Routledge Handbook of Environment and Society in Asia

W0113718

Nowhere is the connection between society and the environment more evident and potentially more harmful for the future of the world than in Asia. In recent decades, rapid development of Asian countries with very large populations has led to an unprecedented increase in environmental problems such as air and water pollution, solid and hazardous wastes, deforestation, depletion of natural resources and extinction of native species.

This handbook provides a comprehensive survey of the cultural, social and policy contexts of environmental change across East Asia. The team of international experts critically examines a wide range of environmental problems related to energy, climate change, air, land, water, fisheries, forests and wildlife.

The editors conclude that, with nearly half of the human population of the planet, and several rapidly growing economies, most notably China, Asian societies will determine much of the future of human impacts on the regional and global environments. As climate change-related threats to society increase, the book strongly argues for increased environmental consciousness and action in Asian societies. This handbook is a very valuable companion for students, scholars, policy makers and researchers working on environmental issues in Asia.

Paul G. Harris is Chair Professor of Global and Environmental Studies at the Hong Kong Institute of Education.

Graeme Lang is a Professor in the Department of Asian and International Studies at the City University of Hong Kong.

Routledge Handbook of Environment and Society in Asia

Edited by
Paul G. Harris
and Graeme Lang

Routledge
Taylor & Francis Group

LONDON AND NEW YORK

First published 2015
by Routledge
2 Park Square, Milton Park, Abingdon, Oxon OX14 4RN

and by Routledge
52 Vanderbilt Avenue, New York, NY 10017

First issued in paperback 2020

Routledge is an imprint of the Taylor & Francis Group, an informa business

British Library Cataloguing in Publication Data
A catalogue record for this book is available from the British Library

Library of Congress Cataloging-in-Publication Data
Routledge handbook of environment and society in Asia/edited by
Paul G. Harris and Graeme Lang.
    pages cm
    Includes bibliographical references and index.
    1. Asia – Environmental conditions. 2. Environmental policy – Asia.
    3. Environmental monitoring – Asia. 4. Environmental impact analysis – Asia.
    5. Climatic changes – Asia. 6. Human ecology – Asia. 7. Nature – Effect of
    human beings on – Asia. I. Harris, Paul G., author, editor of compilation.
    II. Lang, Graeme, author, editor of compilation. III. Title: Handbook
    of environment and society in Asia.
    GE160.A78R68 2014
    304.2095 – dc23
    2014002024

ISBN 13: 978–0–367–66012–3 (pbk)
ISBN 13: 978–0–415–65985–7 (hbk)

Typeset in Bembo and Stone Sans by
by Florence Production Ltd, Stoodleigh, Devon, UK

Contents

Contents

Figures

Figures

Tables

Contributors

Rajesh Basrur is Professor of International Relations at the S. Rajaratnam School of International Studies at Nanyang Technological University, Singapore.

Nitin Bassi is a Senior Researcher at the Institute for Resource Analysis and Policy (IRAP), and is based in Delhi, India.

Bettina Bluemling is an Assistant Professor at the Environmental Policy Group, Wageningen University.

Hans Bruyninckx is the Executive Director of the European Environment Agency, taking office on 1 June 2013.

Faith K.S. Chan is Research Associate at the School of Geography at Leeds University in the UK and is Assistant Professor at the Department of Geographical Sciences, University of Nottingham Ningbo Campus, China.

Youngho Chang is an Assistant Professor of Economics at the Division of Economics and an Adjunct Senior Fellow at the S. Rajaratnam School of International Studies at Nanyang Technological University, Singapore.

Kirsten Conrad is a Singapore-based conservation policy analyst who has been working on the conservation of wild cats in Asia since 1999.

Christopher M. Dent is Professor of East Asia's International Political Economy at the University of Leeds, UK.

Dan A. Exton manages the marine research activities of Operation Wallacea, UK, focusing primarily on the ecology of coral reefs and their conservation.

Maria Francesch-Huidobro, PhD, is Assistant Professor of the Department of Public Policy, City University of Hong Kong and Honorary Assistant Professor of The Kadoorie Institute, the University of Hong Kong.

Daniel A. Friess is Assistant Professor at the Department of Geography at the National University of Singapore.

Notes on contributors

Mary Alice Haddad is an Associate Professor of Government at Wesleyan University.

Paul G. Harris is Chair Professor of Global and Environmental Studies at the Hong Kong Institute of Education.

Gordon M. Heisler retired in 2013 after 41 years as a research meteorologist with the Northern Research Station of the United States Department of Agriculture Forest Service.

Angel Hsu, PhD, is the Director of the Environmental Performance Measurement Program at the Yale Center for Environmental Law & Policy in New Haven, CT.

Swee Lean Collin Koh is an Associate Research Fellow at the Institute of Defence and Strategic Studies at the S. Rajaratnam School of International Studies at Nanyang Technological University, Singapore.

M. Dinesh Kumar is Executive Director of the Institute for Resource Analysis and Policy (IRAP), Hyderabad, India.

Graeme Lang is a Professor in the Department of Asian and International Studies at the City University of Hong Kong.

Toshizo Maeda is an Acting Director/Senior Researcher of the IGES Kitakyushu Urban Centre, specialising in urban environmental management and local environmental policies (P.E.Jp (environmental engineering)).

Darrin Magee is Associate Professor of Environmental Studies at Hobart and William Smith Colleges in Geneva, New York.

Tabitha Grace Mallory is a Research Fellow at Princeton University. Dr Mallory specialises in international relations and Chinese foreign and environmental policy.

Ann Marie Manhart is a consultant working on environmental issues. She is based in Vienna, Austria. She has a MSc in Environmental Technology and International Affairs from the Technical University of Vienna and the Diplomatic Academy of Vienna.

Jack P. Manno is an Associate Professor at the State University of New York College of Environmental Science and Forestry (SUNY-ESF), and a participating faculty member in the Programme for the Advancement of Research on Conflict and Collaboration, Syracuse University.

Bo Miao is an Assistant Professor at the Department of Asian and International Studies at City University of Hong Kong.

Gordon Mitchell is Senior Lecturer at the School of Geography, University of Leeds in the UK.

Hang Ryeol Na is an Adjunct Professor teaching environmental policy at the Rochester Institute of Technology (RIT).

Piya Pangsapa is Head and Senior Lecturer of the Institute for Gender and Development Studies at The University of the West Indies, St Augustine, Trinidad & Tobago.

Dickella Gamaralalage Jagath Premakumara is a Senior Researcher, specialising in participatory planning and urban environmental management at the Kitakyushu Urban Centre of the Institute for Global Environmental Strategies (IGES) in Japan.

Anna Riddell is Publications Editor at the British Institute of International and Comparative Law, and final year PhD candidate studying environmental law and human rights at the European University Institute, Florence, Italy.

Sangbum Shin is an Associate Professor in the Department of International Relations at Yonsei University Wonju Campus Korea.

Paul Simonin is a PhD student at the Department of Natural Resources, Cornell University, Ithaca, New York.

David A. Sonnenfeld is Professor of Sociology and Environmental Policy at the State University of New York College of Environmental Science and Forestry (SUNY-ESF), in Syracuse, and Research Associate, Environmental Policy Group, Wageningen University, the Netherlands.

David J. Smith is a Professor at the Coral Reef Research Unit, University of Essex in the UK.

Benjamin K. Sovacool is Professor of Business and Social Sciences and Director of the Center for Energy Technologies, Aarhus University, Denmark and Associate Professor of Law, Vermont Law School, Institute for Energy and the Environment in America.

Phillip Stalley is an Associate Professor of political science at DePaul University.

Kyle Swan is Assistant Professor of Philosophy at Sacramento State University in California.

James P. Terry is Associate Professor at the Department of Geography at the National University of Singapore.

Karen L. Thornber is Professor and Chair of the Department of Comparative Literature; Chair of Regional Studies East Asia; Professor, East Asian Languages and Civilisations; and Director of Graduate Studies in the Department of Comparative Literature, as well as Walter Channing Cabot Fellow, Harvard University.

Sarah Van Eynde is a PhD Research Fellow at the Department of Political Science (KU Leuven, Belgium), Project Coordinator and Research Fellow at the HIVA-KU Leuven Environment and Sustainable Development Research Group, and Junior Member of the Centre for Global Governance, University of Leuven.

P.K. Viswanathan is Associate Professor at the Gujarat Institute of Development Research (GIDR), Ahmedabad, India.

Bo Wen is the Programme Director of National Geographic Society's Air and Water Conservation Fund and was previously the China Programme Director at Pacific Environment.

Notes on contributors

Fengshi Wu is an Associate Professor at the Rajaratnam School of International Studies, Nanyang Technological University, Singapore.

Amy Zader earned her PhD from the University of Colorado in 2011 where her dissertation focused on changes in rice production and consumption in China.

Part I

Introduction

1

East Asia and the environment

A thematic introduction

Paul G. Harris and Graeme Lang

For those of us who live in economically developed societies, it can be easy to forget that we are completely reliant on the natural environment for our prosperity, wellbeing, and survival. We rely on the environment for the most basic necessities of life, notably air, water, and food. We rely on the environment for material resources and energy that are used, whether directly or indirectly, to produce all of the material things that we take for granted. The environment and the resources that come from it are the raw materials for the majority of economic activity around the world. And, for most of us, the environment is important emotionally: with clean air and water, rich biodiversity, and natural vistas – and knowing that they are likely to thrive in the future – our sense of wellbeing is enhanced. That sense is undermined when air and water are polluted or when nature around us is degraded, overly developed, threatened or simply missing.

Because the resources and benefits from the environment come so easily to most of us – we seldom witness the direct environmental costs of resource extraction, for example – it is easy to forget that the state of the environment – whether it is able to supply human needs and more – is directly related to what all of us do. The environment seldom degrades by itself and resources do not disappear of their own accord. It is human and social behaviour that determines environmental health and, in turn, human and social vitality. Routinely, our behaviour is determined by the cultures and communities in which we live and by the institutions and government policies that largely shape our societies. In short, environment and society – broadly defined to include the actions and interactions of people and the associations and institutions they create – are unified, enormously so in cities and highly developed environments where the impacts of humanity are most obvious, but also in the remotest places on earth. Pollutants and ecological changes, often driven by demand for environmental resources in developed societies or by pollution coming from those places, have insinuated humanity into nature almost everywhere.

Nowhere is the connection between society and the environment more evident and potentially more harmful for the future of world than in Asia. (In this book we focus primarily on the societies and environments of East and Southeast Asia, although other parts of Asia are examined to a lesser extent.) It is, of course, true that Western societies have played at least as large a role in altering the natural environment, the effects of which will be felt for centuries, with climate change being the most obvious and profound example of this. But, arguably, the future of the global environment will be determined in Asia, where countries with very large populations are developing rapidly, in the process adopting many of the environmentally harmful practices of the Western world. The most profound example of this is China. Alone it produces one-third of worldwide greenhouse gases causing global warming and other manifestations of climate change, and this pollution is on the increase. China's demand for natural resources is causing environmental destruction and severe threats to wildlife around the world. And China is using its newfound wealth to exploit other resources around the world to fuel its economy, with profound environmental consequences.

Meanwhile, the billions of people who live in China and other parts of Asia are already suffering the consequences of climate change and environmental pollution, even more severely than most people in the West. This means that the societies of Asia have a profound role not only in the future changes to the global environment, but also in the effects on individuals and societies of those changes and the individual and collective responses that will determine how much the environment declines further and whether it will be possible to cope with the changes that cannot be avoided. Asia is now at the epicentre of environmental change through both its causes and consequences. With this in mind, in this handbook we (the editors and contributors) explore the role of Asia, especially East and Southeast Asia, in shaping relationships between environment and society. We describe the human and policy contexts of environmental change across the region, in the process examining a wide range of environmental problems and their impacts. By describing and analyzing these relationships and contexts, we highlight key environmental problems, locate their causes and help to identify ways of possibly overcoming them in the future.

The handbook is organized into six additional parts. In Part II, we focus on several "human contexts" of environment and society in the region, specifically human rights, justice and literary imagination. Part III takes on questions of politics and policy, considering the roles of policy institutions, democratization, civil society, and corporations. In Part IV, we look at a variety of environmental issues, namely air pollution, solid waste, water and agriculture. Part V is dedicated to understanding the social implications of wilder places, in the process highlighting the region's use of fisheries, reef systems, forests, and animals. Part VI is dedicated to climate change, notably the drivers of the problem, its impacts and how societies are adapting to them, and what is being done to implement alternatives to fossil fuels. In Part VII, we (the editors) draw some conclusions about what the preceding chapters tell us about environment and society in Asia.

Human contexts

We begin our study of environment and society in Asia by looking at the basic unit of society: people. The way that people think, what they believe is just and how this affects behavior are central to the causes of pollution and resource use. The degree to which environment is considered something central to human life is vital for societal and policy change. This is evident in the extent to which the environment – particularly a healthy environment and access to its resources – is considered a right that should be enjoyed by individuals and communities. In Chapter 2, Anna Riddell makes the connections between the environment and human rights. Basing her analysis on the large body of environmental law, she examines the importance of simultaneously reducing the human impact on the environment and improving people's wellbeing. She argues that protecting the environment is one way to fulfill human rights, and indeed doing so may be essential to avoid violating the rights of people to life, health, and livelihoods. This is particularly true in the case of vulnerable groups that are most directly dependent on a healthy environment, notably the world's poor and indigenous communities. At the same time, protecting human rights can bolster environmental protection. Riddell applies these ideas to East Asia, revealing the growing connections between environmental protection and human rights. She finds, however, that progress in the region has been slow. Legally binding provisions to codify human rights and environmental protection, and specifically their connections, are scarce in East Asia. Riddell's chapter suggests that this will have to change if environmental health and human wellbeing are not to be greatly undermined in coming decades.

Riddell's chapter is oriented toward regional relationships between environment and human rights from the perspective of legal justice. In Chapter 3, Piya Pangsapa turns our attention to environmental justice per se. Drawing on case studies in Southeast Asia, especially from Cambodia and Thailand, she explores the different ways in which environmental justice is being fostered by civil society. There is now an understanding among many environmental scholars and even policymakers that social and environmental justice are closely connected. Indeed, new civil society organizations and movements have arisen as a consequence of the injustices that come from environmental pollution and appropriation of local natural resources. This highlights the interdependency of the natural environment and society as well as how this interdependency can be affected by policy decisions. But perceptions of environmental justice may vary from place to place. For example, they are likely to be contingent on local ethical traditions. As Pangsapa notes, in Southeast Asia, discourses about environmental justice are largely premised on Buddhist conceptions of virtue and specifically what they say about the human relationship with the environment. By comparing activist civil society groups concerned about industrial pollution and illegal logging, Pangsapa highlights the connection between social and environmental justice in a rapidly developing region. She shows how these groups increasingly influence environmental policies, although the extent of this influence depends greatly on the willingness of policymakers in the region to let them have it. One of the important capabilities of these organizations is to bring local knowledge into policy processes, not least because this knowledge sometimes contradicts the views of outside experts. This knowledge is coupled with distinctive local traditions and religions that influence views on environmental justice. Traditional respect for nature is often at odds with policymakers' efforts to promote economic growth, thus requiring a better balance between nature and development to achieve environmental justice.

Another human context for looking at the environment can be found in literature, particularly local kinds. In Chapter 4, Karen L. Thornber describes some of the ways in which local literature reflects environmental thought in the region, notably its ambiguities. East Asian literature is often perceived, particularly in the West, as celebrating the natural world. But Thornber points out that it is not this simple; local literature often portrays the way that people and societies have done great harm to the environment. Having said this, she believes that a "planetary consciousness" permeates much of the environmental literature from the region. Very importantly, however, she notes that local authors are not duped into accepting the perception that East Asia is in love with nature. Quite the opposite: they have often challenged official discourse and highlighted environmental challenges facing their societies, the region, and the world. Her objective is to analyze how literature from the region addresses the broader causes and consequences of these challenges. Different writers do this in different ways, through what Thornber calls "ecoambiguity." Writers use a variety of genres, styles, and approaches, but they share many of the same concerns about the human–environment relationship. Thornber's chapter explores the variety of East Asian environmental literature by analyzing a number of the most prominent authors of "ecodegradation," particularly from China, Japan, Korea, Taiwan, and Vietnam. She shows that examining this wide-ranging literature reveals the varied ways in which

East Asian societies have dealt with environmental and pollution issues in the contexts of their own cultures and histories. She also points to how these interpretations, and the manner in which they are expressed through literature, send important messages about environment and society across the region and indeed globally. She shows that local literatures can often negotiate ecological issues in very different ways. This may be the most important message from her chapter – that literature, and perhaps most importantly that from other places, can change the way we think about the environment and ultimately how we behave in our relationships to it.

Politics and policy

Ultimately human–environment contexts develop into, and in turn derive from, politics and the policies of governments. Indeed, it is here where much of the scholarly work on environment and society in Asia is being directed. In Chapter 5, Sangbum Shin looks at environmental policy institutions in comparative perspective. His chapter provides an overview of policymaking processes in East Asian countries, examining how policy institutions have responded to major environmental problems, including climate change. He looks at the roles played by government agencies, legislation and policies, in the process exposing some of the channels by which nongovernmental actors have roles to play as well. In particular, Shin is interested to compare environmental institutions with an eye toward revealing some common patterns across the region. In doing this he tries to identify the extent to which these patterns correlate with economic development. Overall, Shin finds that East Asian governments' environmental institutions are very much top-down entities, with this being most true in Northeast Asia and somewhat less so in Southeast Asia. Generally speaking, government agencies are by far the most important and powerful environmental policy actors across the region, although some nongovernmental actors have significant influence in Southeast Asian contexts and civil society is becoming more environmentally engaged across the whole region. Central governments tend to be more powerful than sub-national ones, undermining the ability of regional institutions and actors to play a greater role in crafting and implementing environmental policies. Shin points out that regional cooperation among countries to address shared environmental challenges is much more active in Southeast Asia, despite some serious issues requiring cooperation across the whole of East Asia. He shows that both domestic and international factors influence environmental policy across the region. Shin argues that more effective policies for addressing environmental problems in East Asia will require institutional reform. In contrast to existing top-down practices, "environmental decentralization" will be needed if countries of the region are to address successfully climate change and other environmental challenges. In short, more cooperation and participation by local actors, both governmental and nongovernmental, from across the region can help to make environmental policy more effective. This would not only have environmental impacts within the region, but might also have significance as the region's role in global environmental problems, not least climate change, grows in importance.

In Chapter 6, Mary Alice Haddad turns more specifically to the politics that underlie environmental policy. As Haddad points out, most countries of the region have experienced simultaneous and interconnected economic development and environmental crises. Sometimes this has resulted in civil unrest in response to environmental pollution and other ecological problems. Haddad argues that the first instinct of regional governments is to suppress any political unrest that might emerge from developmental pressures on the environment. While some of this still occurs, particularly in China, generally speaking in the countries analyzed by Haddad – China, Japan, South Korea, and Taiwan – governments reversed course and have, in recent decades, arguably embraced environmental protection and "green growth." This is somewhat surprising considering the pro-business inclination of these countries' governments and the general weakness of environmental ministries and indeed environmental nongovernmental organizations. The move toward something akin to pro-environment policies is happening across the region despite the significant variety in political regimes. A central question asked by Haddad is whether and how democracy – and democratization – has shaped environmental politics in East Asia. She provides an overview of environmental politics in the four countries, explains the process of political development in each, and attempts to make comparisons. Importantly and significantly, Haddad shows how environmental protection was one of the first issues – indeed, *the* first in most cases – around which civil society activism developed in all four of these countries. She determines that the type of political regime – the amount of democracy in each country – was less important in explaining this than was the timing of the different environmental movements with respect to domestic and international political opportunities. Haddad concludes that democracy matters for environmental movements, but more important for the region, and possibly for other regions, is how well environmental groups are able to time their actions in the process of ongoing political development.

Building on Haddad's analysis, in Chapter 7, Fengshi Wu and Bo Wen examine nongovernmental movements and environmental protests in East Asia, focusing particularly on China, Japan, and South Korea. Wu and Wen argue that since the mid-twentieth century environmental activism stoked by pollution has greatly influenced wider public awareness of the environment and improved environmental governance. Unlike in the West, however, where such developments were largely a function of economic development, in East Asia they have been more closely tied to political development and specifically to the process of democratization. This helps to explain why environmental governance across the three countries examined in the chapter is at different stages, with China lagging Japan and South Korea. Wu and Wen demonstrate that the differences in environmental activism in these countries are generally explained by political culture and each country's progress toward democracy. For example, environmental nongovernmental organizations have had a bigger role in national politics in South Korea than in Japan, despite the latter's longer history of environmentalism. Due to the former country's more radical political transformation in the 1980s, such organizations had a role in creating a democratic and participatory political culture. At the other end of the spectrum, in China the impact of such organizations has been constrained by an authoritarian regime that limits their scope of action and works hard to prevent them from influencing the wider political culture. Wu and Wen also examine how different kinds of environmental activism can result in different environmental policy responses across the country cases. They assess the impacts of environmental activism across three dimensions: public environmental awareness, policy responses (and their actual environmental effects), and long-term institutional reform. As one would expect, Wu and Wen found major differences across the three countries. In China, nongovernmental organizations are not as closely tied to victims of pollution as in the other countries, whereas in Japan and South Korea they have merged with resistance movements. The consequence is that organizations in these last two countries, especially South Korea, have had bigger impacts on raising public awareness and fostering effective environmental policies by government.

A complete picture of environmental politics and policy anywhere in the world must include the role played by corporations and wider industrial interests. In Chapter 8, Phillip Stalley fills in this part of our picture by examining the responses and responsibilities of businesses to environmental challenges in East Asia. As Stalley points out, economic success in the region has come at great environmental cost, with countries across the region becoming some of the largest sources of pollution worldwide. The environment and people have suffered as a consequence. As environmental challenges have increased across East Asia, governments of the region have slowly moved toward striking a balance between economic growth and environmental protection, at least in principle. Stalley shows that a major aspect of doing this involves improving the environmental behaviors of corporations. Strategies have ranged from stronger environmental regulations, market-oriented policies and raising awareness among consumers through increasing environmental disclosure and implementing certification schemes. This attempt to promote "green industry" in the East Asian "factory to the world" has, according to Stalley, generally resulted in more environmental responsibility among the region's corporations. This change is explained by more stringent environmental governance, pressure from publics and the forces of economic globalization. Having said this, the extent to which this has happened varies greatly among countries. Stalley identifies the reasons for the differences, such as inadequate environmental governance – a failure of many governments to implement existing environmental regulations and standards – due to lack of government capacity in some cases or a lack of political will in others. Even where progress seems to be underway, Stalley cautions us to question the reliability of statistics and notes that there may be much by way of industrial behavior that is hidden from view, particularly among the millions of small and medium-sized enterprises spread across the region.

Air, land and water

In Chapters 9 and 10 we look at specific environmental threats in East Asia, starting with air pollution. Air pollution is common across much of East Asia, particularly in its urban areas and industrial zones. This pollution is on the rise, with severe consequences for human and environmental health locally, and growing consequences for those living farther afield as it contributes to acid rain continents away and climate change globally. In Chapter 9, Maria Francesch-Huidobro introduces the problem of air pollution, describing how it is measured before undertaking a case study of two major regional cities – Hong Kong and Singapore – to illustrate the role that institutions play in controlling it. As Francesch-Huidobro notes, stemming the decline in air quality in Asian cities requires addressing a number of issues at the same time, including economic growth, population pressures, demand for transport, and increasing consumption of energy. At least in the two cities that Francesh-Huidobro's chapter examines in detail, air pollution policies are largely influenced by two interlocked factors, namely the balance between governmental and nongovernmental actors' involvement in environmental policymaking, on one hand, and the "ecology of public administration" that determines how environmental issues are governed, on the other. The chapter shows the connections between these factors, especially the ways in which governmental and nongovernmental actors "network" to develop policies for improving air quality. Francesch-Huidobro's case studies reveal that the government of Singapore has adopted orderly forms of collaboration with nongovernmental organizations, in the process co-opting them to a substantial degree without conceding the legitimacy of the state to dictate policy, whereas in Hong Kong policy has been "disarticulated," with much less effective action by the government to learn from the input of nongovernmental actors. Indeed, in the latter city, environmental organizations are considered by government to be hindrances to its overarching goal of economic development, and there is distrust of independent organizations (much like in China more generally). Nevertheless, because it lacks an electoral mandate, the Hong Kong government must at least be seen to be entertaining the ideas of outside groups even if there is very limited evidence that they have significantly influenced policy. For better or worse, this is not an unusual response in East Asian efforts to manage air pollution: governments remain central, with environmental advocates having limited input to policy, and then normally only when governments are already receptive to their ideas and expertise.

In Chapter 10, Angel Hsu builds on the preceding chapter by looking at how the burgeoning problem of smog is being addressed in China, and what this implies for the rest of East Asia. Hsu's chapter describes the causes of air pollution in China, the composition of that pollution and its consequences for the country and its neighbors. It would be an understatement to say that this is an environmental challenge of the greatest magnitude; hundreds of millions of people are affected in China alone, with many millions of them suffering – and more than a few dying early – as a consequence. Indeed, air pollution in most Chinese cities is so bad that the term "air quality" seems inappropriate. Many people in urban and industrial areas routinely wear face masks as they try to avoid inhaling pollutants, and affluent people are doing what they can to avoid the problem, for example by buying air purifiers for their homes or even emigrating to rural areas to escape the problem altogether. Hsu shows that China's smog is an extremely complex environmental problem, involving a wide range of economic and policy actors. Like many other environmental problems, smog is directly associated with economic activity: more economic activity – in short, more burning of polluting fuels and use of polluting chemicals – results in reduced air quality. Consequently, as the chapter reveals, environmental governance to address smog effectively in China requires the involvement of a full range of governmental, industrial and civil society actors, ideally working in concert. Much as with responses to other environmental problems across the region, this is easier said than done, but it is possible. For example, China's urban smog became widely known internationally shortly before the 2008 summer Olympics, held in Beijing. The Chinese government was able to greatly improve air quality in the city during the games, thereby demonstrating the capacity of government to act if needed. Alas, the city's smog returned soon after the Olympics ended, rising to even higher levels than before the games, thus showing what happens when government priorities change. China's air pollution is not just a domestic problem for the Chinese. Much of it is taken by winds to other countries near and far, in turn affecting their air quality. What is more, the Chinese experience with governing air pollution alongside fast paced economic growth offers lessons for other communities in the region undergoing rapid development. As such, smog in China is a classic case of environment and society in Asia, writ large.

Moving on to other aspects of environment and society in Asia, in Chapter 11 Dickella G.J. Premakumara and Toshizo Maeda examine a growing ecological threat: municipal solid waste. This form of waste is one of the most urgent environmental and public health concerns in the region, particularly in urban centers. Population growth, economic development, material consumption, and expansion of cities have resulted in massive increases in the volumes of municipal solid waste in most of East and Southeast Asia. Premakumara and Maeda review this trend, compare a number of cases and use them to propose solutions through regional collective action. While unsustainable production and consumption practices, which do far too little so far to "reduce, reuse, and recycle," are leading causes of municipal waste in the region, Premakumara and Maeda argue that one leading cause of the problem is the rapid rise and expansion of a new "consumer class" whose members want to emulate the high consumption – and thus highly polluting and material-intensive lifestyles – of the world's most economically developed societies, particularly in the West. At the same time, people and industries across the region have greatly increased their demand for natural resources for improved living standards and for industry, the byproduct of which is a much higher level of solid waste byproduct. Premakumara and Maeda point out that systems for managing municipal solid waste face many problems in addition to growing volumes of waste material, including new types of waste, some of it quite hazardous, difficulties in collecting it, and finding suitable locations in which to dispose of it, inadequate regulations, perennial lack of funding, and low public awareness, among other obstacles. Nevertheless, their chapter shows that good practice can take root, as evidenced by policies implemented in urban areas of Japan, Singapore, and South Korea. Less economically developed East Asian societies, including China and most countries of Southeast Asia, have made some progress, too. They can make more if their governments (and citizens) give a higher priority to dealing with solid waste. Importantly, Premakumara and Maeda find that effective policies and measures are most likely to arise from localized solutions supported by help from outside, whether in the form of technical expertise or investment, which builds local capacity and supports solutions that are appropriate for local circumstances.

Like air, water is essential to life. In Chapters 12 and 13 we examine two prominent aspects of the issue in Asia, first by focusing on its scarcity and pollution, and second by looking at how it is controlled via dams. In Chapter 12, M. Dinesh Kumar, P.K. Viswanathan and Nitin Bassi describe and analyze the problems and challenges of water scarcity and pollution in South and Southeast Asia. South Asia is one of the world's regions that often lack water, making irrigation systems essential to its management. Unfortunately, water scarcity in this sub-region is experienced among some of the world's poorest populations. In South Asia, scarcities are exacerbated by shortages of surface water, meaning that much of what is available is extracted from the ground, thereby contributing to long-term shortages for people and agricultural users. Scarcity, in turn, aggravates poverty because people must often pay much of their limited income to those who control wells. What is more, as countries of the region become more urbanized, demands on water systems have increased in terms of both supplies and pollution. In most of the countries of Southeast Asia, water is much more abundant. Nevertheless, much as in South Asia, problems of supply and pollution also exist due to weak regulatory and management institutions. In both sub-regions, availability of water is a function of a multiple socioeconomic factors, including demand by households and industry, dependency of society on agriculture, the degree of urbanization, economic and political structures and of course government capacity. Yet there are also significant differences in the sources of water scarcity between South and Southeast Asia. For example, South Asia has experienced harmful diversions of water from lakes and rivers, creating scarcities, while Southeast Asia is experiencing increasing pollution of ample supplies and sometimes annual problems arising from natural disasters. Both regions are suffering the water-related effects of climate change, not least growing unpredictability of natural water cycles and weather events. With respect to managing common water supplies, specifically those in shared riparian systems, Southeast Asia – with its weak-but-useful Mekong River Commission – offers lessons to countries of South Asia, where cross-border cooperation is less formalized. At the same time, Southeast Asia can learn from experiences in South Asia. For example, poor countries of the former can avoid the uncontrolled exploitation of water and resulting impacts on rural economies that have been experienced in the latter.

In Chapter 13, Darrin Magee examines some of these issues in greater detail by exploring the role of large dams, thousands of which have been constructed across Asia in the last half-century. As he shows, dams can help to solve water problems by controlling and managing supplies, but they can also create problems of their own. Indeed, dams have sometimes resulted in unrest that has contributed to political change. By definition, dams have significant impacts on the environment by altering river courses, affecting water quality and sometimes eliminating seasonal variations in water flow that societies have come to rely on over centuries. Dam construction and flooding of reservoirs can displace entire communities, even cities. In his chapter, Magee looks at these issues in some detail, showcasing the scale of dam building in East Asia and the impacts this has had on social and ecological systems. He finds a common dominant narrative of dams across the region. This narrative entails strong governments working with international development banks and well-connected local financiers to undertake major dam projects officially intended to control floods, irrigate agricultural areas, generate electricity, and provide drinking water. Many of these benefits are realized, at least for some of the population. However, there are also many officially unintended consequences, including disrupted ecosystems, socioeconomic impacts for displaced communities and, paradoxically, flooding, among many other adverse effects. The most affected communities are often those that are already marginalized economically, politically, and ethnically. Consequently, those communities frequently have little say in deciding whether to build dams, where to locate them, and how to compensate those most affected. When compensation is paid, it is almost never considered adequate by recipients, with those being resettled routinely showing declines in wellbeing as a consequence. Significantly, Magee points out that these harms cannot be easily quantified, whereas the benefits that dams can bring are easier to measure in terms of kilowatts of electricity generated, hectares of land irrigated and so forth. This is worth bearing in mind as dams become more attractive as sources of "green" renewable energy to reduce carbon emissions across the region.

Dams are often justified by their potential to enhance agriculture and food production. This is understandable given the importance of these issues for Asia, where they are vital to the livelihoods and survival of billions of people. In Chapter 14, Amy Zader examines food and agriculture. She is especially interested to highlight the implications of security, globalization, and technology in this context. Zader points out that most of East Asia is shifting, or has already shifted, from agrarian to industrial societies, with major consequences for the production and consumption of food. What is more, agriculture is no longer a local pursuit in the region. With globalization, food is being exported and imported in enormous quantities, and multinational corporations now have major operations in the region. As societies in Asia develop, diets are changing, becoming more like those in the West. Drawing lessons from across the region, albeit focusing mostly on China, South Korea, Taiwan, and Japan, Zader explores the options for making agriculture more sustainable in the future. As she points out, food security has been a central concern of governments, arguably more so than in many Western countries. As a consequence, governments of East Asia often have a direct role in agriculture to avoid shortages that might undermine national security. But globalization has made it more difficult for governments to control what happens, even as technology allows increased yields and trade makes it possible to both export and import foodstuffs. Zader argues that agriculture in East Asia faces many challenges. Among these are the need to make food production more ecologically sustainable, coping with burgeoning demands for food among urban residents, improving the livelihoods of subsistence farmers, managing the introduction of new agricultural technologies, and coping with the impacts of climate change. One potentially promising response is the rapid growth in organic agriculture, although it is being held back by pressures on farmers to use chemicals to push up yields. In the longer term, climate change poses the greatest threat to agriculture in the region, in part due to the likelihood that water for irrigation will become scarce. Ultimately, it may be how governments and societies plan for and adapt to climate change that most determines the security of regional agriculture in future decades.

Fisheries, forests and wildlife

Land-based agriculture is the largest source of food across Asia. But food is also taken from the sea, making the marine environment of East Asia a vital resource for all local communities. In Chapter 15, Tabitha Grace Mallory begins three chapters devoted to marine resources by exploring the political, economic and security challenges of fisheries in East Asia. This region's marine fisheries are the most productive anywhere, making them vital to local food security and economic development, and they are significant to local cultures. People in Asia are more reliant on fish as a source of protein than anywhere else, with the increase in fish consumption over the last 50 years being higher than any other region. As a consequence, these resources are under extreme threat. Government programs and organizations to manage fisheries are as diverse as the region itself, with several regional bodies in place to achieve that objective. Civil society organizations also have a major role in the region. However, most fisheries management efforts tend to be fragmented and poorly funded, and attempts by national governments to preserve local fisheries have encouraged illegal fishing, often in disputed maritime territories, such as the South China Sea, where regulation is largely absent. This in turn has exacerbated territorial disputes and has the potential to become a threat to national security of some countries, particularly as China becomes more militarily powerful and enforces its claims to much of the seas around East and Southeast Asia. However, as Mallory points out, the vulnerability of fisheries requires greater cooperation among governments and other actors, not least because fish migrate between different countries' territorial waters. Bilateral and multilateral schemes are in place, but they have been ineffective; fisheries governance in the region has not been up to the task of managing these resources. In addition to improving national and international fisheries management, Mallory argues that other measures – some of them proved to be effective in other world regions – are necessary, such as traceability and certification systems, import bans on illegally caught fish and an the end to economic subsidies paid by governments to fleet owners.

One area where fisheries are most vital is across East Asia's tropical reef systems. In Chapter 16, Dan A. Exton, Paul Simonin and David J. Smith examine artisanal subsistence fishing among these reefs and efforts toward community-based management of reef resources. As was observed in the chapter preceding it, in this chapter we are reminded of the importance of marine resources for the region, not least because it has a vast coastline. It is true that fisheries in the region have become largely industrialized. Nevertheless, millions of people still rely on nearshore artisanal fisheries for livelihoods and often for survival. Among the most important nearshore fisheries are those on coral reefs. Indeed, Southeast Asia has the highest concentration of coral reefs anywhere, equaling over a quarter of the global total. Among these reefs are those found in the Coral Triangle (the subject of Chapter 17; see below). Connected to these reefs are local mangrove forests and sea–grass beds that serve as nurseries for many reef species. Exton, Simonin and Smith focus on the management of local, reef-based fisheries. They describe the status of reef fisheries in Southeast Asia, look at the heavy reliance on them by local people, and analyze options for protecting and sustaining them in the long term. Coral reefs are among the most economically valuable ecological systems on the planet, with their value extending beyond fisheries to services such as coastal protection during storms – something that is becoming more important due to climate change. But these valuable reefs are, like regional fisheries more generally, experiencing unsustainable levels of exploitation, including through highly destructive fishing techniques (e.g., dynamite fishing). Top-down management systems that have worked in other parts of the world to limit overfishing are unlikely to work in Southeast Asia's reef communities, where artisanal practices are often prevalent. Instead, community-based management has been more successful in reducing pressures on reef fisheries. Examples of this kind of management include giving ownership of reef fishing grounds to local communities and restricting access to those areas. More generally, if they are to work management strategies must be geared toward alleviating poverty and ensuring local food security as much as protecting national and regional resources.

Continuing our look at vital marine ecosystems and resources in East Asia, in Chapter 17 Ann Marie Manhart looks at pollution of the marine environment and specifically attempts to protect the Coral Triangle. The Coral Triangle spans an area in the Western Pacific encompassing the reef systems of the Philippines, Malaysia, Indonesia, East Timor, Papua New Guinea, and the Solomon Islands. As Manhart observes, despite covering only 1 percent of the earth's oceans, the triangle contains three-quarters of all reef-building corals and more than one-third of all coral fish species. This makes the Coral Triangle one of the world's epicenters of marine biodiversity. What is more, about 150 million people live in and adjacent to the triangle, with many of them directly dependent on it for food and livelihoods. This becomes truer as populations around the Triangle grow. Despite this, the Triangle and all its species are threatened by overuse and a severe lack of protective institutions. What is more, the Triangle is extremely vulnerable to pollution. Governments of the region have not ignored the plight of the Coral Triangle. For example, they have agreed to the non-binding Coral Triangle Initiative on Coral Reefs, Fisheries, and Food Security. However, preventing pollution is not one of the objectives of this initiative. There is a severe lack of water treatment throughout Southeast Asia, with sewage and waste flowing into the sea, harming Triangle ecosystems. In her chapter, Manhart explains and analyzes these and related pollution issues, examining the Coral Triangle from the perspectives of society, economics, ecology, and governance. She uses these perspectives to highlight issues that need to be considered when developing strategies for reducing pollution. For example, the perspective of society points to the need to consider dynamics of social class in local communities, questions of justice and equity regarding who bears the burdens and who enjoys benefits associated with the Triangle, and whether and how the media in local societies are up to the task of conveying messages about how to protect the Triangle. Similarly, looking at the problem from the perspective of economics highlights the role of corporations, local livelihoods, finance, and the value of environmental services. An ecological perspective focuses attention on the importance of measuring pollution, determining where it comes from and how it affects the Triangle, and ascertaining what works best to mitigate the harm that it does. Governance perspectives highlight the role of environmental regulations and laws, including how they are formulated and whether they are implemented. Manhart concludes that all these perspectives are needed to formulate holistic, multifaceted approaches to reducing society's impact on this vital marine resource.

In Chapter 18, Hang Ryeol Na, Jack P. Manno, David A. Sonnenfeld and Gordon M. Heisler turn our attention away from the sea and back to the land. In their chapter, they describe and assess institutions for the regional governance of forests in East Asia. They argue that forests are social because they provide valuable services to society. Forests are also global because they are linked to global phenomena, for example, they are sources of timber that is traded on the world market and they serve the important role of sinks for carbon that would otherwise contribute to climate change. These aspects of forests imply that their management must involve local, regional, and international governance. In their chapter, Na and colleagues focus on regional governance. Forests in East Asia are unlike those in other parts of the world in some important ways. For example, East Asian forests have a higher human population density than elsewhere, and hundreds of millions of people in the region rely on forests for their livelihoods. This contributes to degradation of forests and conflicts over their use. Very large areas have been – and are being – converted into industrial croplands, contributing to the chronic problem of deforestation, although in some countries reforestation projects have been successful. Regional and multilateral institutions for managing forests have tended to be redundant and to overlap in their initiatives. Drawing on a number of regional cases, Chapter 18 reveals several trends in how forests are being governed. For example, the authors observe that regional forest governance, which started as experimental projects of international institutions, have evolved into collaborative partnerships among local institutions. Regional initiatives have been fostered to some extent by pressure from the West to address illegal logging and to ensure that forest exports are from sustainable sources. Forest governance in the region has slowly evolved to reflect international norms, although longstanding claims of total sovereignty over forest resources have slowed this transformation. The authors argue that forest governance in the region remains poor, but they see some reasons for hoping that it will become more effective with time.

One reason that preserving and protecting forests is so important is because they are the homes of much of Asia's wildlife. Destroying forests destroys biodiversity; preserving forests can preserve biodiversity. A question then becomes whether and how that biodiversity is used by society. In Chapter 19, Kyle Swan and Kristen Conrad look at this relationship. They focus on the ways by which cultural and environmental values affect the consumption of wildlife in China and some Southeast Asian countries sharing similar values. Swan and Conrad begin their chapter by pointing out fundamental ways of thinking about the environment generally and wildlife in particular. Broadly speaking, several Western conceptions suggest that wildlife deserves protection for a variety of reasons; human ends can and should be compromised if that is necessary to preserve environmental goods. Swan and Conrad point out that this "preservationist" view conflicts with traditions common in Chinese-influenced cultures, which put people's needs first, where the consumption of wildlife is seen as fundamental to traditional medicine and religion. If nothing else, this conflict in perceptions about the how to treat wildlife can contribute to difficulties in managing or protecting it. For example, if regulations for wildlife protection fail to consider the values of regional cultures, they may face opposition from those who consume the wildlife. In the countries examined in this chapter – Cambodia, China, Laos, and Vietnam – attitudes about the consumption of wildlife are much more permissive than in the West, with many species being important for cultural rituals and practices. Indeed, consumption of wildlife can give status to people in these societies, which helps to explain why they – and China in particular – are driving much of the global demand for wildlife. National and international policies and programs to regulate the trade in wildlife and its consumption need to reflect such values in their strategies. That is, if they are to be effective, solutions to environmental and related issues in the region will be more likely to succeed if they are compatible with local cultural beliefs. Indeed, Swan and Conrad argue that many of the policies that environmentalists have recommended to protect wildlife are counterproductive because they fail to do this. Preservationist approaches to environmental protection in the region, including with respect to the consumption of wildlife, must attempt to fit local cultures. Swan and Conrad thus conclude that approaches that build on local culture are more likely to be successful.

Energy and climate change

Beginning with Chapter 20, we undertake an in-depth analysis of climate change and society in East Asia, examining its causes and impacts from a regional perspective, and exploring some options for mitigating it. In Chapter 20, Bo Miao describes the drivers of climate change in East Asia and the energy dilemma that is presented by it. Miao focuses especially on China – quite justifiably given its enormous contribution to this problem – comparing its experience with that of two neighboring countries, Japan and South Korea. Miao begins by pointing out that the most important driver of climate change in East Asia is the region's enormous consumption of fossil fuels, especially the burning of coal in China. Indeed, one consequence of China's growing use of fossil fuels is its status as the largest national source of greenhouse gases causing climate change. Its per capita emissions are below those of many Western countries, but it is catching up quickly. Chinese officials are aware of this, and the country arguably has the most ambitious program for reducing reliance on coal, in particular. This includes heavy investments in alternative sources of energy, such as wind farms, hydroelectric dams, and nuclear power stations. However, all of these efforts have not kept pace with increasing demands for inexpensive energy – i.e., coal – as China's economy expands. In other words, even while energy efficiency improves, total emissions of greenhouse gases are increasing. Nearby in Japan and South Korea, national contributions to climate change are also very significant, but the sources of their emissions differ from China. In particular, these countries are less dependent on coal; they rely instead on somewhat less polluting fossil fuels, namely oil and natural gas, which meet about half their energy needs. Nuclear power has until recently also been a major source of energy in Japan, although this may change if reactors shut down after the Fukushima Daiichi nuclear accident are not restarted. Japan and South Korea are planning to move more toward alternative forms of energy, but their plans may not be as ambitious as those of China. All three of these countries have committed to limiting – and in the case of Japan, reducing – their greenhouse gas emissions, but overall this will be difficult because the region will remain largely reliant on coal and other fossil fuels for the bulk of energy well into the future. (In Chapter 23, we look at some of the pathways for increasing the use of renewable, low-carbon forms of energy.)

Asian societies should have a great incentive to reduce reliance on coal and other fossil fuels that contribute to climate change. This is because all of them are vulnerable to its effects, often in very serious ways. Benjamin K. Sovacool looks at some of these vulnerabilities in Chapter 21, in the process considering how countries in the Asia-Pacific region can adapt to the impacts of climate change that cannot be avoided. Asia's vulnerabilities to climate change are numerous, not least from increased and more severe flooding, storms and droughts, and (for coastal countries) sea level rise. These impacts in turn challenge the ability of Asian societies to ensure food security, protect infrastructure and prevent loss of life from natural disasters. It is worth bearing in mind that responding to these vulnerabilities will be very costly. Many countries of the region are already poor, meaning that climate change will inevitably divert resources away from economic development. One of Sovacool's objectives is to examine how these vulnerabilities can be lessened through adaptation measures. The Asia-Pacific region may be the region of the world that is most at risk to the impacts of climate change, in part because so many countries there are relatively poor and still developing economically, but also due to population density, geographic vulnerability, reliance on local biodiversity, and the like. While some countries are more vulnerable than others, Sovacool observes that none of them is immune. Efforts by countries of the region to prepare for and adapt to climate change vary in their approaches, such as planting mangrove trees in Bangladesh, experimenting with climate-hardy crops in Cambodia, and hardening coastal infrastructure in the region's island countries. The key to success, according to Sovacool, is combining infrastructure, organization, and social adaptation to strengthen communities and organizations across the region. Importantly, he finds that success requires that improved local knowledge and assets must be combined with capacity building and improvements in governance. Some forms of adaptation are unique to local areas while others can be useful in many countries. It is likely that small-scale projects that get local communities directly involved – arguably those that are more democratic – will be more successful in fostering genuine adaptation to climate change in countries of the Asia-Pacific region.

As Sovacool notes, one of the most significant impacts of climate change in the region is from flooding. In Chapter 22, Faith K.S. Chan, Daniel A. Friess, James P. Terry and Gordon Mitchell look in detail at the challenges of flooding in coastal East Asia. Over half of Asia's people live in coastal areas. Hundreds of millions of people live in especially vulnerable delta regions and coastal cities where populations are increasing as people move in search of employment and services. These people will be extremely vulnerable to the effects of sea level rise, with livelihoods affected by impacts on fisheries and agriculture, and lives threatened during storms. Arguably, these people are suffering the effects of climate change already, with many coastal areas across Asia being affected by uncommonly severe floods and storms. In their chapter, Chan and colleagues look at these and other vulnerabilities to flooding in East Asia, using China's Pearl River Delta as their case study. They assess coastal vulnerability to flooding and examine whether and how strategies for adaptation to climate change, and specifically flood risk management, can be enhanced. East Asia's delta cities are worth studying both for the lessons they can provide but also because they are now hubs of economic activity that is important for other world regions. This is especially true of the Pearl River Delta, where the growth in both economic output and population has been phenomenal in recent decades, and where flooding has been a recurring problem. Climate change increases the chances of severe rainstorms, which will flood the delta's rivers, and storm surges – which often come alongside heavy rain – that will inundate coastal environments. Unfortunately, the same can also be said of other East Asian deltas, making the experience of the Pearl River Delta worth understanding in detail. Chan and colleagues observe that East Asia's coastal regions lack holistic strategies for flood risk management and more general adaptation to climate change. The region's governments are largely devoted to piecemeal engineering solutions, such as flood protection barriers, which are costly and which the authors of this chapter believe are less effective than land use planning, awareness building and public engagement (like that practiced in some European countries). They also note a lack of collaboration and cooperation to manage climate-related flood risks across East Asia. This will have to change if the region's growing population is not to suffer unavoidably.

Asian societies do not only suffer from climate change; they also contribute to it. The extent to which the region can or cannot limit its use of fossil fuels will have both regional and global impacts. With this in mind, in Chapter 23 Christopher M. Dent details renewable energy strategies and policies in East Asia, focusing on efforts to foster and implement low carbon economic development. According to Dent, the challenge of climate change has combined with that of energy (in)security to compel East Asian governments to develop renewable energy policies and to support "green energy" industries. East Asian demand for energy will almost certainly increase for some time as the region's economies expand. Efforts by local governments to increase the availability and use of renewable energy reflect this likelihood. Indeed, renewable energy has become a defining attribute of the region's industrial policies, part of what Dent calls their "new developmentalism." His aim is to understand the region's actual prospects for a major transition to "low carbon" development, something that is not guaranteed given the region's current reliance on fossil fuels, especially coal. Dent's chapter looks at the evolution of industrial policies and related state capacity, thereby identifying the importance of renewable energy programs and strategies in the region. He draws on case studies from the region, especially China and Japan, to determine how and why specific renewable energy policies have become prominent. Ultimately, his analysis helps us to understand the extent to which these policies are becoming intrinsic aspects of a new development paradigm that will enable the region to become much less reliant on carbon-based fuels. He believes that renewable energy will take hold, but the ways that this will happen will vary from one country to another, even as most countries of the region continue to remain reliant on forms of energy and development that harm the environment.

The transition to truly low-carbon economies will reduce the environmental impact of development, but it will be a long and slow process as societies become more aware of alternatives. Even as low-carbon energy development increases pace in East Asia, economic development will continue to trump environmental protection. In Chapter 24, we explore the notion of how to fuel Asian economies with less carbon-intensive fuels. Sarah Van Eynde, Bettina Bluemling and Hans Bruyninckx do this through a case study of low-carbon development in China. China is a vital case for the region and the world because it has accounted for about three-quarters of the growth in world energy consumption in recent years. Other Asian economies have large and growing appetites for energy, and combined they disproportionately explain why global carbon dioxide emissions continue to increase despite the slowdown in the global economy and the threat of climate change. As observed in the preceding chapter, one response to these trends has been a growing interest in low-carbon development – improving energy efficiency and reducing the amount of fossil fuels necessary to support economic growth. It is important to note that this does not necessarily mean that the upward trend in carbon emissions will be reversed anytime soon. Indeed, one might argue that the continued focus on rapid economic growth dooms the region (and the world) to growing greenhouse gas emissions from carbon for at least some decades. In short, low-carbon development in practice is likely to slow the increase in greenhouse gas emissions, which is welcome, but it will hardly solve the problem of climate-changing pollution. This may be explained by the manner in which low-carbon development is interpreted and implemented in the region. Chapter 24 looks at this in some detail while also describing the emergence, scope, and purpose of low-carbon development.

As noted in Chapter 24, one of the strategies for simultaneously achieving greater energy security and reducing carbon emissions is nuclear power. Japan has until recently relied quite heavily on this form of energy, and other countries of the region have been developing it, with China planning to build many new plants in coming years. In Chapter 25, Rajesh Basrur, Youngho Chang and Swee Lean Collin Koh look at this phenomenon in the context of the Fukushima Daiichi nuclear accident in Japan. How has this event affected what looked to be – and might yet be – a nuclear renaissance in Asia? As Basrur, Chang and Koh point out, the accident has reinforced the thinking of those who were already critical of nuclear power, and it increased the burden for those who have been trying to promote it as an alternative to fossil fuels and almost certainly means that additional safeguards will be demanded of future plants, increasing the costs associated with building and operating them. The accident pointed to flaws in nuclear risk management, plant construction and operation, and political leadership. Nevertheless, it looks likely that nuclear power still has a future in Asia due to the growing demand for energy. Fukushima may turn out to be a temporary setback in the region, at least outside Japan. (Its greater impact may be outside the region. For example, Germany's plans to shut down nuclear power stations were probably accelerated as a consequence of events in Fukushima.) This is partly because other forms of energy are seen to have their own problems, whether environmental or economic. What is most likely to happen is that nuclear energy will become part of a mix of alternatives to fossil fuels that, when combined, help to limit and eventually reduce the use of fuels that make the greatest contribution to climate change (coal, oil, and natural gas). Governments of the region are more likely to recognize that nuclear energy comes with certain dangers. These dangers, if they are to be mitigated, make this form of energy more economically and politically costly than it might have been before the Fukushima accident. As such, even for longstanding supporters of nuclear energy, its attractions may be less bright than before. The search for low-carbon economies in Asian societies will remain a difficult one.

Conclusion

In the final chapter of this handbook, we (the editors) look back at the other chapters, describing many of the lessons that they convey about environment and society in Asia now and in the future. It this stage it is worth reiterating the importance of our topic: for billions of people in Asia, the environment–society nexus is one that certainly affects their wellbeing to some degree, and for millions of them may directly affect their health and even their survival. The seriousness of the topic for Asia cannot be overstated, not least because much of the region is more directly dependent on local environmental conditions and resources than more developed parts of the world, and because its ecosystems and communities are particularly vulnerable to environmental changes. Furthermore, what happens in the region matters for the rest of the world. The relationships between environment and society in Asia will largely determine environmental conditions and developments globally, the region's contribution to future climate change being the most prominent example of this. Additionally, the ability of Asian societies to avert environmental harm and to cope with the harm that cannot be averted will affect their ability to participate in the global economy and provide resources and services to the rest of the world.

Part II

Human contexts

2

Human rights and the environment

Making the connections

Anna Riddell

1 Why connect the environment and human rights?

Environmental law consists of a body of complex interlocking rules, agreements and treaties that operate to regulate the interaction of humanity and the rest of the natural or physical environment. Its goal is to reduce the impact of human activity both on the natural environment and on humanity itself. Similarly, human rights law also strives to create better living conditions for the human population of the earth. Environmental protection can, in fact, be a method of fulfilling human rights standards, given that allowing the degradation of physical environments can contribute directly to infringements of the human rights to life, health and livelihood of their inhabitants. A system of environmental protection thus helps ensure the well-being of future generations, as well as protecting vulnerable groups who depend on natural resources for their livelihoods, including indigenous or economically marginalized groups. Similarly, human rights standards can help to achieve conservation and environmental protection, as will be seen below.

In the twentieth century it became increasingly clear that a clean and healthy environment is the resource base for all life, and that this resource is inherently fragile and in need of protection. As Judge Weeramantry observed in a Separate Opinion in the *Gabčikovo-Nagymaros* case before the International Court of Justice (ICJ), the: 'protection of the environment is . . . a vital part of contemporary human rights doctrine, for it is a *sine qua non* for numerous human rights such as the right to health and the right to life itself' ((*Hungary v Slovakia*) 1997 ICJ Reports 7, 88). Faced with the results of increasing levels of pollution and ecological damage, many international treaties and local laws and regulations on environmental protection were introduced in the second half of the twentieth century. Initially, no connection was made between human rights and environmental protection, but major environmental disasters such as Bhopal and Chernobyl served to highlight the human dimension of a clean environment. Natural environmental disasters too are increasingly having an impact on human lives, and the increasing frequency of such disasters has been attributed to climate change resulting from man's activities on the planet.

2 Environmental considerations in East Asia

When considering the impact of environmental considerations on human rights in East Asia, it must be remembered that the region is particularly vulnerable to natural and climatic hazards such as earthquakes, volcanoes, typhoons, droughts and tsunamis. Examples are the 2004 Indian Ocean tsunami, which killed 230,000 in the 14 countries bordering the Indian Ocean, or the aftermath of Typhoon Utor in 2006–2007, which flooded large areas of Malaysia, Indonesia and Singapore, or the several severe earthquakes in the Philippines in 2012, which affected the lives of thousands, disrupting water and food supplies, destroying homes, and affecting basic infrastructure such as schools and hospitals. A 1998 survey by the Swiss Reinsurance Company (Sigma No 2/1998) highlighted the significance of natural disasters in the East Asian region. From 1970–1997, of the 40 worst catastrophes (assessed in terms of over 1 million fatalities) 30 of these occurred in the Asia–Pacific region.

The impact of natural disasters has sometimes been enhanced by mankind's activities, such as the earthquake and resulting tsunami in Japan in 2011, which caused equipment failures and meltdowns at the Fukushima nuclear plant, severely affecting the basic living conditions and human rights of up to 50,000 people living in the vicinity.

As well as natural disasters, there have been many that have been caused entirely by man's activities. According to the United Nations Economic and Social Commission for Asia and the Pacific, East Asia is the area of the world most gravely affected by poor land management practices causing desertification and loss of livelihood to farmers and starvation in agrarian communities, drought and other related disasters such as large-scale landslides which have claimed many lives ('Emerging issues and development related to natural and man–made disasters', E/ESCAP/1127, 27 January 1999). Forest and bush fires have also had catastrophic effects, such as those in South East Asia in 1997–1998, which were thought to be the worst forest fires in two centuries of recorded history. Much effort was made after this disaster to try and prevent the slash and burn techniques which caused it, but huge-scale fires occurred again in 2005–2006, and June 2013 also saw a state of emergency declared in Singapore and Malaysia as a result of 'hazardous' levels of smog caused by the burning of logged woodland for palm oil plantations over the Strait of Malacca, in the province of Riau in Indonesia. As well as the long-term effects of farming and development practices, individual incidents and accidents have caused environmental disasters, such as industrial accidents involving chemicals or hazardous wastes, often having a devastating effect on human health in the surrounding area.

These examples have made it evident that in East Asia, as in the rest of the world, the question of the environment cannot be effectively addressed without consideration of potential human rights issues.

This chapter will first look at how the connection between human rights and environmental protection has developed at the international and regional level, in order to provide context and comparison for a subsequent examination of how a fusion of environmental law and human rights is beginning to develop within East Asia. This relationship will be examined from both directions – first, looking at how attempts at environmental protection have also resulted in human rights protection, and, second, looking at how the human rights field has offered incidental protection for the environment and redress for environmental harm.

3 Protecting human rights via environmental provisions

Man's relationship to nature has not always been so strained. Early man regarded nature with awe, and worshipped the elements as deities, and in many countries of East Asia, such as Thailand

(Kititasnasorchai & Tasneeyanond 2004: 4), this eco-centric attitude prevailed until relatively recently when industrialization and the encroaching of Western ideas began to emphasize man as the reason for all creation, nature being a resource to use as he pleased. 'Civilized' man lives in a state of alienation from nature, using and abusing the ecological systems of the planet, facing us with a situation where the quality and condition of human life are threatened by our degradation of nature, and the ability of the planet to sustain man is increasingly questioned. It was only with the development of scientific knowledge in the 1960s and 70s that we began to realize that treating the planet in a proprietary manner would have a detrimental effect on human life.

In the past, pollution and environmental degradation were perceived as operating largely on the local level, with effects that were isolated in their impact. With a greater understanding of science effects are now known to be felt on regional and global levels, requiring appropriately scaled responses. Although science has advanced sufficiently to understand the far reaching impacts of industrialization and the release of greenhouse gases, it is still almost impossible to find a causal link between instances of pollution and effects of pollution when many pollutants have global impacts. It is therefore necessary to address the problem of environmental damage at a global level involving all states, but also at a regional level, with focus on the specific problems that region faces. The difficulties faced by East Asia as a region are the distinct geographical differences among the various countries and their varying levels of development. For example, in countries rich in resources and rapidly industrializing, the predominant environmental risks stem from the extractive industries, including logging, and the effects of large-scale construction projects, but in highly developed and urbanized countries, the environmental risks are those stemming from pollution and destruction of the few remaining nature areas. A regional approach is therefore a difficult concept.

The first and most clear connection of human rights and environmental concerns was the Stockholm Declaration of 1972. The very first Principle proclaims: 'Man has the fundamental right to . . . adequate conditions of life, in an environment of a quality that permits a life of dignity and well-being, and he bears a solemn responsibility to protect and improve the environment for present and future generations.' This posits the environment both as a fundamental human right and a responsibility towards future generations. Indeed, at the time this was a very forward-thinking approach to the topic, one that has yet to find its place in a binding international treaty.

The Declaration of the next UN Conference in Rio took somewhat of a backward step, however. It proclaimed in Principle 1 that human beings: 'are entitled to a healthy and productive life in harmony with nature'. The language of 'rights' had disappeared from the Declaration, with the exception of in Principle 10, which detailed procedural rights such as access to information and involvement in environmental decision making, moving away from speaking of a general right relating to the environment.

However, despite the lack of significant political, legal or institutional development at the Rio Conference, it led to a flurry of further treaty making in the area of international environmental law, many containing provisions with an impact on, or relationship to, human rights. These focused mainly on the procedural rights enshrined in Principle 10 of the Rio Declaration, and almost all global and regional environmental treaties since 1991 contain at least some reference to public information, access or remedies.

Of these, the most significant is the legally binding Aarhus Convention on Access to Information, Public Participation in Decision Making and Access to Justice in Environmental Matters,[1] which entered into force in 2001, and expressly establishes a conceptual link between substantive and procedural environmental rights by noting that citizens must have access to environmental information, be entitled to participate in decision making and have access to justice in environmental matters in order to be able to assert their right to live in an environment adequate to their health and well-being. This environmental treaty is significant in its creation of rights for individuals, albeit procedural rights, allowing citizens to become involved in environmental decision making. Forty-five States and the EU are signatories to the Convention, but no East Asian nations have signed.[2]

4 Environmental protection through human rights

Environmental law can be seen as a classic category of international law, regulating the relationship between sovereign States by means of bilateral and multilateral treaties with reciprocal obligations and duties. Human rights law however has broken out of the mould of traditional international law, putting not States, but individuals, at the heart of the system. At the regional as well as the global level, a great number of Conventions have been adopted for the protection of human rights, either in general or focusing on specific rights (against genocide, apartheid, torture, etc.) or on particular categories of human beings (women, children, workers, etc.). Many of these rules protecting human rights have consolidated into customary rules of international law and some have even attained the status of *jus cogens*, binding States whether they have ratified those conventions or not. Human rights law, therefore, has provided individuals with rights under international law, and the means for vindicating those rights on the international plane. In this way, human rights law provides an accessible enforcement mechanism which is often lacking in environmental legislation.

4.1 International connections

There are few explicit references to environmental matters in international human rights instruments, because most human rights treaties were drafted and adopted before environmental protection became a matter of international concern. The International Covenant on Economic, Social and Cultural Rights contains a right to health in Article 12 that expressly calls on States Parties to take steps for 'the improvement of all aspects of environmental and industrial hygiene'. The Convention on the Rights of the Child refers to aspects of environmental protection in Article 24, which provides that States Parties shall take appropriate measures to combat disease and malnutrition 'through the provision of adequate nutritious foods and clean drinking water, taking into consideration the dangers and risks of environmental pollution'.[3]

While environmental rights apply equally to all people, there has also been recognition of the need to provide specific protection for particular groups who are disproportionately affected by environmental degradation and who have a particular cultural connection to land and the environment, and it is perhaps in relation to indigenous peoples and the environment that there has been the most development. International Labour Organisation (ILO) Convention No. 169 concerning Indigenous and Tribal Peoples in Independent Countries contains numerous references to the right to lands, resources and environment of indigenous peoples.[4]

At the UN level, attempts have been made over the years to forge the connection between human rights and the environment in an international treaty, particularly by the Office of the High Commissioner on Human Rights (OHCHR) and the United Nations Environment Programme (UNEP) who have both acknowledged the merit of an interconnected approach and in 2010 even jointly commissioned a group of expert academics and practitioners to draft a Declaration on the Environment and Human Rights. This draft Declaration was very progressive in terms not only of elaborating a substantive right to environmental quality, but also connecting existing rights to environmental issues. It was well drafted and appeared to mark a turning point in the willingness of the UN to push forward an articulated right with the involvement of both its environmental and human rights experts, but despite encouragement by UNEP, little has been done with the draft Declaration since its creation, and it appeared that the OHCHR had lost its enthusiasm for the project, seemingly reflecting the broader UN attitude to the relationship. Although many UN instruments now acknowledge the relationship between human rights and environmental protection, for example, Resolution 64/157 on Promotion of a Democratic and Equitable Order which affirmed the right of every person and all peoples to a healthy environment,[5] there have been no significant moves towards developing a substantive right to the environment.

However, in Resolution 19/10 of March 2012, the Human Rights Council decided to appoint an independent expert on the issue of human rights and the environment. Professor John Knox was selected and delivered his Preliminary Report in December 2012 (A/HRC/22/43). He examined the history of the connection and noted that it is now firmly entrenched in certain fora, but greater conceptual clarity is required in order to assist with more systematic implementation. His appointment demonstrates the importance accorded to a human rights approach to environmental issues and will, it is to be hoped, herald a new era of activity on this topic at the UN level.

4.2 Regional connections

4.2.1 Environmental protection through other human rights

Many human rights tribunals and experts take an approach similar to that of the Stockholm Declaration and understand environmental protection as a precondition to the enjoyment of several internationally guaranteed human rights, especially the rights to life and health. Environmental protection is thus seen as an essential instrument subsumed in, or a prerequisite to, the effort to secure the effective enjoyment of other human rights. In this sense, the General Assembly has called the preservation of nature: 'a prerequisite for the normal life of man'.[6] This approach has most notably been seen in the European Court of Human Rights (ECtHR), which has heard cases in which the environmental harm is such that it interferes with the home, private and family life, as protected by Article 8 of the European Convention on Human Rights (ECHR) and has found states in breach of this article as a result of their environmentally harmful actions. In the seminal judgment in *Lopez-Ostra v Spain* ((1994) 20 EHRR 277), the ECtHR held for the first time that a failure by the state to control industrial pollution was a violation of Article 8 as there was a sufficiently serious interference with the applicants' enjoyment of their home and private life. The Court held that 'severe environmental pollution may affect an individual's well-being and prevent them from enjoying their homes in such a way as to affect their private and family life adversely, without, however, seriously endangering their health'.

This developing vein of jurisprudence linking conditions of the environment to substantive rights under the ECHR has also found violations based on the Article 2 Right to Life (see, e.g., *Öneryıldız v Turkey* (2004) 41 EHRR 20), and Article 6 on the Right to a Fair Hearing (see, e.g., *Kyrtatos v Greece* (2003) EHRR 242), developing some procedural safeguards in environmental cases through the latter. This linking of environmental harm to human rights demonstrates very effectively the utility of such an exercise, as, in many cases, it has resulted in redress of some sort for the victims of environmental harm which they could not have otherwise sought.

4.2.2 Substantive right to a clean environment

In certain fora, the connection between the environment and human rights has been embodied in the development of a substantive human right to a clean environment. Discussion on the need for such a right was particularly prevalent in the nineties (e.g., Boyle & Andersen 1996), many arguing that without such a right, individuals would have no redress for environmental damage caused to their home, livelihood or even health. These arguments seemed pressing in the light of a lack of rules protecting against damage of this kind. However, difficulties plagued the discussion, such as the definition of the right, for example, the question of whether one should use the term 'clean', 'healthy' or 'adequate', the differences between these meanings, and also how to take account of regional differences and inequalities. As a result, many proposals for such a right met with the disapproval of States and thus failed to be adopted.

However, two regional human rights treaties have expressly declared a right to a sound environment. The African Charter on Human and Peoples' Rights, to which more than 50 African nations are parties, states: 'All peoples shall have the right to a general satisfactory environment favourable to their development.'[7] This is a collective or group right – the right of 'all peoples'. For example, in the *Ogoniland* case, a local population in Nigeria complained of environmental damage by the oil extraction industry in the area. The African Commission on Human and Peoples' Rights, in a far reaching judgment (*Social and Economic Rights Action Centre v Nigeria*, Case No ACHPR/COMM/A044/1, OAU Doc. CAB/LEG/67/3 rev 5) said:

> [A]n environment degraded by pollution and defaced by the destruction of all beauty and variety is as contrary to satisfactory living conditions and development as the breakdown of the fundamental ecological equilibria is harmful to physical and moral health.

In contrast to this collective right, the 1988 San Salvador Protocol, joined to date by 25 Latin American and Caribbean nations, contains a right aimed at individuals that proclaims that: 'Everyone shall have the right to live in a healthy environment and to have access to basic public services'[8] and that the States Parties shall promote the protection, preservation and improvement of the environment. However, the Inter-American Court of Human Rights has been progressive in its interpretation of this right and appears to be approximating a collective right in its jurisprudence. For example, in *Mayagna Sumo Awas Tigni Community v Nicaragua* (Inter-Am. Ct. H.R., (Ser. C) No. 79 (2001)) logging concessions that had been awarded by Nicaragua to private investors were held to constitute a violation of the property rights guaranteed in Article 21 of the local tribal community. Despite the fact that Article 21 does not mention a collective right, the Court interpreted it to include the right of indigenous people as a group to use their ancestral land for agriculture and hunting and to be protected from the environmentally and culturally destructive process of commercial logging.

There has been some debate at the Council of Europe over an additional protocol to the ECHR containing a 'Human Right to a Healthy and Clean Environment', with a former president and the Parliamentary Assembly in favour of doing so,[9] but the proposal has met with opposition in the Committee of Ministers and the Steering Committee[10] and it therefore looks unlikely that this progressive step will be taken in the near future.

4.3 National examples of a right to the environment

It is also noteworthy that a right to the environment has been included in many national constitutions around the world either as an individual right or state obligation. Earthjustice summarized the constitutional recognition of environmental rights in a submission to the UN Commission on Human Rights in March 2005:

> Numerous constitutions of the nations of the world guarantee a right to a clean and healthy environment or a related right. Of the approximately 193 countries of the world, there are now 117 whose national constitutions mention the protection of the environment or natural resources. One hundred and nine of them recognise the right to a clean and healthy environment and/or the state's obligation to prevent environmental harm. Of these, 56 constitutions explicitly recognise the right to a clean and healthy environment, and 97 constitutions make it the duty of the national government to prevent harm to the environment.[11]

Jeffords' more recent study notes that '142 out of 198 national constitutions . . . include at least one reference to the environment as of 2010. Out of these 142 constitutions, 125 contain provisions that are explicitly related to environmental human rights' (Jeffords 2011: 2). Several East Asian nations have included a right to a healthy environment in their constitutions, including East Timor, Indonesia, the Philippines, South Korea and Thailand, and several of these have also embodied the right in national environmental legislation.

5 Developing the connection in East Asia

5.1 Environmental protection in East Asia

Noted environmental scholar Benjamin Boer said in 1999 that 'the great revolution of environmental law has only just begun, particularly in the countries of the Asia region' (Boer 1999: 1552). A catalyst for environmental protection in this region has been donations from foreign agencies and development by foreign investors, which often come with environmental expectations or conditions. However, despite the passing of numerous environmental laws, there is still a difficulty in meaningfully internalizing environmental concerns into the governance culture. The relatively youthful environmental consciousness in East Asia has meant that government ministries or agencies dedicated to environmental protection or natural resource management have often faced challenges of enforcing their decisions and asserting their competence as compared with more traditional government departments (Khee-Jin Tan 2002: 891). However, the 'great revolution' has resulted in considerable progress, albeit with a long way to go.

5.1.1 Within the Association of South East Asian Nations

As has been noted earlier, East Asia is one of the most biologically diverse regions in the world, its nations being heterogeneous, each with different environmental concerns and differing regimes for its protection. This has rendered regional cooperation in many fields difficult to achieve and, as such, it is necessary to some extent to study the various countries which comprise this region separately. However, within East Asia one major body for cooperation exists, bringing together the nations of southern East Asia. The Association of South East Asian Nations (ASEAN) was established in 1967 as a political and economic entity and has had a significant impact on governance in the region. Initially formed of Indonesia, Malaysia, the Philippines, Singapore and Thailand, these countries have since been joined by Brunei, Laos, Burma, Vietnam and Cambodia, with East Timor's membership currently pending.

Since the setting up in 1981 of the ASEAN Environment Programme (ASEP), to encourage the establishment of national environmental protection agencies and the 1984 Bangkok Declaration on the ASEAN Environment, which recognized the need to strengthen regional cooperation, the ASEAN countries have worked together on environmental issues, including their response to global environmental problems, but also on a regional approach to trans-boundary pollution, sustainable development, environmental technologies, climate change and protecting biodiversity and marine and coastal areas. This cooperation has resulted in many regional measures on problems specific to or prevalent in the region, such as the ASEAN Agreement on Transboundary Haze Pollution and the setting up of the ASEAN Centre for Biodiversity in 2005.

The ASEAN Vision 2020 adopted in Malaysia in 1997 calls for 'a clean and green ASEAN with fully established mechanisms for sustainable development to ensure the protection of the region's environment, the sustainability of its natural resources and the high quality of life of its peoples'. This suggests that ASEAN policy regarding development is not only environmentally focused but also acknowledges the impact of development on its peoples.

However, even within ASEAN there are disparities in the level of protection for the environment and human rights that this ideology provides. For example, the Philippines is known to be at the forefront of environmental legislation among Asian nations. The 1987 Constitution included a provision requiring the protection and advancement 'of the right of the people to a balanced and healthful ecology' in Section 16, Article II. This 'greening' of the Constitution – i.e. rendering the environment a constitutional concern – paved the way for significant domestic environmental legislation, such as the 1994 Clean Water Act, the Clean Air Act of 1999, the 2001 Ecological Solid Waste Management Act, the National Environmental Awareness and Education Act of 2008 and the Climate Change Act of 2009. The right to a healthy environment is incorporated throughout this legislation.

In 2010 the Philippines, supported by the UNDP, created new 'Rules of Procedure for Environmental Cases', known informally as the 'Writ of Kalikasan (Nature)', a landmark instrument in reform of environmental litigation and protection. They govern all criminal and civil cases in all trial courts regarding environmental issues. Since their adoption, the special remedies they provide for have been applied in several cases, e.g., following a major oil pipeline leak, to make an order for the rehabilitation of Manila Bay, and to issue environmental protection orders, e.g., in a mining case in Suriago and a coal-fired power plant case in Cebu. The Rules have been used creatively to try to recompense past damage and also to prevent potential harm. The former occurred in March 2011 when a Writ of Kalikasan was granted against Placer Dome Inc. and Barric Gold Corporation concerning the 2006 Marcopper mining disaster in Marinduque, the Rules finally helping the petitioners to succeed after five failed attempts to bring the corporation before the courts. The Rules have also been used to prevent potential harm to the environment and indigenous communities by banning mining in the Zamboanga Peninsula. They represent an innovative and forward-thinking approach to environmental justice and clearly embody the judiciary's commitment to upholding the environmental rights and development of the population of the Philippines. The country has also been progressive in a human rights context towards protection of the environment, and these cases will be discussed later.

Burma, in sharp contrast, has lagged behind its neighbours in terms of environmental protection because of the lack of interest in environmental protection shown by the military government. It has not set up a national environmental agency, an early measure taken by other ASEAN nations. However, it has become conscious of the need to pay lip service to environmental protection in order to encourage foreign investment and development, particularly in the field of oil extraction, but appears reluctant to introduce stringent laws, in order to allow investors to operate free of cumbersome environmental regulations. The lack of environmental controls on large-scale development projects has led to serious situations where basic human rights are no longer fulfilled, such as the right to water or to homes and property. Examples are China's construction of two energy pipelines from Western Burma to Yunnan, the Chinese-financed building of a series of hydroelectric dams on the Irrawaddy River in upper Burma which have already forcibly displaced many of the ethnic minority local population and India's construction of major infrastructure on the Kaladan River in Western Burma, which has resulted in land seizures by the Burmese authorities. However, there is hope yet that the government will begin to take environmental considerations more seriously. In 2011 the President Thein Sein announced the Chinese-funded construction of the Myitsone dam at the confluence of the Irrawaddy would be halted during his government, a decision that was applauded worldwide by environmentalists and human rights supporters alike. However, perhaps the economic and diplomatic consequences of this decision have proved too great, as in September 2013 at the sidelines of the 10th ASEAN–China Expo, the Burmese President agreed greater cooperation with China. Whether this will include resuming the dam project has yet to be seen.

5.1.2 Broader East Asia

ASEAN does not include all the countries included in this study, and therefore China, Japan and North and South Korea will now be considered.

South Korea has been progressive in its environmental protection, establishing an environmental agency and developing significant legislation since the incorporation of a right to a healthy environment in its 1988 Constitution, which provides for the right of all citizens 'to a healthy and pleasant environment' (Chapter II, Article 35(1)). The Constitution directs the State and all citizens to 'endeavour to protect the environment' and the State to 'protect the land and natural resources,' and to 'establish a plan necessary for their balanced development and utilization' (Chapter IX, Article 120 (2)). The domestic legislation that has been passed includes the Framework Act on Environmental Policy, the Air Quality Conservation Act and the Toxic Chemicals Control Act. This is further illustration of how granting the right to a healthy environment can prompt action on the part of governments to ensure that they fulfil and protect this right, by introducing environmental protection measures, i.e., the protection of human rights vicariously resulting in the protection of the environment.

Japan has also developed laws on environmental protection, many prompted by protests of the sufferers of several environmental disasters that caused serious illnesses in the late 1950s and 1960s, such as the cadmium poisoning from industrial waste in Toyama Prefecture, methylmercury poisoning from chemical factories in Kumamoto Prefecture, severe air pollution from industrial emissions in Yokkaichi port and arsenic poisoning from dust from arsenic mines in Shimani and Miyazaki Prefectures. The Basic Environmental Law (環境基本法) was adopted in 1993, which restricted emissions, waste, and land utilization, improved energy conservation, promoted recycling and implemented pollution control programmes and relief for victims as well as possible sanctions. In 2001 the Environment Agency became the Ministry of the Environment and a fully fledged part of the government to address environmental problems.

As one of the most rapidly developing countries in the world, China is a big country with a low per capita income that requires a large amount of energy to support its ongoing industrialization and urbanization, relying on huge amounts of coal and oil for its development. It has few laws relating to environmental protection, the main one being the 1989 Environmental Protection Law. A recent proposal to update this inadequate legislation has been criticised for drawing a balance that is too heavily weighted in favour of development with little concern for sustainability. With a 2006 speech revealing that more than 70 percent of the country's two million deaths annually from cancer were pollution related (*The Economist*, 24 January 2008), the human impact of environmental degradation is clear, but progress towards concrete and effective legislation is not yet being made. China is also facing new environmental challenges, such as the development of shale gas fracking. The environmental implications of this practice are not yet fully understood, but it is clear that it has the potential to exacerbate the nation's existing water crisis, as huge volumes of water are required to drill wells and shale gas resources are located in areas already suffering from water shortages. The obvious risk of lack of access to water is compounded by a risk of contamination of precious water resources. China will have to consider these environmental and human impacts in its pursuit of shale gas.

It is difficult to assess environmental protection laws in North Korea because of a lack of access to information. In the 1990s food shortages began due to land degradation caused by loss of forest, droughts, floods and tidal waves and overuse of chemicals. This obvious human impact of the state of the natural environment prompted North Korea to begin to take environmental protection seriously and in 2000 it decided to cooperate with UNEP and gave them information to compile a 'State of the Environment' report in 2003, which revealed a nation in environmental crisis. This process resulted in some environmental legislative progress, such as signing up to some international instruments, e.g., the Convention on Biological Diversity. However, the occasional show of political will in the country appears not to translate into implementing sustainable solutions and little real progress has been made in the environmental field in the past decade.

5.1.3 Conclusion

It is clear that levels of environmental protection in East Asia vary greatly, both within and outside of ASEAN. Those countries that have experienced particularly grave environmental disasters affecting human populations have sometimes been prompted to legislate to prevent further instances, or at least to begin to consider the impact of their activities on the natural environment. Others have not. It is clear, however, that in the countries that have taken environmental protection seriously, there has been a corresponding benefit in protecting and supporting human rights.

5.2 Human rights in East Asia

The protection of human rights and the environment are so intertwined it is difficult to consider the topics separately, particularly with the notion of a right to the environment often being the catalyst for environmental regulation, but the relationship will now be examined from the human rights perspective, particularly with regard to the interesting human rights litigation which has resulted from environmental concerns.

In the 1960s the UN General Assembly tried to encourage regional human rights systems, such as those that developed in Europe, Latin America and Africa. However, one large part of the world remained untouched by this drive, namely, the Asia-Pacific region.

East Asia is one of the most culturally and politically diversified regions in the world. For example, several different religions are followed, each with its own conception of rights; while Thailand, Myanmar, Cambodia, Laos and Vietnam are predominately Buddhist, a considerable proportion of the population of Brunei, Malaysia and Indonesia are Muslim and, in the Philippines, Christianity is the dominant religion. As well as religion, ethnic groups are diverse, across the region and even within nations. From a political perspective, states in the region vary from democratic to military and authoritarian. For example, the Philippines, Indonesia and

Thailand have undergone transition from authoritarian to democratic regimes. Vietnam and Laos are governed by Communist single-state parties and Burma is ruled by its authoritarian military junta. As already noted, the levels of economic development also differ significantly, again also within states as well as between them.

These factors are often considered to have hindered the development of any common conception of human rights. Rather, there is an idea of 'Asian values', which although difficult to define precisely, generally emphasizes communitarian values such as family and community over individual interests (Brun & Jacobsen 2003: 2). This concept is generally favoured by most nations in the region, perhaps because of the ambiguity of its content. Much has been written on the suitability of the concept of human rights for the Asian region, and it is not proposed to rehearse the discussion here; suffice it to say that it is doubtful that a sufficient degree of homogeneity exists which would allow a regional human rights system to function as they have done elsewhere.

5.2.1 A regional human rights treaty?

Attempts, however, have been made to coordinate respect for human rights in the region. The Asia Pacific Forum (APF) has been successful in establishing a network of 17 national human rights institutions (NHRIs) with a mandate to promote and protect human rights, which has been described as 'the closest the Asia-Pacific region has come to a regional arrangement or machinery for the protection and promotion of human rights' (Muntarbhorn 2003: 14).

ASEAN has also managed to achieve some level of cooperation with regard to human rights. Debate and discussion about the viability of, and necessity for, such a mechanism within ASEAN has been described as 'a long and winding road' (Muntarbhorn 2003: 3) characterized by regional meetings and protracted deliberations with scant progress in between (Durbach, Renshaw & Byrnes 2009: 222). Singaporean Ambassador Tommy Koh has famously said: '[There was no] issue that took up more of our time, [no issue] as controversial and which divided the ASEAN family so deeply as human rights' (Koh 2008). There are effectively three groups within ASEAN; those that 'champion' human rights, i.e., Indonesia, Malaysia, the Philippines and Thailand; those that are unenthusiastic, i.e., Cambodia, Laos, Burma and Vietnam; and those that try to bridge the gap between the two, i.e., Singapore and Brunei.

Despite the difficulties in reaching an accord on the issue, there has been some progress. The long and winding road led in 2007 to all 10 ASEAN leaders signing and ratifying the ASEAN Charter, giving ASEAN a legal personality. The Charter calls for the establishment of an ASEAN human rights body for the 'Promotion and protection of human rights and fundamental freedoms of peoples in ASEAN'. A high-powered Working Group for an ASEAN Human Rights Mechanism (Working Group) has been set up, whose primary goal is to establish an inter-governmental human rights commission for ASEAN. The impetus for this development came from the 1993 World Conference on Human Rights, which reiterated 'the need to consider the possibility of establishing regional and sub-regional arrangements for the promotion and protection of human rights where they do not already exist'.

In 2009 the ASEAN Intergovernmental Commission for Human Rights (AICHR) was created. It is faced with a difficult task given the organization's deference to a conservative notion of state sovereignty and its Member States' prioritization of foreign investment fuelled economic growth over human rights. However, it has still been criticised for lacking 'teeth': 'The AICHR has been given very weak terms of reference that limit its mandates, authority and powers to promote and protect human rights', according to Yap Swee Seng, executive director of the Bangkok-based Asian Forum for Human Rights and Development (Bangkok Post Opinion, 28 December 2012).

Further progress towards a treaty was made on 18 November 2012, when all 10 ASEAN nations adopted the ASEAN Human Rights Declaration. It contains nine general principles and lists civil and political rights; economic, social and cultural rights; the right to development and the right to peace. The AICHR was under pressure from human rights and environmental activists to include a right to the environment. Article 28 of the third chapter on economic, social and cultural rights states: 'Every person has the right to an adequate standard of living for himself or herself and his or her family including: [. . .] *The right to a safe, clean and sustainable environment.*' This is a big step forward for ASEAN, but given that it is merely a non-binding declaration, it remains to be seen whether it will have any concrete impact. It is certain that the region is still a long way from creating its own human rights mechanism or court however.

5.2.2 Recognition of the right to the environment in East Asia

The definition of a 'right to the environment' is a fraught task, with many conceptions being possible, and little clarity as to the difference between these conceptions, i.e., the terms 'clean', 'healthy' or 'adequate' and various other phrases used. The ASEAN Article 28 above uses the wording: '*safe, clean and sustainable environment*', but this raises questions as to the meaning of 'sustainable', particularly because most societies in the region depend on non-renewable resources that are fast being depleted. There is no consensus on the appropriate wording to be used, and there does not appear to be a trend in favour of one particular formation. The question remains controversial and is often one of the main criticisms of a substantive right to the environment, opponents using the vagueness of the right to argue against its usefulness. However, in whatever formation it appears, the existence of a right, be it in constitutional, legislative or judicial form, provides a secure platform for consideration of the human rights impacts of environmental matters.

As noted above, the majority of the 193 UN member nations recognize a right to the environment in its various forms through their constitution, environmental legislation, court decisions or ratification of an international agreement. The only countries that have yet to do so are the United States, Canada, Japan, Australia, New Zealand, China, Oman, Afghanistan, Kuwait, Brunei Darussalam, Lebanon, Laos, Myanmar, North Korea, Malaysia and Cambodia. Significantly, of these 16 countries, eight are located in East Asia.

However, as further noted earlier, several East Asian countries do recognize a right to the environment, such as East Timor, Indonesia, the Philippines, South Korea and Thailand. For example, Section 16 Article 2 of the 1987 Philippine Constitution provides that: 'The State shall protect and advance the right of the people to a balanced and healthful ecology in accord with the rhythm and harmony of nature.'

Several East Asian nations have experienced litigation based on their constitutional right to the environment, namely Thailand, the Philippines, Indonesia and South Korea. It is possible that other nations have had such litigation, but a lack of access to court decisions in certain countries, such as East Timor and the Maldives, prevents further discovery, as does a lack of available materials in English, which is a problem for example in China and Burma. It is in these cases that we find the most progressive interpretations of environmental rights and accompanying remedies for those who have suffered harm, and perhaps the most hope for prompting future legislative and policy change.

The famous case of *Minors Oposa v Secretary of Environment and Natural Resources Fulgencio Factoran* in the Philippines is one such example ((1994) 33 ILM 173). This was a lawsuit filed on behalf of children and future generations that sought cancellation of all the timber-harvesting licences in the Philippines, and resulted in a Supreme Court judgment asserting that: '[E]very generation has a responsibility to the next to preserve that rhythm and harmony [of the environment] for the full enjoyment of a balanced and healthful economy', stressing the urgent need to protect the environment on behalf of present and future generations. The judgment did not directly improve environmental conditions, but did result in a fall in the rate of deforestation and a significant reduction in the number of harvesting licences (from 92 when the case was filed to three some 10 years later). In 2008 another significant judgment based on the right to a healthy environment emerged from the Supreme Court in the *Concerned Residents of Manila Bay* case, in which the Court ordered a comprehensive plan to rehabilitate and restore Manila Bay, adopting the remedy of continuing mandamus, a revolutionary concept borrowed from the Supreme Court of India (G.R. Nos. 171947–48, 6 October 2009). This progressive judicial approach, combined with the new powers provided to them with the Writ of Kalikasan, is likely to result in the further successful use of constitutional environmental rights in such cases, providing a ground-breaking example not only for the East Asian region, but also worldwide.

There has also been litigation in Malaysia on the constitutional right to a clean environment despite no explicit mention of this right in the constitution. The Malaysian Court of Appeal has interpreted the right to life broadly as extending beyond mere existence to the quality of life and '[including] the right to live in a reasonably healthy and pollution free environment' (*Tan Tek Seng v Suruhanjaya Perkhidmatan Pendidikan* [1996] 1 MLJ 261). However, in a more recent case, the Malaysian Court of Appeal reinterpreted the right to life in a more restrictive manner (*Pihak Berkuasa Negeri Sabah v Sugumar Balakrishnan & Anor* [2002] 3 MLJ 72), which has been described as the 'final nail' for public interest litigation in Malaysia, concluding that while economic development is the driving force behind Malaysian policy, environmental matters will not receive the attention they have had elsewhere (Sharom 2002: 890).

In Indonesia, there is a constitutional right to a healthy environment and some evidence that citizens are invoking this right and, indeed, some examples of successful cases, e.g., a district court order of restoration of an area particularly affected by a deadly landslide caused by poor forestry practices and compensation to the affected individuals (*Mandalawangi Landslide Class Action Case* No. 49/PDT.G./2003/PN, 4 September (2003)). However, such examples are rare, and there exists an impression that companies are able to control the government and judiciary in environmental disputes (Bedner 2007: 121), leading to a lack of trust in the judicial process. It is clear that there is further work to be done to fully guarantee this constitutional right.

In Australia too, there has been litigation in which a consideration of human rights was imported into an environmental dispute, even though Australia has no constitutional right, its national courts have not recognized a 'right to environment' and there have been no cases where national courts have recognized that environmental harms have violated the rights to life or health or any other human rights.

In this particular case, involving the Port Kembla Copper mine, which had caused high rates of cancer and leukaemia in the area thanks to highly acidic sulphur dioxide and lead fallout from the mine, the issue was never discussed, because the case in the Land and Environment Court brought to review the reopening of the mine (closed due to being unable to meet pollution control standards) was dropped. This was because the New South Wales government passed special legislation declaring the new development consent to be valid despite any decision to the contrary. It is somewhat shocking that such disregard for both the environment and the human rights of a local populace could have occurred as recently as 1997, in one of the most developed countries in East Asia, and a leader within the ASEAN forum. It suffices to demonstrate that although encouraging movements are being made towards protecting human rights and the environment in tandem, there remains an even longer, winding road to travel yet.

6 Conclusion

Despite some very encouraging progress in connecting human rights protection with the protection of the environment in some countries in East Asia, for the most part the region is lagging behind progress in the rest of the world. Even within the sub-regional organization ASEAN, it has not been possible to reach agreement on whether or not legally binding provisions on human rights should exist, let alone whether these would include a substantive right to a clean and healthy environment, though the Declaration is encouraging in this matter. It is further unclear whether the recent Human Rights Declaration will have a positive effect in the region, or whether reaching agreement on such matters among a diverse group of attitudes will actually result in a 'lowest common denominator' and a diminishing effect on the level of protection in some of the more progressive countries.

Considering that the East Asian region contains not only some of the most environmentally hazardous nations, but also some guilty of, or at least turning a blind eye to, egregious human rights abuses, an interconnected approach would benefit both the physical environment and its inhabitants. It would also mean that in countries more attuned to the need for environmental protection, such regulation could be used to foster improved human rights in this field and vice versa in those countries appreciating the need for human rights protection but less concerned about the environment.

Litigation on the constitutional right in countries such as the Philippines and Malaysia, leaders in the environmental 'revolution', will, it is to be hoped, have a persuasive impact on their neighbours, particularly in discussions at the level of ASEAN, and result in a more consistent regional approach to environmental rights, particularly given the cross-border implications of many environmental situations, such as for example, the impact of the Irrawaddy river dams on both China and Burma.

Whether or not a right to the environment develops, it is nevertheless becoming clear to all East Asian nations, even those with little in the way of human rights or environmental protection, that the continued existence of humanity depends on the natural resources of the planet, and therefore that these must be protected accordingly. There are many options for furthering this goal; as explored earlier, it can be done via environmental legislation or via human rights principles. The latter can be both through a substantive recognition of the right to the environment, be it constitutionally or in legislation or from judicial authority or through connecting the need for a safe environment in which to live to other rights, most notably, the right to life or health. Further, procedural rights such as access to information, participation in decision making and access to justice are all capable of persuading governments towards the goal of both environmental and human rights protection. At this time all eyes are on ASEAN following the adoption of the Human Rights Declaration, which has the potential to foster a regional baseline for such protection, and it is hoped that it will be able to meet this challenge head on, with assistance from the more progressive countries such as the Philippines, and will resist pressure from those less amenable to the concept to pursue a lower level of protection. It is hoped that any improvements in ASEAN will have a persuasive impact on the rest of the East Asian nations in fostering better protection for the environment and associated human rights.

Bibliography

Bedner, A. (2007) 'Access to Environmental Justice in Indonesia', in A.J. Harding, *Access to Environmental Justice: A Comparative Study* (Martinus Nijhoff).
Boer, B. (1999) 'The Rise of Environmental Law in the Asian Region', 32 *University of Richmond Law Review* 5, 1503, 1552.

Boyle, A. & Andersen, M. (Eds.) (1996) *Human Rights Approaches to Environmental Protection* (Oxford University Press).

Brun, O. & Jacobsen, M. (Eds.) (2003) *Human Rights and Asian Values: Contesting National Identities and Cultural Representations in Asia* (Curzon Press).

Durbach, A., Renshaw, C. & Byrnes, A. (2009) 'A Tongue but no Teeth? The Emergence of a Regional Human Rights Mechanism in the Asia Pacific Region', 31 *Sydney Law Review* 211, 222.

Jeffords, C. (2011) *Constitutional Environmental Human Rights: A Descriptive Analysis of 142 Constitutions*, 16 Economic Rights Working Paper Series, University of Connecticut. Available at http://www.econ.uconn.edu/working/16.pdf.

Khee-Jin Tan, A. (2002) 'Recent Institutional Developments on the Environment in South East Asia – A Report Card on the Region', 6 *Singapore Journal of International and Comparative Law* 891.

Kititasnasorchai, V. & Tasneeyanond, P. (2004) 'Thai Environment Law', 4 *Singapore Journal of International and Comparative Law* 1, 4.

Koh, T. (2008) Director of the Institute of Policy Studies, Ministry of Foreign Affairs, Singapore, talk at the 7th Workshop on an ASEAN Human Rights Mechanism, 12–13 June, Singapore.

Muntarbhorn, V. (2003) *Roadmap for ASEAN Human Rights*, Publication of the Working Group for an ASEAN Human Rights Mechanism 1, 14. Available at http://www.fnf.org.ph.

Sharom, A. (2002) 'Malaysian Environmental Law: Ten Years After Rio', 6 *Singapore Journal of International and Comparative Law* 855.

Notes

1 Declaration by the Environment Ministers of the region of the United Nations Economic Commission for Europe (UN/ECE), 4th Ministerial Conference 'Environment for Europe', Aarhus, Denmark, 23–25 June 1998.

2 Interestingly, a similar document for the Asian region pre-dates the Aarhus Convention. In 1990 the Ministerial Declaration on Environmentally Sound and Sustainable Development in Asia and the Pacific was adopted in Bangkok, which in Para. 27 affirms the right of individuals and NGOs to be informed of environmental problems relevant to them, to have the necessary access to information, and to participate in the formulation and implementation of decisions likely to affect their environment (16 October 1990), A/CONF.151/PC/38.

3 The Convention on the Rights of the Child (20 Nov. 1989), Art. 24(2)(c).

4 ILO Convention No. 169 concerning Indigenous and Tribal Peoples in Independent Countries (27 June 1989), Arts 2, 6, 7, 15. The UN Declaration on the Rights of Indigenous Peoples, adopted by General Assembly Resolution 61/295 on 13 September 2007 contains similar provisions.

5 UN GA Resolution on Promotion of a Democratic and Equitable International Order, of 8 March 2010, UN Doc. A/RES/64/157.

6 Resolution 35/48 on Drafting of an International Convention against the Recruitment, Use, Financing and Training of Mercenaries, GA Res. 35/48 of 30 October 1980.

7 Organisation of African Unity (OAU) Banjul Charter on Human and Peoples' Rights, OAU Doc. CAB/LEG/67/3/Rev. 5, Article 24.

8 Additional Protocol to the American Convention on Human Rights in the Area of Economic, Cultural and Social Rights, 1989, 28 ILM 156, Article 11.

9 See, e.g., Lluis Maria de Puig, President of the Parliamentary Assembly of the Council of Europe (PACE) on 28 March 2008; Report of the Parliamentary Assembly Committee on the Environment, Agriculture and Local and Regional Affairs, Doc. 12003, 11 September 2009.

10 See, e.g., Steering Committee For Human Rights (CDDH) Report of 69th Meeting, 24–27 November 2009, CDDH(2009)019; Reply from the Committee of Ministers adopted at the 1088th Meeting of Ministers' Deputies (16 June 2010), Doc. 12298.

11 Earthjustice, Environmental Rights Report on Human Rights and the Environment (2005) 37.

3

Environmental justice and civil society

Case studies from Southeast Asia

Piya Pangsapa

One of the most important developments in environmental studies in recent years is the increasing recognition of the close interconnections between social and environmental justice. In conditions where rapid development and modernisation are the central goal, civil society action is even more relevant in pressing reluctant governments to regulate and be more accountable for their actions and for political institutions and private companies to be more responsible. Research and reports of illegal activities, such as the timber trade (Elliott 2007; Fahn 2004; Pangsapa & Smith 2008; Sample 2007; Unger & Siroros 2011) or species trafficking (Davenport 2011; Henry 2004; Karam 2010; Simpson 2012) as well as on legal but irresponsible activities resulting in environmental degradation indicate the range of challenges faced in Southeast Asia (Elliott 2004; Hirsch 2007; Litta 2010; Smith & Pangsapa 2008, 2009c).

The emergence of civil society action and movements in direct response to harmful environmental activities and specific social injustices provides us with an opportunity to consider how social justice issues and the political decisions that ignite social action can be better understood with an eye towards mutual engagement, dialogue and transformation. In addition, by allowing constituencies to become stakeholders (see Smith & Pangsapa 2008), such acts of citizenship (Isin 2008) can generate a set of working principles that acknowledge the complexity, uncertainty and interdependency between society and nature in order to develop flexible strategies for change and cooperation.

When we consider the practices of ecological citizenship, as part of the 'context-specific' ethico-political discourses, we can identify how competing ideas of moral community and political community are articulated. Civil society organisations working in concrete strategic situations inevitably draw on their own ethical traditions that they innovatively and creatively link to political activity and which they use effectively as part of their campaigns (Smith & Pangsapa 2009b: 328). In the context of Southeast Asia, such discourses draw heavily on Buddhist virtues about the environmental relationship with social acts as a basis for ethically informed decisions.

This chapter seeks to underline the connection between social and environmental justice through the comparison and analysis of activist and civil society groups in Southeast Asia in relation to a range of social and environmental problems. Drawing on new case study research of industrial pollution in Thailand and illegal logging in Cambodia, this chapter will examine how social and environmental injustices are connected in the context of rapid development in the Southeast Asian mainland. A particular focus will be how the local and the transnational are configured in these contexts.

New developments on the Southeast Asia mainland

Environmental and community livelihood issues must be understood in the context of regional growth and development. Laos, a country that anticipates becoming the 'battery of Asia', already exports 75 per cent of its electricity to Thailand fuelling rapid export development. Gas pipelines from Burma also satisfy Thailand's energy demands and the opening up of this resource-rich country signal its potential of becoming a major supplier of energy and natural resources in the region. Natural gas generated an estimated US$2.7 billion for the Burmese military junta in 2007, 'constituting almost half of its total exports' (Carroll & Sovacool 2010: 632). The growth of commercial agriculture and the extension of textile manufacturing operations continue to propel Cambodia's growth (generating labour shortages in some provinces) while Vietnam faces energy problems owing to its rapidly increasing urban and industrial expansion. Thailand has undoubtedly been a driving force in economic development in the region often facilitated by Thai industrial investment and trade ties to regional business networks.

The orientation of the Thai economy towards export-led growth (reaching a record high of 11.3 per cent in March 2012)[1] also means that expansion depends on resource grabs in neighbouring countries, whether for raw materials, energy or water. Land designated as forest and woodlands is now at half the level it was in 1960 making Thailand a big importer of wood. As Pangsapa and Smith indicate (2009), increasingly limited forest reserves and fewer suitable locations for large-scale hydroelectric dams since the 1990s have prompted the Thai government to encourage the exportation of environmentally degrading projects into neighbouring countries.

It is therefore crucial to understand the context in which environmental change is taking place. This means we should be thinking of social and environmental issues in terms of regional development. One aspect of this new regionalism is an ambitious plan to create an integrated ASEAN power grid among the countries of the Greater Mekong Region. Carroll and Sovacool examine the trans-ASEAN gas pipeline project (TAGP) as well as the political and economic factors that influence regional cohesion. Their analysis of Southeast Asian regionalism through the energy sector points to what they call 'contested regionalism', which essentially transcends the limits of state-centred analyses and sees regionalism as a process involving the competing interests of state and capital (Carroll & Sovacool 2010: 626).

Collaboration between governments as well as between state and non-state actors has also contributed to regional development in particular when this involves cross-border trade, investment and tourism. It also facilitates the provision of infrastructure such as roads, bridges and energy projects (Hughes 2011: 183). Community livelihood movements outside Thailand find it difficult to make their views heard due to the authoritarian nature of many SEA regimes. It is also important to note that rapid development often generates significant external costs that are borne unequally by some social groups more than others as well as environments on which communities depend for their livelihoods.

From a Thai focus to a regional focus?

China is both an important investor in Southeast Asian economies and a key customer for goods and services derived from these economies. China now figures prominently as both a key import and export partner in the process of regional development, due to its demand for energy and the range of resources including both industrial purpose and luxury timber (Carroll & Sovacool 2010; Cronin 2012; Plokhii 2012b). Cronin has highlighted China's geopolitical interests in the Mekong as part of its perceived trading and strategic influence in the region. However, the policy of pursuing rapid development has resulted in serious environmental degradation and has also endangered human security in the region.

As a major donor to countries such as Laos and Cambodia, China has invested significant sums into a 'network of roads, bridges, dams and power lines' that have affected the millions of inhabitants who live along the Mekong in both positive and negative ways (Cronin 2012). When the environment is damaged in a way that affects how people can maintain their livelihoods, then, in the context of a developing society, that affects their ability to feed themselves, make handicrafts, grow crops for sale, produce a variety of other goods that communities can bring to the market; in short, it impedes their economic independence and that independence is a basis for social activism both here and in other parts of the world. The loss of forest resources for example is one of the most important ways in which communities are affected in a negative way. A joint 2012 UNEP and Interpol study states that illegal logging[2] accounts for up to 'thirty per cent of the global logging trade and contributes to more than fifty per cent of tropical deforestation in Central Africa, the Amazon Basin and South East Asia' (Interpol 2012). Illegal logging also inflicts serious damage on local communities that live in and around the forests through soil erosion, flooding, loss of biodiversity and the release of climate-warming carbon into the atmosphere (Lobe 2012).

It is important to note that when injustices occur they tend to occur in the same place (Smith & Pangsapa 2011), in other words, social and environmental impacts are often conjoined, and have been well documented in the research literature on the countries of the Mekong River Basin that has addressed key aspects of environmental degradation as well as the emergence of social movements, their alliances and civil society actors (Fahn 2004; Forsyth 1999, 2001, 2007; Hirsch 1997, 1998, 2001; Hirsch & Lohmann 1989; Hirsch & Warren 1998; Johnson & Forsyth 2002; Lazarus et al. 2011; Lestrelin et al. 2012; Molle et al. 2009; Pangsapa & Smith 2009; Smith & Pangsapa 2008, 2009a).

Activist networks have been particularly keen to minimise the negative impacts of hydropower development and focus on how benefits derived from hydropower can be improved in the Mekong water systems (Lazarus et al. 2011; Molle et al. 2009). They have concerns about how large-scale developments are proposed and justified as well as the social and environmental consequences of water resource development and management. This raises the important question of who development is oriented towards – is it for citizens, a community or social classes, a nation or a region? This question is particularly important for understanding and explaining the relations and processes that take place within the two case studies explored in this chapter.

Researchers suggest that development-driven agendas can be reconciled with the maintenance of community livelihood provided that the needs and concerns of those communities are recognised through participatory decision making or at least some kind of consultation based on local knowledge(s). As an example, Hirsch investigated forestry as a social movement from the perspective of marginal groups and points out that because community forestry is a contested issue between local people and state authorities, it needs to be approached in a 'more inclusive way' (Hirsch 1997: 16).

Similarly, on rights-based approaches to forest conservation in Thailand, Forsyth points out that efforts to support marginal groups in asserting their rights over livelihood issues are often undermined by state interests (Johnson & Forsyth 2002: 1591). However, in the last decade the focus has shifted more to the complex relationships between rights and duties on environmental impacts as well as their underlying informal relationships between entitlements and obligations (see Smith & Pangsapa 2008: 84). Recognising that environmental NGOs need to go beyond advocacy to stakeholding also implies an awareness of how rights and entitlements relate to duties and obligation.

Johnson & Forsyth (2002) highlight the role played by NGO alliances as advocacy NGOs become increasingly engaged in partnerships with governments as stakeholders in negotiations on policy making and implementation. By the same token, Hirsch (2001) explores resistance through specific case studies such as dam construction in Laos and Thailand. Hirsch points to a 'highly differentiated civil society and complex politics of legitimacy and resistance' associated with development agendas (Hirsch 2001: 239) and poses pertinent questions concerning the relation between environmental politics and the emergence of civil society structures and processes particularly when these actions seek legitimacy across national boundaries.

The recognition of the transnational nature of responsibilities, whether they are described as duties or obligations, implies that a variety of actors and organisations, indeed peoples, share responsibility for particular impacts. So if responsibility is both unevenly distributed and widely dispersed, we need to consider how different institutions and organisations can develop collaborative arrangements or partnerships that address rights and duties on both sides of the equation – in public and private spheres. Similarly, the literature on regional development indicates that we need to be more alert to the increased complexity of environmental problems in Southeast Asia and how responsibility is socially distributed as part of national and regional economic growth and expansion.

When considering the Southeast Asian mainland, it is clearly sensible, in the light of the above development, to develop a perspective that is less state centric and more focused on regional relations and processes, without, of course, forgetting the importance of states in the region. In short, this approach can be designated as *regional-optic* rather than *nation-optic* perspective, to bring these increased complexities into sharper focus. In subsequent sections, this chapter will explore such complexities through a series of cases involving deforestation and industrial pollution on the Southeast Asia mainland.

Our forest in Cambodia's Amazon

> Without forest we would have no access to clean water as the source of life . . . Forest is like the skin covering our body.
>
> (Chut Wutty 2012; Plokhii 2012b)

Forest degradation has long been an issue in Southeast Asia but, in spite of the actions of environmental NGOs, in some contexts it is accelerating rather than slowing down. 'Since 1990, Cambodia has experienced one of the fastest deforestation rates in the world, with some 6,200 square kilometres of old-growth woodlands and endangered timber cut down in the country's most pristine jungles', writes Plokhii (2012b). This kind of problem has a longer history, for in 1992, all the trees in a vast area near Cambodia's border with Thailand were wiped out (Wallace 1992). By the mid-1990s, ministers and officials of the royal Cambodian government were also heavily implicated in the destruction of Cambodia's forests (Gray 1996; see Global Witness 2004; Currey 2001; Currey et al. 2003, for in-depth reporting, investigation and documentation on illegal logging in Cambodia, Indonesia and other countries). By 2012, illegal logging had stripped the Cardamom Mountains considered to be the last intact species-rich forests in Southeast Asia.

As Plokhii identifies, 'In addition to outright illegal logging, land concessions for sugar, rubber, acacia and mining plantations are also a major cause of illegal clear-cutting' (Plokhii 2012b). This continual expansion, often to make way for hydropower dams and other state-supported business ventures, has angered residents whose livelihood depends on the forest. Their daily efforts to safeguard and protect their environment have also led to the risk of harassment, intimidation and violence. On 26 April 2012, the nation's leading environmental activist, Chut (or 'Chuy') Wutty, was killed during a stop at Veal Bei point in Mondul Seima district's Bak Klang commune in Koh Kong province (Schearf 2012; Titthara 2012).

In the absence of adequate ways for civil society movements to ensure demands are heard by government, direct action groups have emerged such as the community-based livelihood movement, the 'Prey Lang' ('Our Forest') Network, which was formed to stop illegal logging in the Prey Lang Forest. In March 2012, 200 members of this movement walked deep into the country's only rainforest 'to discover newly built roads and illegally logged timber' (Noun 2012). Led by Chuy Wutty, the group confiscated and damaged logging equipment and burned the cut wood. The forest, which is also home to more than 200,000 people including the indigenous Kuy people, has lost over 1 million hectares and the capture of an illegal logger by villagers only confirmed the fact that most loggers are doing it for the money. This makes Cambodia a good illustration of where cheap (and often desperate) labour and ample natural resources combined with the compliance of the local political system can lead to environmental destruction, corruption and human rights violations.

The Prey Lang Community Network (PLCN) emerged in 2007 following many years of advocacy and campaigning by localised groups. The PLCN was formed mainly by members of the Prey Lang forest communities who were instrumental in pressuring the government to end large-scale commercial logging in Cambodia.[3] Today the Network is comprised of active members from more than half of Prey Lang's 339 communities across four provinces.[4]

In Cambodian law, all development projects are subject to environmental impact assessments along with consultations with affected communities. PLCN members and forest communities report that no such assessments have been undertaken on any of the major developments that have involved community consultation. Moreover, the disregard of Kuy people's legal rights and entitlement to the forest violates the constitution, which grants collective land ownership rights to indigenous communities. It is also contrary to the rights provided under the UN Declaration on the Rights of Indigenous People (2007), to which Cambodia is a signatory. The royal government of Cambodia recently drafted a sub-decree that would provide for the establishment of a Prey Lang Forestry Protected and Biodiversity Conservation Area but the designated area excludes important areas of forest while cutting off many of the surrounding communities from the forest.

Forests are (our) lives

> We've borrowed the forest from our children. We must protect it for them.
> (Pok Hong, a Preah Vihear mother of five (preylang.com))

Since 2009 the Prey Lang Network has petitioned the government to save the forest and to allow the Network to become co-managers of the forest. The PLCN lobbied the Forest Administration to hold public consultations on the sub-decree and has brought its concerns to media attention. As such, the community has come together to form a citizens' brigade taking its actions to local and national events. Community members would show up with their faces painted in blue inspired by the film *Avatar* and they have staged repeated protests in and around Prey Lang, including a 10-day 'occupation' of the forest to call public attention to illegal logging and land concessions. They have also put up road blocks, collected 30,000 petition signatures, initiated forest patrols with youth groups, mapped various areas of the forest to track government concessions as well as illegal activities and conducted their own biodiversity surveys to document species biodiversity.

The vocal and visual campaign of these 'avatar communities' has enabled the network to develop a strong national following and media coverage, including ardent support from a wide coalition of civil society organisations (CSOs). In June 2012 a PLN coordinator participated in the CSO Green Development Conference that took place alongside the RIO+20 UN Conference on Sustainable Development in Brazil. At the Conference, the coordinator high-lighted how Cambodian communities are mobilising and forming alliances and how they are articulating these demands with the promotion and protection of human rights. The demands of the PLCN on the Cambodian government include halting forest crimes, government trans-parency and accountability, impact assessment of all projects with full community participation, the implementation of regulations concerning land concessions, and punishment of government official who engage in corrupt practices with log traders. These measures would demonstrate support and recognise the claims and demands of the Prey Lang communities.

Prey Lang forests can also be a matter of life and death

> What happened today is meant to be a chilling message to us, the concerned citizens, the rights advocates: You mess with us, you pay with your life.
> (Theary C. Seng, founding president of CIVICUS (Centre for Cambodia Civic Education) on the death of Chuy Wutty (preylang.com))

In October 2012 the Cambodian government made the decision to dismiss a judicial investigation into the April 2012 murder of Chuy Wutty. His position as President of the Phnom Penh-based Natural Resources Conservation Group (NRCG) provided a platform for criticising corruption in Cambodia at a national level and the destruction of the country's natural resources in the Prey Lang forest (Schearf 2012; see also preylang.com):

> Since 2010 we have seized and destroyed many machines that we have found in our Prey Lang Forest . . . When we see the tools used for cutting logs we must destroy them in this area after we take pictures as evidence . . . It is the only action we can do to crack down on the activities of illegal loggers.

Wutty highlighted that their investigation revealed that only 25 per cent of the forest is left and that as a result the government must prevent any further destruction from happening. When the Wutty team and the Prey Lang villagers burned 40 cubic metres of timber as a deliberate act of civil disobedience, they stated that it is 'their only choice' (Noun 2012).

Wutty's campaign against the government's granting of land concessions to develop 7,631 square kilometres (2,946 miles) of land in national parks and wildlife sanctuaries indicated the scale of the problem and the impact on local community livelihoods. This campaign was particularly critical of the role of the military police who were used primarily to protect private business interests (Cambodia Human Rights and Development Association 2012; Thul 2012). The campaign's mission to empower communities to protect livelihoods was documented by

Mathieu Young who highlighted Wutty's two-pronged approach: i) on the ground, documenting evidence and stopping illegal activity as they found it in the forest; ii) in Phnom Penh, working to influence the government through petition and legal counsel through the NRCG (Young 2011).[5] As a result of this combined approach in August 2012, Cambodian Prime Minister Hun Sen cancelled four concessions in Prey Lang, protecting 40,000 hectares (100,000 acres) of forest and signalling a 'rare victory for those battling to preserve the Prey Lang forest' (Earth Action 2012). The PLCN posted a message stating that while it welcomes the RGC's cancellation of concessions in Prey Lang's core area, it remains concerned with the rampant illegal logging that is still going on and the continuation of other concessions that have destroyed many pristine areas of forest (ibid.).

Cambodia's forestry crisis is in part the result of external demand, in particular China's demand, for valuable and rare luxury timber such as rosewood used in furniture as well as specialist material for the manufacture of musical instruments (Plokhii 2012b; Reynolds 2005). The World Bank Report of March 2012, *Justice for Forests: Improving Criminal Justice Efforts to Combat Illegal Logging*, urged local and international enforcement to target organised crime syndicates that profit most from illegal logging. However, few preventive measures had any impact and some environmental groups suggest that the World Bank should consider how *it* contributes to this problem when financing large-scale development projects (Lobe 2012).

Despite the efforts of activists, advocacy groups and the local community, illegal logging is still taking place in Cambodia. It should be stated that this situation highlights how the concepts of rights, obligations, community, ethics and power are interconnected, and are central to understanding what constitutes social and environmental justice. Cultural values are also deeply impacted by social and environmental injustice – the Prey Lang campaign, for instance, is much more linked into Buddhist values especially since the forest is important 'both spiritually and economically' for a large majority of Cambodians (Beebe 2012). Prey Lang is also supported by Svay Phoeun – a Buddhist monk advocating for the protection of the forest. While this case highlights the close connections between loss of community livelihoods and environmental damage, the next one, on Map Ta Phut (MTP), indicates how questions of social and environmental injustice come together in an urban rather than a rural context.

Industrial pollution in urban Thailand

While the previous case study focused on the social environmental injustices that result from resource extraction in particular the effects of deforestation, in this case we explore how the same factors work together in an urban industrial setting in Southeast Thailand. While these two contexts are not far apart within the scope of the region as a whole, they highlight the important differences that occur when developing countries experience rapid change. As with many rapid developing countries and regions, we often witness the combination of both intensive resource exploitation at the same time as increased concentrations of industrial activity within economic zones that have enjoyed a combination of state subsidies, lax regulation and minimal taxation.

One particular province that has specialised in attracting investment in this way is Rayong province. Of particular importance within Rayong is Map Ta Phut (MTP) which has become one of the most intensive sites for industrial investment in the Thai economy. For some financiers, investing in the intensification of production in MTP makes sense because it keeps all the negative effects within that one location. For the people who live there, it is an entirely different matter. What this case study demonstrates is that despite the differences from Prey Lang in Cambodia, similar patterns seem to emerge. In particular, when environmental injustices occur social injustices are present at the same time, as well as the potential for wider violations of human rights and labour standards. As a result, solutions to environmental problems have to take into account the social dimension just as solutions to social injustices often have an environmental dimension. What the MTP suggests is that the disproportionate distribution of environmental and social 'bads' tend to congregate in specific locations, as 'clusters of injustice'.

A swampy area in the late 1970s, MTP soon became a town following industrial development efforts that began to take shape in the early 1980s; by 1990 the MTP Industrial Estate was established, managed by the Industrial Estate Authority of Thailand under the Ministry of Industry. Financed in large part by Japanese investment during the decade 1982 to 1993, the development of the MTP industrial zone involved construction projects to accommodate heavy chemical industry (FOCUS 2012). As early as 1988, residents had started to complain about air pollution. By 1997 cases of severe health problems were being reported along with an incident that involved the hospitalisation of students and teachers after inhaling toxic air. An evaluation study in 2000 noted 'the need to look at measures for limiting environmental impact particularly for MTP area' (ibid.: 3). On 1 May 2009, after years of lobbying, MTP was declared a non-pollution zone. In the same year, the zone's continual expansion was brought to a halt. By this time, at least 2,000 people had died from cancer linked to the pollution generated by the industries in the estate (Kovidhavanij 2012).

A town and sub-district in Rayong Province, MTP is the site of Thailand's largest industrial park and the world's eighth largest petrochemical hub. For this reason, it is a vital part of the Thai economy and the prosperity generated through development for the country as a whole. Few parts of the world have the same degree of concentrated industrial production. The estate contains 45 petrochemical facilities, 12 fertilizer factories, eight coal-fired power plants and two oil refineries (Smith & Pangsapa 2008: 223). Today, the industrial zone also includes a 'high-volume, high-capacity' industrial port built to service the wide range of vessels, equipment and cargo for the area's heavy duty industries (FOCUS 2012). When industrial concentration is so intense without adequate regulatory oversight, the potential dangers are legion.

According to the Pollution Control Department (PCD), 25 chemical-related mishaps have taken place in Rayong since May 2009 that involve illegal dumping of chemical waste, transport-related accidents, as well as fatal accidents and injuries. There are also less frequent but potentially more damaging hazards in MTP. Deboonme reports that volatile organic compounds (VOCs) in the form of toxic gases emitted from certain solids or liquids were discovered by the PCD in 2011 to March 2012. In May 2012 an explosion at a petrochemical factory killed 12 people and injured over 130 bringing to the fore concerns over safety. Residents of the 17 communities who live within a five-kilometre radius of the plant are now concerned about the chemicals in the air that cause breathing problems and skin irritation (Deboonme 2012).

With so many environmental problems affecting people in vulnerable communities there has been growing impetus for citizen action. As early as October 2005, a community campaign against toxic cocktails was launched in a joint effort with Greenpeace Southeast Asia (GPSEA), Campaign for Alternative Industry Network (CAIN) and Global Community Monitor (GCM) to demonstrate the extent of pollution released by the facilities in this 176,000-acre industrial zone (see Smith & Pangsapa 2008: 222). Building alliances between a range of different groups and interests including alliances between business and community groups, has consistently been a key feature of MTP campaigning. Using bucket brigade techniques, the GCM report claimed that the local communities experience airborne toxic chemicals between 60 and 3,000 times higher than EPA standards. This *community-based environmental monitoring technique* allows the communities to develop 'do-it-ourselves pollutant inventories' and test the collected samples.

Environmental activist Penchom Saetang of Ecological Alert and Recovery Thailand (Earth) had spent over 10 years studying and documenting the pollution problems at MTP. She indicates that the root cause of the crisis lies at the feet of government and the companies and points out that Thailand did not have any environmental regulation before 2009. Penchom highlights both the basic levels of the problem and their increased severity: 'Map Ta Phut is a modern industrial estate, but some local communities don't have supply of clean water; they have stopped using rain water because of contamination' (Hariharan, 2010). Penchom adds that locals formed a 'smelling group' and, using their noses, they walk around Map Ta Phut to detect the source of air pollution. The group found seven factories responsible for polluting the air and requested temporary closure of the factories. The residents of the more than 20 communities that live near the estate also have to contend with illegal dumping, which has been going on since 1998, as well as erosion of the coastal area.

It is important to note that MTP is also well known for having problems associated with social injustice in part because of the high level of poverty but also as a result of migrant workers who lack citizenship rights. Feldmann points to the range of social problems in MTP including crime, drug use, the spread of HIV and increasing incidents of suicide as a result of desperation (Feldmann 2012: 2). Suicide rates around the estate are reported to be 11 times higher than the national average and a serious public health concern caused by the problems of industrial expansion. The interviewees in his study expressed their fears about getting cancer, failed crops, personal safety and inadequate medical facilities should they fall ill or be injured through living and working in MTP. This study captures what livelihood means to the different social groups living in this context. For farmers and fishermen, it means the loss of their livelihood through failed crops and declining fish populations. For migrant workers, it means poor living and harmful work conditions. Contrariwise, for industry, it means the generation of revenue without any of the risks associated with having to provide compensation for those negatively affected. As such, MTP exemplifies a lived space in which 'the discrepancies and inequalities manifest themselves within the lives of the people' (Feldmann 2012: 6).

Industrial development has also impacted the local community in other ways. Monks at the local Buddhist temples in MTP have had to conduct their daily meditation separately rather than together and some temples have had to install glass windows and doors, which significantly alter the architectural style of traditional Buddhist temples (Viwatpanich 2010). Buddhist monks have also had to take on the role of consultants providing much needed counsel and spiritual guidance to individuals and their families instead of spreading *dhamma* and Buddhist teachings. Access to the temples is also proving to be difficult especially for elderly residents who are afraid to walk to their church because of heavy commercial traffic linked to the industrial estate.

The Map Ta Phut crisis

In September 2009 the Thai Supreme Court suspended 76 projects at the MTP Industrial Estate due to the absence of health impact assessments required under Article 67 of the 2007 Constitution (FOCUS 2012: 1). Suspension was soon lifted for 11 projects and in the end 74 of the 76 suspended projects were allowed to move ahead. Human Rights Osaka indicates in its June 2012 Focus Report that a 2010 analysis by Silapakorn University of the environmental and health impact studies (conducted by the companies concerned) found that 35 of the 76 industrial plants suspended in 2009 use hazardous chemicals that could cause health ailments and 21 plants use carcinogenic substances in their production. The reasons for the momentous decision and its dramatic reversal are the focus of this subsection.

In 2009 the Thai Supreme Court suspended these developments in MTP, halting projects worth around US$10 million, directly as a result of legal cases initiated by community activists. These cases sought to block additional expansion of the MTP Industrial Estate on the grounds that on top of the existing air pollution and contamination problems these new facilities, particularly petrochemical projects, would intensify these problems even more. The residents of MTP used Paragraphs 2 and 3 of Section 67 on Community Rights of the 2007 Constitution of Thailand that recognises the right of the community to sue the government and other State agencies in their lawsuit (FOCUS 2012: 4). The adamant position of local residents and their desire to focus both on air pollution and illegal dumping in the coastal areas made it difficult for groups such as Ecological Alert and Recovery Thailand to remain neutral or look for compromise.

In 2010 the administration of Prime Minister Abhisit Vejjajiva (2008–2011) ordered a study of the problem through a panel under the auspices of the National Environment Committee (NEC) headed by former PM Anand Panyarachun. The NEC study initially identified 18 types of industrial project as harmful to the environment but inexplicably this list was subsequently shortened to 11 (Sarnsamak 2010). From the community's perspective, the government's amended list angered NGOs and villagers. Subsequently, they threatened to blockade the estate to force a review of this new shorter list proposed by the National Environment Board (NEB) and demanded that the original list should be used instead, taking full account of the range of problems affecting vulnerable communities in Rayong.

According to the chairman of the Environmental and Health Independent Organisation, the NEB accepted only five of the 18 harmful activities from the original list as being legitimate concerns. They then added them to the new list containing six other activities proposed by the board, with no clear explanation or justification (Sarnsamak 2010). Local community activists found it hard to comprehend the intent of those engaged in gathering evidence of the problems of MTP. In September 2010 the Eastern People's Network (EPN) called on the Senate to intervene and recommend that the government review this new list of 11 that would have harmful environmental and health impacts (*The Nation* 2010). The Network coordinator was accompanied by more than 30 residents when making the demand and petitions were submitted to Senate committees on all aspects of rights, protection and good governance. The chair of the senate committee noted that the fact that MTP had already been declared a pollution control zone was not taken into account (ibid.).

Clusters of injustice in Map Ta Phut

The publicity materials of the campaign have primarily focused on the disproportionate distribution of environmental 'bads' in MTP, especially with the increased concentration of industrial development in Rayong following these new projects. Nevertheless, the effects of pollution have impacted on community lives, the health and safety of workers in the industrial facilities and the prospects for local businesses and self-employed citizens. As a result, questions of social and environmental justice are inextricably linked on the ground. In addition, the GCM, which has been active throughout the campaign, has consistently stressed the combination of social and environmental injustices in this industrial zone. The development of community health monitoring and local alliances of citizens, labour unions, businesses and community organisations suggests that there is scope for the effective use of evidence on pollution that is somewhat short of a full public health survey. This campaign provides a concrete example of how communities mobilise against the combined effects of social and environmental injustices in urban and industrial contexts.

Rather than rely on protest and direct action, this alliance has also sought to challenge the increased concentration of industry in MTP through the legal system. On 7 December 2012 114 MTP residents filed a lawsuit against the Industrial Estate Authority of Thailand (IEAT) for 160 million baht (US$5.2 million) in compensation for damages to the environment and public health. The EPN plans to collect evidence that would demonstrate how industrial development on the estate has affected people's health (*The Nation* 2012). So in this particular context, the combination of street protests and marches, alliance-building across a variety of interest groups and legal action has enabled the campaign movement to present obstacles to the implementation of increased industrial development.

Having reviewed the MTP case, authors of the 2012 Human Rights Osaka Focus Report consider whether people's health and wellbeing and protection of the environment can be guaranteed in maintaining this large-scale industrial zone. The authors contend that the link between industrial pollution and environmental and social harm cannot be denied and they emphasise the influential capacity of constitutional provisions on 'community rights, the action of the local residents to protect their rights, and the court decisions' as essential in preventing a worse-case scenario. The authors also point to a new multi-billion dollar deep-sea port and industrial zone being established in Dawei in Southern Burma, dubbed the 'new global gateway of Indo-China' (ibid.: 4–5). The likely environmental impacts in Dawei are likely to match the environmental and social impacts experienced in MTP.

In the Thai case, local residents have come together because their health and wellbeing are at risk and their livelihoods threatened. As a result, they are sensitised to the issues and ready to act when they respond to decisions that affect them directly. Along with support from advocacy and rights groups, environmental movements in Thailand have been successful in embarrassing successive governments on issues of environmental degradation as well as public health.

Forsyth (2001) suggests that the success of campaigns on industrial environmental issues is much more difficult to resolve as their effectiveness depends as much on scientific expertise as it does on activism. Given governments' pressing need for increased power generation, the vested interests of industrial sector funding, and the underlying problems of movements seeking to make a difference in political systems that have histories of graft, corruption and 'money politics' (Pasuk & Baker 1998), it is no surprise that activist groups need to align and work with international organisations to be effective.

Conclusion: towards a more integrated account of social and environmental injustice in Asia

Given the historical and social rootedness of environmental knowledge, it is important to consider how the actors in these kinds of campaign have engaged around specific issues in each location. While in both cases the political organisations of each movement have been successful in highlighting the issues involved, and in causing major difficulties for the businesses and policymakers to achieve their objectives, they have tended to adopt an advocacy group approach. To some extent, this is largely because of the limited political opportunity structures in each context but particularly in Cambodia (McAdam 1982; Tilly 1994). The movements in both cases have operated much more along the lines of the resource mobilisation approach (Tilly 1994).

Considering the successes and failures of both campaigns, each movement needs to become more embedded in the political processes that can secure an improved environment in MTP or to conserve the forest resources that sustain the livelihoods of the communities in Prey Lang.

Advocacy groups tend to focus on specific injustices such as air pollution or contamination to water supplies. Clearly, both the MTP and Prey Lang campaigns go much further than this, drawing on their distinct cultural values to equip them with a more integrated account of the social and environmental problems they face. There are differences. The Prey Lang movement draws on Buddhist values concerning the need for a more harmonious and less exploitative relationship with their natural habitat as compared to their government's instrumental view of forests as resources to be harnessed. However, in MTP, Buddhist values have been relevant to the individuals in the campaign especially religious virtues of courage, sacrifice and care, and their political strategy has been through a combination of civil society alliance building (the aggregation of interest groups) and challenges through the courts. The latter has also involved the support of national and transnational environmental NGOs.

In order to engage in problem solving in the context of 'clusters of injustice', the movements need to become stakeholders in the processes that deliberate on these conjoined issues. Even major private companies involved in Rayong have recognised the need for building stakeholder deliberation. In 2011 industrial companies took the initiative, founding a community partnership association that would provide a forum for discussion and dialogue between MTP business owners, their staff and local residents (*Bangkok Post* 2011).

The two case studies illustrate the emergence of new groups and movements acting in response to distinct environmental problems and their ability to mobilise considerable resources with the support of other organisations and networks of alliances built over the years. It has long been pointed out that environmental activism alone is insufficient to address the concerns of those most affected by environmental impacts but that social movements are effective in representing local livelihood issues (Forsyth 1999: 697). Philip Hirsch indicates that civil society responses have emerged very unevenly in the region and are 'still shaped primarily by the nature of political space within national borders' but that political space is limited within most countries of the Mekong Region (Hirsch 2001: 244–245). However, given the continued push for regional economic integration, we may begin to see a regionalised civil society response to livelihood-related environmental threats (ibid.: 249) in future.

States are still important in environmental negotiations even though they have difficulties in representing the diverse views and interests of their citizens and cannot guarantee the compliance of actors responsible for environmental degradation (Smith & Pangsapa 2008). But as we have seen here, non-state actors and CSOs are increasingly significant in environmental negotiations in terms of influencing, if not also in the formulation of, policy and agreements. Much depends on the scope of civil society action in each location and whether policymakers see the wisdom in drawing such emergent groups and organisations into policymaking. Another important factor here is whether civil society formation is secure enough to allow external NGOs to provide movements with valuable capital.

The solution to environmental issues thus requires that social injustices are addressed as well. All actors – people, families, their communities, states and business – have obligations (sometimes different obligations) to act responsibly because everyone should have access to clean air, water and green spaces.

The two case studies in this chapter have been selected because of the insights they generate in the region more broadly. What is really distinctive about both is their capacity to concentrate some of the effects of regional development in one place (i.e., resource extraction in Cambodia compared to manufacturing and processing in Thailand). Looking further, especially for countries with less effective environmental regulation and weak regulatory frameworks on environmental impacts, such as China, where the local effects of rapid industrialisation have been described as disastrous (Economy 2004), it is still important to look at the empirical evidence and the context-specific values at work in each situation. While the Prey Lang (Cambodia) case highlights the impacts of resource extraction on community livelihoods in a rural context, the Map Ta Phut

(Thailand) case indicates how industrial zones generate similar tensions in urban environments. Environmental groups in Cambodia and Thailand have played an important part in raising awareness for their respective citizens. For countries that may have more limited political opportunity structures such as China, what these two case studies have demonstrated is that local peoples can still seriously and effectively challenge government and government policies by using a variety or combination of tactics to highlight the severity of their issues and generate greater public concern and take action.

One of the distinctive features of environmental movements on the SEA mainland is their capacity to bring in local knowledges in ways that often contradict accounts of outside experts and what we see are community campaigns drawing on local multigenerational knowledge through participatory research and action that involve community members collecting their own data for evidence (Smith & Pangsapa 2008). Moreover, there is variety of cultural beliefs in Southeast Asia – whether Buddhist, Confucian, Hindu etc. – and even within a specific cultural framework such as Buddhism there are quite distinctive traditions that influence practices. What these different religions have at their core are distinctive views on how to be environmentally responsible. While the emphasis they place on respect for nature as part of humanity may be at odds with a country's pursuit of material growth and development, culturally specific values can nevertheless be a great source of leverage in advocacy.

Future research should compare and analyse environmental activism between countries in Asia that may be experiencing similar problems but face differing challenges depending on the issue/s at stake, the political context, the role of environmental NGOs or power that they have. So while activism may be stifled by some states, activists can work towards connecting environmental and livelihood issues as a way of bringing together large numbers of people towards effecting social change. The connections are crucial because injustices are clustered. Effective solutions to Prey Lang, Mae Sot, Lampang, Pearl River Delta, Guiyu (China's recycling city), Guangdong, etc., in fact, throughout Asia, demand that we see environmental and social justice as connected and more often than not, these contexts also include violations of human rights and labour standards (Smith & Pangsapa 2011).

Environmental actors at the grassroots level are increasingly linking local activities to similar activities in other countries and developing networks that are transnational in character. Thus environmental NGOs are also beginning to change both their internal structures and their activities. They are developing new ways to be accountable to their members, transparent to civil society generally, and develop stronger links with other grassroots and community livelihood movements. By reaching out to other NGOs and movements covering environmental and development issues in order to exchange knowledge, they are building co-activist strategies that address the concerns of both partners in a more effective way. Movement effectiveness may be as strongly influenced by the context in which it is working as by the skill of actors and citizens in building and mobilising the resources it commands. In democratising countries, the political opportunities are becoming greater as a result of development. Resource mobilisation can still be seen as a threat to political institutions and when this happens the political opportunity structures are more limited. As such, one way for environmental movements to compensate is to draw on opportunities arising from transnational networking often with the help of international NGOs to place pressure on national governments and private corporations. Whether the latter develops practices and partnerships that embrace environmental responsibility is the next question for researchers in the field but one thing is central, if resources are to be mobilised to take advantage of the political opportunity structures that exist, then movements, community campaigns need to combine their advocacy work with stakeholder activities. While work is being conducted on environmental issues in Asia, connecting that work to all other forms of injustices that are associated with them is the next major task of Asian studies.

Bibliography

Bales, K. (2004) *Disposable People: New Slavery in the Global Economy* (University of California Press).

Bangkok Post (2011) 'Big Business with a Heart: Better Awareness of Local Community Needs is Turning Map Ta Phut Factories into Friends Whereas Once They were Foes', 27/07/2011. Available at http://www.bangkokpost.com/business/economics/ 248965/big-business-with-a-heart.

Beebe, K. (2012) 'Local Communities in Cambodia Protest Logging', 20/04/2012. Available at http://pulitzer center.org/reporting/prey-lang-logging-national-resources-protection-cambodian-center-human-rights.

Cambodian Human Rights and Development Committee (2012) 'Killing of Mr. Chut Wutty' – ADHOC Report, Phnom Penh, 4 May 2012, http://www.adhoc-cambodia.org/?p=1593, accessed 22/10/2012.

Carroll, T. & Sovacool, B. (2010) 'Pipelines, Crisis and Capital: Understanding the Contested Regionalism of Southeast Asia', *Pacific Review* 23(5), 625–647.

Center for People and Forests (RECOFTC) (2012) 'Community Forestry in Cambodia'. Available at http://www.recoftc.org/site/Community-Forestry-in-Cambodia.

Cronin, R.P. (2012) 'China and the Geopolitics of the Mekong River Basin: Part I', 22/03/2012. Available at http://www.worldpoliticsreview.com/articles/11761/china-and-the-geopolitics-of-the-mekong-river-basin-part-i.

Currey, D. (2001) Timber Trafficking—Illegal Logging in Indonesia, South East Asia and International Consumption of Illegally Sourced Timber, Environmental Investigation Agency and Telepak Indonesia.

Currey, D., Doherty, F., Lawson, S., Newman, J. Afianto, M.Y., Hapsoro, M. et al. (2003) Above the Law: Corruption, Collusion, Nepotism, and the Fate of Indonesia's Forests, Environmental Investigation Agency and Telepak Indonesia.

Davenport, J. (2011) 'Demand for Alternative Therapies Fuels Trade in Endangered Animal Products', *London Evening Standard*, 09/13/2011. Available at http://www.standard.co.uk/news/demand-for-alternative-therapies-fuels-trade-in-endangered-animal-products-6442790.html.

Deboonme, A. (2012) 'ANALYSIS: Map Ta Phut Needs to Earn Public Confidence', *The Nation*, 8/05/2012. Available at http://www.nationmultimedia.com/business/Map-Ta-Phut-needs-to-earn-public-confidence-30181476.html.

Dobson, A. (2003) '"Social Justice and Environmental Sustainability" Ne'er the Twain Shall Meet?', in J. Agyeman, R.D. Bullard & B. Evans (Eds.), *Just Sustainabilities: Development in an Unequal World* (Earthscan).

Eagle, W. (2012) 'Billions of Dollars Lost Each Year to Illegal Logging', 05/10/2012. Available at http://www.voanews.com/content/billions-of-dollars-lost-each-year-to-illegal-logging/1521383.html.

Earth Action (2012) 'Victory—40,000 Hectares (100,000 Acres) Protected!', 21/08/2012. Available at http://www.earthaction.org/prey-lang/.

Economy, E.C. (2004) *The River Runs Black: The Environmental Challenge to China's Future* (Cornell University Press).

Elliott, L. (2004) 'Environmental Challenges, Policy Failure and Regional Dynamics in Southeast Asia', in M. Beeson (Ed.), *Contemporary Southeast Asia: Regional Dynamics, National Differences* (Palgrave Macmillan).

Elliott, L. (2007) 'Transnational Environmental Crime in the Asia Pacific: An "Un(der) Securitized" Security Problem?', *Pacific Review* 4(20), 499–522.

Fahn, J. (2004) *A Land on Fire: The Environmental Consequences of the Southeast Asian Boom* (Basic Books).

Feldmann, A. (2012) 'Discourses of Development: The Spatial Struggles of Civil Society in Map Ta Phut', International Conference on International Relations and Development (ICIRD), Towards an ASEAN Economic Community (AEC): Prospects, Challenges and Paradoxes in Development, Governance and Human Security, 26–27 July, Chiang Mai, Thailand. Available at http://www.icird.org/2012/files/papers/Anselm%20Feldmann.pdf.

FOCUS Human Rights Osaka (2012) 'Map Ta Phut: Thailand's Minamata?' Vol. 68. Available at http://www.hurights.or.jp/archives/focus/section2/2012/06/map-ta-phut-thailands-minamata.html.

Forsyth, T. (1999) 'Environmental Activism and the Construction of Risk: Implications for NGO Alliances', *Journal of International Development* (11)5, 687–700.

Forsyth, T. (2001) 'Environmental Social Movements in Thailand: How Important is Class?', *Asian Journal of Social Sciences* 29(1), 35–51.

Forsyth, T. (2007) 'Are Environmental Social Movements Socially Exclusive? An Historical Study from Thailand', *World Development* 35(12), 211–230.

Global Witness (2004) *Taking a Cut: Institutionalized Corruption and Illegal Logging in Cambodia's Aural Wildlife Sanctuary* (Global Witness).

Gray, D.D. (1996) 'Cambodia Selling Remaining Forests', *The Record*, 26/02/1996.

Hariharan, M. (2010) 'Thailand's Map Ta Phut Crisis – the NGO Side of the Story', 18/03/2010. Available at http://www.icis.com/blogs/asian-chemical-connections/ 2010/03/thailands-map-ta-phut-crisis-html.

Henry, L.A. (2004) *A Tale of Two Cities: A Comparative Study of Traditional Chinese Medicine Markets in San Francisco and New York City* (World Wildlife Fund (WWF)).

Hirsch, P. (1997) *Seeing Forest for Trees: Environment and Environmentalism in Thailand* (Silkworm Books).

Hirsch, P. (1998) 'Community Forestry Revisited: Messages from the Periphery', in M. Victor, C. Lang & J. Bornemeier (Eds.), *Community Forestry at a Crossroads: Reflections and Future Directions in the Development of Community Forestry* (RECOFTC).

Hirsch, P. (2001) 'Globalisation, Regionalisation and Local Voices: The Asian Development Bank and Re-scaled Politics of Environment in the Mekong Region', *Singapore Journal of Topical Geography* 22(3), 237–251.

Hirsch, P. (2007) 'Civil Society and Interdependencies: Towards a Regional Political Ecology of Mekong Development', in J. Connell & E.Waddell (Eds.), *Environment, Development and Change in Rural Asia-Pacific: Between Local and Global* (Routledge).

Hirsch, P. & Lohmann, L. (1989) 'The Contemporary Politics of Evironment in Thailand', *Asian Survey* 29(4), 439–451.

Hirsch, P. & Warren, C. (Eds.) (1998) *The Politics of Environment in Southeast Asia: Resources and Resistance* (Routledge).

Hughes, C. (2011) 'Soldiers, Monks, Borders: Violence and Contestation in the Greater Mekong Sub-region', *Journal of Contemporary Asia* 41(2), 181–205.

Interpol (2012) 'Illegal Logging Nets Organized Crime up to 100 Billion Dollars a Year, INTERPOL–UNEP Report Reveals', 27/09/2012. Available at http://www.interpol.int/News-and-media/News-media-releases/2012/PR075.

Isin, E.F. (2008) 'Theorizing Acts of Citizenship', in E.F. Isin & G.M. Nielsen (Eds.), *Acts of Citizenship* (Zed Books).

Johnson, C. & Forsyth, T. (2002) 'In the Eyes of the State: Negotiating a "Rights-based Approach" to Forest Conservation in Thailand', *World Development* 30(9), 1591–1605.

Karam, Z. (2010) 'Trade in Endangered Animals Flourishing in Middle East, Fuelled by Corruption, Weak Legislation', *Star Tribune*, 03/17/2010. Available at http://www.startribune.com/templates/Print_This_Story?sid=88056932.

Kovidhavanij, W. (2012) 'THAILAND: Map Ta Phut, Industry and Environment: Challenge of Sustainable Development', *Asian News*, 10/05/2012. Available at http://www.asianews.it/news-en/Map-Ta-Phut,-industry-and-environment:-challenge-of-sustainable-development-26002.html.

Lazarus, K., Resurreccion, B.P., Dao, N. & Badenoch, N. (Eds.) (2011) *Water Rights and Social Justice in the Mekong Region* (Routledge).

Lestrelin, G., Castella, J. & Bourgoin, J. (2012) 'Territorialising Sustainable Development: The Politics of Land-use Planning in Laos', *Journal of Contemporary Asia* 42(4), 581–602.

Litta, H. (2010) 'Environmental Challenges in Southeast Asia: Why is There so Little Regional Cooperation?', *Asian Journal of Public Affairs* 3(1), 1–16.

Lobe, J. (2012) 'Treat Illegal Logging Like Organised Crime, Urges World Bank', 21/03/2012. Available at http://www.ipsnews.net/2012/03/treat-illegal-logging-like-organised-crime-urges-world-bank/.

McAdam, D. (1982) *Political Process and the Development of the Black Insurgency 1930–1970* (University of Chicago Press).

Molle, F., Foran, T. & Kakonen, M. (2009) *Contested Waterscapes in the Mekong Region: Hydropower, Livelihoods and Governance* (Routledge).

The Nation (2010) 'NGOs, Villagers Angry at Govt Cut to List of Harmful Activities; Demand an Explanation', 08/09/2010. Available at http://www.nhrc.or.th/2012/wb/en/news_detail.php?nid=295&parent_id=1&type=hilight.

The Nation (2012) 'Map Ta Phut Residents Set to Take IEAT to Court', 07/12/2012. Available at http://www.nationmultimedia.com/breakingnews/Map-Ta-Phut-residents-set-to-takeIEAT-to-court-30195768.html.

Noun, B. (2012) 'Cambodia's Amazon Under Threat', 07/03/2012. Available at http://www.asiasentinel.com/index.php?option=com_content&task=view&id=4307&Itemid=20.

Pangsapa, P. (2009) 'When Battlefields Become Marketplaces: Migrant Workers and the Role of Civil Society and NGO Activism in Thailand', *International Migration*, 5 August.

Pangsapa, P. & Smith, M.J. (2008) 'Political Economy of Southeast Asian Borderlands: Migration, Environment, and Developing-country Firms', *Journal of Contemporary Asia* 38(4), 485–514.

Pangsapa, P. & Smith, M.J. (2009) 'When Battlefields Become Marketplaces: migrant workers and the role of civil society and NGO activism in Thailand', *International Migration* (First published online: 5 August 2009).

Pasuk, P. & Baker, C. (1998) *Thailand's Boom and Bust* (Silkworm).

Plokhii, O. (2012a) 'Cambodia Closes Case, Thwarts Justice, in Forester Murder', *Huffington Post*, 10/10/2012. Available at http://www.huffingtonpost.com/olesia-plokhii/cambodia-closes-case-thwa_b_1947199.html.

Plokhii, O. (2012b) 'Murders in the Forest', *Huffington Post*, 20/09/2012. Available at http://blogs.reuters.com/great-debate/2012/09/20/murders-in-the-forest/.

Reynolds, L. (2005) 'Pulping Cambodia: Asia Pulp & Paper and the Threat to Cambodia's Forests', *Multinational Monitor* (March), http://www.multinationalmonitor.org/mm2005/032005/reynolds.html, accessed 10/22/2012.

Sample, I. (2007) 'DNA Kit to Fight Trade in Endangered Animals', *The Guardian*, 06/12/2007. Available at http://www.guardian.co.uk/uk/2007/jun/12/animalwelfare. science.

Sarnsamak, P. (2010) 'Map Ta Phut Blockade Threat', *The Nation*, 07/09/2010. Available at http://www.thaivisa.com/forum/topic/396263-threat-to-blockade-thailands-map-ta-phut-industrial-estate/.

Schearf, D. (2012) 'Attacks on Cambodian Activists Underscore Court Interference', 04/10/2012. Available at http://www.voanews.com/content/attacks-on-cambodian-activists-underscore-interference-in-courts/1520203.html.

Shelley, L. (2010) *Human Trafficking: A Global Perspective* (Cambridge University Press).

Simpson, C. (2012) 'Seizures of Endangered Wild Animal Goods on the Rise in Dubai', *The National*, 15/05/2012. Available at http://www.thenational.ae/news/uae-news/seizures-of-endangered-wild-animal-goods-on-the-rise-in-dubai.

Smith, M.J. & Pangsapa, P. (2008) *Environment and Citizenship: Integrating Justice, Responsibility and Civic Engagement* (Zed Books).

Smith, M.J. & Pangsapa, P. (2009a) 'Buddhist Virtues and Environmental Responsibility in Thailand', in C. Blackmore, M. Reynolds & M.J. Smith (Eds.), *The Environmental Responsibility Reader* (Zed Books).

Smith, M.J. & Pangsapa, P. (2009b) 'Strategic Thinking and the Practices of Ecological Citizenship: Bringing Together the Ties that Bind and Bond', in C. Blackmore, M. Reynolds & M.J. Smith (Eds.), *The Environmental Responsibility Reader* (Zed Books).

Smith, M.J. & Pangsapa, P. (2009c) 'Corporate Environmental Responsibility and Citizenship', in C. Blackmore, M. Reynolds & M.J. Smith (Eds.), *The Environmental Responsibility Reader* (Zed Books).

Smith, M.J. & Pangsapa, P. (2011) 'Clusters of Injustice: Human Rights, Labour Standards and Environmental Sustainability', in A. Voiculescu & H. Yanacopulos (Eds.), *The Business of Human Rights: An Evolving Agenda for Corporate Responsibility* (Zed Books).

Thul, P.C. (2012) 'Top Cambodian Activist Dead after Police Shoot-out', 26/04/2012. Available at http://www.trust.org/alertnet/news/top-cambodian-activist-dead-after-police-shoot-out.

Tilly, C. (1994) 'Social Movements as Historically Specific Clusters of Political Performances', *Berkeley Journal of Sociology* 38, 1–3.

Titthara, M. (2012) 'Verdict in Chut Wutty case delayed', *Phnom Penh Post*, 09/10/2012. Available at http://www.phnompenhpost.com/index.php/2012100959180/National-news/verdict-in-chutwutty-case-delayed.html.

Unger, D.H. & Siroros, P. (2011) 'Trying to Make Decisions Stick: Natural Resource Policy Making in Thailand', *Journal of Contemporary Asia* 41(2), 206–228.

Viwatpanich, K. (20110) 'Suffering from Industrial Estate Development: A Case Study in Map Ta Phut, Thailand', Conference Proceedings of the 1st International Seminar on Population and Development. Available at http://jr.sherubtse.edu.bt/index.php/pd1/ issue/view/4.

Wallace, C.P. (1992) 'Clear-Cut: Rebels Sell Off Cambodia's Forests', *Seattle Times*, 02/12/1992.

World Bank (2012) Justice for Forests: Improving Criminal Justice Efforts to Combat Illegal Logging, Washington, DC.

Young, M. (2011) 'Prey Lang: A Forest on the Brink of Destruction'. Available at http://mathieuyoung.com/blog/?p=866.

Acronyms

ADHOC	Cambodia Human Rights and Development Association
CAIN	Campaign for Alternative Industry Network
CBO	community-based organisation
CIVICUS	Centre for Cambodia Civic Education
CSO	civil society organisation

EARTH	Ecological Alert and Recovery Thailand
EIA	Environmental Investigation Agency
EPA	Environmental Protection Agency
EPN	Eastern People's Network
GCM	Global Community Monitor
GPSEA	Greenpeace Southeast Asia
HRO	Human Rights Osaka
HRW	Human Rights Watch
ICIRD	International Conference on International Relations and Development
IEAT	Industrial Estate Authority of Thailand
INGO	international non-governmental organisation
INTERPOL	International Criminal Police Organization
IO	international organisation
MRC	Mekong River Commission
MTP	Map Ta Phut
NEC	National Environment Committee
NGO	non-governmental organisation
PanNature	People and Nature Reconciliation
PCD	Pollution Control Department
PLCN	Prey Lang Community Network
PLN	Prey Lang Network
RECOFTC	Center for People and Forests
RGC	royal government of Cambodia
SEA	Southeast Asia
TAGP	trans-ASEAN gas pipeline project
TAN	transnational activist networks
TNC	transnational corporation
UN	United Nations
UNEP	United Nations Environmental Programme
VOCs	volatile organic compounds
WB	World Bank

Notes

1   Thailand is a heavily export-dependent country with exports accounting for more than two-thirds of its GDP. http://www.tradingeconomics.com/thailand/gdp-growth and http://data.worldbank.org/indicator/NE.EXP.GNFS.ZS.

2   Illegal logging generates between US$10 and 15 billion a year (World Bank 2012). The UNEP and INTERPOL report estimates between US$30 and 100 billion are lost each year to the illegal timber trade. UNEP says deforestation is responsible for nearly 20 per cent of all global carbon emissions which is 50 per cent more than the amount from shipping, aviation and land transport put together (Eagle 2012).

3   The Cambodian government passed a new law in 2002 which gave the Forestry Administration authority to grant areas of forest to local communities and in 2003 a community forestry sub-decree officially recognised community forestry as a national policy (RECOFTC 2012). This legislation, however, was not legalised and it was only in 2006 that community forestry guidelines, called *prakas*, were introduced, which laid out clear procedural guidelines for identifying, legalising, and managing forests (ibid.).

4   The four provinces are Preah Vihear, Kampong Thom, Kratie and Stung Treng, which are located between the Mekong and Stung Sen Rivers. http://preylang.com/the-forest/.

5   Young posted a picture on his blog of a truck transporting unmarked resin timber out of the forest well after the company's concession had ended. http://mathieuyoung.com/blog/?p=866.

4

Literature and the environment

New approaches to ecocriticism

Karen L. Thornber

'We'd like to cut down the trees with nature in mind.' So declared Suzuki Takehiko, director of Japan's Shōsenkyō Kankō Kyōkai (Shōsen Gorge Tourism Association), in August 2008. Part of Chichibu-Tama-Kai National Park, Shōsen Gorge has for decades been labelled the country's 'most beautiful valley'.[1] Years of deforesting meant that when the park was founded in 1950, little stood between tourists and the majestic rock formations for which the gorge is most famous. But, by the turn of the twenty-first century, visitors were frustrated that trees were now blocking much of the view. The park's laissez-faire approach to the valley's vegetation did not threaten its ecosystems – trees are hardly invasive species there. But this economically disadvantaged part of Japan depended on a steady stream of tourists who wanted to see cliffs, not trees; some even claimed that the trees were depriving the valley of its beauty. So Suzuki argued that 'trees' (part of nature) should be felled so that people could have a better view of 'nature' (the gorge).

This episode encapsulates what I have identified as *ecoambiguity* (environmental ambiguity), the complex, contradictory interactions between people and the nonhuman (Thornber 2012). Environmental ambiguity manifests itself in multiple, intertwined ways, including ambivalent attitudes towards nature; confusion about the actual condition of the nonhuman, often a consequence of ambiguous information or information 'overload'; contradictory behaviours toward ecosystems; and discrepancies among attitudes, conditions, and behaviours that lead to actively downplaying and acquiescing to ecological degradation, as well as to inadvertently harming the very environments one is attempting to protect.

Many parks, although established at least in part to protect ecosystems from human abuse, ultimately depend on the human footprint for their existence; areas that do not attract visitors risk being developed. Likewise, calls to destroy one part of an ecosystem frequently stem from the desire to protect another; deer populations, for instance, are regularly culled so that vegetation can be restored. But the ambiguity of people's relationships with Shōsen Gorge is particularly pronounced. The original requests for deforestation stemmed from the desire not to save but instead to *see* another segment of the landscape; tourists wanted the trees removed not so the cliffs could be protected but so they could be photographed.

Despite their plaint and Suzuki's appeal, most of the trees still stand and, in fact, are highlighted in the park's promotional materials. The Shōsen Gorge Tourism Association's website features images of colourful trees growing beside, and out of, majestic crags; in some pictures, trees effectively obscure the cliffs. A banner running near the top of the website declares Shōsen Gorge the most beautiful in Japan, full of the many wonders of nature; trees remain part of the appeal, their foliage, particularly in autumn, a highlight of visits to Shōsen Gorge.

Environmental degradation, ambiguity and East Asia

Much of East Asia (China, Japan, Korea, Taiwan, Vietnam) has not been so fortunate.[2] East Asia has long been associated with belief systems advocating reverence for nature, especially Buddhism, Confucianism, Daoism and Shinto as well as numerous indigenous philosophies and religions. Popular perceptions both within and outside the region often hold that its environmental degradation began in the late nineteenth century, when East Asian peoples, pressured by Western nations, assimilated the latter's technologies and industries. But actually, East Asian societies have extensive histories of transforming environments. While some of the region's environmental problems have clearly been ameliorated as a result of increased ecological consciousness in the region, others have grown more menacing. In this sense, East Asia is no different from most other parts of the world. Few places today celebrate ecological destruction, instead giving lip service to 'greening' environments, but many promote industries and lifestyles that virtually *ensure* devastation. The separation between practice and environmental protection rhetoric exists practically everywhere; the divergence is now so ingrained it can be taken for granted.

Similarly, East Asian literatures are famous for celebrating the beauties of nature and depicting people as intimately connected with the natural world. But, in fact, like film and popular culture, which have played an important role in awakening environmental consciousness in the region, much modern and even premodern Chinese, Japanese, Korean, Taiwanese and Vietnamese fiction and poetry portray people damaging if not destroying everything from small ponds to the entire planet.[3] References to ecodegradation have appeared regularly in East Asian literatures since the late 1960s (in Japan and Korea), 1980s (in China and Taiwan) and in Vietnam more recently. Some creative works that discuss damage to ecosystems conform to conventional understandings of 'nature writing' or 'environmental/ecological writing', at least in their place of origin, but many others do not.

In the 1980s some Chinese writers explicitly distinguished their 'environmental literature' from the 'pollution literature' of Japan and other nations, arguing that the Chinese strove not only to expose individuals and behaviours that damaged environments but also to extol environmental protection efforts (Wen & Jian 2004: 11–14). But such distinctions are more theoretical than empirical. The degree to which a Chinese text needs to engage with environmental matters for a Chinese critic to consider it 'environmental literature' varies. More significant is the diversity of Chinese texts, especially in the decades since the ecological devastation of the Cultural Revolution, that talk about damaged ecosystems, addressing and often condemning such problems as soil and air pollution, deforestation, desertification, water shortages, flooding, pollution, species extinction and global warming. To be sure, East Asian writers tend to engage primarily with the environmental challenges facing their own nations, rather than those of neighbouring countries. So, in this sense, the precise topics of environmental literature differ across the region. However, a planetary consciousness permeates much of their work; even as East Asian writers discuss devastation close to home, the damage they describe also frequently has counterparts regionally and globally, to which they often at least allude. East Asian authors and other artists have not been deceived by official rhetoric on 'greening' environments. Neither have they been duped by popular discourse on East Asian 'love' of nature. Instead, they have called attention to the many challenges facing East Asia's and the world's environments and the fundamental ambiguities underlying much ecodegradation.

Initially, I had planned to organise *Ecoambiguity: Environmental Crises and East Asian Literatures* (2012), my book on literature and environment in East Asia, around creative treatments of major environmental problems. I had thought of comparing, for instance, Chinese, Japanese, Korean, Taiwanese and other literary engagements with specific environmental problems. After all, we expect area specialists and especially comparatists to highlight cultural and regional differences, differences both among individual (non-)Western societies and between non-Western and Western societies. Separateness is often assumed to be more prevalent, and important, than similarity, not to mention commonality.

But the more I read, the more it became apparent that something quite different was at stake. Throughout history people have routinely damaged both proximate and distant landscapes, despite vast differences in cultures, attitudes toward nature and the resilience of the ecosystems they inhabit. Environmental damage has varied greatly, yet there are few if any places that have not been harmed by the human footprint, either literally or metaphorically. I soon realised that it was important, indeed imperative, to analyse how literature as a form of discourse deals with the causes and consequences of ecodegradation writ large. Once I no longer looked at texts primarily through the lenses of individual societies or environmental problems, but instead examined how creative works from disparate places negotiated more generally with ecological quandaries, their shared environmental ambiguity became unambiguously clear. The authorial, readerly, cultural and environmental circumstances/identities behind the production of a particular text certainly mattered, but not as much as I had presumed. Environmental ambiguity is a hallmark of everything from brief poems to multivolume novels; from the work of writers known globally to those scarcely recognised within their own societies. To be sure, ecoambiguity appears more prevalent in literature from East Asia than in other textual corpuses. And its irony is certainly deeper, considering the region's long cultural history celebrating the intimate ties between humans and nature even as its peoples severely damaged environments. But with several notable exceptions, and especially within East Asia, these disjunctions and their many permutations do not depend as much on specific literary culture or environmental problem as one might anticipate. And so I moved the focus to the concept of environmental ambiguity itself. Languages, genres, styles, and tropes differ within and across cultures, but the concerns raised have much in common.

The tremendous variety of literature in late twentieth- and early twenty-first-century East Asia and throughout the world that addresses ecodegradation – incorporating references that occasionally celebrate, sometimes simply describe, and often condemn harmful changes to environments – testifies to the persistence of damaged environments and to the ecological consciousness, however diaphanous, of literary artistes.[4] In the following pages, I introduce some of the most prominent East Asian writings on environmental crises, with special attention to their articulations of environmental ambiguity: the Chinese writer Jiang Rong's (1946–) novel *Lang tuteng* (Wolf Totem, 2004), the Japanese writer Ishimure Michiko's (1927–) *Kugai jōdo: Waga Minamatabyō* (Sea of Suffering and the Pure Land: Our Minamata Disease, 1969), the Korean writer Cho Sehŭi's (1942–) short story 'Kigye tosi' (City of Machines, 1977), the Taiwanese writer Huang Chunming's (1939–) short story 'Fangsheng' (Set Free, 1987) and the Vietnamese writer Nguyễn Huy Thiệp's (1950–) short story 'Muối cua rừng' (The Salt of the Jungle). These texts not only engage with but also have impacted society, politics, and environmental policies in the region. Huang Chunming's story reveals the ironic ambivalence of even environmental activists toward remediation, Cho Sehŭi's story focuses on the acquiescence and resignation toward environmental crises exhibited among diverse groups and Ishimure's novel demonstrates the ease with which even the most traumatic environmental crises are disavowed, while Jiang Rong's novel goes one step further in portraying those closest

55

to nature as actively harming it. Nguyễn Huy Thiệp's story takes a different approach, highlighting the conflicting attitudes of hunters vis-à-vis nature.

Examining a range of Chinese, Japanese, Korean, Taiwanese and Vietnamese texts that address ecodegradation makes us more aware of the many ways different societies have grappled with phenomena that are grounded in their specific cultures and histories but that also resonate with those of other places and peoples and have widespread regional if not global implications. Readers are invited to consider the particular ways that ecological problems are negotiated in national literatures, while recognising the many commonalities of human relationships with the nonhuman across time and space. I conclude with thoughts on the importance of literature in changing consciousness concerning and ultimately alleviating environmental degradation.

Huang Chunming and the perils of remediation

Published the year Taiwan lifted martial law and just as Taiwanese environmental consciousness was beginning to burgeon, Huang Chunming's 'Set Free' depicts local farmers fighting for their livelihoods amid government-enabled ecological degradation. Significantly, however, the protagonists of this story – Granny Jinzu and her husband Zhuang Awei – are far from relieved when the government announces plans to remediate the ecodegradation for which it is at least partly responsible and against which the couple has long fought. The narrator depicts Jinzu and Awei as simultaneously longing for diverse, prosperous and aesthetically pleasing environments; advocating behaviours that directly harm landscapes; and denouncing policies that would ameliorate existing damage to ecosystems and prevent further occurrences.

'Set Free' takes place in the small town of Dakenggu, at the mouth of the Wulaokeng River in northeast Taiwan. The terrestrial, aquatic and atmospheric environs of Dakenggu have been damaged by the increasing emissions, both wastewater and airborne contaminants, of nearby chemical plants and cement factories. The town's human residents also suffer economically and physically; pollution prevents them from growing crops and catching fish, many persons have become ill, and some have died. People protested, some arguing that authorities had tampered with water samples to conceal the extent of the pollution. But dissenters were quickly silenced, and some, including Jinzu and Awei's son Wentong, were imprisoned.

For the most part, families like the Zhuangs, not to mention local ecosystems, are portrayed as at the mercy of seemingly indomitable, unstoppable and destructive government and industry. Yet the human/nonhuman contacts cited in 'Set Free' in fact are more complex. Several phenomena make it impossible to establish clean divisions between heartless and destructive polluters, on the one hand, and ecologically minded townspeople, on the other. Particularly interesting is Jinzu and Awei's response to their friend Tianying's report on the government's agenda for the region.

To everyone's surprise, after years of showing little concern for anything but corporate profits, the authorities announce plans to transform the coastal area, including Jinzu and Awei's fields, into a 'protected area for birds' (*quanbu guiru niaolei baohuqu*). When Awei asks his friend Tianying what this means, Tianying clarifies that people will not be permitted to catch the teals, goldfinches and swans that migrate to the coast in winter. Awei seems relieved that his life will not be disrupted. Reminding Tianying that ever since polluters came to town there have been few birds to catch, he apparently is not troubled by the idea of a bird sanctuary without birds. But Tianying quickly disabuses Awei of this scenario, telling him, 'After the bird sanctuary is established, the factories will be prohibited from discharging toxic water.' Tianying does not need to state the obvious corollary: paradoxically, an absence of poison, something for which Awei and his neighbours fought for years, will result in an influx of birds, which to Awei is intolerable. He and Jinzu are distressed that they will not be allowed to snare even the birds eating their crops and that in their town 'a bird [now] is more valuable than a person.'

Tianying is not particularly disturbed by this recent development, declaring, 'Who cares if we can't catch sparrows? The factories will no longer be discharging toxic water. Isn't this a good thing?' Awei and his wife disagree, the narrator noting, 'This comment not only did not console them. It instead brought to mind the fact that Wentong had been jailed for several years precisely because the factories had been discharging toxic waste' (114–115).[5] Industrial emissions and the government's resistance to tampering them have brought undue suffering to the Zhuangs and their town. Yet the news that these emissions will be cut brings not joy but fear. A flourishing and untouchable bird population will almost certainly inhibit agricultural output, resulting in a harvest that is no less meagre than when the area was plagued by pollution. The couple resent having their own fields transformed into a space that ironically represents everything for which they fought. 'Set Free' suggests that Awei and Jinzu believe the government has done nothing more than swing from one extreme to another.

This storyline is dropped as quickly as it is introduced. But the couple's immediate reaction to news of the impending conversion of their environs from lifeless cesspool to bird sanctuary highlights their environmental ambivalence. On the one hand, the Zhuangs desire an ecosystem amenable to birds – not only will catching birds increase the family's monthly income, but also the soil of an ecosystem hospitable to birds will almost surely be amenable to crops, the sale of which sustains the family. But, on the other hand, if the family is prohibited from removing birds from their fields, their crops most likely will be destroyed, and they will go bankrupt. In short, what they desire is not so much freedom from pollution as one of the freedoms that pollution curtails, that is to say, freedom to use the nonhuman world to their personal advantage. 'Set Free' makes clear the environmental ambivalence preservation can provoke.

Cho Sehŭi and the pull of resignation

Cho Sehŭi's 'City of Machines,' published in the final years of Park Chung Hee's (Pak Chŏnghŭi, 1917–1979) oppressive regime and part of Cho Sehŭi's legendary and well-travelled linked-story novel *Nanjangi ka ssoa ollin chagŭn kong* (Little Ball Launched by a Dwarf, 1978), is a trenchant exposé of the human-induced environmental demise of the imaginary Korean city of Ŭngang and the abuse of its residents, many of whom have little choice but to work in the city's toxic factories. Reflecting the human and environmental tragedies of Korea and Koreans in the 1960s and 1970s, the narrative – in sharp contrast with Huang Chunming's 'Set Free' – depicts most individuals as trapped within unforgiving political and social systems that make altering the status quo nearly impossible. To be sure, it asserts that when asked, 41.3 percent of Korean workers claim that they believe that in their nation anyone who works diligently, consumes frugally, and saves can live well (*chal sal su itta*); only 3.8 percent believe this 'utterly impossible' (*tojŏhi an doenda*) (145). But in truth, resignation becomes almost a requirement of survival. 'City of Machines' highlights the near universal axiom that people with the most incentive to enact change have the least ability to do so and those with the greatest ability have the least incentive.

Conditions in Ŭngang severely compromise both human and nonhuman lives; abuse of nature is deeply intertwined with harm to people:

Jet black smoke ascends from countless soaring smokestacks, and machines whirl in factories. Workers labor in the factories. So too do the children of the dead dwarf. Mixed into the air of these places are noxious gases and smoke, as well as dust. All the factories spew out a dark to yellowish-brown river of wastewater and waste oil, proportionate to the volume of their production. Factory wastewater emitted upstream is used by other factories. Spewed out again, it flows downstream until it enters the ocean. Ŭngang's inner harbor has festered into a rotten sea. Organisms that live around the factory are gradually dying off. (142)

Responses to pollution vary among the four key players featured in 'City of Machines' – the factory workers (most of whom live in Ŭngang and its environs), residents of Ŭngang who do not work in the factories, Yunho (a young man from a privileged background with considerable interest in the workers' welfare) and those in power (industry leaders). Significantly, although motives differ, acquiescence and resignation dominate throughout. Industry executives appear well aware of the damage their factories have inflicted on their employees, on the other residents of Ŭngang, and on environments. They are the one group that could effectively remediate and prevent future destruction. But they instead actively thwart any such attempts, concerned almost solely with financial gain. As is true of many texts that address human damage to environments, Cho Sehŭi's story depicts those with the greatest discretion to enact change as having the least inclination to upset the status quo.

Feelings of powerlessness prompt the other three (groups of) potential activists in 'City of Machines' to acquiesce, with only brief moments of revolt or thoughts of revolt. These individuals believe the economic, social and political systems behind environmental corruption impossible to reform, and thus the corruption itself impossible to overcome. Variations of the phrase 'nothing can be done' (uriga hal su innŭn il ŭn ŏpsŏ), repeated throughout the text by different characters, haunt this fatalistic narrative. Particularly important in this regard are the labourers and other residents of Ŭngang. Signalling his desperation, the eldest son of one of the workers believes that his only recourse is murdering the director of the Ŭngang Group; he judges his dream of forming a new labour union unattainable and cannot conceive of any other way to improve the lives of the workers. He laments to Yunho, 'There's nothing we can do' before revealing assassination strategies (147). The narrator confirms that this is likely the case, noting that Yunho just has learned that the factories in Ŭngang are managed by the same small group of people who control the economic lives of all Koreans. Petitions, protests and strikes, not to mention the slaying of someone of national importance by a disgruntled employee, would probably result in even harsher working conditions and hardly curtail damage to ecosystems. The labourers do nothing because there seems to be nothing they can do.

More complex is the situation of the people who live in Ŭngang but do not work in the factories. The narrator initially comments on their proclivity for the word 'stifling' (kapkaphada), which accurately describes both the physical position of their town (surrounded by ocean on three sides) and their personalities (plagued by doubts). Living in a socially managed society, these individuals are depicted as having resigned themselves to being smothered by the natural world, by others and by themselves. As the narrator notes: 'There is not a single person there who would voice displeasure at restrictions on individual activity for the sake of maintaining order' (141). In many ways, these people are as helpless as the labourers. Residents of Ŭngang depend on the wind to drive out to sea the toxic gases and smoke that hover above the industrial area. One night the wind unexpectedly changes direction and smog settles over the residential district, inciting chaos. Nearly suffocating inside their homes, people throng the streets, hoping for cleaner air in other parts of town but finding little relief. Although long aware of the factories as polluters, they at last realise that they are living amid 'perilous biological conditions, without precedent in Ŭngang's history'. For the first time they decide to do something to remedy their situation, but not surprisingly they are quickly deterred: '[These people] thought that the following day they would attempt to solve the problem. But they immediately crashed into a large barrier and ended up retreating dejectedly.' The residents of Ŭngang rally their strength, hoping to convene a public meeting or put on a display of force, only to discover that such acts are prohibited. Their response, the narrator indicates, is simply to 'open their mouths' (ibŭl pŏllyŏtta) (143). This phrasing suggests that they have not yet given up but do not know how to proceed; their mouths are open as if to speak, but words have yet to be formed. However, the narrator soon reveals that Ŭngang residents actually have been and likely will continue doing no more than checking on the direction of the wind before turning in for the night: 'The people of Ŭngang stop here. They don't think about the workers of the industrial zone that daily spills more than a hundred thousand tons of wastewater into the ocean. As long as the wind stays over the industrial zone and does not again blow toward the residential district, they will not be awakened from their deep sleep' (144). In other words, as long as they think they personally will not be affected by the poisonous gas and smoke emanating from factories, the people of Ŭngang will do nothing to help the workers, much less the natural world.

Acquiescence is easy to castigate, particularly when the welfare of so many is at stake. Certainly it might be argued that the factory workers could be more imaginative in combatting abuse by their employers, that Ŭngang residents could be more proactive in attempting to curb factory emissions, and that Yunho could take better advantage of his family's high standing to improve conditions for both people and the environment. But 'City of Machines' suggests that even if workers were more imaginative, residents more proactive, and Yunho a better schemer, the tyranny of those in power is so great that it would be almost impossible to effect change. The fleeting calls for action indicate a desire for reform, but the narrative suggests that it cannot be actualised any time soon. Like many creative works concerned with environmental justice, 'City of Machines' depicts the powerlessness of people who have done very little to repair environments not because they do not recognise the dangers of current conditions, not because they do not want conditions to improve, but because there are virtually no opportunities for them even to attempt to do so.

Ishimure Michiko and the politics of repudiation

Writings such as 'City of Machines' highlight ecological degradation that, although disparaged, for various reasons goes almost unchallenged. Other literary works that address human-induced environmental disruption portray disavowing this damage – acquiescing to it by denying responsibility for ecodegradation and/or knowing about but dismissing (potential) ecodegradation – as a common response to and facilitator of compromised ecosystems. This disjuncture between behaviours and irrefutable physical conditions plays a central role in Ishimure's stirring novel *Sea of Suffering* – Japan's most famous literary exposé of Minamata disease, a horrific form of mercury poisoning.[6] *Sea of Suffering* shows disconnects between obvious physical evidence (nonhuman spaces that are clearly polluted; people who are unquestionably disfigured) and the behaviours (disavowals) of many in the Japanese government, the Chisso Corporation (whose poisoning of Minamata Bay caused the disease), and residents of Minamata and surrounding towns.

Although most creative texts concerned with damage to environments address indifference toward and denials of this damage, *Sea of Suffering* is one of a subset that stresses the central role of these behaviours in causing and facilitating environmental degradation. More so than many narratives, it also specifies the reasons behind such disavowals, as well as their consequences. The novel devotes significant attention to alternatives, contrasting denials of Minamata disease with the great compassion for the afflicted demonstrated not only by the families and close friends of Minamata patients but also by the Japanese medical community. Incorporating other instances of industrial pollution both in Japan and abroad, Ishimure's text eloquently exposes denial of environmental degradation as a nearly global phenomenon, one endemic in human societies.

Most reprehensible, according to the narrator, is the Chisso Corporation. In 1959 scientists published a report indicating that Chisso's daily discharges of toxic, mercury-laden wastewater into Minamata Bay were the likely cause of Minamata disease. Yet rather than cooperate in subsequent investigations, for many years the corporation did everything it could to deny its role in propagating this disease, including pumping wastewater under cover of night and prohibiting scientists from taking further samples from the Bay. The narrator describes some Chisso employees as sympathetic to the plight of Minamata patients, even alerting residents of Minamata to Chisso's plans to divert their wastewater channel to another location; similarly, researchers from the Chisso company hospital contribute to efforts to understand the disease better. And at its August 1967 meeting the Chisso First Union issued a declaration condemning its own failure to fight Minamata disease and affirming its commitment to do so in the future. But for the most part, *Sea of Suffering* paints Chisso as an absolute villain, one that denies any connection between factory wastewater and Minamata disease yet prohibits scientists from studying the wastewater; one that does everything it can to avoid paying indemnities and instead continues to discharge poisonous effluent, thus expanding the number of people who may demand compensation; and one that delays dispatching employees to visit hospitalised Minamata patients until 1965, more than a decade after the outbreak of the illness.

Acknowledging Minamata disease belatedly in 1968 and only with great reluctance, the Japanese central government is described as largely responsible for facilitating Chisso's disavowals. This contrasts with local political bodies, which although relatively ineffective, show considerable concern with the spread of Minamata disease and establish various investigative groups. Throughout *Sea of Suffering*, the narrator highlights the tragedy of this situation: the greater and more widespread the traumas of those affected physically or economically (fishers with no market for their contaminated catch, or even with nothing to catch), the greater and more persistent the efforts of those not affected to disregard their suffering, both Chisso and bystanders in the local population. Commenting on the presumably deliberate misperceptions of the local public health department concerning Minamata disease, the narrator notes that 'The strange illness continued to work its way steadily along the coast of the Shiranui Sea, moving from one village to another. The true nature of the strange illness was not officially declared, but the incidents and their ramifications slowly continued to tear apart people's lives and hearts' (179).

Sea of Suffering underscores how national politicians and other government employees downplay if not disavow Minamata disease. The Japanese government for many years did not prohibit Chisso from continuing to deposit outflow, neither did it enact measures to clean polluted waters or to help those stricken with Minamata disease. These disavowals of the significance of this illness marked the beginning of decades of frustrating struggles by Minamata patients and their families with both the central government and Chisso. Like Chisso officials, national politicians and bureaucrats are depicted as disavowing Minamata disease for a variety of reasons: financial dependence of the town, region and nation on industries such as Chisso; inability to appreciate the suffering of Minamata disease patients and the significance of the damage inflicted on local ecosystems; and simple heartlessness, including the belief that because Minamata disease affected such a small, rural and impoverished segment of the Japanese population it did not merit attention. This is particularly true of Japan's central government. In his report on the Minamata Disease Policy Committee's visit to Tokyo in 1957, City Assemblyperson Hirota Sunao recalls that officials in the Welfare Ministry not only had never heard of Minamata but on learning that the disease affected mostly indigent fishers, claimed it too trivial a matter to pursue. Those who listened to their petition did so only to be polite and were eager to see them depart (79).

The meeting in Minamata between Diet representatives and the Municipal Assembly two years later (2 November 1959) is no more productive. The narrator describes this encounter as resembling a 'cross-examination' (76). Diet members take advantage of the recently elected mayor's inexperience with politics and his relative unfamiliarity with Minamata disease and its effects on the town. The narrator laments: 'Both the regional administration and the Diet were supposed to be looking out for the people, but it was inevitable that the meeting between the two sets of officials, with their different agendas, would become a confrontation between the authority of the Diet and the powerless impoverished' (77). The narrator speaks on several occasions of the national government's long history of disavowing industrial pollution, and of its failure to confront much less prevent such occurrences. She reminds readers of the Ashio copper mine incident (1880s) and how the rights of local farmers near Ashio have yet to be recognised nearly a century later, indemnities have yet to be paid and a commission has yet to be established to study Japan's first modern pollution event. And she accuses the Japanese government more generally as having 'a policy of abandoning its people' (*kono kuni no kimin seisaku*) (234).

But the most troubling disavowals of Minamata disease come from residents of the Minamata area who fear that acknowledging both the severity of water pollution and Chisso's culpability in instigating it will further destabilise the region's already precarious economy. Although a number of local government bodies take the disease seriously, many individuals chastise Minamata patients and other activists for threatening the welfare of their town. The narrator includes an article from the 19 October 1968 Kumamoto edition of the *Mainichi shinbun* (Mainichi Newspaper) describing the Development of Minamata City Citizens' Conference. The conference prospectus chastises those residents who have been intent on having Chisso admit its wrongdoing and modify its behaviour; conference participants support those afflicted by Minamata disease but insist on continued cooperation with Chisso.

Sea of Suffering exposes not only the terrible suffering experienced by those stricken with Minamata disease but also the many political, social and economic forces that, in denying this suffering, allow it to proliferate. Ishimure's novel trenchantly reveals that even the most obviously debilitating conditions are repudiated in the name of social stability and commercial profit. People are depicted not only as doing nothing when faced with ecodegradation but also as actively fighting against measures to remediate existing damage and prevent future harm to environments. The novel suggests that with disparities between conditions and behaviours so extreme, with even the most obviously debilitating and painful disease so readily disavowed, there is no real hope of diminishing, much less preempting, environmental crises.

Jiang Rong and the power of reward

Some texts take this conundrum to a logical extreme, featuring damage as being actively accelerated, regardless of predictable and undesirable consequences, even by those who have the closest ties to nature. Powerfully depicting this phenomenon is the Chinese writer Jiang Rong's bestselling *Wolf Totem*, based on his experiences in Inner Mongolia between 1967 and 1978, the Cultural Revolution and its immediate aftermath. The novel follows Chen Zhen, a young Beijing intellectual who during the Cultural Revolution is sent to Inner Mongolia's Olonbulag as part of a production team and becomes enthralled with the region's human and nonhuman occupants. He witnesses an influx of Han Chinese into the Inner Mongolian grasslands, the resulting transformation of the landscape from a space of nomadic herding to grain cropping, and the extermination of the area's wolves.

Throughout *Wolf Totem* the narrator and characters sharply contrast Mongol and Han Chinese attitudes and behaviours toward environments. The Mongols for centuries have respected the delicate ecological balance of the grasslands and engaged in symbiotic relationships with animals and vegetation. In contrast, the Han Chinese do not appreciate the region's ecosystems and recklessly decimate them. Differences between Mongol and Han Chinese attitudes toward and treatment of wolves are particularly striking. The narrator and characters of *Wolf Totem* repeatedly emphasise how the Mongols believe that the wolves are not only responsible for centuries of Mongol glory but are also indispensable in maintaining the homeostasis of the grasslands. The Mongols do hunt wolves; early winter, with its fresh snows that the animals have some difficulty navigating, is dubbed 'funeral season for wolves' (*xinxue chudong shi lang de sangji*) (104). But the Mongols hunt judiciously and are careful to safeguard wolf populations. Han Chinese, by way of contrast, are depicted as fearing and even despising wolves. These attitudes, stemming from baseless yet allegedly inherent prejudices, play a large part in the Han Chinese resolve to eradicate wolves from the Mongolian grasslands.

Yet the relationship of the Mongols with the nonhuman inhabitants of the grasslands changes dramatically the more deeply the Han Chinese advance into Inner Mongolia. The Mongols know all too well that exterminating wolves and converting the rangeland into farms will have devastating outcomes for both the human and nonhuman residents of the region. But they believe themselves helpless to stop or modify the Han Chinese agenda. Many, in fact, collaborate with the Han Chinese, assenting to and even actively participating in transforming the region. The most obvious example is Bao Shungui, 'a Mongol who long ago forgot about his Mongol forebears [and who] hates wolves even more than the Han Chinese' (275). Mongols also participate in the Inner Mongolian Production and Construction Corps, organised by Han Chinese. The first group of cadres from this corps sent into the field are 'half Mongols, half Han Chinese' and their first duty is exterminating wolves (475).

The Mongols greatly revere wolves, recognising the vital role they play in maintaining the health of the region's ecosystems. Yet Han Chinese rhetoric against wolves is so persuasive that most Mongols do not actively protest the extermination of these animals. The Han Chinese argue that removing wolves from the grasslands is necessary if these spaces are to be reclaimed for mechanised agriculture. Converting grazing land to farms, they assert, will 'eradicate damage caused by wolves, diseases, insects, and rats, and greatly strengthen the ability of the grasslands to resist natural disasters such as blizzards and shortfalls of snow [which dry out the land in winter], drought [during other seasons], windstorms, conflagrations, and insect pestilence' (467). This, in turn, will make life easier for the Mongols. One of their top priorities, the Han Chinese claim, is 'allowing the herders, who for thousands of years had lived under difficult conditions and had experienced much hardship and suffering, gradually to settle down to lives of stability and happiness.' To sweeten the deal, the Han Chinese promise to build the Mongols brick houses with tiled roofs, as well as 'roads, schools, hospitals, post offices, auditoriums, stores, movie theaters, etc.,' creating towns that can be lived in (467).

Mesmerised by lists of disasters that will be averted and catalogues of buildings and infrastructure that will be constructed, all the Mongolian educated youth and young herders, and the majority of the women and children, are said to look forward to the arrival of the Han Chinese. In contrast, the majority of middle-aged and elderly herders, described simply as 'keeping silent' (*mo bu zuo sheng*), appear to have resigned themselves to a future on Han Chinese terms (467). The Mongol elder Bilige's lament to Chen Zhen captures the ambivalent feelings of the older men:

We've long hoped for a school for our children and that our sick would no longer have to be taken to the banner alliance hospital by oxcart or horse-drawn wagon. We don't have a hospital, so many have died who shouldn't have. But what's to be done with the grassland? The grassland is too flimsy. The volume of livestock being transported is already too heavy. The grassland is a wooden-wheeled oxcart that can carry only a certain number of people and animals. If more people and machines are added, the cart will overturn. When the grassland is overturned, you Han Chinese can return to your original homes. But what are the herders to do? (467)

Bilige and other herders know that Han Chinese plans for the region are not sustainable, that the stability and happiness they have been promised will be fleeting. At the same time, they succumb to promises of a more comfortable life for themselves and their families. So they not only do not resist the Han Chinese incursion, they also join in transforming the region along Han Chinese guidelines. *Wolf Totem* highlights some of the many contradictions between behaviours and ecological conditions that are implicated in the degradation of environments. It reveals individuals – Han Chinese and Mongols alike – as disregarding incontrovertible physical evidence of the ruinous results of harming ecosystems.

Nguyễn Huy Thiệp and the paradoxes of relationships

The Vietnamese writer Nguyễn Huy Thiệp's partly fantastical 'The Salt of the Jungle' takes a different approach to environmental ambiguity, highlighting the conflicting relationship of a hunter with nature. The story opens with a celebration of the wonders of spring flora and fauna: 'The month after Tet is the most pleasant time in the jungle . . . Nature is both dignified and sentimental . . . If one goes into the jungle during this time, it's extremely delightful . . . All the absurd and ignoble things that one has to encounter every day can be completely dismissed at the sight of a small squirrel leaping onto the branch of a rambai tree.' But then the narrator suddenly alters course, declaring, 'It was during this time that Old Dieu went hunting . . . Old Dieu felt that hunting in the jungle on a spring day with a brand-new rifle would really make life worth living.' Deciding against pursuing bluebirds, which are abundant both in the jungle and closer to home, and mountain goats, which would be difficult to catch because they are few and far between, he decides to look for monkeys; since they are abundant and move in packs, 'shooting one shouldn't be difficult' (130–131). Interesting here is that the scarcity of the mountain goat is what saves it; unlike some hunters, who would embrace the challenge of capturing an endangered species, Dieu desires something more easily harnessed. Also noteworthy is that the glories of spring, far from deterring hunters from entering the jungle, appear to lure them in; part of nature's 'dignity' and 'delight' stems from the possibilities it offers those looking to kill its animals. This sentiment is accentuated shortly thereafter, when Dieu decides that he most desires the 'guard monkey,' which is currently out of sight. So he waits: 'It had been a long time since he had occasion to sit still like this . . . without any worries, and also without any calculations. The stillness and tranquility of the forest penetrated him thoroughly' (132). But this serenity, far from lessening his determination to seize a monkey, only readies him to take aim, albeit not at the animal on which he initially had set his sights.

Interesting as well is Dieu's conviction that 'Nature had saved that very monkey for him and no one else. He knew that even if he . . . was rather careless, it wouldn't affect the outcome.' Nevertheless, after shooting and wounding the animal, he 'trembles with fear,' the narrator declaring, 'He had just done something evil' (133). It is unclear whether the narrator here is stating his own belief, that of Dieu, or – most likely – one that they share. But the contrast is striking between, on the one hand, the idea that the natural world has 'saved' this monkey for Dieu and, on the other hand, the belief that he has committed a crime, particularly since before shooting the animal he had hyperbolically declared it an 'immoral father . . . Decadent lecher! Crude patriarch! Dirty legislator! Wretched despot!' (132). But feelings of remorse are shortlived. Ironically, the monkeys soon gain the upper hand, the wounded animal's female companion spiriting him away to safety and a third monkey grabbing Dieu's rifle and leaping away. After the latter monkey injures himself falling off a cliff, Dieu believes his chance has finally come. Yet observing the animal tremble violently, 'Old Dieu suddenly felt pity for the creature.' And so, amazingly, he constructs a makeshift bandage for it out of plants and his own underwear.[7] But soon thereafter, preparing to return home naked, he consoles himself: 'It was still worth it to be able to shoot an animal like this one' (139). Ironically, his actions in the jungle saved the monkey, which now scratches him with such ferocity that he drops it to the ground, where it lies still; whether it will live is unclear. The narrator draws attention to Dieu's sadness, both at having lost the animal and at its suffering. Yet in the end all would appear to be well. Nguyễn Huy Thiệp's story concludes with Dieu stumbling on flowers that serve as an omen that 'the country would find peace, and the crops would be bountiful'; Dieu vanishes into the mist and the reader is informed that 'In only a few more days, summer would arrive. The weather would get warmer' (140).

The ecosystems depicted in 'The Salt of the Jungle' are for the most part abundant, the story – unlike the other narratives examined in this chapter – sparing its readers scenes of obvious environmental degradation and in fact drawing attention to the jungle's fecundity.[8] Yet time and time again Nguyễn Huy Thiệp's tale reinforces the ambiguous relationships individuals, even those who set out to kill animals, have with nature, relationships that allow them to save, however temporarily, the very animals they target.

Literature and changing societies

For most communities, limiting further ecological degradation and remediating damaged ecosystems of all sizes will require significant cultural change, including 'new learning, a changed ethos, and vigorous action' (Engell 2009: 23). Societies need to reconceptualise the actual and the ideal places of people in ecosystems. Perceptions need to be aligned with actualities, and ideals need to be implemented. Essential to these endeavours is developing deeper, more nuanced understandings of the fluid relationships both among peoples and between peoples and environments in specific places and moments, as well as over time and across spaces. Writing, reading and analysing literature – openly imaginative texts with clear aesthetic ambitions – can perform important roles in this undertaking. Literature has the power to move us profoundly as it exposes how people dominate, damage and destroy one another and the natural world. It also allows us to imagine alternative scenarios. As Lawrence Buell has argued, 'For technological breakthroughs, legislative reforms, and paper covenants about environmental welfare to take effect, or even to be generated in the first place, requires a climate of transformed environmental values, perception, and will. To that end, the power of story, image, and artistic performance and the resources of aesthetics, ethics, and cultural theory are crucial' (2005: vi).

The power of story is particularly significant. Our sense of reality, our understandings of who we are and of our relationships with our surroundings, generally are constructed around stories, not around quantitative data (Akerlof & Shiller 2009: 6). The South African writer Njabulo S. Ndebele's comments on writing hold true for story: '[Writing] has the powerful capability to invade in a very intimate manner the personal world of the reader. Whenever you read, you risk being affected in a manner that can change the course of your life' (1994: 134). Stories, whether of the dangers or benefits of certain behaviours, often surpass data in their power to change people's behaviour. In fact, to become comprehensible, let alone to effect change, data themselves must be translated into narrative, and ultimately stories. Stories have the capacity to awaken, reinforce, and redirect environmental concern and creative thinking about environmental futures. Taking Buell's argument one step further, stories not only help shape legislation that requires changes in behaviours; stories also can cause sweeping changes in behaviours on a large scale in the absence of public policy. Yet the stories we tell ourselves to construct our sense of reality often differ from the stories narrated in creative texts and other art forms: the former tend to impose more logic and unity, or at least what appear to be logic and unity. There are, of course, numerous exceptions. But as George A. Akerlof and Robert J. Shiller have observed, 'The human mind is built to think in terms of narratives, of sequences of events with an internal logic and dynamic that appear as a unified whole' (2009: 51).

Likewise, documentary nonfiction, which frequently translates data into narrative, is committed to precision, or at least to the pretense of precision. Literature's regular and often blatant defiance of logic, precision and unity, by contrast, enables it to grapple more insistently and penetratingly than many other discourses with ambiguities in general and with those arising from interactions among people and ecosystems in particular. More specifically, literature's relative freedom from having to relate a single 'truth' allows it to highlight and negotiate – reveal, (re)interpret, and shape – the ambiguity that has long suffused interactions between people and environments, including those interactions that involve human damage to ecosystems. Ambiguity here emerges not as an ethical or artistic value; in other words, it is not aestheticised. Instead, it is portrayed, usually implicitly, as a deficit of consciousness and/or an implicit confession of the impotence of writers and literary characters.

Future trajectories

Discourse on environmental and disciplinary crises abounds. Many contend that ecological calamities are likely to be the most pressing issues of the twenty-first century. Many also argue that literature scholarship and the humanities more generally are in a state of near crisis due to budgetary constraints and increasing emphasis on science and technology. These dilemmas will not be easily resolved. But scholarship on individual cultures provides vital foundations for comprehending specific contexts of ecological abuse. The fields of comparative and particularly world literature help us appreciate more fully how creative writing and scholarship on creative writing can both reinforce and defy national, cultural, linguistic, geopolitical, and ecological divisions. Ecocriticism and other branches of environmental humanities demonstrate especially clearly the exciting possibilities for humanistic intervention in ecodegradation. Yet there is much work to be done. Without abandoning our time-honoured approaches, humanists need to collaborate more with one another to expand our cultural and disciplinary scopes, incorporating more diverse materials and methodologies even while nurturing expertise in new and specific areas. And ultimately, working with colleagues in the social, physical, and life sciences, we need to develop deeper connections among disciplines with the ultimate aims of embracing more fully the wider world – culturally, geographically and biophysically.

Bibliography

Akerlof, G.A. & Shiller, R.J. (2009) *Animal Spirits: How Human Psychology Drives the Economy, and Why it Matters for Global Capitalism* (Princeton University Press).
Buell, L. (2005) *The Future of Environmental Criticism: Environmental Crisis and Literary Imagination* (Blackwell).

Cho Sehŭi (1978) 'Kigye tosi', in *Nanjangi ka ssoa ollin chagŭn kong: Cho Sehŭi sosŏljip* (Munhak kwa Chisŏngsa).

Engell, J. (2009) 'Plant Beach Grass: Managing the House to Sustain It', Phi Beta Kappa Oration, Harvard University, 2 June.

Huang Chunming (1999) 'Fangsheng', in *Fangsheng* (Taipei: Lianhe Wenxue Chubanshe Youxian Gongsi).

Ishimure Michiko (2004) 'Kugai jodō: waga minamatabyō', in *Ishimura Michiko zenshū* 2 (Tokyo: Fujiwara Shoten).

Jiang Rong (Lü Jiamin) (2005) *Lang tuteng* (Fengyun Shidai).

Mi, Jiayan (2009) 'Framing Ambient *Unheimlich*: Ecoggedon, Ecological Unconscious, and Water Pathology in New Chinese Cinema', in Sheldon H. Lu & Jiayan Mi (Eds.), *Chinese Ecocinema in the Age of Environmental Challenge* (Hong Kong: Hong Kong University Press).

Napier, S. (2001) *Anime: From Akira to Princess Mononoke* (Palgrave).

Ndebele, N.S. (1994) *South African Literature and Culture: Rediscovery of the Ordinary* (New York: Manchester University Press).

Nguyen Huy Thiep (2003) 'The Salt of the Jungle,' in Nguyen Nguyet Cam & D. Sachs (Eds.), *Crossing the River: Short Fiction by Nguyen Huy Thiep* (Curbstone Press).

Paik, P.Y. (2010) *From Utopia to Apocalypse: Science Fiction and the Politics of Catastrophe* (University of Minnesota Press).

Takamine Takeshi (2006) 'Ishimure sakuin o yomu', in Iwaoka Nakamasa (Ed.), *Ishimure no sekai* (Gen Shobō).

Thornber, K.L. (2012) *Ecoambiguity: Environmental Crises and East Asian Literatures* (University of Michigan Press).

Wen Fumin & Jian Rao (2004) 'Zhongguo shengtai wenxue gaishuo', *Shaoguan Xueyuan xuebao (shehui kexue ban)* 25(1), 11–14.

Notes

1 Chichibu-Tama-Kai National Park is mainly in Yamanashi Prefecture, approximately two hours west of Tokyo.

2 Vietnam is geographically part of Southeast Asia, but is culturally part of East Asia.

3 Chen Kaige's (1952–) *Huang tudi* (Yellow Earth, 1984) is often regarded as China's first ecologically oriented film. Recent Chinese cinema addressing damming and water concerns in China includes Tian Zhuangzhuang's (1952–) *Delamu* (The Last Horse Caravan, 2004) and Jia Zhangke's (1970–) *Sanxia haoren* (Good People of the Three Gorges [Still Life], 2006). The documentary filmmaker Wang Bing's (1967–) *Tong dao* (Coal Money, 2008) and epic 14-hour film *Caiyou riji* (Crude Oil, 2008) highlight the difficult lives of miners and oil workers while also revealing the damage these industries inflict on ecosystems. Chinese films dealing with deforestation include Chen Kaige's *Haizi wang* (King of the Children, 1988), Lu Le's (1957–) *Meiren cao* (The Foliage, 2004), and Qi Jian's (1958–) *Tiangou* (The Forest Ranger, 2006) (Mi 2009: 293, n. 42).

   Japanese film has played a vital role as well, from documentaries such as director Satō Makoto's (1957–) *Agano ni ikiru* (Living on the Agano, 1992) and its sequel *Agano no kioku* (Memories of Agano, 2004), both of which dramatise the impact of Minamata disease (mercury poisoning) on a mountain community in Niigata, to the Oscar-winning 12-minute 'Tsumiki no ie' (The House of Small Cubes, 2008), directed by Katō Kunio and depicting an old man attempting to prevent rising water caused by global warming from flooding his house. Even more recent is the Oscar-winning documentary film *The Cove* (2009), which exposes the annual slaughter in Taiji (Southern Japan), of approximately 2,000 dolphins. The Japanese take the lives of about 20,000 dolphins annually, but Taiji is the only place in the country where they are herded and then killed. As shown in the film, in a blatant act of environmental ambiguity, Taiji puts on dolphin shows during which audience members can consume dolphin meat. Environmental degradation occupies an even larger position in Japanese manga and anime. One example is the anime metaseries *Gandamu* (Gundam, 1979–), which features overpopulation and destruction of ecosystems as causing massive armed conflict and migration to outer space. Just as noteworthy is celebrated director Miyazaki Hayao's (1941–) postapocalyptic *Kaze no tani no Naushika* (Nausicaa of the Windy Valley, 1982–94). Even more popular has been Miyazaki's *Mononokehime* (Princess Mononoke, 1997), Japan's highest grossing film of all time, animated or otherwise. This film is 'a wake-up call to human beings in a time of environmental and spiritual crisis that attempts to provoke its audience into realizing how much they have already lost and how much more they stand to lose' (Napier 2001: 180).

Korean films, including Chang Chunhwan's (1970–) *Chigu rŭl chik'yŏra* (*Save the Planet!*, 2003), have also made their mark (Paik 2010: 71–92). Although, in general, Korean cinema has not engaged with environmental degradation to the extent of its Japanese and Chinese counterparts, audiences embraced *Koemul* (The Host, 2006). This film, based on an actual incident and alluding to American use of Agent Orange during the Vietnam War, features an American military pathologist who orders his Korean assistant to dump 200 bottles of formaldehyde down the drain, poisoning the Han River and spawning a dangerous amphibious creature.

4   The first chapter of *Ecoambiguity* overviews East Asia's principal environmental crises and literary works on ecodegradation. The book also introduces several key works of Vietnamese writing on ecological disasters, particularly Agent Orange.

5   All translations from Chinese, Korean and Japanese are my own.

6   The first film on Minamata disease – Tsuchimoto Noriaki's (1928–2008) *Minamatabyō – kanjasan to sono sekai* (Minamata Disease – Patients and Their World, 1971) – was followed by many others on this illness (Takamine 2006).

7   Dieu, having earlier removed his other clothes, is now naked.

8   Although Vietnamese literature does contain a number of texts that expose environmental degradation, most frequently wartime destruction of environments, particularly by Agent Orange, and although many Vietnamese writers are active in environmental groups and have spoken and written eloquently on ecodegradation, they have not made ecological concerns as much a focus of their creative work.

Part III
Politics and policy

5

Environmental policy in East Asia

Institutions in comparative perspective

Sangbum Shin

Introduction

This chapter overviews the processes in which East Asian countries have tried to tackle environmental problems and climate change issues focusing especially on environmental institutions in comparative perspective. Environmental institutions include government environmental organizations, basic environmental legislation, major environmental policies, and channels for non-governmental actors. I shall first briefly introduce economic and environmental profiles of selected East Asian countries. Then, I examine some similarities and differences of the processes and consequences of environmental institutionalization in these countries focusing mostly on the overall patterns of environmental governance such as top down or bottom up. I also see if the patterns overlap with those found in their economic development. In addition, I briefly introduce the climate change policies of these countries and emphasize the environmental decentralization as a key factor for successful climate change policies in this region.

The economy and environment in East Asia

Economic profile

East Asia is diverse in terms of size, degree, speed, and timing of economic development. Countries such as Japan and China are among the largest economies in the world. The size of China's economy alone, measured by gross domestic product (GDP), comprises more than 10 percent of the world's total GDP. By contrast, the size of the economy in Laos or Mongolia is less than 0.1 percent of China's economy. National income measured by GDP per capita also varies from around US$50,000 dollars in Singapore to around US$2,000 dollars in Cambodia. In terms of timing of development, Japan started industrialization as early as the mid-nineteenth century and experienced rapid industrialization from the late 1950s to 1970s. The four East Asian tigers – Singapore, Hong Kong, South Korea, and Taiwan – took off in the 1950s and 1960s and got through the high growth rate period mostly from the 1970s to the early 1990s. China initiated the reform and opening up policies in 1978 and since then maintains high growth rates to date. In Southeast Asia, Indonesia, Malaysia, Thailand, and Philippines are the major early starters and countries such as Vietnam, Laos, and Cambodia are rapidly catching up. Table 5.1 summarizes selected economic indicators of seven major East Asian countries. The combined GDP of these seven countries comprises around 23 percent of world GDP, and it would be slightly more than 25 percent if we included all the other East Asian countries. Although some countries such as Japan and South Korea show relatively low growth rates, overall average growth rates of these countries since 1994 have been always higher than the world average.

Scholars of political economy have paid attention to some institutional factors that enabled the rapid East Asian industrialization and concluded that the government intervention on the market played a significant role in these transformations. The government's key agencies in charge of macroeconomic policies are rational enough to plan and regulate the market effectively, and they initiate and lead the economic development. The developmental orientation of governments comes from the fact that they are late in terms of the timing of economic development compared to the early industrializers. The government not only has willingness but also capability to drive and orchestrate business sectors to catch up with advanced economies (Evans 1995). This type of government is often described as 'developmental state', initially suggested to explain Japan's industrialization, especially focusing on the country's pilot agency, the Ministry of International Trade and Industry (MITI) (Johnson 1982), but soon widely used to analyze other East Asian countries (Wade 1992; Woo-Cumings 1999).

While the term developmental state has brought about some debates for its applicability beyond Japan (Breslin 1996), it should be noted that the key of the institutional arguments is the role of the state. State is itself an entrepreneur and sometimes a banker, which makes decisions to allocate resources, plans, sets up short-term and long-term goals for development, and deploys appropriate development strategies (Amsden 1992; Wade 1992). This type of state-led industrialization is found most clearly in the Northeast Asian countries and soon became influential among the Southeast Asian countries as a desirable model to copy. Due to the critical role of

Table 5.1 Selected economic indicators of seven East Asian countries, 2011

	GDP[1] ($ billion)	GDP growth rate[2] (%)	GDP growth rate, average (1994–2003)[3] (%)	GDP growth rate, average (2004–2011)[4] (%)	GDP per capita, PPP[5] ($)	Urban population (% of total)[6]
China	7,318.4	9.2	9.4	10.8	8,450	43
Japan	5,867.1	–0.8	0.9	1.2	35,530	66
Korea	1,116.2	3.6	5.7	3.9	30,340	81
Indonesia	846.8	6.5	3.1	5.7	4,530	51
Malaysia	278.6	5.1	5.4	4.9	15,190	70
Thailand	345.6	0.1	3.4	3.6	8,390	33
Philippines	224.7	3.9	3.8	5.0	4,160	65
Average	15,997.5*	3.94	4.5	5.0	15,227	–
World	70,160	3.8	3.4	3.9	12,000	50

Notes:
[1] Current USD, World Bank, World Development Indicators.
[2] [3] [4] International Monetary Fund, World Economic Outlook, October 2012.
[5] Current international dollar, World Bank, World Development Indicators.
[6] World Bank, World Development Indicators.
* Total amount (not average).

the government, this model often coincides with political authoritarianism and weak civil society, and often with highly centripetal relationships between the central and local governments.

The state-led industrialization of at least some, if not all, East Asian countries might also have important theoretical and policy implications for environmental politics in the region. Most of all, it has resulted in relatively heavy dependence on the governments in environmental policy initiation and implementation in the region. The overall top-down pattern of environmental governance has persisted even after democratization and rise of civil society in many countries of the region. It also has resulted in relatively centripetal relations between central and local governments, which is not conducive to effective climate change policies.

Environmental profile

The state of the environment in East Asia is as diverse as its economic profile. Northeast Asia is a collection of rapid industrializers with different timings of economic takeoff. Japan is the earliest one followed by South Korea and Taiwan in the 1960s, subsequently China in the 1980s and most recently Mongolia and perhaps North Korea in the near future. One of the important environmental consequences of the different timings of industrialization is that these countries have been experiencing the pollution abatement history in the same order. In other words, early industrializers are also early in terms of abatement of industrial pollution such as air, water, and solid waste problems although the gap between early and late industrializers in terms of environmental policy initiation is much smaller than the gap in terms of timing of industrialization. Therefore, while Japan, South Korea, and Taiwan have steered successfully through the phase of industrial pollution abatement, China is still in the middle of it, and others are at the very initial period of environmental policy.

At the same time, however, there is variation even within a country such as China. Some parts of China are earlier than others in terms of environmental policy initiation presumably due to the region's early industrialization (Economy 2004). Also important is that even though the timing of heavy pollution and its abatement is different among countries, there are some cases where pollution in one country moves to other countries such as transboundary air pollution and yellow dust problems between China and its neighboring countries (Zarsky 1995). In addition, as economies in these countries have become more and more integrated due to economic globalization, some polluting firms move to the countries where environmental regulations are lax in order to save the regulatory cost. Therefore, environmental relations in Northeast Asia have become somewhat complicated and the abatement process also has become more dynamic than before.

The post-colonial development of Southeast Asia has also been remarkable. As the region has been rapidly integrated into the global economic system, it quickly became the center of export-oriented economic development especially based on natural resources such as timber, petroleum, and marine resources. Accordingly, environmental stress and degradation have been especially serious in the resource-related issue areas such as deforestation, soil deterioration, and mining (Hirsch & Warren 1998). Transnational corporations (TNCs) played a significant role in the economic development of the region. At the same time, however, they have been also responsible for rapid deforestation and environmental degradation in the region (Bryant & Parnwell 1996; Ueta & Mori 2007: 173–174). Also, since the mid-1980s, rapid industrialization of some successful economies coupled with rapid urbanization (such as has happened in Indonesia, Malaysia, and Thailand) has generated industrial pollution problems especially in the urban areas.

Table 5.2 illustrates selected energy and environmental indicators of seven East Asian countries. In terms of energy consumption and CO_2 emissions, although China is one of the

Table 5.2 Selected environment and energy indicators of seven East Asian countries, 2010

	Energy use per capita (kg oil equivalent)	Electronic power consumption per capita (kWh)	CO_2 emissions per capita (metric tons)	CO_2 emissions growth (%, 1990–2006)	Particulate matter (μg/cu. m)	Internal freshwater resources per capita (cu. m)
China	1,484	2,332	4.7	152.8	73	2,134
Japan	4,019	8,474	10.1	10.3	30	3,365
Korea	4,586	8,502	9.8	96.7	35	1,338
Indonesia	849	566	1.5	121.7	83	12,632
Malaysia	2,733	3,667	7.2	232.0	23	21,841
Thailand	1,553	2,055	4.1	184.4	71	3,135
Philippines	451	586	0.8	53.5	23	5,399
Average	2,239	3,740	5.5	121.6	–	–
World	1,819	2,846	4.4	34	50	6,617

Source: World Bank (2010).

world's largest energy consumers and the biggest CO_2 emitter, South Korea and Japan are the top two countries of energy consumption and CO_2 emissions per capita in the region. However, China and other middle income countries such as Malaysia and Thailand show extremely high growth rate of CO_2 emissions from 1990 to 2006. In terms of the volume of particulate matter and the internal freshwater resources per capita, respectively a common indicator of air and water pollution, China shows typical records of developing countries while others show some mixed results. For example, Korea's freshwater resource per capita is the lowest among the seven countries and much lower even than the world's average. Therefore, it can be said that while overall industrial pollution is more serious in developing parts of East Asia, results of energy and CO_2-related indicators calculated per capita show some variation among countries.

Environmental institutions in comparative perspective

Historical overview[1]

Two most basic and important indicators of environmental institutionalization for any country might be the passage of basic environmental legislation – enabling the promulgation of environmental regulations and establishment of environmental agencies – and actual formation of environmental regulatory and enforcement agencies (Sonnenfeld & Mol 2006: 122). Therefore, I shall briefly overview the history of environmental institutionalization of selected countries focusing on these two indicators. However, unlike previous literature, I shall suggest that there have been at least two critical periods of environmental institutionalization in almost all the East Asian countries except some early movers such as Japan and Singapore. The first is the period of initiation and the second is the period of substantial development. In general, it can be said that the 1972 United Nations Conference on Human Environment at Stockholm played a significant role in triggering institution building in these countries. However, the follow-up processes of environmental institutionalization vary in different countries according to different international and domestic conditions. The 1992 United Nations Conference on Environment and Development at Rio was certainly a watershed event but other domestic turning points such as democratization and decentralization also became a critical momentum for institutional development. Table 5.3 illustrates the first and second critical periods of environmental institutionalization in major East Asian countries.[2]

As the first industrializer in East Asia, Japan was the first country in the region to build up government environmental organizations and set up environmental legislation. The Basic Law for Environmental Pollution Control enacted in 1967 and the Air Pollution Control Law enacted in 1968 were the first major environmental laws on which various environmental quality standards and regulations were set up. In terms of organizational development, the establishment of the Environmental Agency in 1971 as the comprehensive government organization for environmental administration was also a significant starting point for Japan's environmental institutionalization. Since then, Japan showed steady progress in developing and implementing major environmental regulations and policies for pollution control and natural conservation throughout the 1970s and 1980s (Imura 2005). Singapore was another first mover in the region.

Table 5.3 Establishment of environmental institutions

Year	Environmental agency	Environmental law
1967		Japan
1968		
1969		
1970		
1971	Japan	
1972	Singapore	Singapore
1973		
1974	China 1	Malaysia 1
1975	Malaysia 1, Thailand 1	Thailand 1
1976		
1977	Philippines 1	Philippines 1, Korea 1
1978	Indonesia 1	
1979		China 1, Taiwan 1
1980	Korea 1	
1981		
1982	Taiwan 1	Indonesia 1
1983	Malaysia 2	
1984		
1985		Malaysia 2
1986		
1987	Philippines 2, Taiwan 2	Taiwan 2
1988		
1989		China 2
1990	Korea 2	Korea 2
1991		
1992	Thailand 2	Thailand 2
1993		
1994		
1995		
1996		
1997	Philippines 2	Indonesia 2
1998	China 2	
1999		Philippines 2
2000		

It is said to be the only East Asian newly industrializing economy that did not follow the 'growth first, environment later' strategy (Rock 2002: 23). The government established the Ministry of the Environment in 1972 and in the same year, the Clean Air Standards Regulations were promulgated. Therefore, the starting point of the country's environmental institutionalization might be considered to be 1972.

Other countries in the region show a time gap between the first and the second critical periods ranging from eight to 24 years. In South Korea, the government enacted the Environmental Preservation Act in 1977 and established the Office of Environment as a sub-cabinet agency of the Ministry of Public Health and Social Affairs in 1980 (Chung & Kirkby 2002: 195–196). While these two events were apparently the critical moment of departure in South Korea's environmental institutionalization, the government's reaction to the environmental problems was overall marginal and the citizen awareness was low until the mid-1980s. It was only after the democratization and regime change in 1987 and 1988 that the government began to take environmental issues seriously. The Basic Environmental Policy Act was enacted in 1990 and in the same year the Office of Environment was upgraded to full ministerial level as the Ministry of Environment. Subsequently, a variety of environmental laws and regulations were issued during the 1990s. Similarly, in Taiwan, the Environmental Protection Act in Taiwan Area enacted in 1979 and the Environmental Protection Bureau established in 1982 were the point of departure but the substantial progress in environmental institutionalization came in the late 1980s and early 1990s after martial law was lifted in 1987. The Environmental Protection Bureau was upgraded into Taiwan Environmental Protection Administration in 1987 and the major laws and regulations were issued or significantly revised in the late 1980s and early 1990s.

China shows a relatively long-term and gradual process of environmental institutionalization. After the Chinese delegation participated in the 1972 Stockholm Conference, the government held the first National Environmental Protection Conference in 1973 and the Leading Group on Environmental Protection was established directly under the State Council in 1974. Also, right after the reform and opening up decisions in 1978, the Environmental Protection Law was enacted in 1979 as a trial implementation. However, as was the case in other countries in the region, it has taken many years since the first critical period to make substantial progress. After 10 years of trial period, the Environmental Protection Law was enacted in 1989 and the National Environmental Protection Agency, which was established in 1994 as the main government regulatory organization, was upgraded to sub-cabinet level as the State Environmental Protection Administration in 1998, and again to full ministerial level as the Ministry of Environmental Protection only in 2008. Presumably, China's process of environmental institutionalization is overall slow and gradual due to the size of the economy and lack of radical political change such as democratization. However, as China has been more internationalized and integrated into the global economy, the country has seized more opportunities to strengthen its environmental institutions such as technology transfer and harmonization with various international standards for environmental protection (Shin 2004). Also, it has drawn much attention and support from many international organizations and foreign governments because of its huge impact on global environmental change.

In Southeast Asia, as we mentioned earlier, all countries have experienced two critical periods except Singapore. In Malaysia, the government enacted the Environmental Quality Act in 1974, established the Environmental Division in 1975 and placed it under the Ministry of Science, Technology, and Environment in 1976. Subsequently, the Environmental Division was upgraded to department level as the Department of Environment in 1983. Later, it was placed under the Ministry of Natural Resources and Environment in 2004. Also, the Environmental Quality Act was seriously revised in 1985 and other major laws and regulations were issued in the late 1980s and early 1990s. In the case of Philippines, the first critical period was the late 1970s. The Environmental Policy Decree (Presidential Decree No. 1151) was enacted in 1977 and in the same year the National Environmental Protection Council was established as the first government organization for environmental protection. After the martial law was lifted and the new Aquino administration was inaugurated in 1986, the Department of Environment and Natural Resources was established in 1987 and subsequent legal progress has been made including the 1999 Comprehensive Air Pollution Policy Act.

In Indonesia, the Ministry of Environment and Development Supervision, which was established in 1978, was the first government environmental protection agency. It was a ministry-level organization from the beginning and was renamed as the Ministry of Environment in 1993. Another significant development in the first critical period was that the government passed landmark environmental legislation in 1982, which was the Environmental Management Act. Since then, the country has made steady progress in environmental institutions in a way to allow more citizens to organize environmental NGOs and to participate in the government policy processes. Therefore, it is difficult to find the second critical period of institutional development in Indonesia, but it is also true that there had been much progress in the late 1990s in terms of legal development. Especially in 1997, the 1982 Act was significantly revised to include broader environmental issues and to strengthen various environmental standards. Also, most of the specific environmental regulations were promulgated in 1997. By way of contrast, the two critical periods are identified more clearly in Thailand. The government created the National Environmental Board in 1975 but it played only an advisory role until it was promoted to the ministerial-level organization in 1992. Also, the Environmental Act was enacted in 1975 but was significantly revised in 1992. Therefore the two periods – 1975 and 1992 – are the critical landmarks in the history of Thailand's environmental institutionalization (Harashima 2000: 196; Rock 2002: 115–120).

Patterns of environmental governance

As was described in the introductory section, many successful economies in East Asia have experienced state-led economic development in which the government played a significant role. Therefore, the strong role of the government has become a kind of institutional characteristic of these countries and, at least in some of the countries, still remains as one of the most important factors influencing the politics and society. It has also been one of the most critical factors in determining the patterns of environmental governance in these countries. Overall, it can be said that the pattern of environmental governance in East Asia is still top down rather than bottom up due to the tradition of a strong state. The government is the most important actor in shaping environmental politics, and the role of civil society and environmental NGOs is relatively weak compared with developed countries. However, a closer investigation of each country shows that at least some of the East Asian countries have developed institutional channels for civil society actors to participate in the government policy process and cooperate with the government agencies. Especially in Southeast Asia, countries such as Indonesia and Philippines have a relatively strong environmental civil society although the government is still the most powerful actor in environmental governance. Moreover, even though the government is a powerful actor in this region, the specific contexts in which the government became powerful are different from country to country. This last point is clearly shown especially in the Northeast Asian countries.

In Northeast Asia, Japan's environmental governance shows a typical top-down pattern and perhaps provides a kind of prototype of East Asian environmental governance. In Japan, the environmental consequences of rapid industrialization began to emerge with environmental tragedies in the 1950s and 1960s in major industrial areas such as Minamata, Niigata, and Yokkaichi. In response to these tragedies, the victims and general citizens began to protest and organize environmental movements, and these movements quickly spread into many other local communities. The movements grew in force and number very quickly and they became, along with judicial activism and media coverage, one of the most significant driving forces to push the government to introduce environmental laws and create governmental environmental agencies in the late 1960s and early 1970s. However, these movements did not become institutionalized as occurred in Western advanced countries. Instead, they remained scattered, small, and locally oriented (Schreurs 2002a; 58–59).

The biggest reason why growing environmental activism did not result in creation of national-level environmental NGOs in Japan is that the government's response to the movement was very timely and successful in eliminating the root causes of popular discontent (Upham 1987: 28–67). Whereas the government constrains opportunities for citizen participation and effectively impedes the institutionalization of environmental movements into a Western-style environmental NGO community, it opens various channels for government–industry cooperation in environmental policy. The government's success was possible due to its support of industry in the form of administrative guidance, research and tax incentives. Particularly at the local level, building consensus among industries and local government authorities was an integral part of the environmental policy implementation in Japan. The central government encourages local governments to create extralegal agreements on pollution control with industries, and such agreements result in the creation of more stringent pollution control standards than exist at the national level. During the 1970s and 1980s, these agreements proliferated, and in the late 1990s more than 30,000 polluting facilities had negotiated agreements with local governments (Schreurs 2002b: 69–76). As a result, the government-led environmental policy initiative proved to be quite effective whereas environmental groups lost momentum and left as thousands of neighboring communities took care of local environmental issues such as recycling.

In addition to the government's preemptive strategy, there were some institutional barriers to the formation of environmental civil society in Japan. Scholars have pointed out that Japan had one of the most restrictive regulatory environments governing the non-profit sector of any developed country until the end of the 1990s. All public associations and foundations had to obtain government approval in order to attain incorporation as a non-profit organization, and the government could continually monitor groups that had already been given legal status as public interest foundations or associations and use administrative guidance to decide whether or not a group was financially sound enough to be allowed to incorporate. Therefore, groups often hired former bureaucrats with connections to the ministries in order to facilitate the process. Another institutional barrier is that civil society groups had great difficulty in fundraising because of the limitations on tax benefits accorded to non-profit groups. Unlike Western civil society groups, it was very difficult for them to receive tax-deductible contributions due to the institutional restrictions. Therefore, they had to rely on government subsidies or voluntary activities. Also, the government controlled the environmental information and media coverage, which made it very difficult for civil society groups to effectively respond to major environmental issues (Schreurs 2002a: 59–60).

Similar institutional barriers are found in current China. Although environmental NGOs are becoming an important actor in China's environmental governance, they still have to be registered and tightly controlled by the government. It is certain that activities of Chinese environmental NGOs have become more diversified and the number of environmental NGOs has increased in recent years (Yang 2005). The so-called 'government organized NGOs' (GONGOs) that have dominated China's environmental civil society sector for a long time have gained recently more organizational, financial, and political autonomy from the government (Mol & Carter 2007: 12). However, it is still true that their activities are mostly focused on environmental education, awareness, natural conservation, and endangered species protection, rather than directly confronting and challenging the government by raising politically sensitive issues. The Chinese government is, on the one hand, extremely sensitive and suspicious of the rise of environmental civil activism because it might quickly develop into mass scale movements and contentions and probably result in a spillover to other issue areas. On the other hand, however, the government cannot entirely repress green activism partially because it is under increasing pressure to devolve certain government functions to society and also because by doing so it can reduce the regulatory costs. Therefore, the government cautiously allows green activism to a certain degree (Ho 2001).

It should be noted that China is still an authoritarian, one-party dominated state where the government has enormous power vis-à-vis civil society in all policy areas. Therefore, it is not surprising that green activism in civil society is still relatively weak compared with other countries with competing political parties and elections. Therefore, Western scholars often evaluate recent development of environmental NGOs in China positively because they are occurring under – and in spite of – the authoritarian government. Contrariwise, whereas domestic environmental NGOs are still weak, China has benefited a lot from outside help, i.e., international assistance at various levels and forms. Many international environmental NGOs and international organizations established their offices and began operations in Beijing and other Chinese cities due to China's importance in global environmental politics. The Ford Foundation and the World Wildlife Fund (WWF) opened their Beijing offices as early as in the mid-1980s. Since then, many international environmental NGOs, foundations, research centers, and international organizations have begun to set up branches and offices in China and have become an integral part of China's environmental governance. Although it is still weak, they are increasingly active in establishing networks with Chinese domestic partners (Zusman & Turner 2005: 131–144).

Therefore, the government plays a significant role in China's environmental governance not because it has responded to the environmental crisis readily and effectively but primarily because the green civil society is not yet well developed and environmental activist groups are constrained by the authoritarian government. In the meantime, foreign actors fill the void of a weak green civil society by providing environmental technologies, funding, policy knowhow, and awareness. In addition, more recently, citizen complaints on environmental matters and protests against the government have increased remarkably. The most recent example is the Ningbo city case where the local authority suspended construction at a petrochemical complex after days of protest by residents fearing possible serious pollution by the construction. Similar protests over fears of health risks and environmental damages by industrial projects have occurred frequently within last couple of years.[3] This could be a positive sign for possible changes in China's environmental governance. As income increases, citizens are more concerned with environmental matters as occurred in Western developed countries during the second half of the twentieth century. It can strengthen civil society and result in even larger political changes in the future. In sum, the pattern of China's environmental governance has been overwhelmingly top down but there are certainly some signs of change.

South Korea might be also a unique case in the sense that even after democratization and political development, the government is, overall, still the most important actor in environmental politics. The government had been reasonably successful in addressing industrial pollution, particularly in the 1990s, but at the same time environmental movements grew into national level organizations such as the Korean Federation for Environmental Movements established in

1993 and Green Korea United also founded in 1993. In fact, after the democratization in 1987, the country witnessed rapid growth of civil society, and therefore, the number of interest groups that had been suppressed by the government for almost 25 years also increased rapidly. As a consequence, environmental groups emerged and became an important actor in Korea's environmental politics. However, their role and influence has not been significant consistently. Whenever there was economic downturn and even crisis such as the 1997 East Asian financial crisis, environmental activism quickly lost momentum and the government again dominated environmental policy processes. Similarly, when a pro-growth political leader is elected as the president, civil society participation in environmental processes is severely constrained by government regulations.

It can be said that environmental issues in Korea are closed related to bipartisan political confrontation between pro-growth and pro-(re)distribution forces. Since Korea achieved rapid economic growth based on the political discourse of 'growth first, distribution later,' the pro-growth ideology has become an essential part of Korean politics and society during the rapid development period. However, after the democratization and subsequent regime changes, a new paradigm has emerged in the society to emphasize distributive justice and social welfare. This paradigm began to challenge the existing pro-growth ideology in the presidential elections as well as in street demonstrations. It was in this context that environmental civil society emerged in Korea around the mid-1990s. After the East Asian financial crisis in 1997, however, citizens became less interested in environmental issues due to the economic recession and therefore environmental groups became less influential. As a pro-distribution party leader was elected as the president in 2002, once again environmental groups gained strength and political leverage, but after the 2008 global financial crisis and subsequent regime change in Korea, the pro-growth government of President Lee Myongbak has paid less attention to the voices from the environmental civil society. In sum, although the overall pattern of environmental governance in Korea is top down, it is more dynamic than other Northeast Asian countries mostly due to the political volatility resulting from bipartisan confrontations.

In Southeast Asia, there is more variation in terms of the pattern of environmental governance. Although the government is the most important actor, it can be said that, overall, civil society participation in environmental politics is more active in Southeast Asia than in Northeast Asia. In Indonesia, since the environmental NGOs were legally recognized by the Environmental Management Act (EMA) in 1982, they have played a significant role in environmental policy process both as watchdogs and as partners to cooperate with the government. However, as environmental NGOs gain more leverage and support from citizens, the government began to control them by revising the EMA in 1997 to drop the word 'non-government' and insert specific requirements to establish environmental organizations. This change occurred in the last two years of the Suharto dictatorship, which lasted for 30 years. In some sense, it demonstrates that green civil activism has become powerful enough to challenge the government. In fact, by the 1997 EMA, environmental organizations were granted standing to sue although it was only limited to the purpose of conservation of natural parks and protected species.

Similarly, environmental NGOs have proliferated in the 1980s and 1990s in Thailand and various official and unofficial channels have been created to connect their voice to the government. The notable victories that environmental NGOs and local grassroots activists achieved such the cancellation of the Nam Choan dam in 1988 and banning of logging in 1989 were the landmark events in Thailand's environmental activism (Forsyth 1999: 691). Although environmental NGOs are not powerful enough to challenge the government, they are certainly one of the influential actors in Thailand's environmental politics (Sangchai 2005: 166). In the Philippines, the Department of Environment and Natural Resources has an NGO desk, which was set up to provide the environmental NGOs with information on government plans and programs regarding environmental policy. In 1992, the President created the Philippine Council for Sustainable Development which functions as a major channel for NGOs and people's organizations to participate in the government policy process.

Another difference between Northeast and Southeast Asia is that regional environmental cooperation regimes are more active in Southeast Asia especially by the Association of Southeast Asian Nations (ASEAN). It is still early to conclude that ASEAN has become an effective regime for regional environmental cooperation. However, it is obviously one of the important actors in member countries' environmental politics. It affects domestic environmental policy processes of member countries by organizing various meetings and conferences, making regional agreements, initiating regional action programs, and establishing less official networks (Elliott 2012; Kheng-Lian & Robinson 2002; Nguitragool 2011). It has been engaged in various regional environmental issues especially the forest management and forest fires and haze problems. The Agreement on Transboundary Haze Pollution in 2002 was one of the achievements of ASEAN for regional environmental cooperation. The agreement contains provisions on monitoring, assessment and prevention, technical cooperation and scientific research, mechanisms for coordination, lines of communication and simplified customs and immigration procedures for disaster relief (Nguitragool 2011: 70). Also, ASEAN's most active knowledge networks have been established in the forestry issue area such as the Regional Knowledge Network on Forest Law Enforcement, and the Governance and the Regional Knowledge Network on Forests and Climate Change (Elliott 2012: 50).

In sum, the overall pattern of environmental governance in East Asia is still top down in which government agencies are the most powerful and significant actors in environmental policy. In general, it can be said that the pattern is more top down in the Northeast Asian countries than in the Southeast Asian countries. However, specific reasons why it is top down are different among the Northeast Asian countries. Similarly, while environmental NGOs and civil networks are in general more active in Southeast Asia, there is variation among them in terms of their strength and influence. In addition, regional environmental cooperation is more active in Southeast Asia than Northeast Asia, especially ASEAN's efforts to tackle various regional environmental problems. It is true that there have been attempts at regional environmental cooperation in Northeast Asia such as the Tripartite Environmental Ministers Meeting (TEMM). However, specific cooperation activities and achievements have not been made to date although there have been urgent regional environmental problems such as yellow dust and marine protection.

Climate change and environmental decentralization

Climate change, especially global warming, might have been one of the most serious global challenges during the last two decades. It is not only an environmental issue but a broader challenge that affects every aspect of human life around the world. East Asia is one of the regions most vulnerable to climate change but at the same time it is one of the regions that are *responsible* for climate change, especially global warming. It has large CO_2 emitters such as China, Japan, and South Korea and also many developing countries in the region are either at the beginning of or in the middle of the carbon-intensive stages of economic development. Countries around the world have attempted to tackle climate change and global warming by promising and implementing international cooperation. The United Nations Framework Convention on Climate Change (UNFCCC) in 1992 and the Kyoto Protocol agreed in 1997 at the 3rd conference of the parties (COP) to the UNFCCC are the milestones in these efforts. Since climate change policies in each country directly affect its economic and social policies, governments play a leading role in policy initiation and implementation. They set up national strategies to tackle climate change issues, harmonize climate change policies with other socioeconomic and environmental policies, and negotiate and sometimes cooperate with each other in international regimes such as the annual COPs to the UNFCCC.

The role of the government in climate change policies is particularly prominent in East Asian countries due to its tradition of strong state in economic as well as environmental policies. In Japan, the climate change policy has been predominantly a matter of the government agencies. As the world's fifth largest CO_2 emitter and as the only Annex I country in East Asia, Japan has built up national strategies for climate change mitigation and national plans for emissions reduction of greenhouse gases. The government not only had to consider the industrial interests due to the huge impact of emission reduction plans on the economy but also was concerned about the country's national image and international standing. Therefore, Japan's policy on climate change was largely the result of bargaining among the government bodies – bureaucrats and ruling party – and the industry (Harris 2003: 7). The government has been also very active in clean development mechanism (CDM) investment. Japan has been the fourth largest CDM investor among all Annex I and non-Annex I investor parties. In 2012 Japan's investment comprised 9.38 percent of the total registered CDM projects.[4] Most recently, after the 11 March earthquake and tsunami in 2011, the government has had to review and revise the existing climate change policies such as energy, environment, transportation, and, most of all, its CO_2 emissions reduction plans during the first commitment period of the Kyoto Protocol. For example, prior to the 11 March tragedy, nuclear energy comprised some 30 percent of the country's total electricity generation. However, almost all nuclear power plants have been shut down or suspended after 11 March and the government has had to give up its former energy policy to increase nuclear power dependency from 30 to 40 percent. In all these processes, although citizens and environmental NGOs express their voices, they are largely outside the decision-making framework.

It is similar in China, in that climate change policies are established and implemented mostly by the government agencies. The National Development and Reform Commission (NDRC), one of the most powerful government organizations planning and orchestrating national economic development, is the single most important actor in China's climate change policies. Also, the National Energy Administration (NEA) established in 2008 directly under the NDRC plays a significant role because China's policy objectives and action plans for climate change have been centered on the issues of energy efficiency and renewable energy development. The NDRC chairs the National Leading Group to address Climate Change which is the highest government organ taking charge of China's national climate change strategies. The NDRC also makes key decisions in international negotiations especially in the COPs to the UNFCCC. China has long been arguing for 'common but different obligations' and asked developed countries to reduce greenhouse gas emissions first. However, there is no other voice from the domestic side. In general, Chinese scholars, civil society organizations, industry, and even environmental NGOs strongly support the government position. The NDRC is also the most important organization in China's CDM. China has been always the number one host country of CDM, totaling 65 percent of total certified emissions reductions (CERs).[5]

In other East Asian countries, the role of the government in climate change policies is overall predominant. Governments set up national strategies and action plans to tackle climate changes and these policies strongly reflect their national interests. However, one of the critical factors in climate change policies of East Asian countries might be the devolution of environmental and climate change policies to local authorities, i.e., decentralization of environmental policy.

Local governments in developed countries increasingly play a significant role in climate change policies (Schreurs 2008). It turns out that local initiatives have proved to be more effective especially in the adaptation policies, because they are relatively free from collective action problems of adaptation policies. In other words, they do not have to be concerned about what others might do when they set up their adaptation policies. Also, market incentive policy instruments such as CO_2 emission trading schemes have proved to be more effective when they are initiated and implemented by local governments than central government. Moreover, in some cases, local governments attempt to create regional or global networks to tackle climate change issues and other environmental problems (Nakamura et al. 2011). The International Council for Local Environmental Initiative (ICLEI), Clinton Climate Initiative (CCI), and C40 Cities Climate Leadership Group are the typical examples of global networks of local governments.

In terms of environmental decentralization, however, East Asia falls more or less behind. In Japan, some sub-national governments have begun to set up local climate change policies but still the central government plays a leading role in national climate change policies especially after the 11 March 2011 tsunami tragedy. Korea's 'Low Carbon Green Growth' strategy is almost completely a central government policy. China's local environmental protection bureaus (EPBs) still suffer from the 'dual allegiance problem' in which EPBs depend on local governments of the same level for funding, personnel, and other policy resources but at the same time they receive policy mandates from the central government (Jahiel 2000). In Southeast Asia, countries such as Indonesia and the Philippines have a longer history of environmental decentralization and more cases of local empowerment than Northeast Asia. However, local governments still lack political power and the autonomy to set up and implement local climate change policies.

Conclusions

The environmental institutions in East Asia are, as a whole, government centered. The government is the most powerful and important actor while civil society actors are relatively weak although they have been rapidly growing recently. In terms of central–local relations, environmental policies are still in the hands of central governments, which set limits on the potential development of local climate change policies in these countries. However, a closer examination of these institutions in comparative perspective tells us that there are variations in the region in terms of the history of environmental institutionalization, the degree of government influence, the specific political context that has shaped the pattern of environmental governance, the role of outside actors such as ASEAN, and the degree of environmental decentralization.

The future of East Asian environmental politics lies in institutional reform in which more power is devolved to civil society and sub-national governments so that various actors participate in policy processes at different levels. Similarly, regional environmental cooperation can be facilitated if cooperation is attempted not only between the central governments but also between civil society actors and sub-national governments. In fact, there are already some examples of city-to-city environmental cooperation in Northeast Asia. The success of institutional reform in the region will also have a huge impact on the global environment and global climate change policies.

Bibliography

Amsden, A. (1992) *Asia's Next Giant: South Korea and Late Industrialization* (Oxford University Press).
Breslin, S.G. (1996) 'China: Developmental State or Dysfunctional Development?', *Third World Quarterly* 17(4), 689–706.

Bryant, R. & Parnwell, M. (1996) 'Introduction: Politics, Sustainable Development, and Environmental Change in South-east Asia', in M. Parenwell & R. Bryant (Eds.), *Environmental Change in South-East Asia: People, Politics and Sustainable Development* (Routledge).

Chung, J. & Kirkby, R. (2002) *The Political Economy of Development and Environment in Korea* (Routledge).

Economy, E. (2004) *The River Runs Black: The Environmental Challenge to China's Future* (Cornell University Press).

Elliott, L. (2012) 'ASEAN and Environmental Governance: Strategies of Regionalism in Southeast Asia', *Global Environmental Politics* 12(3), 38–57.

Evans, P. (1995) *Embedded Autonomy: States and Industrial Transformation* (Princeton University Press).

Forsyth, T. (1999) 'Environmental Activism and the Construction of Risk: Implications for NGO Alliances', *Journal of International Development* 11, 687–700.

Harashima, Y. (2000) 'Environmental Governance in Selected Asian Developing Countries', *International Review for Environmental Strategies* 1(1), 193–207.

Harris, P. (2003) 'Introduction: The Politics and Foreign Policy of Global Warming in East Asia', in P. Harris (Ed.), *Global Warming and East Asia: The Domestic and International Politics of Climate Change* (Routledge).

Hirsch, P. & Warren, C. (1998) 'Introduction: Through the Environmental Looking Glass', in P. Hirsch & C. Warren (Eds.), *The Politics of Environment in Southeast Asia: Resources and Resistance* (Routledge).

Ho, P. (2001) 'Greening Without Conflict? Environmentalism, NGOs and Civil Society in China', *Development & Change* 32, 893–921.

Imura, H. (2005) 'Japan's Environmental Policy: Past and Future', in H. Imura & M. Schreurs (Eds.), *Environmental Policy in Japan* (Edward Elgar).

International Monetary Fund (2012) *World Economic Outlook*, Washington, DC, International Monetary Fund.

Jahiel, A. (2000) 'The Organization of Environmental Protection in China', in R. Edmonds (Ed.), *Managing the Chinese Environment* (Oxford University Press).

Johnson, C. (1982) *MITI and the Japanese Miracle* (Stanford University Press).

Kheng-Lian, K. & Robinson, N. (2002) 'Strengthening Sustainable Development in Regional Inter-Governmental Governance: Lessons from "ASEAN Way"', *Singapore Journal of International & Comparative Law* 6, 640–682.

Mol, A. & Carter, N. (2007) 'China's Environmental Governance in Transition', in N. Carter & A. Mol (Eds.), *Environmental Governance in China* (Routledge).

Nakamura, H., Elder, M. & Mori, H. (2011) 'The Surprising Role of Local Governments in International Environmental Cooperation: The Case of Japanese Collaboration with Developing Countries', *Journal of Environment & Development* 20(3), 219–250.

Nguitragool, P. (2011) *Environmental Cooperation in Southeast Asia: ASEAN's Regime for Transboundary Haze Pollution* (Routledge).

Rock, M. (2002) *Pollution Control in East Asia: Lessons From Newly Industrializing Economies* (Resources for the Future).

Sangchai, J. (2005) 'Thailand and the Convention on Biological Diversity: Non-Governmental Organizations Enter the Debate', in P. Harris (Ed.), *Confronting Environmental Challenge in East & Southeast Asia: Eco-Politics, Foreign Policy, and Sustainable Development* (United Nations University Press).

Schreurs, M. (2002a) 'Democratic Transition and Environmental Civil Society: Japan and South Korea Compared', *The Good Society* 11(2), 57–64.

Schreurs, M. (2002b) *Environmental Politics in Japan, Germany, and the United States* (Cambridge University Press).

Schreurs, M. (2008) 'From the Bottom Up: Local and Subnational Climate Change Policies', *Journal of Environment & Development* 17(4), 343–355.

Shin, S. (2004) 'Economic Globalization and the Environment in China: A Comparative Case Study of Shenyang and Dalian', *Journal of Environment & Development* 13(3), 263–294.

Sonnenfeld, D. & Mol, A. (2006) 'Environmental Reform in Asia: Comparisons, Challenges, Next Steps', *Journal of Environment & Development* 15(2), 112–137.

Ueta, K. & Mori, A. (2007) 'Environmental Governance for Sustainable Development in East Asia', *Kyoto Economic Review* 76(2), 165–179.

Upham, F. (1987) *Law and Social Change in Postwar Japan* (Harvard University Press).

Wade, R.H. (1992) 'East Asia's Economic Success: Conflicting Perspectives, Partial Insights, Shaky Evidence', *World Politics* 44(2), 270–320.

Woo-Cumings, M. (Ed.) (1999) *The Developmental State* (Cornell University Press).
World Bank (2010) *The Little Green Data Book 2010*, Washington DC, World Bank.
Yang, G. (2005) 'Environmental NGOs and Institutional Dynamics in China', *China Quarterly* 181, 46–66.
Zarsky, L. (1995) 'The Domain of Environmental Cooperation in Northeast Asia', Paper prepared for the Sixth Annual International Conference on Korea and Future of Northeast Asia: Conflict or Cooperation?, Portland, Oregon, 4–5 May.
Zusman, E. & Turner, J. (2005) 'Beyond the Bureaucracy: Changing China's Policymaking Environment', in K. Day (Ed.), *China's Environment and the Challenge of Sustainable Development* (M.E. Sharpe).

Acronyms

ASEAN	Association of Southeast Asian Nations
CCI	Clinton Climate Initiative
CDM	clean development mechanism
COP	conference of the parties
EMA	Environmental Management Act in Indonesia
EPB	Environmental Protection Bureau
GONGO	government-organized NGO
ICLEI	International Council for Local Environmental Initiative
MITI	Ministry of International Trade and Industry
NDRC	National Development and Reform Commission
NEA	National Energy Administration
NGO	non-governmental organization
TEMM	Tripartite Environmental Ministers Meeting
TNC	transnational corporation
UNFCCC	United Nations Framework Convention on Climate Change

Notes

1　This section is based on the information provided by the official websites of the major government environmental agencies of the countries selected in this research.
2　The first and second critical period in each country are denoted as country 1 and country 2 in Table 5.3.
3　'Chinese protest over chemical factory', *The Guardian*, 28 October 2012. http://www.guardian.co.uk/world/2012/oct/28/chinese-residents-protest-chemical-factory.
4　The information is from the CDM homepage. http://cdm.unfccc.int/Statistics/Registration/RegisteredProjAnnex1PartiesPieChart.html.
5　The information is from the CDM homepage. http://cdm.unfccc.int/Statistics/Registration/AmountOfReductRegisteredProjPieChart.html.

6

Paradoxes of democratization

Environmental politics in East Asia[1]

Mary Alice Haddad

East Asia has jumped onto the environmental bandwagon. Leaders in China, Japan, South Korea, and Taiwan have all made 'green growth' central tenets of their development strategies for the twenty-first century. The entire region, with the glaring exception of North Korea,[2] has enjoyed, but also suffered from, extraordinarily rapid economic growth. Japan, with its 'economic miracle' offered a developmental state model (Johnson 1982; Woo-Cumings 1999), which was promptly followed by others in the region: first the nimble and small 'East Asian Tigers', including South Korea (Korea hereafter) and Taiwan, and then by the much more gigantic China.

As a direct result of their rapid industrial expansions all four countries[3] have faced environmental crises and associated civil unrest. Unsurprisingly, the first instinct of all governments was to suppress the political unrest emerging from the environmental pressures. Quite surprising, given the pro-business orientation of the ruling political regimes, the lack of strong green parties, and the absence of national, professionalized environmental advocacy organizations, all four countries soon reversed their earlier positions and enacted cutting-edge environmental policies. This pattern was common across China, Japan, Korea, and Taiwan even though their political regimes vary dramatically.

This chapter focuses on the question of what role democracy and democratization has played in environmental politics in East Asia. The author recognizes that the particular environmental policies of each government, their capacities and willingness to implement those policies, and the actual environmental outcomes vary dramatically both across and within countries. The goal of this chapter is not to assess policy effectiveness but rather to explain the process of political development. This topic is a broad one, so it is impossible to address all of the nuances found in each place, but the chapter will attempt to offer an overview of the environmental politics in the four countries and draw some parallels found among them. In all four countries, the environment was the first issue area around which grassroots citizen activism began, but the shape and form of the environmental movements varied considerably across the four countries. Surprising the author, regime type had less to do with shaping the environmental movements than the timing of those movements.

Civil society in East Asia

A decade ago there was nearly universal agreement that East Asia had little to no civic participation. Comparative studies all indicated that citizens in East Asia did not join civic organizations, rarely volunteered, and were generally uninvolved politically. Research that relied on statistical surveys found that East Asian states trailed other advanced countries in values and activities associated with political activism. Their citizens have a set of values that are often characterized as 'illiberal' and 'undemocratic': they remain skeptical of individual freedom, have a strong preference for social order, favor an interventionist rather than a limited government, show a reluctance to engage in public protest, etc. (Inglehart & Welzel 2003). Supporting this perspective, academic work focused on the ways in which the heavy hand of the state in regional countries acted to constrain and control civil society (Frolic 1997; Garon 1997; Pekkanen 2006; Schwartz & Pharr 2003; Yamamoto 1998).

Recently, this perspective has begun to change. Beginning with the 'third wave'[4] democracies of Korea and Taiwan, East Asian scholars began to demonstrate that while civil societies in East Asia may not look exactly like their counterparts in Europe and North America, they were still playing increasingly important roles in their country's politics. Robert Weller's *Alternate Civilities* (1999) documents civil society's role in Taiwanese democratization and argues that its success and the expansion of civic, if not necessarily democratic, activity on the mainland suggests that vibrant civic cultures can form in ways that are coherent with non-Western societies. Somewhat more critically, Sunhyuk Kim's *Politics of Democratization in Korea* (2000) and Charles Armstrong's *Korean Society* (2002) both offer detailed accounts of the mixed and varied roles that a wide range of citizen groups have played in Korea's disjointed and lengthy democratization processes.

In Japan, after a brief wave of interest in the movements in the 1960s and 1970s (Krauss & Simcock 1980; Lewis 1980; McKean 1981; Steiner, Krauss & Flanagan 1980; Steinhoff 1989; Upham 1976), academic interest shifted away from citizen activism to economic development. The most recent reexamination of civic activity in Japan may have begun with Jeffrey Broadbent's *Environmental Politics in Japan* (1998) that examined the environmental movements with a focus on Ōita prefecture, to be followed closely by Robin LeBlanc's *Bicycle Citizens* (1999) and Patricia Maclachlan's *Consumer Politics in Postwar Japan* (2002), which both documented the rise of women's participation in consumer groups. All three works focused on particular groups or causes. They were soon augmented by examinations of civil society more broadly in Jeff Kingston's *Japan's Quiet Transformation* (2004), my own *Politics and Volunteering in Japan* (2007a), Yasuo Takao's *Reinventing Japan* (2007), and Kim Reimann's *The Rise of Japanese NGOs* (2009). Now, academics are discussing civic activity in Japan as vibrant rather than dormant, and beginning to study the rise of so-called 'new style' citizen groups that tend to be organized around individual identities and interests rather than local community locations (Haddad 2006, 2007a, 2007b; Reimann 2009; Takao 2007).

Scholarship about China has also experienced an explosion of interest in grassroots political activity aimed at gaining concessions from the state. Kevin O'Brien and Lianjiang Li's *Rightful Resistance in Rural China* (2006) and Elizabeth Perry and Merle Goldman's edited volume *Grassroots Political Reform in Contemporary China* (2007) have both shown ways in which rural protesters have increasingly been able to use divisions within the Chinese state to gain concessions. Peter Ho and Richard Edmonds' *China's Embedded Activism* (2007) discussed not only the constraints but also the expanding opportunities for activists in China. In her innovative book *Accountability without Democracy* (2007), Lily Tsai has demonstrated the positive role that local temple associations have played in improving rural development outcomes, and Andrew Mertha's path breaking *China's Water Warriors* (2008) used case studies of anti-dam protest across China to show the ways that civic groups have been able to mobilize successfully and delay and even cancel state-authorized dam construction projects.

Although the four countries represent very different regime types, there has been a general trend toward more active civil society interacting with a more attentive government. Across the region, environmental organizations are at the forefront of trends toward increasing engagement with governmental policymaking. Citizens and their organizations are becoming increasingly engaged with respect to environmental issues. While a large proportion of that engagement tends to take cooperative forms, more confrontational forms of environmental activism are also present in the region.

Environmental politics in East Asia

Citizen engagement about environmental issues in East Asia emerged directly in response to the negative human health and economic consequences of rapid industrialization. Across the region environmental activism began when local residents living near polluting industrial facilities began to feel the effects of industrial pollution. Residents would organize to try to stop harmful pollution. Initial governmental and corporate responses were predictably resistant to change and were often violent in their attempts to coerce communities to accept the costs of pollution in exchange for a variety of economic and other benefits. When residents refused to be bought off, government and corporate actors engaged in a wide spectrum of responses, ranging from violent suppression, coercion, and co-option, to compromise and even innovation (Aldrich 2008).

While initial movements in the region took the form of classic NIMBY (not in my back yard) (Rabe 1994) struggles that dissolved once the pollution issues were resolved, over time they became more sophisticated. Activists found ways to work with local and national governmental officials as well as corporations to gain better environmental outcomes. They broadened the scope of the issues from strictly pollution to include recycling, conservation, climate change, etc. Now, activists across the region, whether they are private citizens concerned about the environment or NGO (non-governmental organization) professionals or business entrepreneurs or civil servants are able to take advantage of the internet and social networking technologies to connect with others in their own country and abroad who are seeking positive environmental change.

Although regime type has affected the resources available to activists, the most important factor affecting the shape of a country's environmental movement appears to have been timing. The timing of environmental movements mattered in three ways. First, timing mattered with respect to the political opportunities available in domestic politics – e.g., was the ruling party vulnerable and willing to address environmental concerns? Second, timing mattered with respect to the political opportunities available in global politics – e.g., was the environment on the global political agenda? Finally, timing mattered in terms of sequence, because countries that experienced environmental movements later were able to learn from those who had gone before. The next section describes the environmental movements of the four countries in the order in which they occurred: Japan, followed by Taiwan and Korean, and most recently, China.

Japan

Japan's environmental activism began at the turn of the twentieth century as a consequence of the dramatic industrial expansion initiated by Japan's Meiji restoration. During the 1890s and early 1900s village protests erupted first against copper mines: the Ashio mine in Tochigi prefecture, Sumitomo's mine in Ehime prefecture, and Hitachi's mine in Ibaraki prefecture. In all three cases, sulfur gas pollution emitted from smokestacks combined with heavy metals and acid released as wastewater into local streams decimated the livelihoods of nearby farmers and fishermen, causing serious health and livelihood problems for the local residents. Although the government and corporations initially tried to suppress and buy off the protesters, the plants eventually made significant investments to reduce pollution and compensate victims. Once the pollution was reduced and the victims' families were compensated, the protests died off (McKean 1981).

Environmental activism resurfaced again during the early postwar period. High-growth policies favored economic growth over environmental protection resulting in toxic outcomes that threatened life and livelihood across the Japanese archipelago (Walker 2011). By the end of the 1960s environmental pollution had become a major, national political issue. Once again the protests began as residents living near the industrial plants complained to the companies and local governments, and then they began taking more aggressive political action to pressure companies to clean up their plants. Unlike the prewar cases, which were eventually resolved through a combination of government pressure and corporate measures to placate residents, in the postwar cases residents were unsatisfied with government and company responses. Joining counterparts in the United States and Western Europe, pollution victims took their complaints to the public and the corporations to court. To the surprise of many, they won (McKean 1981; Upham 1976).

The government reacted quickly to the widespread and growing concerns about the environmental costs of its growth-first economic policies. Unlike its undemocratic, prewar predecessors, the postwar ruling party, the Liberal Democratic Party (LDP), was sensitive to electoral pressure from opposition parties, and the conservative LDP reacted quickly to pass a sweeping array of environmental legislation in 1970 in what has come to be known as the Pollution Diet. Companies complied, and from that time forward, corporations, especially large corporations, have tried to stay ahead of pollution issues to avoid the commercial consequences of a negative corporate image as well as prevent additional intrusive government regulation.[5]

The Japanese case demonstrates that while the timing of an environmental crisis can put political pressure on ruling parties, it does not always mean the departure of those parties from power. In both the early 1900s and in 1970, Japan's ruling party and major corporations were able to mollify the public, finding policies that enabled them to address environmental health issues without having to give up political power.

After a few decades off the global political agenda, the environment has once again risen to the top of international political discussions. A combination of the end of the Cold War, political instability in the Middle East, rising energy needs, growing concern about energy security, and rising scientific evidence about the multitude of threats related to climate change has brought the environment to international and national political agendas around the world. Citizens have also developed very sophisticated methods of networking with each other within and across borders, creating additional pressure on political and corporate decision makers.

The contemporary form of environmental politics in Japan can be traced to the events leading up to the Third Conference of the Parties of the United Nations Framework Convention on Climate Change (COP 3) held in Kyoto in November 1997. Preceding the conference the Kiko Forum formed in Kyoto to connect Japanese environmental organizations to each other and to groups from around the world. The successful creation of the Kyoto Protocol by participating governments was mirrored in the successful solidification of the global NGO community through the actions of the Kiko Forum and others (Reimann 2003).

These networks grew dramatically as internet technology spread. In 1999 there were only 2,000 internet users in Japan, and just over a decade later, by 2011, there were more than

96 million, with usage rates nearing 100 percent for people between 10 and 50 years old.[6] Examples of environmental networks in the country include Kiko Network (the new name for the Kiko Forum), Japan River Restoration Network, and Climate Action Network Japan (CAN-Japan), to name but three. Commercial networks related to the environment also proliferated, for example Japan Green Purchasing Network, Eco-Networks, GreenBizJapan, etc. These groups usually have connections to international NGOs, but large international environmental organizations, such as Greenpeace, the Nature Conservancy, and World Wildlife Fund, do not usually have offices in Japan, or their offices are a fraction of the size found in other countries.[7]

In the aftermath of the Triple Disaster of 2011, which included a devastating earthquake, tsunami, and nuclear disaster that took the lives of nearly 20,000 people in the northeast region of Japan, these networks of organizations were critical in providing immediate assistance to those in need in disaster areas. They also helped focus the Japanese public on the importance of a wide variety of environmental concerns, especially those related to energy, conservation, and community resilience (Kingston 2012; Samuels 2013).

Environmental politics reached a fevered pitch following the disaster, with all segments of society getting involved. Japanese government officials have been working with global organizations and governments to improve international nuclear safety protocols and standards, disaster management, and eco-city development.[8] The corporate sector has also responded with innovative plans. For example, within weeks of the disaster, billionaire Masayoshi Son gained the cooperation of nearly all of Japan's governors for a bold new plan that would dramatically enhance Japan's renewable power infrastructure and expand renewable energy across the region.[9]

Private citizens and civic groups have also formed innovative groups. Safecast.org established an interactive map just days after the disaster that serves as a crowd-sourced platform where individuals from around the world upload and view radiation readings from individual Geiger counters in an effort to offer a non-governmental source of radiation information to the public. Networks of anti-nuclear groups such as Sayonara Nukes have taken advantage of public outrage as well as new social networking technology to organize regular, simultaneous protests all over Japan with the numbers of participants often reaching into the tens of thousands.[10]

Contemporary environmental politics in Japan is sophisticated and complex. Millions of ordinary citizens are working through local organizations to improve their local environments. Hundreds of small organizations help link Japanese citizens with local policymakers and with international environmental organizations around the world. Dozens of groups work closely with counterparts around the world and with high-level government officials in Japan. However, for the most part, environmental politics in Japan, as in the rest of East Asia, remains somewhat bifurcated. At an elite level there is a small group of technocrats from the government, civil society, and business working on high-level issues of national and international environmental policy. At the grassroots, millions of Japanese are working in local groups to improve the environment in their communities. There is not much happening in the middle. While thousands of Japanese are taking to the streets regularly, we do not yet see sustained efforts to influence environmental politics through political parties or the electoral process, although the coming elections may change that situation.[11]

Taiwan

As was the case in Japan, Taiwanese environmental organizations began as community protests against instances of local environmental pollution. Rapid industrialization under a developmental state prioritizing economic growth had the expected result of high levels of pollution. When farmers and fishers found their livelihoods threatened by contaminated soil and water, and when residents found themselves and their loved ones getting sick at unusually high rates, they began to protest.

The beginning of Taiwan's environmental movement strongly resembled the early stages of environmental organizing in Japan. Disgruntled and largely disempowered residents appealed to their local government officials and directly to the factories themselves. Usually they found a sympathetic ear to listen, but their complaints were largely met with a variety of appeasement measures that were intended to make the political problem go away for the local elites but not to take significant steps to address the pollution problems.

Of critical difference in the Taiwanese and Korean cases as compared to Japan was their timing. The timing of their movements mattered in all three dimensions mentioned already. The domestic political context at the time of the emergence of environmental movements meant that they became inexorably entangled with democratization movements and strongly linked with liberal political parties. The international context at the time meant that groups could take advantage of the rising global concern with environmental issues, the political openings brought about by the end of the Cold War, and the increasing international support for both environmental activism as well as third wave democratization. Finally, because Taiwanese and Korean environmental movements followed those in the United States, Europe, and Japan, they were able to learn tactical and strategic lessons about grassroots organizing and political advocacy from those who had gone before.

Taiwan's first environmental protesters were victims of industrial pollution: farmers, fishers, and local residents who had their health and livelihoods threatened by nearby industrial production. From 1980–1987, 97 percent of environmental protests were reactive – victims seeking redress against damage that had already occurred (Tang & Tang 1997). Examples include a lawsuit in 1981, when villagers in Hua-t'an village demanded compensation from local brick manufacturers for damage to their nearby rice paddies. They eventually won NT$1.5 million (US$375,000).

In a context in which newspapers were increasingly reporting on local pollution issues, Taiwanese citizens were also affected by international events. Although not as geographically widespread as the Chernobyl explosion that would follow two years later, in 1984 a disastrous gas leak from a Union Carbide factory in Bhopal, India killed 2,000 people and made major headlines across the globe. Taiwanese were already feeling sensitive to the damaging effects of chemical pollution in their own communities, and they immediately recognized that they were similarly vulnerable. The very next year, protests and threats of violence against pesticide companies in Hsin-chu and T'ai-chung forced the closing of both factories for cleanup (Reardon-Anderson 1997: 11–12).

The largest turning point for Taiwan's environmental movement occurred in 1986, when the sleepy fishing village of Lukang began a protest against a planned titanium dioxide plant. The protest represented a shift away from a strategy of reactive protests against damage already done to proactive political action aimed at preventing damage that had not yet occurred. According to James Reardon-Anderson, who wrote an excellent book about the protest in Lukang, if the protest had happened earlier, it would have been crushed; if it had happened later, it would not have been noticed. 'But it came just at the time when environmental consciousness in Taiwan had reached a critical mass and as the government was introducing political reforms that gave unprecedented scope to new forms of civic action.' In that context, a small group of determined local activists 'focused the attention of the entire island on this sleepy provincial town, raised the national consciousness about threats to the natural environment, and challenged the rules that government officials and industrial leaders in Taiwan had come to take for granted' (Reardon-Anderson 1997: x).

As in Japan, most of the early movements were often led by women and conducted in ways that kept them from being noticed by authorities until they had gained considerable momentum. As Robert Weller has eloquently explained: 'Women's networks often rely less on patrilineal kinship than men's, they have stronger local roots, and are often less formally organized and therefore less visible to the state. This has put them at the forefront of informal civil associations' (Weller 1999).

Different from Japan, these local networks also took advantage of religious institutions, organizational linkages, symbolism, and rituals to spread their messages and gain supporters. Buddhist temple associations were critical for providing both the institutional framework that facilitated political organizing and also for offering the opportunities for gathering in 'apolitical' ways in public that could then be transformed into more political action (Weller 1999).

All of these activities coincided with the rise of the middle class and the beginning of the third wave of democratizations that would sweep the world in the next decade. Taiwan ended martial law in 1987, and a number of groups that would lead Taiwan's environmental and democratization movement, such as the New Environment Foundation, Taiwan Greenpeace, the Taiwan Environmental Protection Union (TEPU), and the Homemaker's Union Environmental Protection Foundation were founded shortly thereafter.

As was the case in Japan, their origin fighting local pollution problems made it difficult for early environmental movements in Taiwan to grow beyond NIMBY protests. Groups tended to focus primarily on local pollution issues and would disband once a particular battle was over (Tang & Tang 1997). Throughout the 1990s the Taiwanese continued to grow and political reform spread such that by the end of the decade citizens were well placed to hold their governments accountable for poor decisions in the wake of Asian (and global) economic crisis of 1997. In 2000 Taiwanese ousted the KMT, which had ruled the island for nearly 40 years, electing native-born Chen Shui-bian of the Democratic Progressive Party (DPP). Because environmental organizations and their leaders were closely linked with liberal political parties, activists enjoyed considerable access to policymaking during the eight years the DPP was in power.

Even though they had access, however, many activists felt like their interests were pushed aside in favor of big business once the parties they had supported gained power. As one Taiwanese activist phrased it, 'The DPP changed when it took power. When in power then it didn't like the environment any more.' From another, 'They want the votes, but they don't want to hear the voices.'[12] However, even their limited access was severely curtailed when voters returned conservatives to power, electing Ma Ying-jeou of the Kuomintang (KMT) in 2008. When the global financial crisis hit at the end of the year, Ma was quick to put 'green growth' at the top of his strategies for economic recovery, making significant public investments to promote green technology and green industry related to the information technology (IT) industry, renewable energy, eco-tourism, and the like.

Although the conservative governments and businesses were promoting environmentally friendly economic development, environmental activists felt completely shut out. Several suggested that the election brought back the very same people who were in charge under the military government. One Taiwanese activist was very blunt in assessing the situation, '[In Taiwan] Corporations are a shadow government. Our government is their puppet.'[13]

Responding to what they perceive as a lack of effectiveness and access, Taiwanese activists are beginning to shift their tactics away from protests and partisan politics. Environmental leaders talk about how the public and the policymakers have become anesthetized to public protests, such that they are no longer effective as a mode of advocacy. Instead, scientific and policy reports that give policymakers new information about an environmental problem and create an opportunity for dialog about solutions appear to be more effective.[14] As a result, Taiwan's environmental politics may be moving closer to the model found on the Chinese mainland, where advocacy has been aimed more at working with the government rather than against it. It also exhibits the same bifurcation of environmental advocacy found in Japan: many small, local grassroots groups working on a volunteer basis to improve their communities and a few elite organizations working to influence policy, but not much in between.

Korea

Of the four countries under examination in this chapter, Korea is an outlier because unlike the other three countries which either have no professional environmental advocacy organizations or only very small ones, Korea boasts the largest in the region: the Korean Federation for Environmental Movements (KFEM) has more than 50 local branches, nearly 100,000 members, and it is deeply involved in local and national politics.[15] The highly politicized and well-organized nature of environmental politics in Korea makes it somewhat distinct from that found in the other three countries.

Korea's environmental movement has grown out of and has contributed to a long history of protest politics. Korea's tradition of mass protests can be traced back more than a century, beginning with the 1894 Donghak Peasant Movement protesting government corruption. Other famous mass movements in Korea include the March 1st Movement protesting Japanese rule in 1919, the April Revolution when student and labor organizations successfully ended the autocratic rule of Syngman Rhee in 1960, the failed student-led pro-democracy protests in May 1980, and the successful democracy movement of June 1987. This long history helped normalize protest, even violent protest, as a regular method through which civil society organizations would engage the state. Although considerably less violent than their predecessors, contemporary Korean civic organizations, of which environmental organizations are one group, continue to favor confrontational modes of political engagement (Kim 2009; Oh 2012).

Scholar and environmental activist See-Jae Lee argues that Korea's environmental movement has passed through four stages: negation (1960s and 1970s) – where neither the state nor civil society viewed the other side as legitimate, resistance (1980s) – where the state acknowledges the existence of civic actors but does not view them as partners for dialog and seeks to suppress them, negotiation (1990s) – where both sides recognize one another and struggle against one another for policy influence and public support, and participation (2000–present) – where environmental organizations are incorporated into the state's decision-making process and participated in jointly developed policy projects (Lee 2000).

Early environmental organizations in Korea were generally located in churches and universities and were focused primarily on raising environmental consciousness among the population (Lee 1999). However, as was the case for Japan, Taiwan, and now also China, high-speed growth policies led to rapidly deteriorating environmental conditions that threatened human health and livelihoods. Early environmental protest movements in Korea, like their counterparts elsewhere in the region, began with residents' demands for compensation for damages. The first major case to generate national attention was in the Ulsan and Onsan areas of Gyeongsangnam-do province. Construction on the government-approved Ulsan Industrial Complex began in 1962, and in 1967 farmers began to demand compensation for agricultural losses. In 1971 they formed a pollution countermeasures committee, and 1978 they increased their demands to include financial aid for relocation as well as compensation for residents experiencing health damages. Urban residents in large cities such as Seoul and Inch'on also began to protest against noise and air pollution (Lee 2000).

Responding to growing public concern, a small number of environmentalists, religious leaders, and pro-democracy activists formed the Pollution Research Institute (PRI) in 1982 in order to conduct pollution-related research independent of the government (Ku 2002). In contrast to the local NIMBY protesters, who were usually farmers, fishers, and local residents living near factories, PRI had connections to national church leaders, leading academics, and student groups. It was the first professional environmental organization in Korea with dedicated staff and office space (Lee 1999). Because of its close connection with student groups involved in the democracy movement, the anti-pollution movement was also seen as an anti-government movement, so the government tried to prevent connections between local grassroots NIMBY groups and the professional environmental organization (Ku 2002: 76).

In 1983 heavy metal pollution made the water near the Ulsan complex so toxic that the government suspended fishing rights. Immediately the people took to the streets to demand financial compensation for lost revenue, and the PRI began an independent investigation into the situation. In 1985 they released a very detailed report that claimed that more than 500 people in Onsan suffered from cadmium contamination. The press offered considerable public exposure to the findings, and the issue became one of national interest. Soon after the PRI report, the Environment Administration conducted its own tests and reported that the illness spreading among the Onsan population was not a pollution-related disease. Residents and the PRI refuted the official test results and engaged in a series of public protests. Eventually the government was forced to concede to the growing pressure from environmental groups and the public and resettle about forty thousand residents to new areas (Ku 2011: 211–213; Lee 1999: 215).

Although martial law was lifted in 1981, it was not until the creation of a new constitution that guaranteed basic civil rights in 1987 that the legal protections needed for the creation of environmental organizations were established. Soon afterwards key organizations, such as Citizen Alliance for Economic Justice, Green Korea United, and the Korean Federation for Environmental Movements (KFEM), were established and became important organizations for the combined environmental and democratization movements.

Throughout the 2000s, Korean environmental organizations expanded their membership and broadened the scope of their issues, moving beyond merely responding to situations of environmental degradation to initiating preventive campaigns. The anti-Doggang Dam Campaign was one such campaign, led by KFEM, which aimed to prevent the construction of a new dam and protect the natural ecosystem of the river. The campaign had a sophisticated strategy that included promoting tourism to the river. It also utilized cultural symbols, such as the annual Jeongseon Airirang festival, as well as ecological symbols, displaying rare species of fish and otters that live in and along the river. In 1999 Kim Dae-jung's government formed a citizen government joint investigation panel to research the dam, and a year later the committee recommended that construction be canceled. On Environment Day, June 5, 2000, President Kim announced a New Millennium Vision for the Environment, and pledged to repeal construction plans (Lee 2010: 215).

In the 2000s, receiving a considerable political boost with the victory of Kim Dae-jung in the 1997 presidential elections, Korea's environmental groups grew increasingly politically involved and enjoyed closer connections to the government, serving on joint panels and being appointed to key government positions. They joined several other NGOs in forming the Citizens' Alliance for the 2000 General Elections, which was a pro-transparency, anti-corruption campaign that blacklisted candidates with records of political corruption or illegal activities and succeeded in defeating 56 of the 89 candidates it targeted. President Roh Moo-hyun of the Millennium Democratic Party pledged to create a 'participatory democracy' in Korea, appointing many civil society activists to government positions and expanding NGO participation in policymaking (Choi 2010: 17–22; Kim 2004; Oh 2012: 547).

Just as had been the case in Taiwan, the liberal renaissance ended by the end of the decade, bringing the environmental movement's access to policymaking to a wrenching end. In late 2007 Korean voters elected Myung-bak Lee of the conservative Grand National Party, and he assumed his post as President in February 2008. Once again paralleling Taiwan, when the global financial crisis hit at the end of the year, Lee promoted 'green growth' as a key component of a strategy for economic recovery. The 'Four Rivers Project,' a massive plan to re-engineer the country's major rivers became the core of Korea's Green New Deal, an economic stimulus package that pledged $40 billion dollars (equivalent to 4 percent of total GDP) for four years to promote sustainable economic growth (Chang, Han & Kim 2011).

Environmental activists in Korea faced the same problem as their counterparts in Taiwan. Although the conservative government and businesses were publicly promoting environmentally friendly economic development, activists felt completely shut out. As one Korean activist stated in an interview in 2010, 'The ex-government valued governance and wanted to hear civil society. It didn't decide everything on its own but always had channels with civil society. This government doesn't.'

However, in contrast to the Taiwanese activists who are shifting strategies and trying to find more ways to work with the government, Korean activists are returning to the protest repertoires of the past. Mere months after the new conservative government had assumed office, it reversed the decision of the earlier administration and re-allowed the import of US beef, which had been banned since 2003 after evidence of bovine spongiform encephalopathy (BSE or mad cow disease) had been identified in the United States. Playing on food safety fears spurred by BSE and the concurrent tainted milk scandal in China, building on rising anti-US nationalist sentiment among the youth, and capitalizing on national discontent with the new conservative leadership, KFEM joined other environmental and social groups to support public protests against the national government. The protests, which came to be called the Candlelight Protests because they were usually held at night and protesters brought candles, grew to be national in scope and attracted hundreds of thousands of protesters from late May through August. Interestingly, although Taiwan experienced a similar reversal in its ban on US beef, its protests remained isolated and small, failing to grow as they did in Korea (Oh 2012).

Korea's environmental movement shares with China, Japan, and Taiwan its origins as NIMBY protests against pollution that was harming human health and livelihood. It shares with Taiwan its close links to the democracy movement. However, it remains unique in the region for its highly political organizational structure. Just as Korean activists were able to learn from Japanese and Taiwanese, Chinese government officials and environmental advocates have been paying very close attention to the Korean experience and are working to ensure that their movement takes a different path.

China

Environmental policy did not feature prominently in China's politics until Mao, and then it was largely as anti-environmental policy. Judith Shapiro argues Mao's policies were not just pro-industrialization but collectively they amounted to a 'war against nature' (Shapiro 2001). Not until Deng Xiaoping's reforms to introduce some market reforms in the late 1970s did China's economy start to take off. Reforms that began as experiments in special economic zones were expanded to include much of the country by the late 1980s, spreading wealth and pollution in their wake.

By the 1990s, with the end of the Cold War hostilities and the rise of the other economies of East Asia, foreign investment was flowing into China and new policies made it easier for Chinese to travel and study abroad. By the time that Beijing hosted the United Nations' Fourth World Conference on Women in 1995, China had begun the groundwork to enter the World Trade Organization (which it joined in 2001) and was becoming more economically as well as politically and socially integrated into the rest of the world. According to many activists in China, the World Conference on Women was a very eye-opening experience for the Chinese political leadership. In addition to the official representatives from governments around the world, more than 5,000 members of a variety of non-governmental organizations poured into the city for the event. Their large numbers and the productive role that they played at the conference helped make the Chinese leadership aware of the rising importance and usefulness of the nonprofit sector.[16] This impression was solidified for the Chinese Communist Party (CCP) in 2008 after a devastating earthquake in Sichuan killed nearly 70,000 people. Just as they had in Japan after the 1995 and 2011 earthquakes, international and domestic NGOs rushed to the scene to assist with the rescue and reconstruction efforts. Their relative successes helped to demonstrate the usefulness of the NGO sector to the Chinese leadership.[17]

As was the case for the other countries in the region environmental activism began primarily at the local level with citizens protesting pollution by particular factories in their villages. Because of the size of the country and the scope of the problems, local protests are much more widespread in China than elsewhere, with official statistics recording tens of thousands of separate environment-related protests ever year (Zissis & Bajoria 2008). The Chinese government's response to the rising environmental concerns and public dissent that have emerged from its rapid development policies have been highly diverse and erratic. As explained in great detail by Elizabeth Economy in *The River Runs Black* (2004) the Chinese government has used all of the tools at its disposal to deal with environmental concerns, ranging from violent suppression of environmental protesters to collaboration with academics and others concerned with climate change to craft some of the world's most stringent environmental policies.

As in the other three countries, the central government is the primary crafter of environmental policy, and it does this in close consultation, both formal and informal, with a wide variety of stakeholders including academic experts, who often also represent NGOs, as well as the business community (Wells-Dang 2011; Xie & Ho 2008). One of the main institutional mechanisms for this consultation in all four countries is the use of government-organized NGOs (GONGOs). In China as elsewhere, GONGOs provide an institutional location in which environmental activists, technical experts, and policymakers work together to address environmental issues (Ho & Edmonds 2007; Wu 2002).

Unlike the other countries in the region, China's environmental movement came of age at a point in time when it could take advantage of the experiences in other countries and draw on the enormous resources of the international NGO sector.[18] These international environmental NGOs see themselves as utilizing a wide array of delicately balanced political tools that provide negative and positive pressure, with many NGOs in all fields tending to favor the latter type of activity to the former. As one activist phrased it, 'Big confrontational actions don't work in China. We try to push the boundaries of what civil society can do slowly . . . We have and need to establish ourselves as a constructive partner.'[19]

Starting from the late 1990s and accelerating through the 2000s, the number of international NGOs in China proliferated, and they have also expanded their staffs and offices. The primary reason for their involvement is the scale and scope of the environmental problems in the country. China is now the largest emitter of greenhouse gasses in the world,[20] and there is no doubt that the fate of the global environment is heavily influenced by what happens in China.

While environmental activists in China benefited from the timing of their environmental movement by being able to draw on the experience and resources of environmental movements in other countries, the Chinese government also learned valuable lessons from the Japanese, Korean, and Taiwanese experiences. From Taiwan and Korea, the CCP leadership learned that local environmental protests can transform into national democracy movements if allowed to grow. As a result, the Chinese government has been working at multiple levels to prevent the formation of national environmental movements that might be transformed into democratization movements. From the Japanese experience the CCP learned that if the party in power can accommodate or even exceed the public's demands for environmental action, the party can remain in power and even enhance its legitimacy. In response to spreading protests, the central government is ramping up its efforts to improve the environment in the country. In fact, it recently launched a new smart phone app that gives subscribers live pollution data for China's major cities.[21]

Environmental activism and its paradoxical relationship to democracy

There is no doubt that democracy affects environmental politics. Activists operating in authoritarian contexts have to worry about their personal safety being affected by things that they say, while those in mature democracies can be fairly confident that although their advocacy may anger officials and will likely not effect the change that they would like, they will be able to go home to their families at night without concern for safety of themselves or their loved ones. In newly democratic countries, the picture is somewhat more mixed. Most activists I interviewed in Korea and Taiwan were willing to be quoted. Kuang-Jung Hsu, a Professor at National Taiwan University and former Chair of Taiwan Environmental Protection Union (TEPU), expressed a common sentiment when she said, 'Yes you can quote me. I need to take responsibility for my own words.' Or, even more enthusiastically, from Echo Lin of Taiwan Environmental Action Network, 'Sure you can quote me. Make it public!' Many others were more cautious, asking that I check with them if I wanted to publish their names. The response from one activist in Korea made it clear that not everything in that new democracy is as free and open as one might want. Responding to a standard question about willingness to be identified in my publications, the activist said: 'Better to be anonymous. The government monitors people here. I know that my name is on a blacklist.'

The findings and arguments in this chapter are not intended to belittle or ignore the very real differences in the ways in which democracy affects environmental activism. However, this chapter does argue that focusing primarily on the constraints that authoritarianism puts on activism can cause one to miss the extensive and creative ways that citizens, even in those nondemocratic states, are finding to engage policy actors and effect change. Activists everywhere have to be very conscious not to cross 'the line' that will cause public officials to shut down their operations or turn public sentiment against them. Where that 'line' exists in the political space and the consequences of crossing it are profoundly different in democratic and authoritarian states, and the 'line' is constantly shifting no matter where you are. Certainly the political space in which activists may engage policymakers, the public, and the international community is significantly more constrained in authoritarian contexts. However, in most places that space does exist. Furthermore, as Andrew Mertha has shown in *Water Warriors* (2008), many of the political strategies related to issue framing and savvy political entrepreneurism can work in authoritarian contexts as well as democratic ones.

This chapter has three primary findings. First, the democratic status of the country affected the policymaking process less than the author expected. In all four countries environmental policies were made primarily by central government bureaucrats who gathered advice from experts outside the government, and the experts were often academics affiliated with environmental NGOs. In no country were there mass environmental movements, although Korea does have a powerful environmental advocacy organization, unlike the other three countries. In none of the countries were there electorally viable green parties. Activists in all of the countries used the courts in innovative, interesting, and different ways, but in no country was a legal strategy a major component of the environmental movement (Tang & Tang 1997; Upham 1976; Wang & Gao 2010).[22]

Perhaps most striking, but not surprising given the common themes of 'connectedness' and 'embeddedness' in their civil society literatures, in each country the boundaries between government, civil society, academia, and business were very blurred. In many cases, single individuals could have identities that placed them as part of several communities simultaneously. For example, a professor who is on the governing board of a renewable energy NGO, frequently advises the government about energy policy, and is part-owner and patent holder for solar technology, is a member of academia, civil society, the government, and the corporate sector simultaneously. It would be erroneous to suggest that when she is advising the government she has 'dropped' her affiliation with the NGO. Indeed, her connection with civil society and the corporate sector may be the very reason she was asked to join the policymaking process.

In these countries, at least with respect to the issue area of the environment, it is frequently very difficult to determine what the 'interest' of a key policymaker is because he or she has several 'interests' at the same time. In most cases, analyses that assume that policymakers have single identities shaped by their affiliation with particular organizations and that they are always trying to maximize those organizations' interests would be woefully inadequate and misleading because most of the individuals involved in the process have multiple identities and interests, and they are trying to accommodate all of those interests simultaneously. The broad similarity in the policymaking process across all regime types, especially this last point that key policy players often had multiple interests that they were accommodating simultaneously, was a surprise to the author.

Second, when environmental movements become politicized as part of a democratization process, as they did in Korea and Taiwan, it can have negative as well as positive consequences. On the positive side, the issue of the environment gains national attention from the government, the public, and corporations. Furthermore, environmental activists are likely to get unusually good access to government policymaking while their party is in power. On the negative side, when their party leaves power, they are likely to have worse access than they did before. Furthermore, because their issues, organizations, and personnel have become politicized, they become subject to political pressures that have nothing to do with the environment. Although I have been unable to find a good way to measure access, my impression from interviews was that during the 2000s when the liberal parties were in power activists in Korea and Taiwan enjoyed greater access to government policymaking than those in either Japan or China. However, once the conservatives returned to power and took political revenge on the activists who had favored the opposition, Korean and Taiwanese activists had less access to policymakers than counterparts in Japan or China. In fact, many of them were struggling to find ways to 'depoliticize' their organizations to return the focus onto the environment more centrally, rather than on which political party was in power.[23]

Finally, the timing of rapid industrialization and the attendant environmental movement significantly influenced the shape of environmental politics in the four countries, more than their democratic status. In Japan, the movement began at a time that coincided with movements elsewhere in the world and during a period during which Japan was under a single-party (democratic) government, and prior to the introduction of the internet, or the expansion of the global civil society sector. The result has been that Japan's environmental NGOs tend to be much smaller and almost completely bifurcated between research/think-tank GONGOs and very small grassroots groups that focus primarily on local environmental issues and education, although this may be shifting with the explosion of technology and the reinvigoration of the movement after the 3–11 Triple Disaster.

The benefits of this tradition of relatively nonpartisan environmentalism is that environmental activists have long been brought into the policymaking process at both the local and national levels of government. The corporate community as well is, on the whole, environmentally conscious and all large corporations are actively working toward both improving internal conservation efforts, their carbon footprint, their impact on the local environments around their facilities, as well as creating new environmentally friendly products. Networks of activists in Japan are numerous, large, and highly diverse. They include individuals in the NGO, government, academic, and corporate communities as well as organizations, companies, and local governments. Although members, both organizations and individuals, vary in their influence and power, the networks themselves are horizontally organized and information flows freely across state–civil society–corporate divisions because many people are involved in all three sectors simultaneously.[24]

The main cost of this organizational arrangement is that it encourages cooperative relationships among all parties and discourages confrontation. While this can be conducive to creating positive change, it can also mean that it is difficult for the NGO community to act as a significant 'check' on either the government or the corporations. When an NGO community remains underfunded, volunteer rather than professionally oriented, and closely linked to government, it is nearly impossible for it to engage in independent initiatives or undertake to challenge governmental and corporate agendas in any kind of meaningful way (Haddad 2011; Pekkanen 2006).

In China the timing of the expansion of the civil society sector, including the segment related to the environment, occurred at a moment when the domestic political environment was constrained but opening up. As was the case in the other countries, the environment was a relatively 'apolitical' issue area because it centered around addressing human health concerns and was not aimed at toppling the regime. In China, in particular, most activists have been careful to keep their advocacy focused primarily on public education efforts, and to the extent that they are talking about government policy, they tend to focus more on the positive, high-lighting examples of positive change. When they do focus on negative environmental issues, they present scientific data to show the problem (e.g., toxins in the Yangze),[25] but do not blame particular companies or governments for causing the problem. All sides are focused on *solutions*.

Because China is so large, its environmental problems so vast and of such global consequence, and because its environmental movement emerged at a time when the global civil society and advanced communications technologies had already been developed, it is the only country of the four where there is a significant presence of international environmental NGOs. The primary benefit of their presence is that they bring with them significant resources both in terms of funding and also, perhaps more importantly, in terms of expertise. There are many more NGO professionals in China than in any of the other countries. Since the international organizations usually partner with local groups for their projects, their knowledge and funding resources are being transferred, slowly, to the domestic NGOs. The other benefit of having many international NGOs active in China is that these groups are often granted slightly more political space in which to operate, which means that environmental advocacy in China can reach beyond the narrow confines of what is permitted domestic NGOs.

The main cost of the large role of international NGOs is the asymmetrical power dynamic that it creates. Chinese environmental networks, like those in the other three countries, are highly diverse, crossing state–civil society–academic–corporate boundaries. Formal as well as informal networks encompass both individual as well as organization/corporate/government members. However, unlike the networks in Japan, Korea, and Taiwan, these networks are not horizontal in their power relations. In many cases the networks are organized around a funding organization. So, the organization with the money, often an international NGO or a GONGO, operates at the top or the center of a large web of organizations and individuals, but since one network member has the money and the rest are dependent on that money, the relationships cannot be horizontal. In the same way, although this is beginning to change, international NGOs have a least part of their staff that are highly paid, well educated professionals. Most domestic NGOs are run by volunteers with perhaps one or two underpaid staff members. So, there is an asymmetry in knowledge and training in addition to finances. As a result, there is a complex web of dependency, where the international NGOs rely on the local ones for local knowledge and labor to carry out projects, the local NGOs rely on the international NGOs for monetary and technical support, and they both rely on each other to get access to the political connections that make their advocacy possible. The network structure itself helps protect the smaller groups since it is easy for the government to shut down a particular organization, but it is difficult to shut down a whole network.[26]

Across the region, and in China especially, the shape of the NGO sector generally and the environmental NGO sector in particular is shifting rapidly, and it is difficult to predict where it will end up. When I visited China in November 2010, most people I spoke with were very optimistic. One long-term activist told me: 'The more I work here the less I self-censor. Sometime I think we've crossed the line, but it ends up being OK, so we tend toward being more bold.' When I returned in April 2011, mere weeks after the Jasmine Spring had toppled governments across the Middle East and a crackdown was taking place in China, activists were more wary. I found it more difficult to get interviews, and the political situation was the first topic of conversation that people raised when they sat down with me. Several people brought up the artist/activist Ai WeiWei as an example of how the regime had become less tolerant than it used to be and how the political space for activism had closed. Professional activists who spoke with me were not worried for themselves personally, but the relative optimism that I had felt months earlier had shifted to a sense of watchful caution. The future felt quite uncertain.

Conclusion

In a region known for its strong governments and businesses, it is somewhat surprising that environmental policy has progressed so far. This chapter has investigated the ways in which democracy and democratization have affected environmental politics in the region. To the author's surprise, the level of democracy in a country, while it did matter, was not the defining difference that explained the variations in the environmental organizations and their role in policymaking. The primary explanatory factor for the shape of the environmental movements had to do with their timing. Timing mattered in three ways: (a) with respect to domestic political opportunities, (b) with respect to international political opportunities, and (c) with respect to sequence, governments and organizations were able to learn strategies from movements that had come before.

In particular, the timing of the movements in each country affected: (1) the ability and willingness of the ruling party and corporations to respond to the demands of the public, (2) the funding availability from the international community and especially international NGOs to support local environmental organizations, and (3) access to internet technology, which facilitated communication and coordination among activists.

In Japan in the early 1900s and again in the 1960s and in China today we see single-party states and large corporations responding to citizen unhappiness with high pollution levels. In Japan, not only were the political parties able to retain political power, but they discovered that their willingness to take bold action on environmental issues helped raise their popularity with the public. We do not yet see the same positive public response in China on the part of the CCP, but we do see significant action at the pinnacle of Chinese government as leaders in that country strive to repeat the Japanese success. Activists in Korea and Taiwan linked their environmental activism to issues of poor governance, and threw their ruling parties out of power. Now we see a divergence in the political strategies of both activists and governmental actors in these two countries. While Korean environmental politics has become more confrontational and more politicized, Taiwan's is shifting more toward a cooperative model that has been successful for Chinese and Japanese activists. In the aftermath of the Triple Disaster in March 2011, the environmental movement in Japan appears to have split into three groups. An elite level branch continues to work with government, corporations, and international organizations to shape a better energy and environmental future for Japan. A grassroots level branch located in local neighborhoods continues to focus on community beautification, recycling, and conservation projects. A third branch, which has existed since the 1960s, is growing rapidly. This branch of the movement has become more confrontational, organizing mass street protests, linking environmental issues to failures of governance, and making the issue of the environment more political. The three groups are linked to each other through personal, professional, and geographic networks, although they do not appear to be coordinating activities with one another. Japanese politics has diversified and is likely to remain so into the foreseeable future (Haddad 2012).

Democracy and democratization have had profound and important effects on the development of environmental movements in East Asia. However, the effects of democracy have been more paradoxical and complex than expected, and the timing of the rise of environmentalism appears to be more important than democracy for explaining the variation in environmental politics in East Asia. The experience of these East Asian countries offers important lessons for both environmental activists as well as democracy promoters around the world who are working in authoritarian, transitional, or democratic political environments.

Bibliography

Aldrich, D.P. (2008) *Site Fights: Divisive Facilities and Civil Society in Japan and the West* (Cornell University Press).

Armstrong, C. (Ed.) (2002) *Korean Society: Civil Society, Democracy and the State* (Taylor & Francis).

Broadbent, J. (1998) *Environmental Politics in Japan: Networks of Power and Protest* (Cambridge University Press).

Chang, Y.-B., Han, J.-K. and Kim, W.-H. (2011) 'Green Growth and Green New Deal Policies in Korea: Are They Creating Decent Green Jobs', In *GURN/ITUC workshop on 'A green Economy that Woks for Social Progress'*. Brussels, Belgium.

Choi, J.J. (2010) 'The Democratic State Engulfing Civil Society: The Ironies of Korean Democracy', *Korean Studies* 34, 1–24.

Economy, E. (2004) *The River Runs Black: The Environmental Challenge to China's Future* (Cornell University Press).

Frolic, B.M. (1997) 'State-Led Civil Society', in T. Brook & B.M. Frolic (Eds.), *Civil Society in China* (M.E. Sharpe).

Garon, S. (1997) *Molding Japanese Minds: The State in Everyday Life* (Princeton University Press).

Haddad, M.A. (2006) 'Civic Responsibility and Patterns of Voluntary Participation Around the World', *Comparative Political Studies* 39, 1220–1242.

Haddad, M.A. (2007a) *Politics and Volunteering in Japan: A Global Perspective* (Cambridge University Press).

Haddad, M.A. (2007b) 'Transformation of Japan's Civil Society Landscape', *Journal of East Asian Studies* 7, 413–437.

Haddad, M.A. (2011) 'Contribute to Renewal, Not Just Recovery', *Asahi Shinbun*.

Haddad, M.A. (2012) *Building Democracy in Japan*. New York: Cambridge University Press.

Ho, P. & Edmonds, R. (Eds.) (2007) *China's Embedded Activism: Opportunities and Constraints of a Social Movement* (Routledge).

Inglehart, R. and Welzel, C. (2003). 'Political Culture and Democracy: Analyzing Cross-Level Linkages', *Comparative Politics* 36(1), 61–79.

Johnson, C. (1982) *MITI and the Japanese Miracle: The Growth of Industrial Policy 1925–1975* (Stanford University Press).

Kim, E. (2009) 'The Limits of NGO–Government Relations in South Korea', *Asian Survey* 49, 873–894.

Kim, H.-R. (2004) 'Dilemmas in the Making of Civil Society in Korean Political Reform', *Journal of Contemporary Asia* 34(1), 55–69.

Kim, S. (2000) *Politics of Democratization in Korea: The Role of Civil Society* (University of Pittsburg Press).

Kingston, J. (2004) *Japan's Quiet Transformation: Social Change and Civil Society in the Twenty-first Century* (RoutledgeCurzon).

Kingston, J. (ed.) (2012. *Natural Disaster and Nuclear Crisis in Japan: Response and Recovery after Japan's 3/11.* New York: Routledge.

Krauss, E. (1974) *Japanese Radicals Revisited: Student Protest in Postwar Japan* (University of California Press).

Krauss, E. & Simcock, B. (1980) 'Citizens' Movements: The Growth and Impact of Environmental Protest in Japan', in K. Steiner, E. Krauss & S. Flanagan (Eds.), *Political Opposition and Local Politics in Japan* (Princeton University Press).

Ku, D. (2002) 'Environmental Movement and Policies During High Economic Growth in Korea', In Y. Arayama (Ed.) *Environment and Our Sustainability in the 21st Century: Understanding and Cooperation between Developed and Developing Countries*, Nagoya, Japan: Nagoya University.

Ku, D. (2011) 'The Korean Environmental Movement: Green Politics Through Social Movement', In J. Broadbent and V. Brockman (Eds.) *East Asian Social Movements*, New York: Springer.

LeBlanc, R.M. (1999) *Bicycle Citizens: The Political World of the Japanese Housewife* (University of California Press).

Lee, J.-H. and Yun S.-J. (2010) 'A Comparative Study on Governance in Korea: Focused on a Comparison between the Roh Moo-hyun Government and the Lee Myung-bak Government.' Seoul.

Lee, S.-J. (2000) 'The Environmental Movement and Its Political Empowerment', *Korea Journal* 40(3), 131–60.

Lee, S.-H. (1999) 'Environmental Movements in Korea', In Y.-s. F. Lee and A. Y. So (Eds.) *Asia's Environmental Movements: Comparative Perspectives*.

Lewis, J. (1980) 'Civic Protest in Mishima: Citizens' Movement and the Politics of the Environment in Contemporary Japan', in K. Steiner, E. Krauss & S. Flanagan (Eds.), *Political Opposition and Local Politics in Japan* (Princeton University Press).

Maclachlan, P.L. (2002) *Consumer Politics in Postwar Japan: The Institutional Boundaries of Citizen Activism* (Columbia University Press).

McKean, M. (1980) 'Political Socialization through Citizens' Movements', in K. Steiner, E. Krauss & S. Flanagan (Eds.), *Political Opposition and Local Politics in Japan* (Princeton University Press).

McKean, M. (1981) *Environmental Protest and Citizen Politics in Japan* (University of California Press).

Mertha, A. (2008) *China's Water Warriors: Citizen Action and Policy Change* (Cornell University Press).

O'Brien, K. (1996) 'Rightful Resistance', *World Politics* 49, 31–55.

O'Brien, K. and Li, L. (2006) *Rightful Resistance in Rural China*. New York: Cambridge University Press.

Oh, J.S. (2012) 'Strong State and Strong Civil Society in Contemporary South Korea Challenges to Democratic Governance', *Asian Survey* 52(3), 528–49.

Pekkanen, R. (2006) *Japan's Dual Civil Society: Members Without Advocates* (Stanford University Press).

Perry, E. and Goldman, M. (eds.) (2007) *Grassroots Political Reform in Contemporary China.* Cambridge: Harvard University Press.

Rabe, B. (1994) *Beyond NIMBY: Hazardous Waste Siting in Canada and the United States* (Brookings).

Reardon-Anderson, J. (1997) *Pollution, Politics, and Foreign Investment in Taiwan: The Lukang Rebellion.*

Reimann, K. (2003) 'Building Global Civil Society from the Outside In? Japanese International Development NGOs, the State, and International Norms', in F. Schwartz & S. Pharr (Eds.), *The State of Civil Society in Japan* (Cambridge University Press).

Reimann, K. (2009) *The Rise of Japanese NGOs* (Routledge).

Samuels, R. (2013) *3.11: Disaster and Change in Japan* (Cornell University Press).

Schreurs, M. (2002) *Environmental Politics in Japan, Germany, and the United States* (Cambridge University Press).

Schwartz, F. & Pharr, S. (Eds.) (2003) *The State of Civil Society in Japan* (Cambridge University Press).

Shapiro, J. (2001) *Mao's War against Nature: Politics and the Environment in Revolutionary China* (Cambridge University Press).

Steiner, K., Krauss, E. & Flanagan, S. (Eds.) (1980) *Political Opposition and Local Politics in Japan* (Princeton University Press).

Steinhoff, P. (1989) 'Protest in Japan', in Ishida Takeshi & E.S. Krauss (Eds.), *Democracy in Japan* (University of Pittsburg Press).

Steinhoff, P. (2011) 'Transforming Invisible Civil Society into Alternative Politics', in *Politics of Popular Culture* (Temple University Press).

Takao, Y. (2007) *Reinventing Japan: From Merchant Nation to Civic Nation* (Palgrave Macmillan).

Tang, D.T.-C. (1997) 'New Developments in Environmental Law and Policy in Taiwan', *Pacific Rim Law & Policy Journal* 6, 245–246.

Tang, S.-Y. & Tang, C.-P. (1997) 'Democratization and Environmental Politics in Taiwan', *Asian Survey* 37, 281–294.

Teets, J.C. (2009) 'Post-Earthquake Relief and Reconstruction Efforts: The Emergence of Civil Society in China?', *China Quarterly* 198, 330–347.

Tsai, L. (2007) *Accountability without Democracy: How Solidary Groups Provide Public Goods in Rural China* (Cambridge University Press).

Upham, F. (1976) 'Litigation and Moral Consciousness in Japan: An Interpretive Analysis of Four Japanese Pollution Suits', *Law and Society Review* 10, 579–619.

Walker, B. (2011) *Toxic Archipelago: A History of Industrial Disease in Japan* (University of Washington Press).

Wang, A. & Gao, J. (2010) 'Environmental Courts and the Development of Environmental Public Interest Litigation in China', *Journal of Court Innovation* 3, 37–51.

Weller, R.P. (1999) *Alternate Civilities: Democracy and Culture in China and Taiwan* (Westview Press).

Wells-Dang, A. (2011) *Civil Society Networks in China and Vietnam* (University of Birmingham Press).

Woo-Cumings, M. (Ed.) (1999) *The Developmental State* (Cornell University Press).

Wu, F. (2002) 'New Partners or Old Brothers? GONGOs in Transnational Environmental Advocacy in China', *China Environment Series* 45–58.

Xie, L. & Ho, P. (2008) 'Urban Environmentalism and Activists' Networks in China: The Cases of Xiangfan and Shanghai', *Conservation and Society* 6, 141–153.

Yamamoto, T. (Ed.) (1998) *The Nonprofit Sector in Japan* (Manchester University Press).

Zissis, C. & Bajoria, J. (2008) China's Environmental Crisis. Washington, DC, Council on Foreign Relations.

Acronyms

CCP	Chinese Communist Party (China)
DPP	Democratic Progressive Party (Taiwan)
GONGO	government-organized non-governmental organization
KFEM	Korean Federation for Environmental Movements
KMT	Kuomintang (Taiwan)
LDP	Liberal Democratic Party (Japan)
NGO	non-governmental organization
NIMBY	not in my back yard
TEPU	Taiwan Environmental Protection Union

Notes

1 Research for this project was made possible by an Abe Fellowship from the Japan Foundation and the Social Science Research Council, and a fellowship from the East Asian Institute.

2 North Korea is not included in this study. There may be environmental movements that are active in the North; certainly there are those in the North who are concerned with environmental issues, such as those working to create a nature reserve out of the demilitarized zone. However, I have been unable to get access to sufficient information to include North Korea in this analysis.

3 I will refer to all four political units – mainland China, Japan, South Korea, and Taiwan – as countries even though there is considerable dispute about whether Taiwan should 'count' as an independent state. It has most of the features of an independent state, such as printing its own currency, a national military, and sovereign control of its territory. However, it is missing some key symbols of contemporary statehood, such as recognition by the United Nations as an independent state, and the People's Republic of China also claims that it is one of China's territories. Therefore, I am using the label of 'country' in this chapter largely for rhetorical ease and not as a political statement coming down on one side or the other of the debate about Taiwanese statehood.

4 The third wave of democracy usually refers to those that occurred in the late 1980s and early 1990s concurrent with and immediately following the dissolution of the Soviet Union. Huntington, S.P. (1991) 'Democracy's Third Wave', *Journal of Democracy* 2(2), 12–34.

5 Interviews with senior managers from Keidanren, Toyota, and Hitachi in Tokyo, 2011.

6 1999 figure from Historical Statistics of Japan. http://www.stat.go.jp/data/chouki/zuhyou/11-05.xls; 2011 figures are from *Statistical Handbook of Japan 2012*. http://www.stat.go.jp/english/data/handbook/c08cont.htm.

7 Interviews in Beijing, Hong Kong, Taipei, Seoul, and Tokyo, 2010–2011. For example, in 2010 Greenpeace had 50 staff in their Beijing office, 40 staff in Hong Kong, and only 15 in Tokyo (interviews in Beijing, Hong Kong offices of Greenpeace, November 2010, and email communication with the Tokyo office, October 2012).

8 Interviews with Ministry of Environment staff in Tokyo, June 2011. See also the IAEA website about its Nuclear Safety Action Plan (http://www.iaea.org/newscenter/focus/actionplan/); the UN Environment Program's disaster management (http://www.unep.org/pdf/UNEP_Japan_post-tsunami_debris.pdf); the UN Centre for Regional Development and Japan's future city initiative (http://www.uncrd.or.jp/env/docs/120621_FutureCitiesWeWant.pdf).

9 For the Japan Renewable Energy Foundation that Son created in August 2011, see http://jref.or.jp/en/; for news coverage, see the *Wall Street Journal* (http://online.wsj.com/article/SB10001424052702304371504577404343259051300.html).

10 No Nukes Japan (http://nonukes.jp/wordpress/) (Japanese); Sayonara (Goodbye) Nukes has an English website (http://sayonara-nukes.org/english/). For coverage of one of their biggest rallies, see *The Economist* (http://www.economist.com/node/21559364).

11 For example, a new Green Party was established in the summer of 2012 with the intention of competing in the 2013 Lower House elections. http://ajw.asahi.com/article/0311disaster/fukushima/AJ201207300010.

12 Interviews in Taipei, November 2010.

13 Interviews in Taipei, November 2010.

14 Interviews with Echo Lin of Taiwan Environmental Action Network (TEAN) and Kuang-Jung Hsu, Professor of Atmospheric Sciences of Taiwan National University and former chair of Taiwan Environmental Protection Union (TEPU) in Taipei, November 2010.

15 Homepage in Korean (http://www.kfem.or.kr/). Interview with a KEFM leader in Seoul, November 2010.

16 Interviews in Beijing, November 2010 and April 2011.

17 Interviews in Beijing, 2010 and 2011. Please note that one activist noted the different responses to NGO involvement during a similar earthquake that occurred only months later in the politically sensitive Tibetan region of Quinghai. In that instance, the government attempted to react quickly, claim responsibility, and keep the NGOs out.

18 For example, in 2012 the Nature Conservancy has assets in excess of $6 million (http://www.nature.org/about-us/our-accountability/annual-report/2012-financial-report-with-report-of-independent-auditors.pdf); in 2012 WWF collected more than 500 million euros in income (http://awsassets.panda.org/downloads/int_ar_2010.pdf).

19 Interview in Beijing, November 2010.

20 http://www.ucsusa.org/global_warming/science_and_impacts/science/each-countrys-share-of-co2.html.

21 https://itunes.apple.com/us/app/china-air-pollution-index/id477700080?mt=8.

22 Of the 338 environmental organizations in the region studied by the author for this project, only 21, or 7%, appeared to have a legal strategy as a component of their activism.

23 Interviews in Beijing, Seoul, Taipei, and Tokyo, 2010 and 2011.

24 Interviews in Tokyo, 2011.

25 See Greenpeace Report, 'Swimming in Poison'. http://www.greenpeace.org/eastasia/publications/reports/toxics/2010/swimming-in-poison-yangtze-fish/.

26 Interviews in Beijing, 2010 and 2011.

7

Nongovernmental organizations and environmental protests

Impacts in East Asia[1]

Fengshi Wu and Bo Wen

Activism and mass movements are integral parts of the rise of modern environmentalism in Western democracies that demands 'a radical transformation in the values and structures of society' (Carter 2007: 7). Comparative history of environmental politics in post-World War II East Asia confirms such observations. Activism driven by nongovernmental organizations (NGOs) and protests by pollution victims have provided important impetus for changes in environmental governance and public awareness in most East Asian societies to different degrees since the 1950s. However, differently from the Western Europe and Northern America experiences, the dynamics of environmental movements is not only shaped by the trajectory of industrialization and state–market structures, but also, if not more importantly, by the overall political development and the process of democratization. In fact, environmental activists and NGOs face fundamental challenges of political repression and lack of participatory mechanisms. This chapter examines the rise and impact of environmentalism in East Asia with evidence from Japan, South Korea, and mainland China. It endeavors to compare and explain the relationship between the trajectory of environmental activism and its effects on public opinions and policy outcomes, simultaneously shaped by specific political contexts.

Since the 1950s, these three countries in the region have embarked on radically different journeys of economic development and political transformation. South Korean politics has experienced a sea change from authoritarian rule to a relatively consolidated democracy. With its first female president in history sworn in, the country is arguably one of the most successful cases of the 'third wave' democratic transition. In contrast, Japanese society has not been surprised by fundamental changes in politics in spite of frequent replacements of prime ministers and recent decline in public support for the Liberal Democratic Party. The People's Republic of China is by far an outlier here, even though it shares much of the Confucian cultural and other historical legacies with the two neighboring countries. Over three decades of miraculous economic growth and intensification of social discontent have not propelled comprehensive political reform on the mainland.

Environmental activism and protests in these three cases emerged at different times and went through divergent trajectories. Chronologically, victims' collective resistance and mass protests against pollution broke out first in Japan in the 1960s. The 'Onsan illness' movement, considered the beginning of mass protest against industrial pollution and policy failure in modern Korea, started in the mid–1980s. Only after the 2007 anti-PX peaceful marching in Xiamen, did protests against pollution and development projects at local levels begin to catch nationwide publicity and generate policy impact in China.

Unlike their South Korean peers, forming NGOs or NGO coalitions has never been a central strategy for Japanese environmentalists. Even with the passing of the Non-Profit Organization Law (hereinafter, the NPO Law) in 1998, there has not emerged a national umbrella social organization comparable to the Korean Federation for Environmental Movement (KFEM) or the Taiwan Environmental Protection Union. However, KFEM, together with many other NGOs with broad membership base, has been critical in forging public campaigns and shaping national environmental agenda in South Korea since the 1990s. In China, environmental NGOs have emerged in major cities since the mid–1990s. They have been argued as pioneers in reviving civil society, expanding the scope of social activism and experimenting policy advocacy. But, due to the overall political constraints, even the most influential and capable environmental NGOs have restrained themselves from openly organizing protests by specific constituencies or victim groups.

The main divergence in the overall development of environmental activism in the three cases can be in general explained by key features of political culture and progress or lack of democratic reforms. However, what remains to be found out is whether different kinds of environmental activism have generated uneven socio-political impact in environmental protection. The parallel and intertwining relationship between the two main routes of environmental activism – NGO-centered policy advocacy and mass-based protests – is worthy of particular attention and can potentially affect the results of activism. For each case, the analysis starts with a brief history of the environmental movement, leading NGOs and important public campaigns. Then, the sequence and interplay between NGOs and mass-based collective actions are examined in order to find more contextualized differences across the cases. The impact of environmental activism is assessed and compared along three dimensions: educational effects and the rise of public environmental awareness, input in specific policy change and pollution reduction, and direct and indirect contribution to long-term institutional reform. The last section of the chapter intends to sort out the possible explanations for the varying impact of environmental movements in the three countries.

Japan: locally rooted environmental activism and a 'soft elite'

In postwar East Asia, industrialization and economic development took off first in Japan, as did mass protests against pollution and large construction projects. Four major public campaigns broke out in the 1960s against the construction of the Narita Airport, and demanded compensation for the victims of Itai-Itai cadmium poisoning, Minamata and Niigata mercury disease, and Yokkaichi asthma incidents. These campaigns not only awakened environmental awareness at the time, but also left their impact on the emergence of civil society and environmental governance in Japan. The most important direct result is the 1971 'Pollution Diet', which produced 14 environmental related national-level laws and established the State Environmental Agency. However, nationwide protests have since subsided and gradually been replaced by localized activism and community-based conservation groups. In addition, a pro-environmentalism 'soft elite' – like-minded technocrats and academicians – gradually emerged and mobilized for policy changes across the divide between the state and society without sensational public presence. Such a loosely coordinated, often low-profile and cross-sectoral coalition can be extremely effective in pushing for policy change under specific circumstances in Japanese political culture.

As Robert Mason once observed (1999: 188), Japan's environmental movement at the national level is 'politically marginalized' and 'weakly consolidated', although it is 'far from impotent'.

Japan's environmental movement has gone through considerable ups and downs (McKean 1981). The first wave of public awareness awakening and protests surged in the 1960s and lasted until nearly the end of the 1970s, highlighted by the four major campaigns mentioned above. The postwar Japanese regime at that time was run by a 'ruling triad' – the Liberal Democratic Party (LDP), a highly organized business community and the administration – that largely ignored the mounting pollution problems along with a rapidly growing economy. Sustained protests by pollution victims and not-in-my-back-yard (NIMBY) activists in various locations eventually staged real challenges to the triad (Broadbent 1998). The LDP suffered a decline in popular vote from 58% in 1960 to 48% in 1969, and their seats in the Diet reduced from 63% in 1960 to 57% in 1967 accordingly. The big businesses allied with the LDP also softened their attitudes towards environmentalists and victims after the campaigners changed their strategies and began to file law suits against the specific companies responsible for the pollution.

The 1980s saw environmental activism hit bottom in two decades. The rebound took place after the 1992 Rio United Nations Earth Summit. The passing of the Kyoto Protocol on climate change in 1997, for which the Japanese government's initiative and leadership in international affairs was widely recognized, and the establishment of the NPO Law in 1998 both contributed to the improving system for activism and policy advocacy. However, the improvement was limited and the general institutional environment for activism was still 'paradoxical' (Mason 1999: 190–193). The NPO Law allows social groups to gain a corporate entity status – more commonly known as NGOs in Western democracies – without having to meet the onerous requirements that were formerly in place. Yet, they are provided with few entry points into the national policymaking processes. 'NGOs must find their way within a system that is ostensibly open and democratic, yet in reality is unreceptive to direct citizen participation in national affairs. The implicit argument is that professional bureaucrats can best manage Japan, and there is scant place in this industrial-bureaucratic venture for environmental NGOs' (Mason 1999: 196).

Compared with neighboring South Korea and Taiwan, environmental activism in today's Japan is far less unified and visible as a collective social force at the national level (Hasegawa 2010; Pekkanen 2004). In 1999, there were 4500 environment-related NGOs in Japan, most of which are locally based, and only 9.5% nationally active. By 2009, over 10,000 registered NGOs have included environmental issues in their overall missions, but few touched on national policies. Japanese green NGOs tend to be small – operating with limited, and sometimes even insufficient, financial and human resources, and only focus on clearly defined local problems. The three largest environmental NGOs in Japan, Wild Bird Society (WBS), World Wildlife Fund Japan, and Nature Conservation of Japan, all focus on specific local conservation projects. By comparing them with their counterparts in Western democracies, South Korea and Taiwan, Koichi Hasegawa (2010: 86) pointed out that most of Japan's environmental NGOs and movements had kept 'strong puritanical tendencies' – self-depoliticizing and with a 'lack of critical perspectives' on both internal organizational and external public affairs. He even warned that overemphasizing the 'emotional ties among members' would turn these NGOs into 'cozy clubs' and eventually alienate them from their original social missions.

Along with the decline of environmental protests and victim resistance, a 'soft elite' has emerged quietly promoting environmental values, principles, and practices that are suitable for the Japanese context. They are circles and networks of like-minded governmental officials, academics, and environmental experts who collaborate via various formal and informal channels to push forward pollution prevention policies and practices. As a whole, the 'soft elite' 'stand against (and to some extent between) a "hard elite" (or the ruling Triad) of business leaders, politicians and bureaucrats and a small band of "hard" campaigners against dams and similar construction projects' (Waley 2005: 196). These are the main reasons for explaining the gradual yet consistent improvement in environmental protection since the 1980s without a national-level unified NGO alliance. Among the 'soft elite', governmental officials and technocrats are most interesting and challenging in terms of conceptualization. They have to act wisely along legitimate lines, advocating for environmentally informed ideas without bluntly infringing state authority. They should not be simply confused with 'policy entrepreneurs' that are commonly known to scholars of public administration. For they do not merely push for policy adjustment or innovative administrative measures via established institutional channels; rather, they act both as a governmental official during daytime and a concerned citizen when off duty. They would use strategies of social mobilization outside and beyond the state apparatus when necessary.

The main component of the 'soft elite' is the first generation of environmental technocrats who were in university during the peak of the 'four major campaigns' and rose to positions of influence from the mid-1980s to mid-1990s. In a sense, they carried on the legacy of the early environmental protests in quiet yet effective ways. For example, in the case of river conservation, the movement of 'home country rivers' was essentially led by an official of the Ministry of Construction (MoC), Seki Masakazu. Facing entrenched opposition from within the state, he first mobilized within the MoC and the bureaucratic system and made an effort to modify the ideas that underpinned river landscaping and planning among the technocrats. The movement then started to set up government-affiliated bodies to host conferences and study trips to disseminate new information and principles, and conducted two important pilot projects in Hino waterway and Tsurumi River in order to find and promote 'good practices'. After the mid-1990s, further diffusion of 'good practices' throughout the country was carried out beyond Masakazu and his immediate allies by local 'lay activism' featuring recreational activities, school outreach and community-based groups (Waley 2005: 205).

Such a seemingly fit and organic solution to transform the practices of environmental protection that mostly rests on the shoulder of a 'soft elite' could be a double-edged sword for environmentalism as a social movement in Japan. The elite's conscious effort to balance ideology and empirical rigor, environmental cause and political reality, and radical departure in principle and gradual change in practice has introduced many specific corrections of pollution, but also contributed to the fragmented, partially isolated and overly community-tied nature of environmental activism in the Japanese society of today. Simon Avenell's detailed study (2012) of the Pollution Research Committee established in the 1970s, particularly the part on individual committee members who were both influential public intellectuals and state affiliated specialists, their Marxist perspectives and pragmatic-minded leadership in changing Japan's environmental governance, convincingly explained that the gradual approach adopted by prominent environmentalists and the 'soft elite' in fact had delayed more proactive resistance and movements. Due to the lack of more critical and independent social forces, there have been 'blind spots' in environmental activism in Japan such as the nuclear issue. Despite local and regional campaigns to support victims of the atomic bombings in Hiroshima and Nagasaki since the 1950s, national-level NGOs and policy advocacy coalitions in this field did not emerge until after the 2011 Fukushima Daiichi nuclear disaster.

South Korea: national spread and coalitions of environmental NGOs

Environmental activism and campaigns have been an integral part of the democratic transition in South Korea since the 1980s. Most of the first generation of environmental activists were university students and at the center of the early pro-democracy protests. At that time, environmental activism was mainly embodied by staging protests by pollution victims, campaigning for compensation and making the government accountable for the established laws on pollution control. During the presidencies of Kim Dae-jung and Roh Moo-hyun, environmental NGOs blossomed, and, more importantly, NGO coalitions across regions emerged and became a critical social force to push for agenda change and policy reform at the national level. Entering the new millennium, South Korean NGOs are the most active in regional and international environmental politics compared with their peer organizations in neighboring countries.

South Korea's environmental movements in the early years were highlighted by the Goldman Environmental Prize Laureate, Choi Yul, who chaired the Korea Anti-Pollution Movement since 1982 and founded the first environmental NGO in the country, the Korean Research Institute of Environmental Problems (KRIEP), to support the residents who were affected by the toxic wastes from a nonferrous metal industrial complex in Onsan. The Onsan campaign became the landmark event in the history of environmentalism in South Korea. In 1993, together with several other environmentalists, Choi Yul established the first national coalition of environmental NGOs, the KFEM. This umbrella organization led environmental activism at the national level in the coming decade and has become the largest environmental NGO in Asia, with 46 local branches across the country and 150,000 registered members.

Since its inception, KFEM adopted Greenpeace-style campaign tactics – staging public campaigns and symbolic acts to trigger dramatic public reaction and push for policy change. One well-known case was masking the face of the Admiral Yi Sun-Sin statue in downtown Seoul to call for public awareness of and protest against air pollution in the city. It has also led long-term campaigns against the damming of the Donggang River and for preserving the Saemangeum costal tidal flat. With successful domestic campaigns such as these, KFEM extended its work to other regions and began actively engaging environmental NGOs in neighboring countries. For example, it worked with bird-watching groups in Japan, Taiwan, and Hong Kong to form a Northeast Asia Black-faced Spoonbill Network for information exchange and coordination of conservation efforts. The three governments of China, Japan, and South Korea launched the Tripartite Environment Ministers Meeting in 1999, and KFEM was entrusted with funds to coordinate nongovernmental cooperation across borders. In 2002, KFEM joined the international environmental federation, Friends of the Earth, and became its Korean chapter, and has since been visible on the stage of global environmental politics.

Besides KFEM, there were three other NGOs that emerged in early 1990s and played an important role in fomenting environmental movements in South Korea: People's Solidarity for Participatory Democracy (PSPD), Citizen's Movement for Environmental Justice (CMEJ), and Green Korea United (GKU, aka Green Korea). All of them demonstrate the close linkage between environmental activism and democratic transition. Park Won-Soon, another veteran student activist of the 1970s and 1980s, founded the nonprofit watchdog organization PSPD in 1994 responding to the need for a comprehensive movement to oversee and challenge abuses of power. Different from KFEM which is more specialized in environment related public affairs, PSPD focuses on working with labor unions, monitoring governmental behaviors and regulatory practices, and promoting direct forms of public participation in a wider range of issues including yet not limited to environmental protection.

Founded in 1989 right after the first waves of student protests, the Citizens' Coalition for Economic Justice (CCEJ) was the largest social organization taking up the role of public monitoring in South Korea during the processes of political transition. In 1992, CCEJ founded the Center for Environment and Development as its environmental research arm. Over the years, staff at this center became increasingly convinced that environmental protection could best be achieved through social mobilization and activism rather than research. Eventually in 1999, the center separated itself from CCEJ and changed its name to the CMEJ. CMEJ became an independent network of environmental activists focusing on social mobilization and public campaigns. Its major achievements include preventing development projects in Daejisan Mountains in 2001, launching the Food Safety Campaign against the use of toxic chemicals and genetically engineered food in 2002 and bringing together some 40 organizations to form a 'Saving Rivers Network' in the mid-2000s.

In 1991, two youth-led environmental groups joined hands and formed GKU. GKU now has some 15,000 members and, like KFEM, a number of local chapters. It has utilized various strategies to achieve environmental goals, such as running public education programs to preserve the Baekdudaegan region and marine life, filing litigations against environmental damage caused by U.S. military bases, and monitoring governmental policies related to nuclear energy. Different from other leading environmental NGOs, leadership of GKU has remained among university students and relatively younger activists. Therefore, it has maintained particularly close contact with the youth in society and is capable of initiating more progressive and forward-looking public campaigns.

Despite setbacks in policies towards civic organizations during Lee Myung-bak's administration, environmental activism continues to gain vigor in South Korea mainly due to NGOs' professional development. A new generation of environmental groups has started to sprout such as the Energy Justice Actions (EJA). This NGO evolved from a student environmental club, Korea Eco Center, founded in 1999. In 2008, the center held a public campaign to raise public awareness of sustainable energy. Using pinwheels as a symbol of green energy, the campaign quickly received a highly positive reception from the public and became one of the highlights in environmental activism after the democratic transition. EJA was later established as a direct fruit of the successful campaign, and has since grown into an effective force in national politics surrounding nuclear power, energy efficiency and energy-related social justice matters.

Nevertheless, observers are concerned with the actual impact of environmental protests and NGOs in formulating and implementing pollution control policies. Excessive usage of confrontational strategy can make decision making impossible and turn public consultations into deadlock situations. Some even argue that South Korea's environmental activism has hit a structural turning point after the consolidation of democratic institutions. Whether NGOs should modify their conventional strategies and take up a less confrontational role is a pertinent question for South Korean environmental activists (Ju & Tang 2011; Ku 1996; Zusman 2007). Despite scholars' concerns, there is little doubt that environmental NGOs in South Korea have influenced national policies in explicit ways: First, they submit policy proposals directly to the National Assembly and the Blue House; second, they have their voice heard through a regular mechanism, the Environmental Policy Council established by the late President Kim Dae-jung; third, they submit research findings and policy recommendations to relevant ministries; last but not least, they act against environment endangering policies by directly reaching the public, supporting (potential) victims, mobilizing media campaigns, and staging mass protests (Kim 2007).

In the case of nuclear energy, all of these strategies have been applied by various environmental NGOs. After South Korean companies won a US$20 billion contract to build civil nuclear power plants in the United Arab Emirates, the South Korean government announced plans to draw more than 50% of domestic energy needs from the nuclear sector by 2020. With years' experience of campaigning against nuclear power plants, KFEM has since organized several 'No Nuke Asia Forums' to gather international support against the government's plans. The Climate Justice Actions also has been conducting public education programs on nuclear energy-related issues for over a decade. After the 2011 Fukushima nuclear accident, EJA hosted internet live shows everyday giving detailed accounts of the incident to the public. Moreover, South Korean environmental NGOs have reached a high level of consensus to guide the public's attention to go beyond Fukushima or Japan and to reflect on South Korea's own energy policy and nuclear export strategy.

As illustrated by all these cases, environmental NGOs in South Korea usually utilize a wide range of strategies. Professionalized NGOs are keen to work with communities, victim groups, experts and external peer organizations. Social mobilization and direct resistance has been accompanied by consistent development of NGOs and institutionalized advocacy strategies. In addition, many leading environmental NGOs are not restricted to environmental causes; instead, they aim to bring about more wide-ranging political and social changes in South Korea. Embedding environmental activism in the comprehensive political democratization is what distinguishes South Korea from the other two cases studied here, in particular, and most developing countries in general (Alagappa 2004; Schreurs 2002).

China: emerging convergence between environmental NGOs and NIMBY protesters

Modern environmentalism emerged in China only after the end of the Cultural Revolution, and more precisely, since the controversy over the Three Gorge Dam in the mid-1980s. The 1980s once saw rapid revival of civil society in China, yet ended with the Tian'anmen Square crackdown. Citizens' actions against environmental degradation gradually regained momentum by the end of 1990s, and have since grown steadily driven by committed individuals, grass-roots groups and NGOs (Economy 2004: 129–176; Hildebrandt 2013). Incidents of pollution victims' resistance have increased and become more confrontational in recent years. However, environmental NGOs have strategically chosen not to be directly involved in open collective actions against the authorities in China (Yang 2005; Peng & Wu 2013).

The Chinese environmental NGO sector has gone through three main phases (Wu 2009). Pioneer Chinese environmental activists included the scientists, professors, journalists, and members of the National People's Congress who participated in the debate against the proposal of the Three Gorges Dam. Even though most of the debate and policy advocacy took place within the academic and high-level policymaking circles, this prototype movement still had evident political implications. For the first time, it alerted the general public about potential environmental problems hidden beneath grand proposals of economic development (Dai et al. 1994; Economy 2004: 142–145; Heggelund 1993; Khagram 2004: 170–176).

The second period started from the mid-1990s, which saw significant growth of NGOs, public education programs and cross-regional campaigns. The establishment of the Friends of Nature (FoN) in Beijing in 1994 marked a turning point in environmental activism in China. It is a first successful attempt by citizens to establish a formal organization fully devoted to environmental public education and advocacy. Almost all other 'social organizations' (*shehui tuanti*) at that time in China were either professional associations or government-affiliated organizations (otherwise known in scholarship as GONGOs – government-organized NGOs) (Wu 2004). The founding board members of FoN consisted of Liang Congjie – an established intellectual with a distinguished family background – and three younger public intellectuals and writers Liang Xiaoyan, Wang Lixiong and Yang Dongping, who all participated in the Tian'anmen movement with different roles and continue to be extremely influential in civil society development in China beyond the field of environmental protection. Liang has been regarded by the environmentalist community as a father figure and later became the first Chinese environmentalist who won an international award, the Ramon Magsaysay Award of 2000.

Since FoN, environmental activism in China entered a new phase, and 'NGO' gradually became the main institutional venue and strategy for environmentalists to operationalize their ideas. Soon after FoN, a dozen environmental NGOs emerged and obtained formal status in Beijing and other major cities in spite of bureaucratic hurdles. Examples include Global Village of Beijing, Chongqing Green Volunteers, Green Civil Association of Weihai City (Shandong province), Farmers' Association for the Protection of Biodiversity of the Gaoligong Mountains (Yunnan province), Green Earth Volunteers in Beijing, and Chengdu Green Rivers (Sichuan province). The most common types of activity conducted by environmental NGOs in the 1990s were public education activities, media campaigns for a green lifestyle and providing policy recommendations to local environmental agencies. Very few leading organizations started to experiment with promoting new norms related to environmental justice and building stronger ties with the communities that faced industrial pollution.

Despite mixed signals from different state agencies and the harsh crackdown of Falung Gong exercisers and sympathizers in 1998, the environmental activism community has not only survived but also continued to grow in the 2000s. Researchers have found that at least 128 NGOs had been established nationwide by 2004, and the number rose to 300 by 2012 (Ru & Ortolano 2009; Wu 2013). In May 2004, Dai Qing, a veteran environmentalist of the anti-Three Gorges Dam campaign, a prominent political dissident and the first Chinese winner of the Goldman Prize, attended a meeting in Beijing against damming the Nu River in southwest China. She reflected on changes in environmental activism and commented that the timing for a new anti-large dam campaign was much more mature by now than in the 1980s, particularly because the NGO community has strengthened its capacity for policy advocacy and public campaigning.

Over 50 environmental NGOs responded to the Sichuan earthquake swiftly in May 2008 and joined other grassroots groups in turning a major natural disaster into the 'Year of Civil Society' in China (Xu 2008). This set off the nascent beginning of a third phase of environmental activism in China marked by a younger generation of NGO leaders with considerably diverse social backgrounds. Mostly born in the 1980s, these young environmentalists, on the one hand, do not enjoy the social capital accumulated throughout lifelong professional careers as did the first generation of NGO leaders such as Liang Congjie and Dai Qing, yet, on the other hand, they are much more innovative in public campaigning taking full advantage of the development of social media and Web 2.0. They have developed new strategies of environmental activism merging policy advocacy with entertainment and festivity. For example, Biker Guangzhou, a new environmental NGO run by young people in their twenties, set off a campaign to promote public awareness of carbon emission and climate change by openly giving a bicycle to the city mayor and inviting him to bike together (Wu 2013). Soon after, Biker used its own website and most popular social networking sites in China to spread the news and images of Guangzhou's mayor receiving the bicycle. Such a non-contentious approach has been proved to be highly effective as it is both accepted by the authorities and able to arouse public enthusiasm about the issues. By the first of half of 2013, this organization has successfully influenced the municipal government's decision and policymaking in urban planning and the creation of more dedicated traffic lanes for bikers.

Compared with Japan and South Korea, the case of Chinese environmental NGOs and activism is peculiar not only in the general patterns but also in a specific aspect: that there has been a clear disconnection between NGO-centered advocacy and mass-based protests. It is not rare for rural or urban residents to take collective and drastic actions against local authorities in post-Mao China, and such actions range from peaceful sit-in, petition to more radical forms including burning governmental building and open vandalism (O'Brien 2008). Recently, an official term, *qunti xing shijian* ('collective event'), has been coined to describe these actions, among which cases against industrial pollution or large-scale infrastructure construction count as a significant percentage (van Rooij 2010). Official data indicate that the number of protests against pollution, environmental accidents and large construction projects has grown by a rate of 29% per annum since 1996 (Feng & Wang 2012). In urban regions and more recent years, precautionary protests organized by home owners against potential damage have even emerged. For example, 2007 anti-PX (*para-xylene*) chemical plant in Xiamen, 2008 anti-magnet railway in Shanghai, 2011 anti-PX in Dalian and 2012 anti-chemical industries in Ningbo. These protests seem to resemble the ones in Japan and South Korea in earlier decades; however, environmental NGOs have been missing in the entire process of community mobilization, campaign organizing and policy advocacy follow-up (Johnson 2010; Stern 2011). Chinese environmental activists and grassroots NGOs have been cautious and made conscious decisions not to use the strategy of mobilizing collective resistance due to daunting political pressure, prevalent surveillance and hidden constraints they endure on a daily basis (Deng 2010; Wu & Chan 2012). Also, the campaigners from residential neighborhoods, who may have succeeded in protecting their own interests, usually recognize the political danger of becoming a 'professional activist' and choose not to commit themselves to long-term policy changes (Stern 2008; Zhu & Ho 2008). This feature of clear separation between the protesters and the NGO sector is critical to understand environmental movements as a whole in China.

In the year 2012 alone, at least three mass protests against chemical factories in different cities (i.e., Shifang, Chongming, and Ningbo) took local authorities by surprise. The year 2013 began with hundreds of thousands marching in peace in Kunming, capital of Yunnan province, against a proposal of building a PX chemical plant adjacent to the city. This is the first time anti-pollution protest has happened in a provincial capital city and against a provincial authority, at that. These waves of mass protests have generated pressure on the state in the face of mounting industrial pollution and environmental accidents, which in turn creates more political space for environmental NGOs. Therefore, in the most recent years or months, some environmental NGOs are beginning to get involved in collective resistance of industrial pollution, and more importantly, to search for methods to maintain the fruit of successful protests and turn the temporary resolution into some form of local policy that will have long-term effects.

The most developed case of this nascent convergence of NGO-centered activism and mass-based resistance is the anti-incinerator campaign and policy advocacy in Guangdong province. Since 2009, residents of a suburban neighborhood in Panyu district, Guangzhou city, started petitioning, peaceful sit-ins and other forms of direct resistance to municipal government's plan of building an incinerator in their neighborhood to solve the problem of increasing solid waste for the whole city. Activists from FoN in Beijing paid close attention to these local initiatives from the beginning, and provided important assistance in information sharing and brought in technical expertise along the way. For a while, FoN activists maintained a very low profile to protect both their own organization and the campaign itself from unnecessary political scrutiny.[2] But, after the initial success of pressuring the government to withdraw the project, FoN encouraged and helped community leaders to establish a formal NGO in Guangzhou in 2012. FoN staff sit on the board of this new NGO and will continue to push for long-term policy changes related to solid waste treatment in the province.

To a large extent, NGO-centered activism in the environmental field is much more developed than other policy areas in China. However, like their peer organizations in other fields, environmental NGOs are consistently under surveillance by the state and therefore are greatly constrained in what they can accomplish in terms of grassroots outreach, social mobilization and staging mass protests. The anti-incinerator example indicates recent positive changes in this respect and it is possible that veteran NGOs may be able to exert some influence and play a direct role in collective resistance in order to push for policy changes in addition to one-time compensation or cancellation of a construction project. Yet it is not certain whether such will become a general trend in China's environmental politics. The chapter will further discuss the impact of this factor in the comparative section and explain how it has differentiated environmental activism in China from other East Asian countries.

Comparative analysis: the impact of environmental activism

In this section, the impact of environmental activism and movements in East Asian societies is examined and compared in three aspects: public education and raising environmental awareness, provisions of solutions to specific pollution problems, and participation in national-level policymaking. It is evident that activists and NGOs have contributed substantially to the rise of public environmental awareness in East Asian societies, as in the rest of the world. Comparing environmental activism within the region, the Chinese case is still lagging behind in this regard. Despite the fact that Chinese environmental NGOs have launched a number of successful campaigns to promote a green lifestyle and environmental values since the 1990s, the public remained indifferent to environmental degradation and industrial pollution until very recently (Chinese Academy of Social Sciences 2007; Li 2006; Wong 2003, 2005). Continuous heavy smog in the Pearl River Delta and Beijing has led to citizens' open criticism of the government and voluntary reporting of $PM_{2.5}$ since 2012. Tens of thousands of people in Beijing decided to use the U.S. Embassy's AQI feed to check the air quality of their city, and found the readings completely differently from those released by the Chinese government (Sebag-Montefiore 2012). The outbreak of dreadful, even toxic, smog in Beijing in the spring of 2013 further increased the public's resentment and called for governmental responses. Ironically, NGOs have not played significant roles in this recent wave of public outcry for environmental protection. In contrast, environmental NGOs are usually ahead of the public in exposing governmental failures in controlling pollution and calling for public attention in Japan and South Korea.

To a great extent, NGOs depend on their own capacity in solving problems and supporting victims at the grassroots level to enhance their accountability and legitimacy in the long run. In this respect, both South Korean and Japanese environmental NGOs have achieved substantial accomplishments, while in China, precisely due to the low level of professional capacity in addition to the lack of institutional space for public participation in environmental protection, most environmental NGOs have not grown roots in local communities and therefore have accomplished less in preventing degradation on the ground.

In South Korea, the most noteworthy examples of NGOs' success in water and river conservation include the campaigns against damming the Donggang River, land reclamation in Sammunkuen and the Four Rivers Restoration Project. The Donggang River flows through the Gangwon-do district of Seoul, and is a tributary to the South Hangang River. Starting in 1998, KFEM took up the campaign and gained widespread public support for preserving the river for endangered otters and bird species, landscapes and the culture of bordering communities. In 2000, after a one-year joint investigation by both citizen groups and governmental agencies, the Kim Dae-jung government decided to cancel the project.

Saemangeum is a large tidal flat that sits on the Yellow Sea coast of the Korean Peninsula. It is one of the most important habitats for migratory birds and marine species. In 1991, the South Korean government started to build Saemangeum seawall and aimed to create new land for farms and potentially, for an industrial complex. The campaign against the Sammunkuen reclamation project ran from 1998 to 2006. South Korean NGOs formed a wide-reaching alliance and accumulated religious, political and international support. They also utilized a variety of strategies such as a 305km 'three-steps-and-a-bow' (*sambo ilbae*) march from Seoul to Saemangeum during March 28 to May 31, 2003, and environmental litigation against the government. On July 15, 2003, Seoul Administration Court ruled against the construction of the Saemangeum project, but the ruling was revoked by Seoul High Court in 2004 and again by the Supreme Court in 2006. Despite the disappointing result, this movement set up new examples for environmental activism and prepared the NGO community to face even more massive state-led projects later.

Initiated by the then President Lee Myung-Bak in 2009, the Four Rivers Restoration project was to construct a series of canals linking the Han, Nakdong, Kum and Youngsan rivers across the entire country. The plan encountered strong resistance from civil society. With media and public campaigns driven by environmental NGOs, it became highly unpopular among the public and led to considerable debates within the National Assembly. On February 10, 2012, the Busan High Court declared the project illegal.

In a way, Japanese environmentalists have conducted similar campaigns to prevent and halt state-led projects as their peers in South Korea, except that most were at community levels. Moreover, environmentalism has been further embraced into local governance without national-level mobilization or policy shifts due to at least two factors. First, since the 1970s, progressive politicians, among whom many were environmental activists involved in the first wave of protests, have been able to run successful elections and were elected as mayors in a number of cities. By the end of the 1970s, 38% of the Japanese population including major cities such as Tokyo, Kyoto and Osaka were governed by progressive mayors (Steiner et al. 1980: 326). Second, 'good practices', advanced technologies and new models of sustainable development and environmental protection have been disseminated by non-contentious and within-system mechanisms; and, more importantly, the 'soft elite' preferred such a gradual and technical approach instead of the strategy of social movement.

As observed by the long-term practitioner and scholar, Koichi Hasegawa, currently environmentalist-oriented groups at local levels are becoming more proactive in introducing innovative methods without waiting for policy improvement at the national level (Hasegawa 2010). Examples include the Hokkaido Green Fund and the promotion of communal wind power plants. Japan's first citizens' communal wind power project was introduced by the Hokkaido branch of the Seikatsu Club Consumer Co-op established in 1965 in Tokyo. The Hokkaido branch was established in 1982 and has been working on radiation, nuclear energy and alternative solutions since then. A Hokkaido Green Fund to establish a wind power plant based on regular contributions by members was launched in March 1999. The importance of this communal wind power plant is that it links the movement and community business through small investments and small shareholdings by citizens. Its success is now widely reported in the country and hopefully will be copied by more communities.

The third indicator to examine the effect of environmental movements focuses on direct policy and institutional input at the national level. Only with reform of relevant policies and governing structures, environmental NGOs can turn short-term successes into long-term mechanisms of environmental protection. In general, this is a very challenging task for any environmental NGO across the world. Because the environmental movement has always been an integral part of the democratic transition in South Korea, and most environmental NGOs have a wider agenda and embedded environmental causes in broader discourses of social justice and public participation, they have been able to achieve more observable accomplishments in this respect than their peers in Japan and China. In contrast, Japanese NGOs have not been able to form a consolidated coalition at the national level to pursue a comprehensive environmental agenda after the 1971 Environmental Diet.

In China, existing structures provide least space for environmental NGOs to take part in policymaking at any level. However, in recent years, some more experienced NGOs have begun to gain sufficient support and built up their capacity to advocate for policy changes at the national level. The most promising example here is the Institute of the Public and Environment (IPE) based in Beijing. Founded by an internationally known and award winning environmentalist, Ma Jun, in 2004, IPE has been collecting, analyzing and publishing data of pollution in China independently (Wu 2013). Ma Jun, author of the book *China's Water Crisis* first published in Chinese in 1999 and later published internationally in English, was named one of *Time* magazine's '100 People Who Shape Our World' in 2006 and awarded the Goldman Prize in 2012. IPE has recruited a team of young, capable and committed college graduates to conduct reliable research, and collaborated with a network of grassroots NGOs across the country that have an interest in pollution-related work. Its interactive website provides numerous maps, charts and user-friendly data sets free for the public, media and even governmental agencies to use. Each year, IPE publishes a report ranking foreign and Chinese corporations' waste water discharge using publicly available information. This form of 'information campaign' has generated evident pressure on both the businesses and regulatory agencies since its first publication. Because it strictly uses official data released by the central government, the authorities have not been able to find legitimate reasons to force IPE to stop. Instead, environmental bureaus find it actually lends them some help when enforcing environmental regulations. Collaborating with the China Environmental Law Project of America's National Resource Defense Council, IPE has jointly published the annual report of Pollution Information Transparency Index (PITI) since 2009. The PITI project evaluates and ranks 113 municipal environmental protection bureaus (EPBs)' implementation of the Environmental Information Disclosure Measures. PITI in a way has similar effects as the 'information campaign' of waste water management. The Chinese government, particularly the Ministry of Environmental Protection, has endorsed and even used the results of PITI to modify EPBs' behaviors in releasing environmental data to the public. These changes may not seem to be comparable to what South Korean NGOs have achieved, but in the context of contemporary Chinese politics they are noteworthy and have the potential to be replicated in other fields.

Conclusion

Comparative evidence from Japan, South Korea and China first and foremost indicates that the impact of environmental activism is heavily modified by the domestic political context. Even though mass protests, pollution victim groups and environmental legislation have a longer history in Japan, South Korean environmental NGOs have been able to play a more visible and effective role in national-level policymaking and politics. This is precisely due to their crucial participation in the entire processes of the political transition since the 1980s, and therefore, they have been part of the creation of a participatory culture of the newly established democratic regime in South Korea, while in China, the post-authoritarian rule still determines the general institutional settings for any social activism and policy advocacy, in spite of the fact that environmental NGOs have been highly self-motivating and innovative in expanding the boundary of organizational autonomy and capacity.

Table 7.1 both summarizes the key characteristics of environmental activism and compares their effects in the three countries. The most important finding is that the level of integration of mass-based protests and NGOs varies across the cases. China's NGO-centered environmental activism almost has taken a separate trajectory from community-based or victim-organized collective resistance, and thus its public visibility, local outreach and capacity to provide specific

Table 7.1 Comparing environmental movements in East Asia

	Effects			Features
	Raising public environmental awareness	Solving specific environmental problems	National-level environmental policymaking	
South Korea	+ + +	+ +	+ + +	Integration of mass protests and NGO activism and the formation of NGO coalitions at national level
Japan	+ + +	+ + +	+ +	Integration of mass protests and NGO activism at local levels and a circle of 'soft elites'
China	+	+	+	Separate development of victim protests and environmental NGOs

Note: + is a relative measurement to indicate the ranking among the three cases of this study.

solutions on the ground remain very limited. However, the other two cases imply that the merging of resistance movements and NGO professionalism can produce positive outcomes in raising public awareness and preventing degradation in specific locations.

Moreover, it is mainly because South Korean environmentalists have been able to embed their effort in a broader public agenda and always use both strategies of mass mobilization and professional policy advocacy at the national level, their participation and impact on national-level environmental governance is more visible than their peers in Japan. In Japan's national-level environmental policymaking, NGOs are still marginalized. Their voices and pursuit of policy change are usually channeled via sympathetic or activist-minded technocrats and academicians, which is explainable by the salient political culture of the country.

Entering 2013, China faced even more daunting environmental challenges and rising social resentment marked by anti-PX protests in multiple cities, angry 'netizens' posting data of $PM_{2.5}$ and pollution victims' collective litigation against the government. That year coincidentally was also the 20th anniversary of the first and leading Chinese environmental NGO, FoN. With two decades of experience, whether Chinese environmental activists can navigate through the perplexing political systems in the country to not only build up their self-capacity in campaigning and policy advocacy, but also reach out to victims and communities affected by environmental pollution and articulate their concerns remains to be seen. As demonstrated by the history of environmentalism in neighboring Japan and South Korea, the possible convergence of increasingly professionalized NGOs and radicalized anti-pollution protests will have significant implications for Chinese environmental politics in the coming years.

Bibliography

Alagappa, M. (Ed.) (2004) *Civil Society and Political Change in Asia: Expanding and Contracting Democratic Space* (Stanford University Press).

Avenell, S. (2012) 'From Fearsome Pollution to Fukushima: Environmental Activism and the Nuclear Blind Spot in Contemporary Japan', *Environmental History* 17(2), 244–176.

Broadbent, J. (1998) *Environmental Politics in Japan. Networks of Power and Protest* (Cambridge University Press).

Carter, N. (2007) *The Politics of the Environment* (Cambridge University Press).

Chinese Academy of Social Sciences (2007) '2007 National Report on Public Environmental Awareness', Beijing.

Dai, Q., Adams, P. & Thibodeau, T. (1994) *Yangtze! Yangtze!* (Earthscan Canada).

Deng, G. (2010) 'The Hidden Rules Governing China's Unregistered NGOs: Management and Consequences', *China Review* 10(1), 183–206.

Economy, E. (2004) 'The New Politics of the Environment', in *The River Runs Black* (Cornell University Press).

Feng, J. & Tao Wang (2012) 'Open Window: Search for Environmental Dispute Resolution' (in Chinese), *South Weekly* 29 November. Available at http://www.infzm.com/content/83316.

Hasegawa, K. (2010) 'Collaborative environmentalism in Japan', in H. Vinken, Y. Nishimura, B.L.J. White & M. Deguchi (Eds.), *Civic Engagement in Contemporary Japan* (Springer Science+Business Media).

Heggelund, G. (1993) *China's Environmental Crisis: The Battle of Sanxia* (Norwegian Institute of International Affairs).

Hildebrandt, T. (2013) *Social Organizations and the Authoritarian State in China* (Cambridge University Press).

Johnson, T. (2010) 'Environmentalism and Nimbyism in China: Promoting a Rules-Based Approach to Public Participation', *Environmental Politics* 19(3), 430–448.

Ju, C.B. & Shui-Yan Tang (2011) 'Path Dependence, Critical Junctures, and Political Contestation: The Developmental Trajectories of Environmental NGOs in South Korea', *Nonprofit and Voluntary Sector Quarterly* 40(6), 1048–1072.

Khagram, S. (2004) *Dams and Development: Transnational Struggles for Water and Power* (Cornell University Press).

Kim, E. (2009) 'The Limits of NGO-Government Relations in South Korea', *Asia Survey* 49(5), 873–894.

Ku, D.-W. (1996) 'The Structural Change of the Korean Environmental Movement', *Korea Journal of Population and Development* 25(1), 155–177.

Li, X. (2006) 'Environmental Concerns in China: Problems, Policies, and Global Implications', *International Social Science Review* 81(1/2), 43–57.

McKean, M.A. (1981) *Environmental Protest and Citizen Politics in Japan* (University of California Press).

Mason, R. (1999) 'Whither Japan's Environmental Movement? An Assessment of Problems and Prospects at the National Level', *Pacific Affairs* 72(2), 187–202.

O'Brien, K.J. (Ed.) (2008) *Popular Protest in China* (Harvard University Press).

Pekkanen, R. (2004) 'Japan: Social Capital without Advocacy', in M. Alagappa (Ed.), *Civil Society and Political Change in Asia: Expanding and Contracting Democratic Space* (Stanford University Press).

Peng, L. & Fengshi Wu (2013) 'Beyond NIMBYism: China's Environment Social Mobilization in 2012', China Policy Institute Blog, the University of Nottingham, April. Available at http://blogs.nottingham.ac.uk/chinapolicyinstitute/.

Ru, J. & Ortolano, L. (2009) 'Development of Citizen-Organized Environmental NGOs in China', *Voluntas: International Journal of Voluntary & Nonprofit Organizations* 20(2), 141–168.

Schreurs, M. (2002) 'Democratic Transition and Environmental Civil Society: Japan and South Korea', *The Good Society* 11(2), 57–64.

Sebag-Montefiore, C. (2012) 'The Great Smog of China', *New York Times*, Latitude-Views from around the World, 28 December. Available at http://latitude.blogs.nytimes.com/2012/12/28/the-great-smog-of-china.

Steiner, K., Krauss, E.S. & Flanagan, S.C. (Eds.) (1980) *Political Opposition and Local Politics in Japan* (Princeton University Press).

Stern, R.E. (2008) 'From Rhetoric to Action: Talking About Environmental Rights in Contemporary China', Paper presented at the ISA's 49th Annual Convention, San Francisco, 26 March.

Stern, R.E. (2011) 'From Dispute to Decision: Suing Polluters in China', *China Quarterly* 206, 294–312.

van Rooij, B. (2010) 'The People vs. Pollution: Understanding Citizen Action against Pollution in China', *Journal of Contemporary China* 19(63), 55–77.

Waley, P. (2005) 'Ruining and Restoring Rivers: The State and Civil Society in Japan', *Pacific Affairs* 78(2), 195–215.

Wong, K.-K. (2003) 'The Environmental Awareness of University Students in Beijing, China', *Journal of Contemporary China* 12(36), 519–536.

Wong, K.-K. (2005) 'Greening of the Chinese Mind: Environmentalism with Chinese Characteristics', *Asia-Pacific Review* 12(2), 39–57.

Wu, F. (2004) 'Environmental GONGO Autonomy: Unintended Consequences of State Strategies in China', *Journal of the Good Society (a PEGS journal)* 12(1), 35–45.

Wu, F. (2009) 'Environmental Activism and Civil Society Development in China', Harvard-Yenching Institute Working Paper Series. Available at http://www.harvard-yenching.org/sites/Harvardyenching.

org/files/featurefiles/WU%20Fengshi_Environmental%20Civil%20Society%20in%20China2.pdf.

Wu, F. (2013) 'Environmental Activism in Provincial China', *Journal of Environmental Policy & Planning (Special Issue: China's Local Environmental Politics)* 15(1), 89–108.

Wu, F. & Kinman Chan (2012) 'Graduated Control and Beyond: The Evolving Governance over Social Organizations in China', *China Perspectives* 3, 9–17.

Xu, Y. (2008) 'New Era of China's Civil Society' (in Chinese), *People's Net*, June 3. Available at http://politics.people.com.cn/GB/1026/7336201.

Yang, G. (2005) 'Environmental NGOs and Institutional Dynamics in China', *China Quarterly* 182, 46–66.

Zhu, J. & Ho, P. (2008) 'Not against the State, Just Protecting Residents' Interests: An Urban Movement in a Shanghai Neighborhood', in P. Ho & R. Edmonds (Eds.), *China's Embedded Activism* (Routledge).

Zusman, E. (2007) *What Makes Dragons and Tigers Brown? A Comparative Institutional Study of Air Pollution Regulation in East Asia* (University College of Los Angeles Press).

Notes

1   The authors wish to thank professors Graeme Lang (City University of Hong Kong) and Esook Yoon (Kwangwoon University, Korea) for their valuable comments.
2   Interviews with Guangzhou local activist, Mr. Ba Suo (internet pseudonym), and FoN staff, Mr. Zhang, in Guangzhou city, 20 February 2013.

8

Corporations and the environment in East Asia

Responsibilities and responses of businesses

Phillip Stalley

Introduction

Over the last three decades the nations of East Asia have made tremendous strides in promoting economic growth.[1] Between 1990 and 2005 the number of people living in poverty, defined as less than $1.25 per day, declined 70 percent (Preneuf & Kitt 2010). The Asian Development Bank estimates that during this period 521 million Asians escaped poverty (ADB 2011: 1). However, economic success has come at a steep environmental cost, particularly in the form of industrial pollution. The World Bank estimated that at the end of the 1980s Asian developing economies had already surpassed the OECD nations as the leading producer of industrial wastewater. By the turn of the century, Asia's water pollution, as measured by suspended solids, was more than four times the world average. Air pollution, as measured by particulate matter in the air, was twice the world average and more than five times that of the developed world (Rock & Angel 2007: 10, Table 1). In 2005, more than half of Asia's cities reported total suspended particulates and nitrous dioxide concentrations above World Health Organization standard limits (ADB 2011: 1). In 2002, there were more than 500,000 deaths per year attributed to outdoor air pollution in East and South Asia, roughly 60 percent of the global total (Zhang 2008: 3906, Table 1). In 2007, the World Bank stated that air pollution led to 750,000 premature deaths in China alone (World Bank 2007).

Economic growth has also contributed to a rapid rise of greenhouse gases (GHG) emissions coming from East Asia. Today the Asia-Pacific region is the fastest growing source of new GHG (ADB 2011: 8). In 1971, Asian countries accounted for just less than 21 percent of the world's energy-related carbon dioxide emissions. By 2004, that number was up to more than 50 percent (IPCC 2007: Table 7.1). This increase was driven largely by the developing countries of Asia whose energy-related carbon emissions increased more than 470 percent. In terms of total carbon emissions, in 1988 East Asia accounted for approximately 20 percent of global emissions. Twenty years later, that number was more than 33 percent. This increase was driven by a 200 percent increase in CO_2 emissions by the developing countries of East Asia.[2]

The situation is similar if one looks at other contributors to climate change (e.g., methane). In 1990, developing Asian countries accounted for 7.2 percent of the world's industrial emissions of non-CO_2 GHGs. By 2000, the figure had tripled to more than 21 percent and by 2030 it is projected to be more than 34 percent (IPCC 2007: Table 7.3). Although this emissions increase comes from a variety of sources such as deforestation and urbanization, carbon-intensive industrial production is a major driver of Asia's growing contribution to climate change.

Faced with resource constraints and a growing pollution problem, over the last two decades the governments of East Asia have dedicated considerable resources to achieve a better balance between economic growth and environmental protection (ADB 2011: 17–18). A critical component of this effort has been improving the environmental behavior of industry. Across Asia, governments have not only strengthened traditional regulations such as pollution limits, but also put forward a variety of market-oriented policies that promote public participation such as green label standards, environmental management certification schemes, corporate rating programs, and environmental information disclosure rules for businesses. The focus on corporate practices is sensible given that manufacturing comprises a comparatively high portion of East Asian economies. In 2010, manufacturing accounted for 26 percent of GDP in East Asia (measured as value added). This compares with a global average of less than 17 percent (World Bank 2011c). As of 2005, China alone made 70 percent of the world's kitchen appliances, while Asia produced 75 percent of footwear, 90 percent of sporting goods, and 80 percent of the world's ceramics (Welford 2005).

Given that East Asia is the factory of the world, it is hardly surprising that industrial manufacturing accounts for a high proportion of Asia's pollution and consumption of resources. In China, by the mid-1990s industrial pollution accounted for over 70 percent of the total. This included more than 70 percent of SO_2 emissions and 70 percent of chemical oxygen demand, a measure of organic water pollution (Wang & Wheeler 2005: 177). Today, industry accounts for 70 percent of China's primary energy use (CSEP 2012). More broadly, as of 2008 manufacturing contributed 28 percent of Asia's CO_2 emissions, which is roughly 50 percent higher than the world average of 19 percent (World Bank 2011a, 2011b). After agriculture, industry is the biggest consumer of water in Asia (ADB 2012: 27, Box 2.2). The Asian region accounts for approximately a quarter of the world's GDP, but half of the world's material consumption (ADB 2011: 7). In short, it is hardly controversial to assert that making industry more environmentally friendly is a critical element in the battle against pollution in East Asia.

How successful have East Asian nations been in their quest to green industry? How has business responded to the pressure coming not only from state regulators, but also markets and society? And what explains variance in the corporate environmental behavior across East Asian nations? This chapter answers these questions and in the process offers a sense of the recent achievements and remaining challenges in Asia's quest to green its corporations. I start with an overview of the evolving approach to corporate environmental management among East Asian firms. I argue that, although there is considerable divergence across and within countries, the general trend is towards greater corporate environmental responsibility. The second half of the chapter focuses on the variance in industrial environmental performance and describes the factors that explain the wide range of corporate behavior one sees in East Asia. I conclude with a brief summary, a caveat, and a suggestion for future research.

The greening of East Asian companies

Although East Asia still faces substantial challenges in reducing pollution, over the last two decades there has been clear progress in terms of corporate environmental management. A little more than a generation ago, environmental protection was barely on the agenda of Asian companies and non-compliance, where regulation even existed, was common well into the 1990s. Today, the situation has changed considerably. In general, shameless polluters represent the minority of firms and the majority of companies feature at least a modicum of environmental concern in their corporate policies. In the more economically developed countries such as Singapore, Korea, and Japan, leading firms display environmental practices and technologies on par with advanced firms in OECD countries. Schreurs' (2004: 88) portrayal of Japan captures this evolution well as she states: 'In the 1960s, Japan was arguably among the most polluted countries in the world as a result of rapid industrialization and minimal pollution control. In contrast, today Japan is providing technological and scientific expertise for pollution control to developing countries.'

The progress of Asian firms is evident in their increasing participating in various international sustainability initiatives. For instance, Figure 8.1 shows the dramatic increase in ISO certification by Asian companies. Under the ISO 14001 program a licensed third party certifies that a company's environmental management system (EMS) is in accordance with guidelines established by the International Standards Organization (ISO). Put another way, ISO certification is a signal that a company's EMS is on par with international best practices. In 1999, Asian firms accounted for approximately 30 percent of global ISO 14001 certifications. By 2010, the figure was almost 50 percent. China, with almost 75,000 certified companies, is the world's leader, while Japan ranked second and Korea sixth in terms of the number of companies certified (ISO 2010).

East Asian firms have also increased significantly their participation in the Global Reporting Initiative (GRI). Similar to ISO 14001, GRI is a voluntary program focusing on a company's environmental management program. Under the GRI, companies report environmental practices and policies according to a common framework established by an international nonprofit organization working in collaboration with United Nations Environment Program. The aim is to encourage transparency, comparability, and accountability in corporate environmental governance. As seen in Table 8.1, in 2005, East Asia comprised 9.3 percent of all firms utilizing the GRI. Just five years later, the percentage had doubled to 18.6 percent. China went from having only seven participating companies to more than 160; the number of Japanese firms increased from 19 to 197.

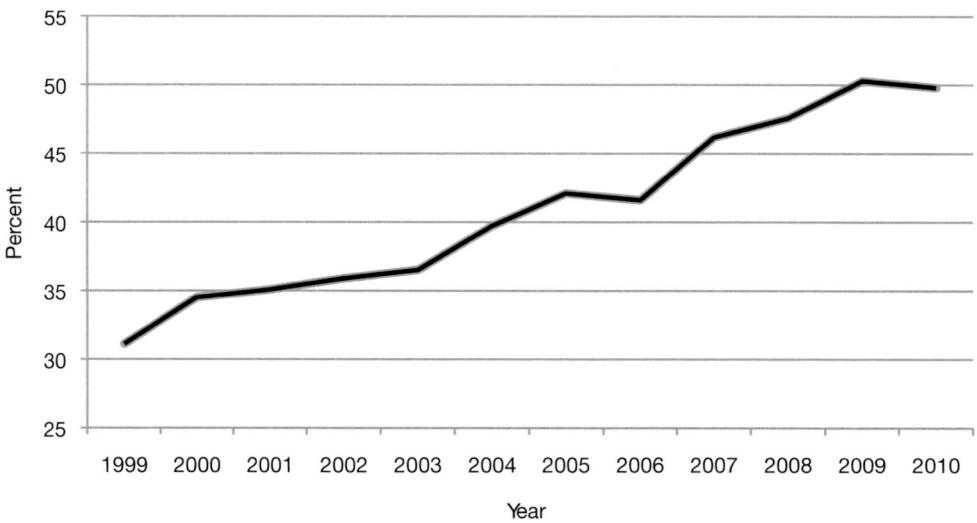

Figure 8.1 Asia's percentage of global ISO 14001 certifications

Source: ISO (2010). ISO uses the term 'Far East' rather than 'Asia' in its data tables.

Table 8.1 Number of East Asian companies participating in the Global Reporting Initiative, 2005 compared with 2010

Country	2005	2010
China	7	161
Indonesia	4	3
Japan	19	197
Malaysia	2	9
Philippines	1	13
ROK	16	97
Singapore	0	21
Thailand	0	25
Other		2
Total Asia	49	528
Asia: percent of global total	9.3%	18.6%

Source: GRI (2012).

In addition to GRI, more than 50 companies from East Asia are listed on the Dow Jones Sustainability World Index (DJSWI). The DJSWI aims to assist investors who want to put their money in companies that have sustainable and ethical corporate governance practices. Dow Jones evaluates the sustainability performance of the 2,500 largest companies listed on the Dow Jones. Approximately one-third of the evaluation is based on a company's environmental practices and performance. As seen in Figure 8.2, in 2011 East Asian companies comprised 15 percent of firms listed on the DJSWI.

Domestically, East Asian governments have been busy building green label programs, which provide a set of standards that allow companies to market their products as superior in terms of the impact on environmental and/or human health. As Table 8.2 shows, Asian countries in general, and the countries of Northeast Asia in particular, have developed a large number of

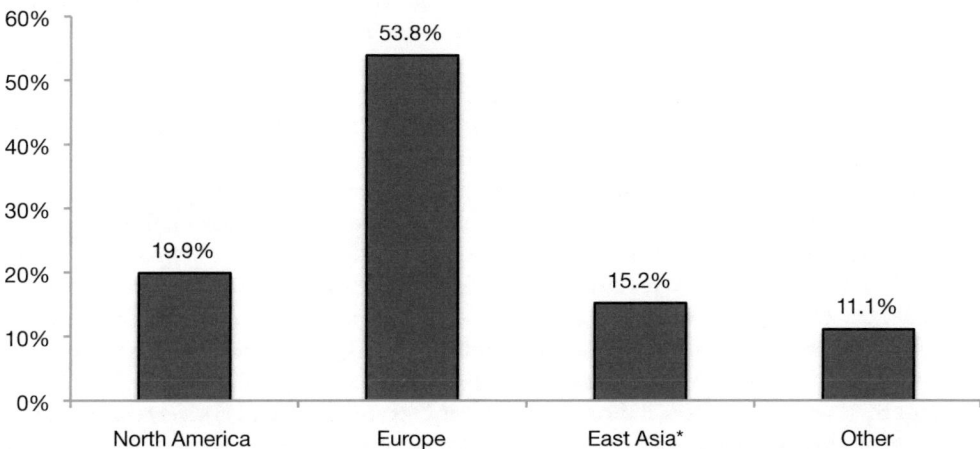

Figure 8.2 Regional distribution of companies listed on the Dow Jones Sustainability World Index, 2011

Source: Dow Jones, data provided directly to the author. * Includes: China, Hong Kong, Japan, Korea, Singapore, Taiwan, and Thailand.

Table 8.2 Eco-labelling programmes in East Asia, 2010

Country	Number of standards	Number of products and services covered
China	81	40,000
Hong Kong	76	573
Taiwan	112	6,000
Indonesia	n/a	n/a
Japan	47	4,904
Korea	148	8,012
Malaysia	23	7
Philippines	36	20
Singapore	66	936
Thailand	48	461

Source: GEN (2010).

standards covering a wide range of products and services. China, Taiwan, and Japan have eco-label standards for more than 50,000 products.

The discussion in the previous paragraphs indicates there is a significant and growing number of prominent firms in East Asia that demonstrate a strong commitment to environmental management. While this is an important development, it should be noted that the companies participating in ISO 14000, GRI, Dow Jones Sustainability Index, as well as those selling green label products, represent a small minority of all firms in the region.[3] The majority of firms in East Asia demonstrate a lower level of ambition and appear generally satisfied with basic compliance. This is evident in several surveys of corporate environmental management. For instance, Lee and Ball (2003) conducted detailed case studies of 15 Korean chemical companies and found that eight of the 15 firms fall into what they term a 'defensive compliance' cluster, while another three are 'environmentally sensitive'. Both categories consist of firms that comply with the law of the land, but do not consistently invest resources to promote environmental protection. Another three firms qualify as genuinely green firms – 'proactive catalysts' in the authors' terminology – while the remaining firm is an environmental laggard. Focusing on the Korean construction sector, Park and Ahn (2012) observe a similar distribution of corporate environmental behavior. Of the 99 firms they examine, 87 are deemed to have 'passive' environmental strategies with modest environmental budgets and little effort to incorporate environmental considerations into overall corporate strategy.

Scholars focusing on other countries present similar findings. For example, Stalley (2010: 151–154) compiles data on the environmental behavior of more than 500 manufacturing companies in three cities of Jiangsu province, China. Approximately 11 percent of his sample shows strong, 'beyond compliance' environmental management practices, while another 11 percent have a history of serious regulatory fines and virtually no internal environmental management systems. The remaining 78 percent are divided roughly equally between those that either have no history of environmental violations or have received a modest penalty.

What these and other studies indicate is that across East Asia there is a relatively small fraction of firms that possess little to no sense of environmental responsibility, a perhaps slightly larger percentage that demonstrate a deep commitment to corporate environmental management, and a large middle that meets basic regulatory requirements but adopts a largely passive environmental strategy. The firms occupying this middle territory may make investments to improve efficiency and reduce waste (e.g., acquiring new technology), but the investments are aimed primarily at lowering costs and environmental improvement is an ancillary consideration. These middle-range companies establish basic environmental management systems and generally observe local regulations in order to avoid legal action or loss of customer base. But they do not go 'beyond compliance' by having their EMS externally certified, adopting environmental quality standards where local regulations are inadequate or nonexistent, or vigorously communicating and promoting their environmental commitment to their employees, suppliers, and local community. However, to say most firms adopt a passive or reactive basic compliance strategy should not detract from the fact that the overall trend has been a greening of the firm. As studies have shown (Angel & Rock 2009), there has been a steady tightening of industrial environmental standards across the region. While this progress has not been universal across all jurisdictions and there remain considerable gaps in enforcement, particularly in the developing countries such as China, overall the trend is toward more stringent environmental governance. Thus, even if most firms are content merely to comply, basic adherence with local environmental standards in 2013 equates with stronger environmental practices than it did in 1983, 1993, or even 2003.

While the general trend is toward a greener corporate Asia, there is nonetheless a wide divergence of corporate environmental performance within the countries of East Asia, let alone across the region. Indeed, it is variance in firm behavior that drives much of the academic research on corporate environmental performance as scholars seek to understand the factors that explain divergent practices. This variance is evident in a number of ways. For instance, Baughn, Bodie, and McIntosh (2007: 192) review a set of studies that demonstrate companies from South Korea and Japan are far more likely to have corporate responsibility policies (CSR) and/or conduct sustainable development reporting than their counterparts in Thailand, Malaysia, and Singapore. In the case of Japan, more than 60 percent of companies engage in CSR.

Variance in corporate behavior is also seen in the aforementioned Dow Jones Sustainability Index where there is a notable lack of firms from Mainland China. Only one Chinese company is listed on the 2011 World Sustainability Index. (In 2011, 59 Chinese companies were invited by Dow Jones to submit information for evaluation and possible inclusion on the Global Index.) By contrast, there are 25 Japanese and 16 South Korean firms listed; these two countries alone account for almost 60 percent of all East Asian firms listed on the Global Index. Figure 8.3,

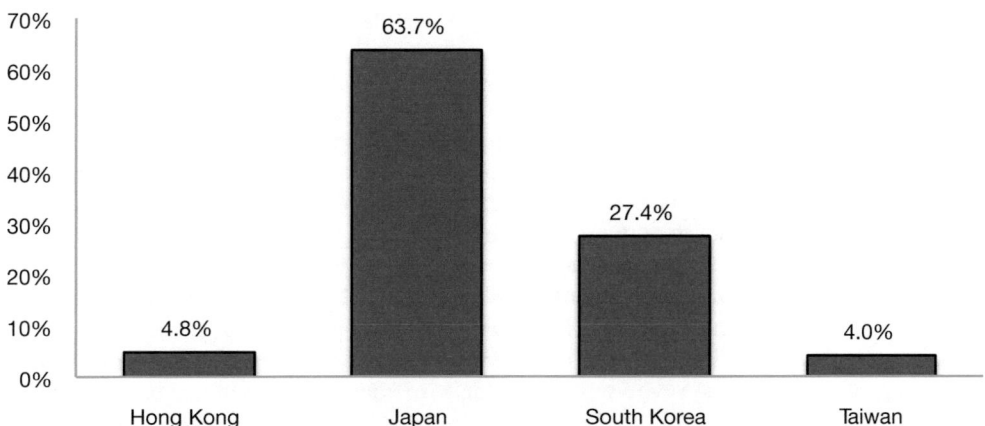

Figure 8.3 Distribution of companies listed on the Dow Jones Sustainability Asia-Pacific Index, 2011
Source: Dow Jones, data provided directly to the author.

which displays a country breakdown of the Dow Jones Sustainability Asia-Pacific Index, further demonstrates the high degree of national variation. Again, there are no companies from Mainland China and South Korea and Japan comprise the overwhelming majority of listed firms (more than 90 percent).

Drivers of corporate environmental behavior in East Asia

The previous section pointed out that, although the broad trend is toward greater corporate environmental responsibility, there is substantial disparity both within and across the countries of East Asia. What accounts for this variance? Put another way, what factors exert the greatest influence on a firm's environmental strategy? Scholars with an interest in corporate environmental responsibility have dedicated a great deal of effort to answering questions such as these. Drawing on their work, we can divide the drivers of East Asian firms' environmental performance into two broad categories: factors related to the external environment and those stemming from the internal characteristics of the firm itself.

Features of the institutional environment

Enhanced regulatory pressure

It is virtually axiomatic that stronger environmental governance positively influences a firm's environmental strategy. Schreurs (2004: 93) asserts that the about-face of Japanese industry, from notorious despoiler of the environment to ecological leader, was in large part a function of the flurry of multilateral efforts in the late 1980s and early 1990s aimed at addressing problems like ozone depletion, climate change, and biodiversity loss. While international negotiations may help raise corporate awareness and enhance the saliency of environmental protection among government regulators and community stakeholders, ultimately domestic environmental governance has the most pronounced effect on company behavior. Governance includes not only the ratcheting up of standards, but also enhanced supervision of business as well as support for voluntary programs that provide incentives for corporate participation.

Several studies have found that, where there is greater regulatory action and capacity, East Asian firms respond. Wang and Wheeler (2005: 190) gather data on the more than 3,000 firms in China and find that, the greater the number of regulators per firm, the higher the pollution levies collected (both water and air). They further discover that higher levy rates are associated with lower emissions. Aden, Kyu-hong, and Rock (1999: 1205–1206) attribute South Korea's success in controlling industrial pollution in the 1990s directly to a dramatic increase in regulatory activity. Between 1988 and 1993, wastewater inspections tripled; inspections of facilities emitting air pollution almost doubled. Wang et al. (2011) show that an increased strictness in discharge standards in Jiangsu (China) led to industrial efficiency improvements in water usage and for most pollution outputs, although the impact on energy efficiency was less positive.

This is not to say that a command and control regime is the only way for governments to promote stronger corporate practices. East Asian nations have been quite willing to experiment with environmental governance approaches that rely on market incentives or public participation. For instance, several East Asian countries have achieved modest success with 'shaming lists' that rank and publish companies' environmental practices in order to bring external pressure on the firms to improve their behavior. This includes Korea, China, Philippines, and Indonesia (Hua et al. 2004; Liu et al. 2012; Zhang 2008). Others have made use of voluntary public–private programs. Matsuno (2007) shows that in Japan more than 30,000 voluntary pollution control agreements (PCAs) have been signed between businesses and local governments. These PCAs typically require a company to accept more stringent standards than the existing national or local regulations. Mori and Welch (2008: 430) explore the factors that lead Japanese companies to pursue ISO certification. Conducting a survey of more than 1,700 businesses in four sectors, they find that the existence of a previous local agreement with the government was highly associated with ISO certification. They report that over 50 percent of Japanese companies that certified early reported having a previous local agreement.

It should be pointed out that the strength of environmental governance varies noticeably across Asia and can be weak to non-existent in areas that are relatively poor. China, in particular, has a well-established gap between its environmental standards as they exist on paper and implementation on the ground, especially as one moves away from urban, coastal areas (Wang et al. 2008). One survey of more than 500 Chinese cities revealed that only 23 percent of factories properly treat sewage before disposing of it (Economy 2007). Kasim (2007: 683) studies the environmental practices of the hotel industry in Malaysia and argues that the 'rather complacent business attitude towards environmental responsibility' is in part a function of a lack of enforcement threat. In short, despite advances in environmental governance there are many jurisdictions in Asia where regulation is lax and companies possess a relatively free hand when it comes to pollution control.

Stakeholder and community pressure

Along with attention from government regulators, pressure from the local community has also proved to be a significant influence on the environmental behavior of industry in East Asia. This pressure can take a variety of forms such as complaints through officially designated channels, environmental litigation, mass protests, customer demands for green products, or organized social movements involving domestic and international NGOs. Citizen demand for cleaner industry is often facilitated by the media, which plays an increasingly important role in spotlighting industry practices even in authoritarian settings like China. Citizen activism not only sways businesses directly, it also can strengthen the hand of environmental regulators vis-à-vis local companies and other government agencies that may be more prone to protecting business interests. Because community pressure generally increases concomitantly with income, East Asia's rapid economic growth is a major contributor to better corporate environmental management. Not surprisingly, studies have found that, where income per GDP is higher, companies exhibit stronger environmental management practices (Baughn, Bodie & McIntosh 2007).

The emerging importance of community pressure is evident across East Asia. China has in recent years witnessed a number of NIMBY (not-in-my-backyard) incidents that have been driven by concerns about industrial pollution. In 2012 citizens from Ningbo (Zhejiang) donned chemical masks and held banners saying 'we want to survive' to protest the expansion of a petrochemical plant (Jacobs 2012). A few months earlier protestors in Shifang (Sichuan) clashed with the local government over plans to construct a copper alloy plant (BBC 2012). A few weeks after the Shifang protests, there was another uprising in Qidong (Jiangsu) about a planned pipeline that citizens feared would pollute local fisheries. In both cases, the demonstrations led to the cancellation of the projects. In 2011 protests in Shanghai led to the closing of a solar panel plant and demonstrations in Dalian resulted in the shuttering and relocation of a plant manufacturing paraxylene, a chemical used to make polyester products (*Economist* 2011). These successes followed in the wake of similar, high-profile protests aimed at industrial polluters in Xiamen and Chengdu (Wong 2008). Zeng and Easton (2011: 149) show how environmental NGOs played a crucial role in inducing an Indonesian pulp and paper company to clean up its practices in China. Child and Tsai (2005: 110), focusing on rising pressure on the environmental practices of multinationals in both China and Taiwan, conclude:

[P]oliticians, NGOs and the concerned public played a more prominent role than the formal regulators. In a green era, demonstrations over environmental protection have become hot news for the mass media. It is arguable that informal regulators have become more powerful and influential than formal government regulators.

Mori and Welch (2008: 435) examine drivers of ISO certification in Japanese companies and conclude that 'broad social pressure was one of the most important reasons for their pursuit of ISO 14001 certification'. Nishitani (2009) looks at more than 400 firms listed on the Tokyo Stock Exchange and demonstrates that a variety of stakeholders exert leverage on corporate environmental management. Companies that manufacture end products directly for consumers, as opposed to producing components for other manufacturers, tend to exhibit stronger environmental management. Aden, Kyu-hong, and Rock (1999) conducted a survey of 99 Korean firms in the textile, plastics, and petrochemical industries that shows community pressure induces manufacturers to increase pollution abatement expenditures. One study of Korea reported that companies indicated it is harder to 'satisfy the demands of the local community than to comply with regulations' (Lee & Rhee 2005: 393).

Outside Northeast Asia, one sees similar evidence of growing community pressure. Sonnenfeld (2002: 1) explores the impact of social movements in Indonesia, Malaysia, and Thailand and finds that they were 'instrumental in the transformation of the pulp industry', although their influence was much more apparent in Indonesia and Thailand than Malaysia. O'Rourke (2004) shows how, even in the absence of a vibrant NGO movement or a free press, citizens in Vietnam were able to influence the practices of local industry. Rock (2002) argues that Taiwan's growing protest movement in the early 1990s compelled the government to increase its efforts at industrial environmental regulation. Kasim (2007: 686) also cites the importance of community pressure, arguing that the *lack of* NGOs focusing on the Malaysian tourism industry, and more broadly the 'general lack of community concern', translates into business complacency since there are no reputational benefits from investing in stronger environmental management.

Kasim's study serves as a reminder that, although citizen activism and media attention have generally increased across the region, there are still many areas where citizens are unwilling, or more likely unable, to voice their concerns about industrial pollution. As is the case with environmental governance more broadly, the influence of citizen activity on corporate environmental behavior varies widely. Moreover, citizens' capacity to influence corporate practices depends on a variety of exogenous factors (e.g., strength of the legal system). This observation is perhaps nowhere more relevant than in China where, despite the large number of environmental protests, the impact of citizens and environmental activists on environmental protection remains uncertain and highly contingent on local political and economic factors (Grano 2012; Mertha 2008: 65–93; Van Rooij et al. 2012).

Features of the firm

A corporation's environmental practices are not only a function of the institutional environment in which it operates. As might be expected, a number of internal features of a company influence its environmental responsibility. Two critical factors are the company's size and profitability. For a variety of reasons large, profitable companies tend to demonstrate a higher degree of environmental responsibility. First, large firms are more likely to appear on the radar screens of regulators, consumers, and the local community. As the more likely targets of enforcement action, they typically pay greater attention to environmental protection (Nishitani 2009). In addition, creating and sustaining an environmental management system and associated company culture requires significant initial investment and long-term maintenance costs. Small firms, companies with high debt, or those operating at a loss, are typically not well positioned to make such investments. Cole, Elliott, and Shimamoto (2006) explore the environmental performance of more than 800 Japanese firms. They find better environmental performance among firms that are larger, have lower debt ratios, and higher total factor productivity. Similar findings about the importance of size, profitability, and/or productivity are offered by scholars focusing on Japan (Mori & Welch 2008; Nishitani 2009), Korea (Aden, Kyu-hong & Rock 1999; Cho & Voss 2012), and China (Studer, Welford & Hills 2006).

If it is relatively straightforward that large, profitable firms demonstrate greater environmental responsibility, the impact of other factors is less clear. Intuitively, one would expect that individual leadership and company culture are important. Studies from outside Asia have found that a company whose top leaders possess a strong sense of environmental responsibility can transform a company's culture, which typically translates into stronger pollution abatement practices and proactive environmental management (Cho & Voss 2012). A corporate leader's sense of responsibility toward the environment and society may itself depend on a number of factors such as the particular leader's educational background, experience, and training. Data on variables such as management attitude and background are relatively hard to gather and measure and are less studied than more objective and observable characteristics such as firm profitability. Those studies that have explored management attitudes, however, have found mixed support. Mori and Welch's (2008: 433–434) interviews with firms seeking ISO certification in Japan, for instance, reveal that only a tiny fraction of companies were driven by upper management's notions of environmental consciousness.

There are also questions about the impact of ownership structure. Do private firms exhibit stronger environmental management? How do the environmental practices of foreign-invested companies compare with domestic ones? This is a particularly pertinent issue in East Asia where foreign investment is widespread and where in some countries (e.g., China) state-ownership is common. The findings here are mixed. Some have argued that private companies, because they are less able to draw on a close relationship with the government than are state-owned or collectively-owned enterprises, are pushed towards environmental compliance. Chun (2009), for instance, looks at corporate environmentalism in Chinese energy companies and finds that state ownership is associated with weaker environmental attitudes among employees. Wang and Jin (2007) gather data on 800 industrial firms in three Chinese cities and offer similar findings that state-owned enterprises display poor environmental behavior when compared to private firms. They argue that foreign-invested companies have the strongest environmental performance. Liu et al. (2012) support this conclusion as they show that wholly foreign-owned firms in China respond more proactively to an environmental ratings scheme than do private firms. Yu and Wong (2011), by contrast, find little evidence of corporate environmentalism among foreign-invested enterprises in Southern China. Aden, Kyu-hong, and Rock (1999) also see less benefit to foreign ownership. They find that domestic firms in Korea have higher pollution abatement expenditures than do foreign-invested companies. Cole, Elliott, and Shimamoto (2006) find that membership in a *keiretsu* has a negative impact on environmental performance in Japan. Their explanation is that membership potentially provides the firm with greater resources to resist government regulation.

Another area where there remains some debate is the influence of a firm's integration into the global economy. Like ownership structure, the impact on corporate environmental responsibility of participation in the global economy is particularly pertinent for East Asia where so many nations have promoted growth through a combination of producing for export and attracting foreign direct investment. In Taiwan, for instance, exports account for 60 percent of

GDP (Teng 2010: 383). In the Pearl River Delta region of China, between 10 and 40 percent of air pollution is related to export-related activities (Yu & Wong 2011: 80). Nationally, it is estimated that exported products account for 30–35 percent of China's emissions (Lin & Yang 2012).

Some have expressed concern that, as developed countries shift their manufacturing bases to take advantage of cheap labor and low environmental standards, portions of East Asia are turning into pollution havens (Cole, Elliott & Okubo 2010; Mukhopadhyay 2009). In the 1970s and 1980s, Japan achieved a degree of notoriety in this regard as its companies, as well as the government's official development assistance, were associated with environmental destruction in SE Asia (e.g., tropical deforestation) (Schreurs 2004). More recently, China's quenchless thirst for natural resources, combined with Chinese companies' apparent lack of attention to environmental sustainability in their overseas ventures, has drawn criticism and renewed concern about the potential for foreign investors to exert a negative influence on environmental protection (Hagglund 2009; Turner & Wu 2012).

While there are certainly plenty of examples of foreign companies engaging in unscrupulous environmental practices in East Asia, particularly in the extractive industries, studies have by and large found that concerns about an East Asian pollution haven are unwarranted. Integration into the global economy is associated with better corporate environmental performance and stronger regulation. Asian companies that are foreign-invested, that supply to multinationals, or produce for export face a range of pressures and opportunities to improve environmental performance. Participation in the global economy can provide access to advanced technologies, which is typically more energy efficient and environmentally friendly. Asian companies producing for export may also improve their EMS, for instance by pursuing ISO certification, in order to enhance their reputation and assure access to overseas markets. Asian firms selling to foreign multinationals may also be subject to 'green supply chain' pressures as when Wal-Mart announced it would work with its top 200 Chinese suppliers to improve energy efficiency by 20 percent over a five-year period.

Park and Ahn (2012: 164) offer empirical support for the benefits of internationalization as they find that Korean construction firms that engaged in global activity became 'more aggressive in responding to environmental management . . . as they realized the necessity of incorporating strategic environmental management policies'. Rock and Angel (2007: 12) introduce two case studies involving Thai manufacturers and conclude that 'technological learning and interaction with OECD multinationals can contribute to substantial environmental technology effects'. Cole, Elliott, and Shimamoto (2006) construct a dataset on the environmental management practices of approximately 400 Japanese manufacturing firms. Drawing on a survey of firm managers, they find that variables associated with globalization, such as whether a Japanese firm invests abroad or produces for export, are positively associated with 'beyond compliance' environmental management systems, the use of environmental accounting, and the release of an environmental disclosure statement. They conclude that participation in multinational activity is making firms increasingly aware of their environmental obligations. Similarly, Nishitani (2009) finds that the export ratio of a Japanese firm is positively associated with ISO 14001 adoption. Firms that sell to foreign customers also exhibited better environmental management. Focusing on China, Stalley (2009) argues that participation in the global economy has a bifurcated effect. It provides incentives and opportunities to improve environmental management to a small percentage of Chinese exporters, i.e., those that sell to big foreign multinational corporations or export to Europe. But for other companies, particularly those that compete on price or serve as secondary subcontractors, globalization provides a motivation to skirt environmental regulation. A recent study by five environmental NGOs of the textile industry in China lends support to this notion. It found that some foreign companies worked closely with Chinese contractors on improving environmental practices. However, the suppliers for many, including well-known foreign brands such as Disney, Polo Ralph Lauren, and Marks & Spencer, had a history of pollution violations or complaints from the local community (Jun et al. 2012).

As Stalley's study indicates, the impact of integration into the global economy is not identical across all countries and firms and is dependent on other factors. The impact of internationalization on an Asian firm's environmental management strategy depends in large part on the firm's particular customers and markets. Interaction with customers in developed countries, which have rigorous environmental standards, tends to impact positively an Asian firm's environmental performance. Cho and Voss (2012) argue that Korean companies whose customers are Western European or North American are likely to place greater emphasis on environmental management. Zeng and Easton (2011) conduct a survey of 50 business executives from several Chinese cities and find that selling to European and Japanese customers is associated with better corporate environmental performance. This is particularly the case when the Chinese firm exports a high percentage of its product to Japan. They also find that foreign ownership contributes positively to environmental performance. Dean, Lovely, and Wang (2009) wade into the pollution haven debate by addressing the question of whether foreign firms seek out jurisdictions with weak environmental standards. Looking at almost 3,000 joint venture projects in the 1990s, they find that companies in pollution-intensive sectors coming from Taiwan, Hong Kong, or Macau are drawn by lax environmental standards. However, companies coming from other parts of the world do not appear attracted by low pollution levies. The importance of firm sector is seen in other studies as well. In their survey, Cho and Voss (2012) point out that the impact of environmental regulation varies across sectors. Sectors that are pollution intensive (e.g., chemicals), and thus draw increased regulatory attention, tend to have firms with the most elaborate and proactive environmental management systems. Thus, both the country of origin of an Asian firm's customers and partners, as well as the company's own industrial sector, influence its environmental management practices.

To summarize, there are a variety of firm-specific characteristics that influence where an East Asian company falls on the spectrum from environmental laggard to eco-friendly leader. All things equal, one would expect that a large profitable multinational company that produces for multiple markets in the developed world or that has a close relationship with other well-known multinationals will display the strongest environmental behavior. One can expect a much lower level of environmental responsibility in small companies that fly under the radar of regulators, compete on price, operate in lax regulatory settings (e.g., rural areas), and/or produce for the domestic market. Likewise, those whose ownership structure inures them from government and stakeholder pressure or that are cut off from the global economy typically have worse environmental performance.

Conclusion

This chapter has provided a survey of the broad trends in corporate environmental performance, as well as the factors influencing the environmental strategy of business in East Asia. It has argued that a combination of more stringent environmental governance, community pressure, and participation in the global economy has resulted in the greening of business in East Asia. This is particularly the case in the most developed nations of East Asia. The about-face of Japanese companies in terms of environmental responsibility is cause for optimism for all of Asia.

However, to say business *is greening* is not to say it *has greened*. There remains much to be done in improving the environmental responsibility of Asian firms and many obstacles to overcome. The most notable of these obstacles, particularly in the developing countries of Asia, is inadequate environmental governance. While Asian governments have demonstrated an impressive readiness to experiment with market-based and public participation-based environmental initiatives, many jurisdictions still fail to enforce existing pollution standards. Whether this is a lack of capacity among government regulators, a lack of political will, or some combination of both depends on the particular jurisdiction, but the fact remains industrial environmental governance is not as strong as it needs to be to confront the dual challenges posed by resource scarcity and pollution.

Much of the analysis presented in this chapter has to be interpreted with a measure of caution. Discerning trends in corporate environmental performance in developing countries is an inherently difficult endeavor. For obvious reasons, heavily polluting companies or those with a history of non-compliance prefer to remain obscure, as do the regulators that often shield them from greater scrutiny. This poses a formidable challenge for researchers in their quest to gather reliable data that capture trends in corporate environmental responsibility.

This challenge is made all more acute by the fact that there is often a lack of institutional capacity to gather information and enforce environmental law in developing countries. Thus, one cannot necessarily draw on reliable data produced by government regulators. While scholars of American environmental politics, for instance, can utilize the Toxic Release Inventory to get a sense of firms' environmental performance, there are few such resources available to scholars interested in Asian countries. For that reason, it is hardly surprising that there are far more studies about corporate environmental *management* than about corporate environmental *performance*. We know a great deal more about the extent to which Asian companies are reorienting their internal structures to incorporate environmental issues than we do about firms' actual pollution emissions and propensity to comply with environmental regulation. While there are several studies from the industrialized world that indicate a firm's environmental management practices are, in fact, correlated with its overall environmental performance in terms of pollution emissions, some studies have cast doubt on the notion that internal management leads directly to actual environmental benefits (Mori & Welch 2008: 436).

Along the same lines, much of the literature on business and the environment tends to focus on firms that go beyond compliance and to rely on firm self-reporting via surveys of top executives. Inevitably, this raises questions about whether we are really seeing the whole picture or, like the drunk man who drops his keys in the dark alley but looks for them under the street lamp where he can see, observing only those firms that are willing to step out of the shadows. This is particularly the case in developing Asian countries where enforcement is weaker, industry scattered, and there is an abundance of small and medium-sized enterprises that can fly under the radar of both regulators and scholars. It is worth bearing in mind that few scholars are able to directly observe the outcome of greatest interest. Going forward, scholars would be well served by gathering data that directly capture outcomes in corporate environmental behavior (e.g., abatement expenditures, pollution emissions, regulatory history), rather than rely on self-reported descriptions of internal environmental management structures and company practices.

Bibliography

ADB (2011) *Environment Program: Greening Growth in Asia and the Pacific* (Asian Development Bank).
ADB (2012) *Green Growth, Resources, and Resilience: Environmental Sustainability in Asia and the Pacific* (United Nations and Asian Development Bank).
Aden, J., Kyu-hong, A. & Rock, M.T. (1999) 'What is Driving the Pollution Abatement Expenditure Behavior of Manufacturing Plants in Korea?', *World Development* 27(7), 1203–1214.
Angel, D. & Rock, M.T. (2009) 'Environmental Rationalities and the Development State in East Asia:Prospects for a Sustainability Transition', *Technological Forecasting & Social Change* 76, 229–240.

Baughn, C.C., Bodie, N.L. & McIntosh, J.C. (2007) 'Corporate Social and Environmental Responsibility in Asian Countries and Other Geographical Regions', *Corporate Social Responsibility and Environmental Management* 14, 189–205.

BBC (2012) 'China Factory Construction Halted Amid Violent Protests', *BBC News*, 3 July.

Child, J. & Tsai, T. (2005) 'The Dynamic Between Firms' Environmental Strategies and Institutional Constraints in Emerging Economies: Evidence from China and Taiwan', *Journal of Management Studies* 42(1), 95–125.

Cho, E.H. & Voss, H. (2012) 'Determinants of International Environmental Strategies of Korean Firms: An Explorative Case-Study Approach', *Asian Business & Management* 10(3), 357–380.

Chun, R. (2009) 'Ethical Values and Environmentalism in China: Comparing Employees from State-Owned and Private Firms', *Journal of Business Ethics* 84, 341–348.

Cole, M.A., Elliott, R.J.R. & Shimamoto, K. (2006) 'Globalization, Firm-level Characteristics and Environmental Management: A Study of Japan', *Ecological Economics* 59, 312–323.

Cole, M.A., Elliott, R.J.R. & Okubo, T. (2010) 'Trade, Environmental Regulations and Industrial Mobility: An Industry-Level Study', *Ecological Economics* 69, 1995–2002.

CSEP (2012) *Fact Sheet: China Emerging as New Leader in Clean Energy Policies*. Available at http://www.efchina.org.

Dean, J.M., Lovely, M.E. & Wang, H. (2009) 'Are Foreign Investors Attracted to Weak Environmental Regulations? Evaluating the Evidence from China', *Journal of Development Economics* 90, 1–13.

Economist (2011) 'Poison Protests', *Economist*, 20 August.

Economy, E. (2007) 'The Great Leap Backward?', *Foreign Affairs* 86(5), 38–59.

GEN (2010) *Global Ecolabelling Network 2010 Annual Report*. Available at http://www.globalecolabelling.net/docs/annual_reports/2010annual_report.pdf

Grano, S.A. (2012) 'Green Activism in Red China: The Role of Shanghai's ENGOs in Influencing Environmental Politics', *Journal of Civil Society* 8(1), 39–61.

GRI (2012) *Global Report Inititative, Sustainability Disclosure Database*. Available at http://database.globalreporting.org/.

Hagglund, D. (2009) 'In It for the Long Term? Governance and Learning Among Chinese Investors in Zambia's Copper Sector', *China Quarterly* 199, 627–646.

Hua, W., Bi, J., Wheeler, D., Wang, J., Cao, D., Lu, G. et al. (2004) 'Environmental Performance Rating and Disclosure: China's GreenWatch Program', *Journal of Environmental Management* 71(2), 123–133.

IPCC (2007) *Intergovernmental Panel on Climate Change, Fourth Assessment Report: Climate Change*. Available at http://www.ipcc.ch/publications_and_data/ar4/wg3/en/ch7s7-1-3.html.

ISO (2010) *ISO Survey, 2010*. Available at http://www.iso.org/iso/iso-survey2010.pdf.

Jacobs, A. (2012) 'Protests in China Against Refinery Reach Third Day', *New York Times*, 28 October.

Jun, M., Jingjing, W., Collins, M., Malei, W., Orlins, S. & Jie, L. (2012) *Sustainable Apparel's Critical Blind Spot*, October. Available at http://www.ipe.org.cn/Upload/Report-Textiles-Phase-II-EN.pdf.

Kasim, A. (2007) 'Corporate Environmentalism in the Hotel Sector: Evidence of Drivers and Barriers in Penang, Malaysia', *Journal of Sustainable Tourism* 15(6), 680–699.

Lee, S.-Y. & Rhee, S.-K. (2005) 'From End-of-Pipe Technology Towards Pollution Preventive Approach: The Evolution of Corporate Environmentalism in Korea', *Journal of Cleaner Production* 13, 387–395.

Lee, K.-H. & Ball, R. (2003) 'Achieving Sustainable Corporate Competitiveness: Strategic Link Between Top Management's (Green) Commitment and Corporate Environmental Strategy', *Greener Management International* 44, 89–104.

Lin, A. & Yang, F. (2012) *China's Carbon Tax is Very Real, China Dialogue*, 12 January. Available at http://www.chinadialogue.net/article/show/single/en/4742-China-s-carbon-tax-is-very-real.

Liu, B., Yu, Q., Bi, J., Zhang, B., Ge, J. & Bu, M. (2012) 'A Study on the Short-Term and Long-Term Corporate Responses to the Greenwatch Program: Evidence From Jiangsu, China', *Journal of Cleaner Production* 24, 132–140.

Matsuno, Y. (2007) 'Pollution Control Agreements in Japan: Conditions for their Success', *Environmental Economics and Policy Studies* 8, 103–141.

Mertha, A. (2008) *China's Water Warriors: Citizen Action and Policy Change* (Columbia University Press).

Mori, Y. & Welch, E.W. (2008) 'The ISO 14001 Environmental Management Standard in Japan: Results From a National Survey of Facilities in Four Industries', *Journal of Environmental Planning and Management* 51(3), 421–445.

Mukhopadhyay, K. (2009) 'Trade and the Environment: Implications for Climate Change', *Decision* 36(3), 83–102.

Nishitani, K. (2009) 'An Empirical Study of the Initial Adoption of ISO 14001 in Japanese Manufacturing Firms', *Ecological Economics* 68, 669–679.

O'Rourke, D. (2004) *Community-Driven Regulation: Balancing Development and the Environment in Vietnam* (MIT Press).

Park, J.-H. & Ahn, Y.-G. (2012) 'Strategic Environmental Management of Korean Construction Industry in the Context of Typology Models', *Journal of Cleaner Production* 23, 158–166.

Preneuf, F.D & Kitt, F. (2010) *A Climate For Change in East Asia and the Pacific* (World Bank).

Rock, M.T. (2002) *Pollution Control in East Asia: Lessons From Newly Industrializing Economies* (Resources for the Future Press).

Rock, M.T. & Angel, D.P. (2007) 'Grow First, Clean Up Later? Industrial Transformation in East Asia', *Environment* 49(4), 8–19.

Schreurs, M. (2004) 'Assessing Japan's Role as a Global Environmental Leader', *Policy and Society* 23(1), 88–110.

Sonnenfeld, D.A. (2002) 'Social Movements and Ecological Modernization: the Transformation of Pulp and Paper Manufacturing', *Development and Change* 1–27.

Stalley, P. (2009) 'Can Trade Green China? Participation in the Global Economy and the Environmental Performance of Chinese Firms', *Journal of Contemporary China* 18(61), 567–590.

Stalley, P. (2010) *Foreign Firms, Investment, and Environmental Regulation in the People's Republic of China* (Stanford University Press).

Studer, S., Welford, R. & Hills, P. (2006) 'Engaging Hong Kong Businesses in Environmental Change: Drivers and Barriers', *Business Strategy and the Environment* 15, 416–431.

Teng, M.-J. (2010) 'The Effects of an Environmental Management System on Intangible Assets and Corporate Value: Evidence From Taiwan's Manufacturing Firms', *Asian Business & Management* 10(3), 381–404.

Turner, J. & Wu, A. (2012) *Step Lightly: China's Ecological Impact on Southeast Asia* (China Environment Forum, Woodrow Wilson Center).

Van Rooij, B., Fryxell, G.E.L., Wing-Hung, C. & Wang, W. (2012) *Decentered Authoritarian Regulation in China: Changes in Social and Political Influences on the Effectiveness of Environmental Law Enforcement in Guangzhou City, Amsterdam Law School Research Paper No. 2012–77*, 18 July. Available at http://ssrn.com/abstract=2111965 or http://dx.doi.org/10.2139/ssrn.2111965.

Wang, H. & Wheeler, D. (2005) 'Financial Incentives and Endogenous Enforcement in China's Pollution Levy System', *Journal of Environmental Economics and Management* 49, 174–196.

Wang, H. & Jin, Y. (2007) 'Industrial Ownership and Environmental Performance: Evidence from China', *Environmental & Resource Economics* 36, 255–273.

Wang, M., Webber, M., Finlayson, B. & Barnett, J. (2008) 'Rural Industries and Water Pollution in China', *Journal of Environmental Management* 86, 648–659.

Wang, Y., Liu, J., Hansson, L., Zhang, K. & Wang, R. (2011) 'Implementing Stricter Environmental Regulation to Enhance Eco-efficiency', *Journal of Cleaner Production* 19, 303–311.

Welford, R. (2005) 'Editorial', *Corporate Social Responsibility and Environmental Management* 12(6), 1.

Wong, E. (2008) 'In China City, Protesters See Pollution Risk of New Plant ', *New York Times*, 6 May.

World Bank (2007) Cost of Pollution in China: Economic Estimates of Physical Damage (World Bank).

World Bank (2011a) *World Bank Databank, CO2 Emissions (kt)*. Available at http://data.worldbank.org/indicator/EN.ATM.CO2E.KT.

World Bank (2011b) *World Bank Databank, CO2 Emissions from Manufacturing Industries and Construction (million metric tons)*. Available at http://data.worldbank.org/indicator/EN.CO2.MANF.MT.

World Bank (2011c) *World Bank, Manufacturing, Value Added (% of GDP)*. Available at http://search.worldbank.org/quickview?view_url=http%3A%2F%2Fdatabanksearch.worldbank.org%2FDataSearch%2FLoadReport.aspx%3Fdb%3D2%26cntrycode%3D%26sercode%3DNV.IND.MANF.ZS%26yrcode%3D.

Yu, X. & Wong, K.K. (2011) 'Environmental Performance of Foreign Direct Investment (FDI) Companies in the Pearl River Delta Region (PRDR): A Case Study of Dongguan City', *Australian Geographer* 42(1), 79–93.

Zeng, K. & Easton, J. (2011) *Greening China: The Benefits of Trade and Foreign Direct Investment* (University of Michigan Press).

Zhang, Z. (2008) 'Asian Energy and Environmental Policy: Promoting Growth While Preserving the Environment', *Energy Policy* 36, 3905–3924.

Acronyms

CSR	corporate social responsibility
DJSWI	Dow Jones Sustainability World Index
EMS	environmental management system
GHG	greenhouse gas
GRI	Global Reporting Initiative
ISO	International Standards Organization

Notes

1   In this chapter, I use the term 'East Asia' broadly to include the nations of Northeast Asia (i.e., China, Japan, and the Koreas) as well as Southeast Asia.
2   These figures were derived by combining information from two separate data tables from the World Bank (World Bank 2011a, 2011b).
3   There are also serious questions about whether voluntary initiatives focusing on corporate environmental management, as opposed to official government programs regulating corporate performance, are sufficiently effective to change the environmental practices of companies. This is particularly pertinent for manufacturing companies that compete primarily on price (as opposed to brand) and may therefore be tempted to skirt environmental regulation in order to keep production costs low. Scholars have also expressed misgivings about the reliability of third-party certifications (e.g., ISO 14000, green label) and the quality of information provided by companies in environmental disclosure schemes (e.g., GRI). For a more extended discussion of these issues, see Stalley 2010.

Part IV

Air, land and water

9

Air pollution

Inventories, regional control and institutions

Maria Francesch-Huidobro

Air pollution in Asia

Air pollution is neither new nor exclusive to Asia. From antiquity, Greek physicians warned of the diseases caused by stale air and Romans yearned to escape the smoke, the wealth and the noise of the *civitas*. At the start of the twentieth century, smog – the cloud of fumes formed by carbon particles and sulfur dioxide (SO_2) from extensive coal burning mixing with fog – lingered above London. During the 1950s and 1960s, Los Angeles suffered from photochemical smog, a mixture of hydrocarbons (HCs) and nitrogen oxides (NO_x) from vehicular emissions reacting in the presence of sunlight. Air quality management has changed that, and smog in London and Los Angeles is today a rare occurrence. However, in most megacities in Asia a combination of population growth, industrialization, urbanization, and increased vehicle use is causing increasing levels of air pollution. At the start of 2013, pollution levels in Beijing have consistently risen above an Air Quality Index (AQI) of 200 in a range of 0–500 points measuring concentrations of particulate matter (PM) of 10 and 2.5 microns in diameter. An AQI of 50 is considered good while levels above 100 are found to be hazardous to health. Dry and calm weather conditions, the emissions ensuing from factories operating in neighboring Hebei province, and the number of cars on Beijing's streets, a city of 20 million people, seemed to be blamed. And Beijing sits on a plain flanked by hills that can trap pollution on days with little wind. Meanwhile, one person hiking at the Great Wall in the hills at Mutianyu, north of Beijing, took photographs of crisp blue skies there (*New York Times* 2013; *Wall Street Journal* 2013). On 8 March 2013, the Causeway Bay district of Hong Kong recorded 291 micrograms of NO_x per cubic meter of air, nearly 50 per cent more than the World Health Organization (WHO) safety guideline of 200 (*South China Morning Post* 2013).

The main cause of pollution in Asia comes from both stationary (power generation, cooling/heating and cooking) and mobile (traffic) sources. Stationary sources increase emissions such as sulfur oxides (SO_x), PM and NO_x that result from the combustion of high (e.g., coal and biomass), medium (e.g., gasoline/diesel and kerosene) and low emitting fuels (e.g., liquefied petroleum gas (LPG) and natural gas (NG) in inefficient combustion devices with limited flue gas control. Residential cooling and heating, the open burning of agricultural waste, slash and burning farming practices and cooking also emit high amounts of carbon monoxide (CO), HCs, and SO_x. Mobile sources, such as poorly maintained, diesel-powered and two-stroke engine vehicles, emit pollutants such as volatile organic compounds (VOCs), CO, NO_x and PM. Concentrations of pollutants are exacerbated by dense urban settings with poor dispersion conditions.

These air pollutants are not only hazardous to the health of Asia's populations contributing to conditions such as asthma, emphysema, bronchitis, cardiovascular and respiratory diseases, eye and throat soreness (Hedley et al. 2008; WHO 2006: 8–17) but are also impacting on climate change, acidification and ozone depletion. For example, carbon dioxide (CO_2), methane (CH_4), nitrous oxide (N_2O) and fluorinated gases (e.g., sulfur hexafluoride (SF6), HCs and perfluorocarbons (PFCs) are major greenhouse gases (GHGs) included in the Kyoto Protocol. These indirectly impact on health through increased morbidity and mortality caused by extreme weather events (e.g., heatwaves, cold spells, floods, etc.), which in turn cause vector-borne diseases, mental stress, etc. (Hedley et al. 2008).

These health and climate hazards are not confined to the localities and populations where they are emitted as the transboundary movement of air pollution across borders causes adverse effects in countries other than the country of origin. In Asia, pollutants with a potential for regional transport include: PM, acidifying substances (SO_2, NO_x), ozone (O_3) and its precursors (VOCs and NO_x), heavy metals (mercury) and persistent organic pollutants (POPs). The yellow dust from the Gobi Desert in Mongolia has been reported to affect pollution levels in other parts of China, Hong Kong, Taiwan, Korea, Japan and even the Pacific coast of North America. Indonesian forest fires caused by the burning of farmland and forest clearance have caused haze pollution in Singapore and Malaysia (Francesch-Huidobro 2007, 2008). Atmospheric brown cloud (ABC), regional dust and acid deposition are other transboundary air pollution issues of concern.

To improve air quality, Asian cities must respond to the combined pressures of rapid growth in population, transport, economic development, and energy consumption. This chapter focuses on analysing the pollution inventories, control measures, and modes of air pollution govern-ance with special focus on two Asian megacities: Singapore and Hong Kong. It argues that air pollution policies and programs are influenced by two mutually related factors: the two localities' modes of governance regarding state and non-state actors' relations and their participation in policy decisions; and, what Gaus (1947, cited in Olsen 2004) has termed the ecology of public administration – that is, the contextual, sectoral and organizational factors surrounding and affecting these governance modes.

The chapter is organized as follows: the section that follows provides a brief presentation on pollution inventories and modelling; on regional control, monitoring and impacts; and on air quality management (AQM) principles and practices. The third section analyses how notions of governance can contribute to the study of the dynamic relations between state and non-state actors for effective air quality management by examining the main approaches of such relations. Through two case studies (the fourth and fifth sections), the chapter then considers the applica-tion of these governance approaches to air pollution problems in Singapore and Hong Kong as a way of probing the ability of government and non-government actors to network and make decisions for sustainable air quality. Finally, the chapter considers the governance modes ensuing from and explaining these relations and draws conclusions about their adequacy in dealing with acute (Singapore) and chronic (Hong Kong) air pollution situations in order to ensure good air quality. Reflections are also drawn for air quality in Asia as a whole.

Inventories, control standards and air quality management

Choosing the *method* to create emissions inventories among those that are commonly used (e.g., WHO–Rapid Inventory Assessment Technique (RIAS), European Environment Agency Core Inventory of Air Emissions (EEA–CORINAIR)), compiling an *inventory* of pollutants from different pollution sources (e.g., stationary, area, mobile, regional and transboundary), selecting *emissions control devices* (e.g., for particulate matter, for gaseous emissions, etc.) and implementing a clear *inspection and maintenance* program, are the pillars of air quality modelling and an essential part of air quality management (AQM) strategies and plans (Haq & Schewela 2008; WHO 2006).

In Asia, regular estimation and reporting of emissions *inventories* of sufficient quality is absent, not publicly disclosed, or only available in a few cities. Yet, reliable emissions inventories are the basis on which to develop strategic plans to deal with air pollution and to monitor the effectiveness of such plans. Hong Kong has compiled as of 2010 and makes publicly available an inventory of SO_2, NO_x, respirable suspended particulate (RSP), VOC and CO by source (Table 9.1). The table shows that in Hong Kong electricity generation and navigation are the main contributors to SO_2 at 50 per cent and 48 per cent, respectively, while road transportation contributes to high emissions of NO_x (30 per cent), RSP (21 per cent) and CO (68 per cent).

Table 9.1 Hong Kong air pollutants emissions inventory, 2010

Pollutant source categories	(Unit: tonnes)				
	SO_2	NO_x	RSP	VOC[3]	CO
Public electricity generation	17,800	27,000	1,010	413	3,310
Road transport	286	32,700	1,340	7,900	47,600
Navigation	16,900	35,000	2,260	3,660	11,400
Civil aviation	299	4,350	54	396	2,530
Other fuel combustion[1]	268	9,520	778	849	5,100
Non-combustion[2]	N/A	N/A	898	20,500	N/A
Total	35,500	109,000	6,340	33,700	70,000
Pollutant source categories	(Unit: %)				
	SO_2	NO_x	RSP	VOC[3]	CO
Public electricity generation	50%	25%	16%	1%	5%
Road transport	< 1%	30%	21%	23%	68%
Navigation	48%	32%	36%	11%	16%
Civil aviation	< 1%	4%	< 1%	1%	4%
Other fuel combustion[1]	< 1%	9%	12%	3%	7%
Non-combustion[2]	N/A	N/A	14%	61%	N/A
Total	100%	100%	100%	100%	100%

Notes:
For all pollutants: data subject to revision when more information are available. Data are rounded to three significant figures, which may not sum up to the total due to rounding.
N/A = not applicable.
Updated: October 2012.
[1] Other fuel combustion sources include industrial, commercial and domestic applications including off-road mobile equipment.
[2] RSP emission sources include quarrying, cooking fumes, construction site dust, tyre, brake and road surface wear.
[3] VOC emission sources mainly consist of consumer products, paints and printing.

Singapore only makes available a 2010 inventory of its SO_2 emissions by source (see Table 9.2) on the basis that SO_2 is one of the its major pollutants from sources such as refineries (51 per cent), power stations (27 per cent), shipping (18 per cent), and other industries and vehicles (about 4 per cent). In the Singapore data, emitters are identified by name (e.g., Shell, ExxonMobil, etc.), likely following a political culture that increasingly supports greater disclosure.

Asian cities have been generally more effective in establishing *control mechanisms* and implementing legislation to reduce vehicle emissions than in compiling inventories. Table 9.3 summarizes the standards in Hong Kong and Singapore in contrast with the EU. The two countries have adopted Euro emissions standards that regulate emissions of NO_x, HC, CO, and PM. For example, Singapore has adopted Euro 2 standards for all vehicles since 2001, while Hong Kong adopted Euro 3 standards in 2001 and Euro 4 in 2006.

Most Asian countries have followed the Asian Development Bank (ADB) recommendations to set: (1) emissions standards and adopt new vehicle technology and green driving assistance aids. For example, rooftop solar panels that help keep a vehicle's battery charged and/or help with the air-conditioning/heating system of the car. The 2010 'Toyota Prius' adopted this technology. Honda's new 2010 'insight hybrid' uses an 'EcoAssist dashboard' that changes color

Table 9.2 Singapore sulfur dioxide (SO_2) emissions inventory, 2010

Sources	Emitters	SO_2 emissions (tonnes)	Contribution of SO_2 emissions
Refineries	Shell	28,278	51%
	Singapore Refining Company	26,754	
	ExxonMobil	22,421	
Power stations	Power Seraya	14,194	27%
	Tuas Power	18,382	
	Senoko Power	7,754	
Shipping	—	28,026	18%
Other industries	Sembcorp Utilities and Terminals	748	3%
	ExxonMobil Petrochemical	722	
	Linde Syngas	546	
	Mitsui Phenol	436	
	Petrochemical Corporation of Singapore	208	
	Invista	3	
	Other fuel oil users	1,300	
	Diesel users	46	
Motor vehicles	Petrol vehicles	891	1%
	Diesel vehicles	101	
Total		150,810	100%

Source: National Environment Agency.

Table 9.3 Vehicle emissions standards in the European Union (EU), Hong Kong and Singapore (gasoline and diesel), 1995–2010

Country/year	95	96	97	98	99	00	01	02	03	04	05	06	07	08	09	10
EU#	1	2	2	2	2	2	3	3	3	3	4	4	4	5	5/6*	5/6*
HK	1	1	2	2	2	2	3	3	3	3	3	4	4	4	4	4
S'pore a	1	1	1	1	1	1	2	2	2	2	2	2	2	2	2	2
S'pore b	1	1	1	1	1	1	2	2	2	2	2	3	3	3	3	3

Notes: #: The 'Euro standards' regulate emissions of nitrogen oxides (NOx), hydrocarbons (HC), carbon monoxide (CO), particulate matter (PM) and particle numbers (PN). There are separate regulations for light vehicles (under 3.5 tonnes) and heavy-duty vehicles. The standards for both light and heavy vehicles are designated 'Euro' and followed by a number (usually Arabic numerals for light vehicles: Euro 1, 2, 3 etc. and Roman numerals for heavy vehicles: Euro I, II, III and so on). Compliance is determined by running the vehicle or the engine in a standardised test cycle. Non-compliant vehicles may not be sold in the EU, but new standards do not apply to vehicles already on the road.

Legend: Euro 1 (red) standards (also known as EC 93): Directives 91/441/EEC (passenger cars only) or 93/59/EEC (passenger cars and light trucks). Euro 2 (blue) standards (EC 96): Directives 94/12/EC or 96/69/EC. Euro 3 (green)/4 (orange) standards (2000/2005): Directive 98/69/EC, further amendments in 2002/80/EC. Euro 5/6 standards (2009/2014/2017): Regulation 715/2007 ('political' legislation) and several other regulations.

Source: CAI-Asia (2006); European Commission, Environment.

from green to blue to teach the driver how to be more efficient. Ford uses a similar system in the 2010 'fusion hybrid' called the 'SmartGauge cluster' that uses dual 4.3-inch liquid crystal display (LCDs) screens with a graphic that shows leaves growing on a tree when you are driving efficiently (Green Autoblog 2013); (2) emphasize inspection and maintenance; and (3) implement transportation planning and demand management.

All Asian countries have phased out leaded gasoline. Table 9.4 shows sulfur content in diesel in Hong Kong and Singapore in contrast with the EU and the United States. While sulfur is a naturally occurring compound in crude oil, when fuel is burned, sulfur combines with oxygen as SOx creating emissions that contribute to decreased air quality. High sulfur content also decreases the catalytic conversion capacity of a system, thus increasing the emissions of NOx, CO, HCs, and VOCs. The main environmental concerns of sulfur emissions are acid rain and the formation of PM. From 2000, regulations have been changed in most developed countries to use ultra low sulfur diesel (ULSD) of 50 parts per million (ppm) and 15–10 ppm. Hong Kong and Singapore have made the shift at different paces and levels.

Table 9.5 presents Hong Kong's and Singapore's specifications of gasoline linked to Euro 3 and Euro 4 standards.

As suggested by Haq and Schwela (2008: 54), key issues for emissions standards, vehicle technology and cleaner fuels for Asian cities include establishing a roadmap for emissions standards and fuel quality; developing a comprehensive fuel quality monitoring system; promoting fuel efficiency standards; introducing alternative fuels for public transportation and low carbon footprint fuels with lower GHGs emissions, and improving efficiency by improving engines and fuel technology.

Many Asian countries have also introduced *inspection and maintenance (I&M)* requirements for vehicles mainly focusing on increasing public awareness; linking vehicle registration and emission testing; establishing testing facilities; steeping up information management and analysis; routine checking; adequate repair facilities and mechanics; and effective enforcement (Haq & Schwela 2008: 55–59).

Table 9.4 Sulfur content in diesel in the European Union (EU), United States, Hong Kong and Singapore, 1996–2011

Country/year	96	97	98	99	00	01	02	03	04	05	06	07	08	09	10	11
EU					500					50/10#*						
US	500											15				
HK		500					50#					10*				
Singapore	3000			500					50#							

Notes: The figures indicate parts per million (ppm) content of sulphur in diesel.

\# Euro 4 standard vehicles are required to have sulphur contents not higher than 50ppm. Hong Kong has adopted this standard from 2002.

* 10–15ppm or ultra-low sulphur in diesel (ULSD) is the standard to be applied from 2005. Hong Kong had adopted ULSD from 2007.

Source: ADB (2003).

Table 9.5 Specification of gasoline in the European Union (EU), Hong Kong and Singapore

	Lead	Sulphur ppm#	Benzene %v/v, max	Aromatics %	Olefins %	Oxygen % m/m, max	RVP summer kPa, max
Linked to Euro 3, effective 2000	Lead free	150	1.0	42	18	2.7	60
Linked to Euro 4, effective 2005	Lead free	50	1.0	35	18	2.7	60
HK	Lead free	5	1	42	18	2.7	60
Singapore	Lead free	—	—	—	—	—	—

Source: ADB (2003).

With inventories, control standards, and inspection and maintenance in place, an AQM program can be developed. AQM sets the basis for the effective governance of air pollution through the participation of the relevant actors, the clear implementation of plans, and monitoring of laws and regulations relating to air pollution and GHGs with special emphasis on the co-benefits of GHG mitigation and air pollutants reduction. Hong Kong has recently revised its existing Air Quality Objectives (AQO), last promulgated in 1987, to align them with WHO recommendations and international best practices (Table 9.6). For example, for PM_{10} the existing allowed emissions are 55 micrograms per cubic meter ($\mu g/m^3$) annually. The proposed objective is to achieve one of the WHO interim targets (ITs) of 50 $\mu g/m^3$ on the basis that the WHO accepts governments to set standards according to their circumstances. The WHO PM_{10} standard is 20 $\mu g/m^3$. The new AQOs will also see targets set for $PM_{2.5}$ at 35 $\mu g/m^3$ annually. The WHO target for $PM_{2.5}$ is 10 $\mu g/m^3$. The legislative process to transform the proposal into law has begun and the objectives are expected to be implemented by 2014.

Singapore also announced in its Sustainable Singapore Blueprint (SSB) of 2009, a target to achieve an annual mean of 15 $\mu g/m^3$ of SO_2 and 12 $\mu g/m^3$ of $PM_{2.5}$ by 2020: 'We will be adopting the World Health Organization (WHO) Air Quality Guidelines (AQG) for particulate

Table 9.6 Existing and proposed new air quality objectives (AQOs) for Hong Kong

Pollutants	Avg. time	Existing AQOs		Proposed AQOs				
		Micrograms per cubic meter ($\mu g/m^3$)	No. of surplus allowed	WHO IT-1[3] ($\mu g/m^3$)	WHO IT-2[3] ($\mu g/m^3$)	WHO IT-3[3] ($\mu g/m^3$)	WHO AQO ($\mu g/m^3$)	Surplus allowed
Sulfur	10 min.	—	—	—	—	—	**500**	3
dioxide	24 hour	130	1	**125**	50	—	20	3
Respirable	24 hour	180	1	150	**100**	75	50	9
suspended particulates (PM$_{10}$)	Annual	55	NA	70	**50**	30	20	NA
Fine	24 hour	—	—	**75**	50	57.5	25	9
suspended particulates (PM$_{2.5}$)	Annual	—	—	**35**	25	15	10	NA
Nitrogen	1 hour	300	3	—	—	—	200	18
dioxide	Annual	80	NA	—	—	—	40	NA
Ozone	8 hour	240[1]	3	**160**	—	—	100	9
Carbon	1 hour	30,000	3	—	—	—	30,000	0
monoxide	8 hour	10,000	1	—	—	—	10,000	0
Lead	Annual	1.5[2]	NA	—	—	—	0.5	NA

Notes:
Bold = proposed new AQO.
[1] There is no existing 8-hour AQO for ozone in Hong Kong. The figure presented here is the 1-hour AQO.
[2] There is no annual AQO for lead in Hong Kong. The figure presented here is the 3-month AQO.
[3] The WHO accepts the need for governments to set national standards according to their own particular circumstances. The WHO guidelines therefore also suggest interim targets (IT) on SO$_2$, PM$_{10}$, PM$_{2.5}$ and O$_3$ to facilitate a progressive approach for achieving the ultimate AQOs in achieving better air quality.

matter 10 (PM$_{10}$), Nitrogen Dioxide, Carbon Monoxide and Ozone, and the WHO AQG's Interim Targets for PM$_{2.5}$ and Sulfur Dioxide, as Singapore's air quality targets for 2020' (National Environment Agency 2012: n.p.) (see Table 9.7).

We now turn to an institutional analysis of how notions of governance can contribute to the study of the dynamic relations between state and non-state actors for the effective implementation of these air quality management programs.

Governance, stakeholders and policies

Governance refers to the 'inability of a single actor to have sufficient action potential to dominate unilaterally in a particular governing model' (Kooiman 1993: 4). Thus, there has been a growing recognition that real progress comes about when state and non-state actors engage in meaningful networking (Wettenhall 2003: 236). The traditional public administration approach to governing – the Weberian model – is characterized by the use of agencies, by a mode of organization that is hierarchical, by a clear delineation of what should be done by the public and private sectors, by a mode of management that follows a command and control

Table 9.7 Singapore ambient air quality targets, 2013

Pollutant	Singapore targets by 2020	Long-term targets#
Sulphur dioxide (SO$_2$)	24-hour mean: 50µg/m^3 (WHO interim target)	24-hour mean: 20µg/m^3 (WHO final)
	Annual mean: 15µg/m^3 (Sustainable Singapore blueprint target)	
Particulate matter (PM$_{2.5}$)	Annual mean: 12µg/m^3 (Sustainable Singapore blueprint target)[1]	Annual mean: 10µg/m^3
	24-hour mean: 37.5µg/m^3 (WHO interim target)	
		24-hour mean: 25µg/m^3 (WHO final)
Particulate matter (PM$_{10}$)	Annual mean: 20µg/m^3 24-hour mean: 50µg/m^3 (WHO final)	
Ozone	8-hour mean: 100µg/m^3 (WHO final)	
Nitrogen dioxide (NO$_2$)	Annual mean: 40µg/m^3 1-hour mean: 200µg/m^3 (WHO final)	
Carbon monoxide (CO)	8-hour mean: 10mg/m^3 1-hour mean: 30mg/m^3 (WHO final)	

Notes:
[1] Sustainable Singapore blueprint annual target for PM$_{2.5}$ of 12 micrograms per cubic meter (µg/m^3) will be retained and aligned with WHO interim target of 37.5 µg/m^3 for 24-hour mean.
NEA's document does not define 'long-term' targets except that are targets aligned with WHO final targets as opposed to interim targets (ITs).

Source: National Environment Agency (2013).

approach and by an emphasis on management skills. In contrast, the governance approach prefers a redesigned agency that is organized as a network, acknowledges the need for public and private partnerships, prefers negotiation, persuasion and collaboration as modes of governance, and emphasizes enabling rather than management skills (Salamon 2002: 9).

Thus, governance is conceptualized as the patterns of interaction between the state and civil society, including the various processes of coordination and collaboration between government and NGOs for policymaking, as well as the participation of nonstate actors in the policymaking process (Atkinson & Coleman 1992; Francesch-Huidobro 2008; Kooiman 1993; Peters 2002; Rosenau 1992). Unlike other existing theories and approaches to governing, governance acknowledges that state adaptations to the external environment have altered the overall steering capacity of states (Burns 2004; Peters 1996, 2002, 2004: 5; Peters & Pierre 2003: 4; Pierre &

Peters 2000; Rhodes 1997; Rosenau & Czempiel 1992). Part of the realization that governing is not only the government's purview is that the running of a modern state requires governments to develop new and evolving relationships with the civil society. Models have been proposed that identify the nature and evolution of public–private relationships and show how these may suit the necessities of different policies and policy sectors (Head & Ryan 2003).

The commonality of these models is the type of relationships they identify. Some relationships are identified as extensions of government. Other relationships are identified as affiliations. Such is the 'partnership model', which closely resembles a corporatist policy community model characterized by power sharing. There are relationships that are identified as adversarial that focus on the political relationship between government and civil society organizations. In such relationships, the government and NGOs act independently and in opposition (Young 2000: 153). The classical explanatory theories of a governance trend of greater non-government participation and closer government – NGO relations are: (1) the market contract and government failure (service delivery failure); (2) the voluntary failure; and (3) the political failure (policy failure) (Hannsman 1987; Lipsky and Rathgeb-Smith 1989; Salamon et al. 1996; Weisboard 1977). Here, we are especially interested in cases of political failure – that is, instances in which the government falls short when responding to citizens. Such a focus helps clarify the relationship between the availability of political and policy space, and the place at which NGOs can enter the process.

In general, the relationship between government and NGOs impacts on the efficiency and effectiveness of service delivery, on the quality of the responsiveness of public policies, on the degree of social exclusion, and on the expression of public values and the building of social capital. Such a relationship may be participatory, usually taking the form of institutionalized channels of co-operation, or through more confrontational tactics that create lobby pressure. The focus in such instances is on co-operation. Four main factors, which constitute the ecology of these relations, are identified in existing research as impacting co-operation: the political and socioeconomic environment; the substantive policy in question; the characteristics of NGOs; and the networking of actors (Casey 1998; Shigetomi 2002).

We now begin discussing two case studies in which these notions are explored. The criteria used for selecting the two cases were that the NGOs under consideration had the highest level of interaction with government and the maximum local, regional and international impact. The rationale behind this was that, since the co-ordination of societal actors is very complex, there is a need to enhance our understanding of forms of co-ordination through explorative in-depth case studies that seek to locate these forms of co-ordination. The case studies seek to clarify the roles, relations, influencing factors and interplay of power between the government and NGOs seeking to improve air quality. The criterion used for selecting the specific problem addressed was that it should be central to the internationalization of environmental problems and, thus, require effective co-operation across sectors and political boundaries (Hills & Roberts 2001). Locally chronic and transboundary acute air pollution successfully met this criterion.

Case 1: Acute air pollution: Singapore and the 1997–1998 forest fires and haze crisis in Southeast Asia

Fighting the haze originating from burning fires in Indonesia, palliating its effects, and stopping the burning altogether were not steps that the Singapore NGOs initiated either through proactive persuasion or reactive protest. The issue was entrusted to NGOs by the Singapore government, largely as a consequence of the nonconventional *modus operandi* adopted in 1994 by the Association of Southeast Asian Nations (ASEAN) as part of its Strategic Plan of Action on the Environment (Cotton 1999: 342). Eventually, however, NGOS became more involved over time.

From August 1997 to March 1998, the Indonesian provinces of South Sumatra, Jambi, Riau and most of Kalimantan were enveloped in smoke and haze originating from palm and rubber plantations, forests, and scrubland burning out of control. Between 800,000 and 4.5 million hectares of forest and bush were lost in Sumatra and Kalimantan alone (Tay 1998: 1). The smoke from the fires combined with pollution from the cities produced haze. Friends of the Earth estimated that up to 70 million people were affected by haze pollution and health experts warned that up to 20 per cent of all deaths in the region could have been related to haze (Friends of the Earth 2004). Regional health, tourism and airline losses were estimated at US$1.3 billion (Tay 1998: 2). Press reports at the time estimated that the economy of Sarawak, Malaysia, was under threat from hazardous haze levels with daily losses estimated to be Malaysian Ringgit 30 million (*Straits Times* 1997). Singapore alone lost S$104 million in 1997 (*Straits Times* 1998a). Air pollution indexes reached 650 points, in a range of 0–500 points measuring concentrations of particulate matter (PM) of 10 and 2.5 microns in diameter well over the hazardous level of 300–400 points (*Straits Times* 1997).

Institutional and collaborative arrangements were already in place before the 1997–1998 crisis struck. The Action Plan on the Haze of 1997 was the basis of the ASEAN Agreement on Transboundary Haze Pollution (ASEAN Haze Agreement) promulgated in June 2002 and ratified in November 2003. A major drawback, however, is that Indonesia has not yet ratified the Treaty (ASEAN 2003; Channel News Asia 2012; Environmental News Network 2003; *New Straits Times* 2003; SIIA 2006; *Straits Times* 2003). Confronted with the latest incidence of forest fires and haze affecting the region in 2012, the Indonesian President, Susilo Bambang Yudhoyono, has pledged to ratify the Agreement. Only confirmation by the legislature and due implementation by local officials will give teeth to his pledge (Channel News Asia 2013; *Straits Times* 2006).

From a governance perspective what characterized the 1997–1998 forest fires and haze crisis were the Singapore NGOs' proposals after the issue had been entrusted to them by the Singapore government, and the Singapore government's disciplined manner in handling the crisis and its approach towards Indonesia. The latter was clearly shown in the Singapore government's outright demand that the NGOs should intervene. The NGOs' action climaxed when the Singapore Environment Council (SEC), prompted by the government, organized the International Policy Dialogue on the Southeast Asian Fires on 4–5 June 1998 that brought together representatives of international, regional and local NGOs; businesses; and public agencies (Francesch-Huidobro 2007, 2008; Ruland 2002; *Straits Times* 1998c, 1998d).

After two days of discussion, the participants resolved to call on Indonesia and to prevent similar episodes in the future. NGOs working in Indonesia were encouraged to pursue greater co-ordination, in the region and in the international arena (*Straits Times* 1998c, 1998d). After the dialogue, the Singapore Environment Council began to act and implement recommendations. Recognition of the need for NGOs to establish a dialogue with ASEAN senior officials for the environment materialized when Simon Tay, nominated member of the Singapore Parliament and Director of the Singapore Environment Council, met with ASEAN senior officials for the environment at their Haze Taskforce meeting on June 18, 1998 (SEC 1998). Tay conveyed the message that ASEAN ought to review the compliance of member states as laid down in the 1998 Haze Plan and explore how NGOs might supplement official efforts. This was the first time NGOs had participated in an ASEAN meeting (SEC 1998; *Straits Times* 1998c). Tay also brought the issue to Parliament by moving a motion on June 30. In his speech, he urged the government to consider what else it could do to deal with the fires and haze (Interview 2003; SEC 1998).

In relation to ASEAN, Tay suggested that Singapore should work from the existing plans towards a binding treaty that would accept the principle of state responsibility and create a system of dispute resolution. He also suggested focusing on the breach of international law Principle 2[1] incurred by Indonesia as it had failed to take responsibility to ensure that activities within its jurisdiction do not cause damage to the environment of other States or of areas beyond the limits of its jurisdiction. He also suggested not to use water bombing when ground efforts were insufficient as it will not make a long-term difference to stop the slash and burning of forests (Interview 2003; SEC 1998). In relation to bilateral measures, Tay urged the Singapore authorities to take a harder stand with Indonesia as the culprits were big businesses, while those suffering the consequences were ordinary people. He also proposed setting up a special court between countries to allow special access to courts in both countries. In continuing aid to Indonesia, Tay suggested for Singapore to promote certification of environmentally sound products, such as timber and palm oil; advising Singaporean firms conducting business in Indonesia to observe Indonesian laws; and assisting, when needed, in investigating Singaporean companies (Interview 2003).

In the aftermath of the International Policy Dialogue when the haze and fires had abated, the issue was taken over from SEC by the Singapore Institute of International Affairs (SIIA). From 1998 to 2000, the matter was continuously debated in Parliament (Singapore Parliamentary Reports System, 9, 1, 68, 2, Jan. 15, 1998: cols. 158–159; 9, 1, 68, 3, Feb. 19, 1998; 9, 1, 68, 10, March 16, 1998; 9, 1, 70, 12, May 4, 1999; 9, 2, 71, 17, March 13, 2000). SIIA has afterwards taken four initiatives in regard to this issue.

As a result of SIIA's work in Indonesia, a report was released based on a public lecture delivered by a panel of Indonesia-based environmental experts (SIIA 2002a). This was followed by the publication, in February 2003, of Preventing Indonesia's Future Fires – Special Report (SIIA 2003) describing several 'governance' approaches recommended to governments by various lending agencies and donors for dealing with local communities that cause small-scale fires and also for dealing with private companies that cause large-scale fires that threaten the survival of the ecosystem. One recommended approach was community-based fire management, which gives communities a sense of ownership over the land since ownership elicits protection. The other more cost-effective method was to recognize and support native knowledge and practices.

The second initiative taken by SIIA was to support the 2001 campaign of World Wide Fund for Nature (WWF)-Europe and WWF-Indonesia that addressed the Indonesian fires and haze crisis by focusing on existing practices in the palm oil industry. The third initiative was a Cross-Sectoral Dialog on Sustainable Development in Southeast Asia jointly organized by SIIA and the University of British Columbia, which was part of SIIA's efforts to draw regional and international attention to the fires and haze crisis (SIIA 2003).

In addition to tapping NGOs' expertise, the Singapore government also sought to tackle the haze problem by introducing measures to lessen negative health impacts on Singaporeans. Thus, the government sought the help of a taskforce, set up back in 1994, to create a Haze Plan of Action. The government also responded to individual citizens' and NGOs' calls to provide information about pollution levels by frequently showing the Pollutants Standard Index (PSI)[2] in various media. Regionally, the Malaysian government took the initiative as early as 1995 to hold an ASEAN meeting to formulate a strategy to fight cross-border pollution including haze (*New Straits Times* 1997). In 1998, the Regional Haze Action Plan was agreed among ASEAN environment ministers.

The ASEAN plan and haze-related issues were often debated in the Singapore Parliament. In an Oral Answers to Questions session in January 1998, then Nominated Member of Parliament (NMP) Simon Tay asked what the government was doing to ensure that member states fulfilled their obligations under the Action Plan and how much the government was planning to spend on local activities and on supporting regional activities to curb fires and the resulting haze. The issue of the involvement of Singapore companies in activities causing or contributing to forest fires in Indonesia was also debated on the same day. Debates continued through 1998, 1999 and 2000, with the landmark ASEAN Agreement among 10 Southeast Asian nations (Brunei, Cambodia, Indonesia, Laos, Malaysia, Myanmar, the Philippines, Singapore, Thailand, and Vietnam) on preventing forest fires coming into force on November 24, 2003. The Agreement, which allows for better monitoring, assessment of prevention, and technical and scientific research, was acclaimed by the United Nations as a possible global model (*Straits Times* 2003), and was a clear victory for those governments and NGOs engaged in the process of environmental protection policy (ASEAN 2003). As mentioned earlier, the battle that is being fought to date is to get Indonesia to ratify it.

How were the above-mentioned collaborative arrangements influenced by Singapore's mode of governance and how did they transform it?

An independent republic since 1965, Singapore has been ruled for the past 40 years by the People's Action Party (PAP) although its monolithic rule was challenged in the general elections of 2012 and in a by-election early 2013. Although the system follows that of a procedural democracy[3] questions of legitimacy have been raised given the lack of viable political opposition, the hegemonic position of the executive, and the weakness of civil society (Francesch-Huidobro 2007, 2008; Rodan 1996; Worthington 2003). Government capacity has changed through Singapore's history. During the colonial period (1819–1965), the state often relied on civil society without the government diminishing its capacity (Gillis 2005). The years of nation building that followed under Lee Kuan Yew's rule (1965–1990) saw the dismantling of civil society and the consequential overpowering position of the state that nevertheless managed to sustain its hegemony without social unrest. A change of leadership in 1990 and subsequently in 2004 saw the beginning of the reconstitution of a civil society under Goh Chok Tong and Lee Hsien Loong with government capacity still remaining strong but increasingly sharing its potential to act with non-state actors (Francesch-Huidobro 2005, 2008). The process is a work in progress that needs to be regularly monitored in various areas of policy to be meaningfully conclusive.

This emerging mode of governance may be termed disciplined governance (Francesch-Huidobro 2008) characterized by a state that plays a mixed rowing–steering function, uses traditional agencies and instruments, and allows networks but maintains tight control over them. Public and private co-operation is sought but not to the extent of forming partnerships. A bureaucratic workforce that possesses management rather than enabling skills is preferred. The case of the 1997–1998 forest fires and haze fits the collaborative towards co-optation dimension of the disciplined governance approach. This case is also a clear landmark in 'transboundary' governance, setting the precedent of the government devolving power to ASEAN and Singapore NGOs; that is, of 'passing the buck'. It is also an example of regional and national commitment to curb air pollution through collaborative arrangements.

Despite the limited form of governance of the disciplined governance approach, the institutional arrangements that have emerged cannot be dismissed. Public dialogs and collaborative programs were created to facilitate partnerships, which ultimately resulted in the 2002 ASEAN Agreement on Transboundary Haze Pollution. Although, as can be deduced from other environmental protection and nature conservation cases (Francesch-Huidobro 2005), the Singapore government remains the agenda setter, implementer and regulator of environmental policy, a more strategic structure of co-operation with national and regional NGOs appears to be emerging. The case of forest fires and haze has shown that talent and money are not wanting. A lack of political will may be the problem. Finally, it is difficult to quantify whether these collaborative arrangements have curbed transboundary air pollution originating from forest fires in the region. To date, the problem persists.

What was the ecology, that is, the political and organizational issues influencing this mode of governance?

The global governance *context* at the end of the 1990s may have influenced Singapore's approach to governing. Consultative modes of governing, which began to characterize the late 1980s, may have influenced the way the government approached decision making into the 1990s. But the positive effects a governance approach could have had on the protection of forest resources were greatly hampered by regional events. Regionally, an economic crisis had struck and its impacts were compounded by political crises and ecological challenges in Indonesia, the source of the fires. In 1997, Indonesia, having achieved remarkable economic development over the previous three decades, fell into recession. The chief connection between the economic crisis and the forest fires is that Indonesia found it was partially able to overcome the economic downturn by expanding agricultural production done through rapid land clearing which was carried out by slash-and-burn methods.

A second connection between the economy and the clearing of forests was the policy reforms attached to the US$43 billion International Monetary Fund (IMF) rescue package offered to Indonesia. From January to March 1998, the demand for Indonesian plywood from China was increasing as a new Chinese policy of reducing logging by 60 per cent had been put in place. The reduced supply had raised Chinese plywood prices and China had found the importation of Indonesian plywood to be the preferred option. This rescued the Indonesian timber sector. But the increased demand led companies to fell more trees than their annual quota. The problem was compounded by an actual shortage of plywood due to previous over-harvesting and loss of timber to devastating fires (CIFOR 1998).

Another policy reform required by the IMF rescue package was agricultural expansion. As many as 2.4 million hectares of palm oil were cultivated when the crisis struck and 1.5 million were later added. This conversion of forest land into plantations together with export policy changes imposed by the IMF also had an effect on Indonesian forests. These included removing restrictions on foreign investment in palm oil plantations implemented in 1998 and replacing the ban on exporting palm oil products with an export tax of 40 per cent (CIFOR 1998). The combined effects of these two policies clearly increased the interest in, and therefore the threat to, forest resources.

If the context influences government–NGO relations, the *policy* in question is an important determinant of such relations. Global, regional, and national environmental policies at the end of the 1990s continued to be characterized by growing environmental awareness alongside growing damage (Gilbert 2000: 847). The 1997 Singapore Ministry of the Environment (ENV), now Ministry of Environment and Water Resources (MEWR), report also highlighted the fact that Singapore was severely affected by haze. The 1998 report indicated that the various agreements and meetings of ASEAN environment ministers and senior officials had produced the positive result of the implementation of the Regional Haze Action Plan in December 1997 (ENV 1998).

While *context* and *sector* are important ecological factors affecting government–NGO relations in particular events, the characteristics of the *organizations* themselves are equally relevant. Organizationally, the involvement of ENV in the forest fires and haze crisis was because it was responsible for air quality. At the time of the crisis, ENV was directly involved in haze monitoring because the Meteorological Services Division (MSD) was under its portfolio (*Straits Times* 1998b). The MSD, after a brief spell within the Transport Ministry, has moved to the newly created National Environment Agency (NEA) in clear recognition of the relationship that climate conditions and crises have with air quality now being monitored by the NEA (NEA 2002/03: 25, 2012).

As mentioned earlier, the SEC was involved in this case as the NGO which initiated the 1998 Policy Dialogue to tackle the forest fires crisis. Today, SEC maintains an umbrella status under which other NGOs continue to pursue the abatement of fires and their consequences. SIIA's involvement in this case originated when it participated in the 1998 Policy Dialog initiated by the SEC. Since then, SIIA has taken a leading role in bringing the various national, regional and international stakeholders together to tackle the recurring problem of forest fires and haze. The ASEAN-Institute of Strategic and International Studies (ASEAN-ISIS) was not involved in the 1997–1998 crisis but it has since then provided a forum for discussion as the problem re-emerged.

The analysis of this case demonstrates a shift in Singapore's ever changing dynamic relations between state and non-state actors for better air quality. Although the Singapore government still sets the agenda and acts as implementer and regulator of the process, a more strategic structure of co-operation with national and regional NGOs appears to be emerging. In this case, it may be argued, the fact that Indonesia was involved called for avoiding direct confrontation between the two states and involving SEC. Nevertheless, what originally was an act of 'passing the buck' to NGOs so as not to interfere with Indonesia's sovereignty while still offering support in a non-political way, has resulted, in the long term, in mutual engagement.

Case 2: Chronic air pollution: Hong Kong and the 1995–2006 smog crisis in the Pearl River Delta of southern China

This case addresses the issue of government and environmental NGOs' relations in the regional effort to ameliorate haze and smog pollution in the Pearl River Delta (PRD).[4] In contrast to the Singapore case, the Hong Kong NGOs' involvement was not directly advocated either by the Hong Kong government or the local governments across the PRD. That is, governments on both sides of the border have neither 'passed the buck' to NGOs nor have relied on them for expertise but are very much in the driver's seat without being able to orchestrate a concerted effort between themselves, the NGOs and the business sector.

In the past 15 years, Hong Kong's air pollution has been seen as a serious and growing threat to the health and well-being of its people, the populations of the PRD, the promotion of tourism and the presence of foreign corporations. In recent years, the problem has come to the top of the government's agenda, as shown by the extensive coverage it received in the Policy Addresses since 1995 (e.g., Policy Address 1999: 89–105, 2001: 89–92, 2005, 2006: 51–66, 2013: 135). The latter suggests new AQOs will be in place by 2020 after having agreed general stations air reduction targets with Guangdong Province for 2015 and 2010.

There are several compelling reasons for this change in priority. First, the air quality as measured by the Air Pollution Index (API) has gradually deteriorated since 1991. Second, the findings of a study conducted by the Community Medicine Department of the University of Hong Kong indicated strong short-term effects of air pollutants on hospital admissions and hospital deaths (Hedley et al. 2008; Wong et al. 2002). Third, as Hong Kong loses its competitive advantage in terms of costs of business, foreign investors increasingly complain about poor air quality and consider it a factor in determining whether to continue to invest (Business Environment Council 2005, 2012; Wah 2000). Fourth, the tourist industry suffered a downturn when the local economy was affected by the Asian economic crisis of 1997–1998 and again in the 2003 Severe Acute

Respiratory Syndrome (SARS) crisis. Tourism has traditionally acted as a major boost to the economy but air pollution is seen as having a significant negative impact on the industry.

All these factors have made air pollution not only an environmental problem, but also a socioeconomic and political problem. The present positioning of the problem has determined the way the government agenda has been set and the choice of alternative policies. For example, the Third Comprehensive Transport Study released in October 1999 and still in force, proposed measuring levels of nitrogen oxide (NOx) and respirable suspended particulates (RSPs) for fifteen roadside stations across Hong Kong (Wilbur Smith Associates Ltd 1999). The high levels of NOx were found across the Kowloon Peninsula, the northern coast of Hong Kong Island and the North West New Territories. The daily average of air quality objectives (AQOs) for RSPs was breached 38 times in one month at 10 monitoring stations, while levels of sulfur dioxide (SO_2) had significantly decreased. The control framework put in place to tackle the situation is guided by the Air Pollution Control Ordinance (Cap 311) which requires industries to reduce harmful and excessive air pollutants.

In 1989, after the publication of a white paper, the entire Hong Kong territory was declared an air control zone with a set of AQOs for seven pollutants currently under review (Table 9.6). Besides setting up a control framework, the Environmental Protection Department (EPD) also set up control programs as specific actions to reduce air pollution. The effect was that SO_2 concentrations fell by up to 80 per cent in industrial areas. This, combined with a reduction in industrial activity, resulted in a total fall from 46,616 tonnes in 1989 to 16,688 tonnes in 1997. SO_2 emissions in 2010 were 22,100 (Environmental Protection Department 1999b, 2010). The government also aimed at cutting emissions from power generation. SO_2 emissions from power generation came down from 131,600 tonnes in 1991 to 52,659 tonnes in 1997 and to 17,800 in 2010 (Environmental Protection Department 1999a, 2010).

On the vehicular front, unleaded petrol became available in 1991 and a complete ban on leaded petrol was issued on April 1, 1999. Moreover, three-way catalytic converters were made available and 75 per cent of the total petrol car fleet was eventually fitted with them. In order to reduce emissions from diesel vehicles, the EPD adopted Euro I and Euro II standards in 1995 and 1997 respectively and Euro 4 and 5 in 2005 and 2008 (Table 9.3). The effect was that RSPs were reduced by 80 per cent and NO_x was reduced by 20 per cent for vehicles complying with the latest standards. Finally, inspection and enforcement programs were set up for smoky vehicles with reductions reported at 30 per cent from 1993 to 1998.

The handling of the chronic air pollution affecting Hong Kong and the PRD was not initiated by local or mainland China NGOs. Unlike Indonesia, where environmental NGOs thrive, China's NGO activity is greatly limited by its political system and underdeveloped civil society. Hong Kong NGOs are more autonomous and independent than those of Singapore, but their resources are limited (Francesch-Huidobro 2005, 2008). Thus, the initiatives and the involvement of the Hong Kong and PRD governments are more important than those of NGOs.

In reaction to the continuing deterioration of the regional air quality, the Hong Kong and Guangdong governments carried out a Joint Study of Air Quality in the Pearl River Delta Region (henceforth the 'Joint Study') (CH2M Hill (China) Ltd 2002; Civic Exchange 2004: 2–3). This aimed at building an emissions index based on existing emissions inventories, mostly modeled on US inventories that fitted local characteristics. The Joint Study found that around 80 to 95 per cent of regional pollutants were generated in the PRD with the remainder originating in Hong Kong. A caveat is necessary here, however. On a per capita basis, emissions were found to be comparable in Hong Kong and the PRD; for example, total emissions per capita in the PRD weighed 50 kilograms while in Hong Kong the total weighed 40 kilograms.

Furthermore, the Hong Kong and Guangdong Joint Working Group on Sustainable Development and Environmental Protection has set up a PRD Air Quality Management and

Monitoring Special Panel (Special Panel) that in turn has put in place a Regional Air Quality Management Plan listing control measures needed to achieve regional targets (see Civic Exchange 2004: 2–3).

Groups such as Clear the Air, Friends of the Earth (HK), Green Power and Greenpeace (China) are involved in campaigns ranging from lobbying Hong Kong-owned transport and power plants generating pollution in the PRD (Clear the Air 2004) to advocating renewable energy solutions (Friends of the Earth 2004; Green Power 2004; Greenpeace 2004).

How were the above-mentioned collaborative arrangements influenced by Hong Kong's mode of governance and how did they transform it?

A former colony of Britain (1841–1997), Hong Kong's retrocession of sovereignty to the People's Republic of China (PRC) marked the beginning of the Hong Kong Special Administrative Region (HKSAR), a local government of the PRC ruled by the Chinese Communist Party (CCP). The Basic Law, its constitutional framework under the slogan 'one country, two systems', establishes an executive-led system and limits power for the legislature while ignoring the political effects of the transition and the higher expectations of the population in terms of demands for representative government (Scott 2005). Rapid economic growth, public–private sector co-operation, a rich and autonomous civil society, respect for the rule of law and political stability had facilitated the management of state–society relations until the retrocession of sovereignty (Burns 2004). Yet after 1997 a sharp decline in economic growth plus an inefficient and outdated bureaucracy generated blunder after blunder, making apparent a lack of co-ordination in government bureaucracy that negatively affected state–society relations by undermining the government performance-based legitimacy (Burns 2004; Lam 2005).

This has resulted in a form of governance one can term disarticulated governance characterized by a state that neither rows nor steers but governs reactively. Its agencies and instruments are redesigned but are highly compartmentalized, lacking overall co-ordination. Networking is favored but only when windows of opportunity open for interaction. The public and private sectors compete rather than co-operate. A bureaucracy that possesses crisis management skills to react to recurrent policy emergencies is favored.

What was the ecology, that is, the political and organizational issues influencing this mode of governance?

Hong Kong's ecology of government–NGO relations in relation to tackling air pollution is identified in that although air pollution has been on the government agenda for some time and in the more recent Policy Addresses of 2013: 135, a series of ideological, political and social factors has brought about further changes in the ecology of government–NGOs relations. These contextual and policy specific factors have had an impact on the actors involved in this policy area.

Hong Kong has reached a level of development that has led it to pay attention to a particular set of problems experienced by cities at similar levels of development. High levels of economic development and wealth create similar problems and opportunities that are dealt with in the same manner regardless of differences in the socio-political structure (Dye 1972). In Hong Kong, this sort of determinism is not so clear because government agencies continue to have their own agendas and there is little joint thinking and responsiveness to outside factors. Nevertheless, the latest developments in air pollution policies have shown that outside factors can modify these agendas.

Thus, in the past few years, several factors have resulted in a new agenda. The establishment of a dedicated Environment Bureau (EB) was conducive to better policy co-ordination.

Moreover, the EB seemed to have the resources and willpower to improve the overall environmental quality situation. Second, the then Legislative Councillor Christine Loh (now Undersecretary for Environment), who had been a leading critic of the government's environmental policies, announced her decision not to run for the September 2000 elections and set up a think tank to monitor environmental quality in the city. This probably prompted other legislators such as Choy So-yuk (Democratic Alliance for the Betterment of Hong Kong) and Audrey Eu Yutmee (former Civic Party) to become more actively involved in environmental protection issues. Third, the API reached a record high of 174 on March 29, 2000, but this was broken by a 'severe' API reading of 212 in Central in August 2012. Readings of 198 API were recorded as recent as March 8, 2013, prompting the government to issue health warnings to those with respiratory illnesses (*South China Morning Post* 2013). Fourth, from 2000, the Citizens Party, Clear the Air and other green groups have initiated a well-articulated lobbying of the government through the internet to persuade it to tackle the air pollution problem and have since then proposed very specific targets. Fifth, the 2000 Asian Development Bank Report predicted that economic recovery in Hong Kong would be set back unless it cleaned its air. Finally, Hong Kong hosted the 2006 Asian Games, which prompted the government to show that it was cleaning up its act in earnest.

All these factors led the government to propose more stringent measures to tackle air pollution. Thus, the agenda had set the policy direction. After the critical air pollution problem was recognized and defined, solutions began to appear and go through a selection process. Particulate trap trials, for example, were completed and a scheme was implemented as soon as funding became available. The administration also supported the use of ultra-low sulfur diesel (ULSD) for diesel vehicles and a trial using government vehicles began in July 2000.

The above proposals were chosen from among the various possible solutions because they were feasible, the result of current political values and preferences, and highly acceptable to the public. At this point in time, it seems that problems, policies and politics have met, forming a window of opportunity (Kingdon 1994). Yet, unlike in the Singapore case, this process seems to resemble an impromptu mode of governance that is referred to here as disarticulated governance.

In Hong Kong, the way the problem has been addressed seems to indicate that the government is working on a strategy. The aim appears to be to ensure that polluting emissions do not cause harm to human health or the environment. The Administration seems to have begun to acknowledge that such harm has social and economic costs. However, while they can be regarded as a positive step forward, the measures taken so far appear to depict what I have termed a disarticulated governance approach to governing (Francesch-Huidobro 2007). There are other measures that the government needs to adopt to assist long-term and sustainable planning. Since 1998, the government has opened up debate on various policy issues. Fully engaging the community to assist government decision making is a fundamental condition for sustainability. It may be argued that although this is a good start, such a synergistic form of joint thinking is still in its infancy (Francesch-Huidobro 2012). At this point in time, it looks as if Hong Kong needs a political system that makes people feel they are governing. Article 68 of the Basic Law, Hong Kong's constitution, makes provisions for Hong Kong to eventually base its political system on universal suffrage. To advance the process and overcome problems that affect the way decisions are made, such as the part-time nature of the Executive Council or the lack of expertise of generalist decision makers, a series of measures could be considered.

First, is the importance of access to information as a condition for participation. On the one hand, the new politics of the internet provide for more direct contact among people and with policymakers. On the other hand, governments need to actively open up the channels of discussion to demystify administrative processes and steer interaction. Although the various government bureaux and departments are great collectors of data, some of which are posted on the various agencies' homepages, data are not being properly analyzed and collated, resulting in deficient information provision. Better information will help to improve the quality of public discourse, which in turn will help the government to make better policy (Loh 1999).

Second, proper institutional support of sustainability is needed to support policy goals. There are essentially three institutional problems that have hindered the development of more sustainable policies in Hong Kong: the size and diversity of the tasks of some policy bureaux; the lack of multidisciplinary expertise and management structures to tackle specific issues and problems; and the conspicuous lack of participation of the community in the development of their city (Francesch-Huidobro 2012). The way that air pollution policy has developed over the past fifteen years may be indicative of environmental policy integration – a vital prerequisite of sustainable development.

Conclusion: governance for air quality in Asia

The analysis of the way air pollution is dealt with in the two cities highlights two types in the continuum of the governance model that, I argue, depict Singapore and Hong Kong. While the Singapore government is recently approaching relations with NGOs by fostering limited forms of collaborative arrangements and co-opting them in instances such as the 1997–1998 forest fires crisis for reasons of political correctness while still believing that election by the majority gives a leader a blank cheque to mold society and make an exclusive claim on wisdom and power, the Hong Kong government governs in a disarticulated manner.

Without fostering a concerted effort among government, businesses and NGOs the sustainability of Hong Kong's governance and the quality of its air remain hampered. Hong Kong suffers from the Chinese government's mistrust of NGOs, hence the current stand-off. Many of the Hong Kong environmental NGOs have an expatriate element that disengages them from the mainstream population and vice versa. The Hong Kong government does welcome NGO activity but mainly for social services; on the environmental front there is plain distrust because of the planning and development agenda that the government is pursuing. Environmental NGOs are regarded as hindrances to development. Yet, because of the lack of a political mandate, the government has to entertain their proposals and respond to their lobbying. The NGOs, for their part, have limited themselves to proposing initiatives that the government ignores. Hong Kong has missed a march here by not harnessing its NGOs. Singapore, by way of contrast, has its NGOs in its pocket and has used them for its own ends.

Urban air pollution in Asia is closely linked with urbanization, industrialization and car ownership. Asian countries, most significantly China, suffer from two major sources of pollution: coal smoke and vehicle emissions. In implementing AQM programs, Asia has many challenges. First, many Asian countries are developing nations where economic development is a primary consideration with the consequential 'develop first, clean later' *modus operandi*. Second, the dominance of coal to supply energy cannot be lessened in the near future. Third, pollution-curbing measures have only been recently introduced in Asian countries and their effects have not been felt. And also there are problems with the management and control of pollution sources which frustrate the implementation of pollution reduction strategies. Control and mitigation of urban pollution is an arduous task that Asia will have to deal with for years to come. Science, technology, management and political will are the pillars on which an effective air quality strategy rests.

Bibliography

Asian Development Bank (ADB) (2003) *Reducing Vehicles Emissions in Asia*, Manila, Asian Development Bank.

ASEAN (2003) Press Release on the ASEAN Ministerial Meeting on Haze, March 4 (Indonesia: ASEAN).

Atkinson, M. & Coleman, W. (1992) 'Policy Networks, Policy Communities, and the Problems of Governance', *Governance: An International Journal of Public Administration* 5(2), 154–180.

Burns, J.P. (2004) *Government Capacity and the Hong Kong Civil Service* (Oxford University Press).

Business Environment Council (2005, 2012 (rev.)) *Living Under Blue Skies: A Review of Air Pollution in Hong Kong and the Pearl River Delta* (Business Environment Council).

CAI-Asia (2006) Country Synthesis Reports on Urban Quality Management in Asia, Manila, Clean Air Initiative for Asian Cities.

Casey, J. (1998) 'Non-government Organisations as Policy Actors: the Case of Immigration Policies in Spain', Unpublished doctoral thesis, Barcelona.

Centre for International Forestry Reserves (CIFOR) (1998) 'The Economic Crisis and Indonesia's Forest Fires Sector', May 18.

Channel News Asia (2013) Available at http://www.channelnewsasia.com/stories/singaporelocalnews/view/1228144/1/.html.

CH2M Hill (China) Limited (2002) Final Report Agreement No. C/E 106/98 Study of Air Quality in the Pearl River Delta Region, Hong Kong, CH2M Hill (China) Limited.

Citizens Party (2000) *Clearing the Air – Still a Long Way to Go. A Comprehensive Review of Air Pollution Problems and Solutions* (Citizens Party).

Civic Exchange (2004) White Paper on Air Quality Management Issues in the Pearl River Delta Region, Hong Kong, Civic Exchange.

Clear the Air (2004) 'Campaigns, Hong Kong: Clear the Air'. Available at http://www.cleartheair.org.hk.

Cotton, J. (1999) 'The "Haze" over Southeast Asia: Challenging the ASEAN Mode of Regional Engagement', *Pacific Affairs* 72(3), 331–351.

Dye, T. (1972) *Understanding Public Policy* (Prentice-Hall).

Environmental News Network (2003) ASEAN Pact on Fighting Forest Fires takes Effect, November 25.

Environmental Protection Department (1999a) Environmental Protection in Hong Kong, Hong Kong, Environmental Protection Department.

Environmental Protection Department (1999b) Short-term Effects of Ambient Air Pollution, Hong Kong, Environmental Protection Department.

Environmental Protection Department (2012) Environmental Protection in Hong Kong, Hong Kong, Environmental Protection Department.

Francesch-Huidobro, M. (2005) 'A Disciplined Governance Approach to Government–NGOs Relations: the Structures and Dynamics of Environmental Politics and Management in Singapore', PhD thesis, University of Hong Kong.

Francesch-Huidobro, M. (2007) 'Impact of Government-NGO Relations on Sustainable Air Quality in Singapore and Hong Kong Compared', *Journal of Comparative Policy Analysis: Research and Practice* 9(4), 383–404.

Francesch-Huidobro, M. (2008) *Governance, Politics and the Environment: A Singapore Study* (Institute of Southeast Asian Studies Publishing).

Francesch-Huidobro, M. (2012) 'Institutional Deficit and Lack of Legitimacy: the Challenges of Climate Change Governance in Hong Kong', *Special Issue: Climate Change, National Politics and Grassroots Action, Environmental Politics* 21(5), 791–810.

Friends of the Earth (HK) (2004) Campaigns (Hong Kong: Friends of the Earth (HK)). Available at http://www.foe.org.hk.

Gilbert, M. (2000) *A History of the Twentieth Century, Volume 3: 1952–1999* (Perennial).

Gillis, E.K. (2005) *Singapore Civil Society and British Power* (Talisman).

Green Autoblog (2013) http://green.autoblog.com/2009/04/17/aaa-picks-10-best-new-vehicle-technologies-thinks-green/.

Green Power (2004) Campaigns (Hong Kong: Green Power). Available at http://www.greenpower.org.hk.

Greenpeace (China) (2004) Campaigns (Hong Kong: Greenpeace (China)). Available at http://www.gpchina.org.

Hannsman, H. (1987) 'Economic Theories of Non-profit Organizations', in W.W. Powell (Ed.), *The Non-Profit Sector – A Research Handbook* (Yale University Press).

Haq, G. & Schwela, D. (2008) *Foundation Course on Air Quality Management in Asia* (Environment Institute).

Head, B.W. & Ryan, N. (2003) 'Working with Non-government Organisations: A Sustainable Development Perspective', *Asian Journal of Public Administration, Symposium on the Third Sector in Transition: Non-profit Policy, Management Theory* 25(1), 31–56.

Hedley, A.J., McGhee, S.M., Lai, H.K., Chau, J., Chau, P.Y.K., Chung, K.W.Y. et al. (2008) Report on Air Quality and the State of Public Health in Southern China, Hong Kong, Civic Exchange.

Hills, P. & Roberts, P. (2001) 'Political Integration, Transboundary Pollution and Sustainability: Challenges for Environmental Policy in the Pearl River Delta Region', *Journal of Environmental Planning and Management* 44(4), 455–473.

Interview (2003) Interview with Chairman, Singapore Environment Council, Singapore Institute of International Affairs, February 20.

Kingdon, J. (1984) *Agendas, Alternatives and Public Policies* (Little, Brown, & Company).

Kooiman, J. (1993) 'Social-political Governance: Introduction', in J. Kooiman (Ed.), *Modern Governance: New Government-Society Interactions* (Sage).

Lam, W.F. (2005) 'Coordinating the Government Bureaucracy in Hong Kong: An Institutional Analysis', *Governance: An International Journal of Policy, Administration, and Institutions* 18(4), 633–654.

Lipsky, M. & Rathgeb-Smith, S. (1989) 'Non-profit Organisations, Government and the Welfare State', *Political Science Quarterly* 104(4), 625–648.

Loh, C. (1999) *Alternative Policy Address: Hong Kong, a Sustainable World City of the Millennium* (Citizens Party).

Ministry of the Environment (ENV) (1998) Annual Report 1998, Singapore, Ministry of the Environment.

National Environment Agency (NEA) (2002/3) Annual Report 2002–2003, Singapore, National Environment Agency.

National Environment Agency (NEA) (2012) Annual Report, Singapore, National Environment Agency.

New Straits Times (1997) 'Transboundary Pollution: Jakarta Urged to Spearhead Action Plan', October 7.

New Straits Times (2003) 'Haze Pollution Control Agreement to be Enforced by November 25', October 4.

New York Times (2013) 'On Scale of 0 to 500, Beijing's Air Quality Tops "Crazy Bad" at 755', January 12.

Olsen, J.P. (2004) 'Citizens, Public Administration and the Search for Theoretical Foundations, The 2003 John Gaus Lecture to the American Political Science Association, PSOnline'. Available at http://www.apsanet.org.

Peters, B.G. (1996) *The Future of Governing: Four Emerging Models* (University Press of Kansas).

Peters, B.G. (2002) 'The Changing Nature of Public Administration: From Easy Answers to Hard Questions', *Asian Journal of Public Administration* 24(2), 153–183.

Peters, B.G. (2004) 'Governance and Public Bureaucracy: New Forms of Democracy or New Forms of Control?', *Asia Pacific Journal of Public Administration* 26(1), 3–15.

Peters, B.G. & Pierre, J. (2003) 'Introduction: the Role of the Public Administration in Governing', in B.G. Peters & J. Pierre (Eds.), *Handbook of Public Administration* (Sage Publications).

Pierre, J. & Peters, B.G. (2000) *Governance, Politics and the State* (Macmillan Press).

Rhodes, R.A.W. (1997) *Understanding Governance: Policy Networks, Governance, Reflexivity and Accountability* (Open University Press).

Rodan, G. (1996) 'State Society Relations and Political Opposition in Singapore', in G. Rodan (Ed.), *Political Oppositions in Industrialising Asia* (Routledge).

Rosenau, J.N. (1992) 'Citizenship in a Changing Global Order', in J.N. Rosenau & E.-O. Czempiel (Eds.), *Governance without Government: Order in Change in World Politics* (Cambridge University Press).

Rosenau, J.N. & Czempiel, E.-O. (Eds.) (1992) *Governance without Government: Order and Change in World Politics* (Cambridge University Press).

Ruland, J. (2002) 'The Contribution of Track Two Dialogue towards Crisis Prevention', *ASIEN* 85, 84–96.

Salamon, L. (2002) 'The New Governance and the Tools of Public Action: An Introduction', in L.M. Salamon (Ed.), *The Tools of Government: A Guide to the New Governance* (Oxford University Press).

Salamon, L., Anheier, H. & Sokolowski, W. (1996) 'The Emerging Sector: A Statistical Supplement', in *Papers in Johns Hopkins Comparative Non-profit Sector Project* (Johns Hopkins Institute for Policy Studies).

Scott, I. (2005) *Public Administration in Hong Kong: Regime Change and Its Impact on the Public Sector* (Marshall Cavendish International).

Shigetomi, S. (Ed.) (2002) *The State and NGOs: Perspective from Asia* (Institute of Southeast Asian Studies).

Singapore Environment Council (SEC) (1998) Press Release, Singapore Environment Council Presents NGOs Concerns about SEA Fires to ASEAN Senior Officials, June 19.

Singapore Institute of International Affairs (SIIA) (2002a) Press Release, Public Lecture on Indonesia Fire and Haze, August 16.

Singapore Institute of International Affairs (SIIA) (2002b) Press Briefing, Indonesia Vegetation Fires and Transboundary Haze: More Meaningful Dialogue between Scientific and Policy Communities is Needed, August 16.

Singapore Institute of International Affairs (SIIA) (2003) Preventing Indonesia's Future Fires – Special Report, Singapore, Singapore Institute of International Affairs.

Singapore Institute of International Affairs (SIIA) (2006) Chairman's Statement, Dialogue on Southeast Asian Fires and Haze, October 19.

Singapore Parliamentary Reports System, 9, 1, 68, 2, Jan. 15, 1998: cols. 158–159; 9, 1, 68, 3, Feb. 19, 1998; 9, 1, 68, 10, March 16, 1998; 9, 1, 70, 12, May 4, 1999; 9, 2, 71, 17, March 13, 2000. Available at http://www.parliament.gov.sg/publications-singapore-parliament-reports.

South China Morning Post (2013) 'Air Pollution Soars above WHO Safety Guidelines', March 9.

Straits Times (1997) 'Hazardous Haze May Cost Sarawak RM 30 Million a Day', September 20.

Straits Times (1998a) 'Haze Costs S'pore at least $104 m, Say Environmentalists', March 16.

Straits Times (1998b) 'Haze May Return in May', March 21.

Straits Times (1998c) 'NGOs Want to Be at ASEAN Meet on Haze', June 8.

Straits Times (1998d) 'Boycott of Haze Fire Firms Next?', June 25.

Straits Times (2003) 'ASEAN Pact to Stop Forest Fires Takes Effect', November 26.

Straits Times (2006) 'ASEAN Must Mind the Credibility Gap', October 14.

Tay, S. (1998) 'Southeast Asian Fires: Haze over ASEAN and International Environmental Law', *Review of European Community and International Environmental Law (RECIEL)* 7(2), 199.

Wah, S. (2000) 'Air Pollution and Vehicles Emissions', Unpublished MPA Coursework, Hong Kong University.

Wall Street Journal (2013) 'Beijing's Murky Air Pollution Measures', February 1.

Weisboard, B.A. (1977) *The Voluntary Non-profit Sector: An Economic Analysis* (Heath).

Wettenhall, R. (2003) 'Exploring Types of Public Sector Organizations: Past Experiences and Current Issues', *Public Organization Review: A Global Journal* 3, 219–245.

WHO (2006) Use of Air Quality Guidelines in Protecting Public Health: A Global Update. Factsheet No. 313, October 2006, Geneva, World Health Organization.

Wilbur Smith Associates Ltd (1999) Third Comprehensive Transport Study for the HKSAR Hong Kong, Wilbur Smith Associates Ltd.

Wong, C.M., Atkinson, R.W., Anderson, H.R., Hedley, A.J., Ma, S.L.S., Chan, Y.K. et al. (2002) 'A Tale of Two Cities: Effects of Air Pollution on Hospital Admissions in Hong Kong and London Compared', *Environmental Health Perspectives* 110, 67–77.

Worthington, R. (2003) *Governance in Singapore* (RoutledgeCurzon).

Yin, R.K. (1994) *Case Study Research Design and Methods*, 2nd edn (Sage Publications).

Young, D. (2000) 'Alternative Model of Government and Non-profit Relations: Theoretical International Perspectives', *Non-Profit and Voluntary Sector Quarterly* 29(1), 149–172.

Notes

1 Principle 2 of the Rio Declaration on Environment and Development (1992) states that 'States have the right to exploit their own resources pursuant to their own environmental and development policies and the responsibility to ensure that activities within their jurisdiction or control do not cause damage to the environment of other States or of areas beyond the limits of national jurisdiction' (United Nations Environment Program 2005).

2 Air Pollution Indexes (APIs) are a series of numbers expressing the relative levels of air pollution taking into account various contaminants. They protect health by triggering control actions designed to curb air pollution episodes associated with these contaminants (Ministry of the Environment, Ontario 2012; http://www.ene.gov.on.ca). The Environmental Protection Agency (EPA-US) developed an index to measure pollution levels for the major air pollutants regulated under Title I of the Clean Air Act (Environmental Protection Agency 2004; http://www.epa.gov/air).

3 These are democracies that follow electoral due process but where the rights, interests and involvement of citizens are not taken into consideration by the officials-to-be in between elections.

4 The immediate Pearl River Delta Region comprises the cities of Zhaoqing, Guangzhou, Huizhou, Foshan, Dongguan, Jiangmen, Zhongshan, Shenzhen, Zhuhai, Hong Kong and Macau.

10

Seeing through the smog

China's air pollution challenge for East Asia

Angel Hsu

Introduction

Air pollution is one of the most pressing environmental concerns facing Asia, and in particular, China today. In 2011 populations in more than 160 Asian cities were routinely exposed to air pollution levels above those deemed safe by the World Health Organization (WHO). Many of these cities experienced air pollution levels more than double recommended levels (CAI-Asia 2012). The recent Global Burden of Disease project estimates that ~3.4 million deaths worldwide each year can be attributed to outdoor air pollution (Lim et al. 2012). As cities within Asia continue to industrialize and urbanize, air pollution and its social, economic, political, and environmental consequences will pose major regional challenges.

China's air pollution first rose to international attention prior to the 2008 Olympic Games, when high levels of pollution and poor air quality threatened to endanger athletes (Macur 2007). To reduce air pollution in a relatively short period of time before the Olympics, the Chinese government took to more stringent measures to tackle pollution from industry, vehicles, and construction. Beginning two months prior to the opening ceremony, Beijing started to drastically reduce the number of allowable vehicles on its roads, first removing around 300,000 heavily polluting vehicles and then cutting the number of vehicles by half through an odd–even license plate system, which alternated between which cars were allowed on roadways on 'odd' or 'even' numbered days. Construction within Beijing was halted, power plants were instructed to use cleaner fuels, and many polluting factories, including those in neighboring provinces, were ordered to reduce production or in some cases shut down (Wang et al. 2010). While these policies and heavy-handed measures resulted in a notable reduction in air pollution during the Olympic Games and immediately following, pollution levels rose soon after to surpass their previous levels (Chen et al. 2011).

Presently, the issue of air pollution has been a chief concern to citizens both within China and abroad, garnering continuous international media attention over the last half-decade. The staggering statistics of China's environmental problems paint an alarming challenge of the reality of life citizens must face in China and the enormity of the task at hand for Chinese leaders to address. To name a few of these oft cited figures: the World Bank in 2007 listed 16 out of 20 of the world's most polluted cities as being located in China; environmental pollution and degradation were estimated to cost around 4 percent of China's GDP in 2008 (CAEP 2010); and an estimated

750,000 premature deaths in China are thought to be attributed to air and water pollution alone (World Bank & SEPA 2007). Death rates from chronic respiratory disease and infections in children are four to 44 times higher in China than other parts of the world (He et al. 2012). China's air pollution in urban areas regularly exceeds recommended safe levels by the WHO. In the first half of this year, China's Ministry of Environmental Protection (MEP) reported levels of fine particulate matter pollution ($PM_{2.5}$) on average three times higher than the WHO recommendations (MEP 2013).

As economic development and per capita incomes are on the rise, citizens within China are demanding improved environmental quality. Much of the citizen demand for improved air quality has occurred within the last few years. In October 2011 Chinese netizens took to the web to demand explanations as to misleading air quality statistics that bore discrepancies to rival readings atop the U.S. Embassy (AFP 2011). In January 2013 several days of 'beyond index' air pollution was deemed an 'airpocalypse' by media and triggered the Chinese government to consider a series of drastic policy measures to address its acute air pollution. In October 2013 air pollution levels in Harbin in northeastern China were so severe that officials shut down airports and schools in the city of 11 million (Reuters 2013). Off-the-chart air pollution levels once again triggered a national state of emergency for China, inciting Beijing to develop a 'Heavy Air Pollution Contingency Plan' in preparation for impending intense air pollution during winter months, which are particularly prone to high air pollution events (Xinhua 2013b).

What is the cause of these 'airpocalpytic' levels of pollution in China, and what is being done to address it? Is China's air pollution an inevitable consequence of its rapid industrialization and economic development, or does this represent a case of exceptionalism? What are the consequences – in terms of human health, economy, politics, and society – of air pollution in China, both in the short and long term? How is the government working to address air pollution control in China, and what are the remaining challenges ahead? What are the regional and global consequences of China's air pollution? How China chooses to address its air pollution will have regional and global consequences, all of which are a concern for neighboring countries within East and South Asia. Due to its transboundary nature, how China deals with air pollution internally has consequences beyond its borders.

This chapter will introduce China's air pollution challenge – its causes and consequences – both within China and abroad. By presenting an analysis of the multiple sources and factors that contribute to both the complexity of the air pollution problem as well as the various actors and institutions that are required to implement a solution, this chapter will facilitate a deeper understanding of the complex forces that make environmental governance in China challenging. In doing so, China's experience of managing environmental impacts amid rapid industrial growth provides salient lessons for other countries and cities within Asia that are undertaking similar growth trajectories.

Understanding what comprises air pollution in China

The composition of air pollution within China is made complicated by the variation in the sources, climatic conditions, lifestyles, and populations that are unevenly distributed throughout the country. Obscured by natural sources and anthropogenic sources, China's air pollution in the broadest sense is comprised of both urban and rural particulate pollution due to fossil fuel (primarily coal) combustion and biomass burning, respectively. Frequent dust storms and natural sources of pollution also contribute to larger air particles, particularly in northeastern China, which is prone to frequent influxes of dust from the Gobi Desert (Ma et al. 2004). Many terms have been applied to articulate China's air pollution, including 'complex air pollution' – a term coined to refer to the mixture of coal combustion, vehicular emissions, and biomass burning that can address the difficulty of isolating sources and even the scale at which pollution occurs (Fang et al. 2009). For many years, citizens and officials within China simply referred to overcast, smoggy skies by the word *wu*, meaning fog, rather than the term for pollution, *wu ran*, suggesting confusion as to the nature and cause of urban smog and haze.

For human health, the dominant relevant health pollutants are by and large particulate matter (PM). Particulate matter pollution in populated regions originates from primary particles, such as soot and dust that result from combustion processes, and secondary particles, such as sulfates and nitrates, that form in atmospheric reactions (Brauer et al. 2012). Coarse and fine particulate matter (PM_{10} and $PM_{2.5}$), which have diameters of under 10 micrometers and 2.5 micrometers, respectively, are the current focus of air pollution control policies worldwide, although many developing countries are still primarily focused on regulation of PM_{10}. Long-term exposure to $PM_{2.5}$ has been found to cause serious negative health effects, such as increased mortality from chronic cardiovascular and respiratory disease and lung cancer (WHO 2006). Globally, $PM_{2.5}$ causes an estimated 1.2 percent of premature deaths per year, with 65 percent of these occurring in developing countries in Asia alone (Cohen et al. 2005). In most parts of China, $PM_{2.5}$ has been found to account for more than half of air particulate pollution, whereas in many other parts of the world with less severe pollution, $PM_{2.5}$ comprises a smaller proportion (Zhang et al. 2004). In some locations within China, $PM_{2.5}$ has been found to constitute higher than 70 percent of particulate pollution (Hopke et al., 2008).

Addressing fine particulate matter is now the primary focus of China's air pollution control efforts. What makes $PM_{2.5}$ particularly difficult to tackle is its formation as a secondary particle, meaning that measures taken to reduce $PM_{2.5}$ must also focus on other primary pollutants that react to form it. Primary emissions of sulfur dioxide (SO_2), nitrous oxides (NOx), volatile organic compounds (VOCs), black carbon, and ammonia (NH_3) react in the atmosphere to form $PM_{2.5}$. During winter months when the most extreme, staggering air pollution events typically occur, $PM_{2.5}$ is mainly formed as secondary – as opposed to – primary pollutants. Therefore, policies must tackle PM as both primary particulates, in addition to the other pollutants that react to form it as secondary pollutants. China has only started to grapple with control of NOx, ozone and VOCs. Recognizing its key role in producing secondary $PM_{2.5}$, the central government adopted pollutant reduction targets for NOx in the 12th Five-Year Plan, which was widely lauded as the greenest national policy plan to date (Seligsohn & Hsu 2011).

Sources of air pollution in China

China's air pollution arises from a variety of sources. Its composition varies depending on a number of factors, including location, geography, industrialization, and population. From a policy and governance perspective, there are natural conditions related to geography and climate that have dealt many cities in China and throughout East Asia a difficult hand with respect to managing environmental impacts with development. At the same time, these natural factors likely are not the dominant forces at play with respect to high levels of air pollution blanketing many Chinese cities. The following section will elaborate and explain some of the common sources of China's severe air pollution.

Weather conditions and temperature inversions

Cold weather and temperature inversions that occur during winter months can prevent the dispersal of air pollutions, leading to unusual buildups of air pollutants and the extreme 'air pollution events' that occur almost predictably in megacities such as Beijing. Inversions occur when cold air becomes trapped beneath a warmer layer of air above the earth's surface (Beard et al. 2012). During winter months, as pollution accumulates above the surface of megacities such as Beijing, cold air can form ice particles that serve as catalysts for the gases to react to produce more $PM_{2.5}$. In the 'airpocalypse' events in January 2013, secondary particulate pollution increased by as much as 80 percent, leading to levels of air pollution hazardously above WHO-recommended levels (Seligsohn 2013). However, cities outside China and developing countries are likewise not immune to periods of unusually high local air pollution, which occur when pollutants are trapped by temperature inversions. For example, Salt Lake City in the United States experienced inversions on 57 percent of winter days from 1994 through 2008, which have led to spikes in particulate matter pollution that exceed National Ambient Air Quality Standards (Bailey et al. 2011).

Heavy reliance on coal in the overall energy mix

At the end of 2011 70 percent of China's energy was generated from coal. In terms of electricity generation, around 80 percent is produced from coal (EIA 2012a). This proportion is much higher than other East Asian countries, such as Japan (65 percent) and South Korea (69 percent) (EIA 2012b, 2012c). During winter months, provinces in northeastern China suffer worse air quality because more coal is used to generate central heating in buildings. This phenomenon affecting primarily northeastern China is a relic of China's central planning period from 1950–1980, when the government provided free coal for boilers to heat homes and offices (Chen et al. 2013). Many cities, including Beijing, have plans to eliminate coal-fired boilers from the city center to reduce air pollution from these significant sources (*China Daily* 2013). The central government has also made plans to reduce the percentage of coal consumption in the overall energy mix. These policies include a limit on coal consumption to less than 65 percent of the overall energy mix, increasing the percentage of non-fossil fuels to 11.4 percent by 2015 and 13 percent by 2017 (State Council 2013). Plans for new coal-fired power plants are also being restricted from major metropolitan and industrial centers, including Beijing, Shanghai, and the Yangtze and Pearl River Delta regions in eastern and southern China, which are major manufacturing centers.

Despite these policies aimed at major urban and industrial areas, which will have a significant impact on air pollution in these locations, coal will still remain a significant contributor to China's energy mix for the foreseeable future. What is perhaps more worrisome is the shift of coal development from the Eastern and coastal regions in China to the interior and Western, less developed areas. More than 80 percent of proposed coal projects that were approved in 2012 are not covered by policies to eliminate coal-fired power plants (Yang & Cui 2013) (see Figure 10.1). What this suggests is the short-sightedness of national policies targeted singularly at present-day urban areas that fail to take into consideration future growth in less developed regions.

Increasing car ownership

Within urban areas in China, vehicle emissions are the fastest growing source of atmospheric particulates and the largest source of human exposure to $PM_{2.5}$ (Fang et al. 2009). At the end of 2012, passenger car ownership reached 88 million, while the total vehicle population numbered over 100 million (China National Statistical Yearbook 2012). While the percentage of the population owning personal vehicles is still low at 6.5 percent, this number is growing,

Angel Hsu

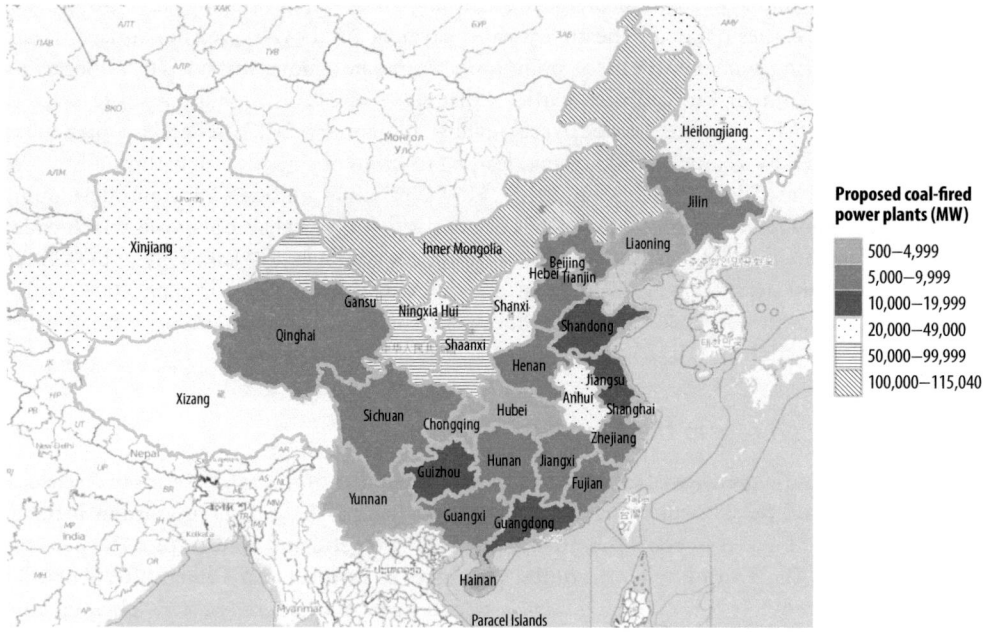

Figure 10.1 Map of planned coal-fired power plants

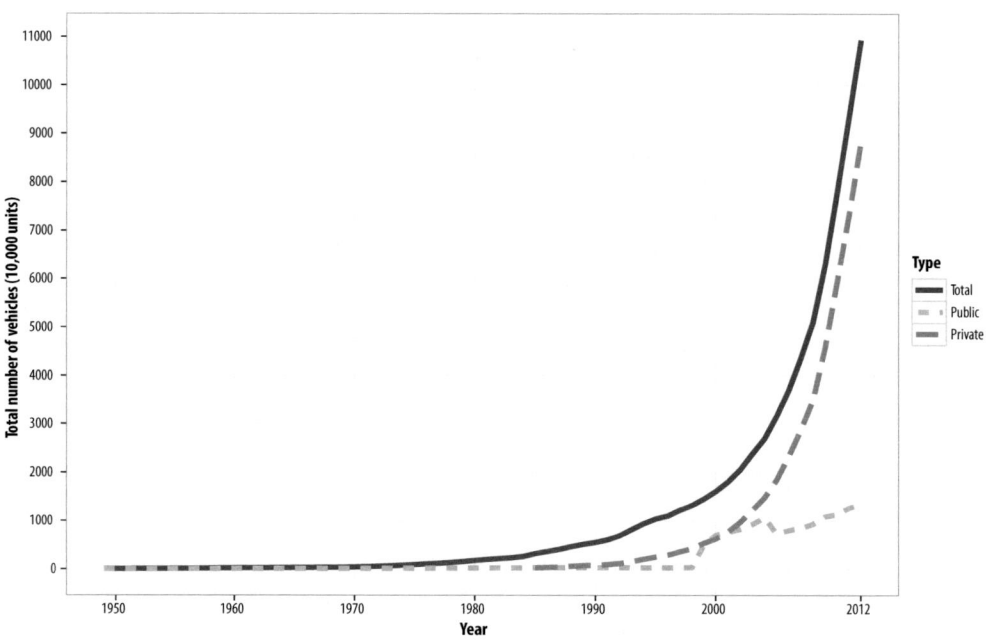

Figure 10.2 Trends in vehicle ownership in China

as China has become the world's largest car manufacturer and the largest new car market (*China Daily* 2010) (see Figure 10.2).

What is the more significant statistic in evaluating the impact of vehicle emissions, which can contribute between 20 and 50 percent of particulate matter pollution in cities in China (Larson 2013), is the population being exposed to vehicular emissions. For example, the exposure impact of roadside emissions is greater for cities such as Beijing, where about 75 percent of the population live near a roadway (Jerrett et al. 2010). Consequently, authorities are now beginning to think of more aggressive measures to curb emissions from vehicles. Several major cities within China have policies to regulate car use within metropolitan areas to address traffic congestion and air pollution, including Beijing, Nanchang, Changchun, Lanzhou, Guiyang, Hangzhou, and Chengdu (Zeng 2013). Several cities, including Shanghai, Beijing, Guiyang, and Guangzhou, have implemented additional restrictions such as the above-mentioned odd–even license plate regulations and vehicle quota systems to cap the growth of private vehicle ownership.

The elimination of heavy-polluting, yellow-label vehicles within urban centers will also reduce vehicular emissions and sources of pollution. Yellow-label or high-emitting vehicles only accounted for 28 percent of vehicles in 2009 but were responsible for around 75 percent of emissions (Xinhua 2009). To address this critical component of vehicular emissions, China's State Council in September 2013 approved a massive US\$275 billion air pollution control plan that included a pledge to remove of all of these yellow-label vehicles (State Council 2013). To provide more context for how ambitious this goal is, the number of yellow-label vehicles in China in 2011 totaled around 15 million – a substantial number of vehicles to completely remove from roadways (MEP 2011). Shanghai plans to eliminate 70,000 of these vehicles by the end of next year, while Beijing has already scrapped around 150,000 yellow-label vehicles (Beijing EPB 2013; Jiang 2013).

The government has also targeted the adoption of higher fuel quality standards to improve air quality within urban areas. The Chinese government in 2013 announced a timeline for the adoption of more stringent fuel quality standards that would bring the fuel quality for motor vehicles on par with the European Union and Japan by 2018 and more stringent than the United States (Wagner & He 2013). Residents in the three key industrial corridors (Beijing corridor, Yangtze and Pearl River Delta regions) will see the benefit of higher fuel quality standards by the end of 2015, which the 2013 State Council Air Pollution Control Plan establishes as a new goal. While the new 'China V' standards for ultra-low sulfur fuel will reduce vehicular sources particulate matter, SOx and NOx emissions, the latest State Council-approved air pollution plan lacks vehicle emission standards that would support other efforts to scrap older, more polluting vehicles and replace them with cleaner vehicles that can fully maximize the air pollution benefits from higher fuel quality standards,

Regional impacts of industrial activity

China's rapid industrialization, with average annual growth rates of 10 percent over the last two decades, is a major contributing factor to poor air quality. Because air pollution that affects major urban areas is largely regional rather than local in nature, industrial activities that occur outside of urban centers still influence air quality within metropolitan areas. Streets et al. (2007) estimated that, on average, 34 percent of $PM_{2.5}$ pollution and 35–60 percent of ozone pollution originate from areas outside of Beijing. Neighboring province Hebei contributes between 50 and 70 percent of Beijing's particulate pollution (Streets et al. 2007).

Inefficient, polluting industrial enterprises are a primary culprit in China's regional air pollution problems. China's industrial sector traditionally has not been characterized by a high degree of concentration, meaning most manufacturing occurs in small facilities due to market demand and the high cost of energy intensive goods (Price et al. 2011). For example, in 2007 steel production from the top 10 largest iron and steel companies in China represented only a little more than a third of total production, suggesting that two-thirds of production was occurring in small and medium-sized enterprises (SMEs) (Shan & Wang 2007). During the 11th Five-Year Plan (2005–2010), China established a Structural Adjustment/Small Plant Closure program that targeted 14 high energy-consuming industries, including electric power, iron and steel, chemicals, cement, coal, aluminum, and pulp and paper for elimination and shutdown of small plants and outdated facilities (Price et al. 2011). Through this program, China was able to meet most of its 20 percent reduction in energy-intensity goal – a target that was carried over to the 12th Five-Year Plan period (2011–2015). However, it is recognized within the Chinese leadership that most of the low-hanging fruit with respect to energy savings were achieved through these SME plant and facility closures during the 11th Five-Year Plan. The challenge to achieve even greater returns in the face of economic slowdown and necessary economic restructuring is a task China's new leadership has inherited.

Agricultural biomass burning

In peri-urban and rural areas, biomass burning of agricultural waste is a major source of air pollution that also contributes to poor urban air quality (Duan et al. 2004). In May 2012 high levels of air pollution in Wuhan, the capital of Hubei province, was the worst pollution experienced by the city in a decade (Bloomberg 2012). Satellite imagery revealed multiple biomass incineration fires that occurred in the perimeter of Wuhan were largely responsible for spikes of pollution that resulted in thick, orange-brown smog covering the area (Hsu 2012).

Some countries, including India, have managed to provide farmers alternatives to agricultural waste burning in an effort to improve air quality. During harvest season, when agricultural waste burning peaks, villagers in Ghanaur, located in Punjab, collect 120,000 pounds of straw and other agricultural waste products to generate 12 megawatts of electricity that is then fed back into the grid (Yee 2013). Such alternatives could address gaps in education and public awareness that environmental protection officials in China say are ineffective in terms of curbing biomass incineration (Hsu 2013). Providing alternatives such as waste-to-power plants would provide benefits for air quality and reduce the carbon intensity of electricity generation.

Urban growth

China's population has rapidly urbanized since the 1980s, and by 2011, more Chinese officially resided in urban than rural areas for the first time in the nation's history, with 690 million residing in urban areas and 656 million in rural areas (CASS 2012). The number of urban dwellers in China is only expected to increase, with an additional 400 million predicted to live in cities by 2025 (Woetzel et al. 2009). The influx of new urban residents will not only drive the construction of additional built infrastructure, but massive urban migration will also increase the demand for vehicles and public transport – all sources of air pollution described in this section. The November 2013 meetings of the 18th National People's Congress once again emphasized increased urbanization as a high-level policy focus, to encourage the growth of domestic consumption that is needed for China to shift from a primarily export-led economy. The patterns and shapes of urban development – for example, whether growth sprawls or is densely compact – affect infrastructure choices, transportation, and ultimately energy consumption and air quality (Ewing & Rong 2008). Infrastructure, once developed, can lead to a form of path dependency with respect to energy use and may constrain or lock urban residents into poor air quality for decades to come.

Consequences of China's air pollution

China's severe air pollution has led to a range of 'pressure points' in the domains of health, society, economy, and politics. Such pressure points, if not addressed, could have serious implications for the future legitimacy of China's leadership, which are discussed in this section.

The health effects of toxic air pollution are one of the most apparent consequences of China's industrialization. The latest Global Burden of Disease Study estimates that air pollution resulted in 1.2 million premature deaths in China in 2010 (GBD 2012). Another study demonstrates that residents living north of the Huai River – the arbitrary line the central Chinese government set as a boundary between areas receiving free coal for the purpose of winter heating – lived on average 6 years less than people living south of the river (Chen et al. 2013). In an aggregate sense, Chen et al. (2013) estimate that the 500 million residents living in northern China during the 1990s experienced a collective loss of more than 2.5 billion life years due to increased total suspended particulates and pollution originating from greater coal consumption tied to heating. Higher asthma rates and hospitalizations, particularly among young children, have been reported throughout cities in China. In some areas, asthma affects up to 11 percent of children (Yangzong et al. 2012). Not only is asthma, which can be fatal, the leading cause of hospitalization among children, but it can also result in other economic and societal losses due to school absenteeism and the costs of medical care (Wilkinson 2013).

More controversial are economic estimates of the costs of environmental degradation and pollution. Matus et al. (2012) estimate the loss of labor and the increased demand for health care due to air pollution cost the Chinese economy $112 billion of its GDP in 2005, compared to $22 billion in 1975. Poor air quality advertised through international media outlets has discouraged tourists from visiting Beijing, which reported a 50 percent reduction in the number of tourists in 2013 (Deng 2013). The financial investments required to address air pollution also represent substantial costs. China's comprehensive air pollution control plan, approved in 2013 by the State Council, has a price tag of around US$275 billion. In comparison, the 1990 U.S. Clean Air Act Amendments only cost US$85 billion (USEPA 2013).

But perhaps the greatest pressure point with respect to China's air pollution is the potential ramifications for continued performance legitimacy, which the leadership relies on to maintain social stability. The shift from a form of legitimacy and authority based on ideology to one based on performance is contingent on continued achievement and results (Huntington 1991). High economic growth rates provided for sustained performance legitimacy since market reforms began in 1978. However, as standards of living and income increase on a per capita basis, citizens are making more demands with respect to environmental quality. One Chinese official cited a growing number of environmental protests by 29 percent per year from 1996 (Jie & Tao 2012). After 2012 saw several high-profile citizen environmental protests, including one at Shifang in Sichuan province that turned violent, the Chinese government issued mandatory 'social risk' assessments to prevent any potential citizen conflict for any major industrial project (Bradsher 2012a, 2012b).

The desire to maintain performance legitimacy and social stability explains the swift response on the part of the government to address citizen grievances on opaque air pollution statistics and misleading information. The Chinese citizen frustrations and demands for nationwide $PM_{2.5}$ data vocalized via social media did not go unnoticed by the government. Figure 10.3 shows the near exponential increase in mentions of '$PM_{2.5}$' from January 2011 to January 2013 on the popular microblog platform Sina Weibo. In response to the heightened citizen concern for air quality, the MEP announced within weeks following the initial heightened concern that it would release trial $PM_{2.5}$ data in January 2012, immediately prior to the Lunar New Year holidays (Xinhua 2012). The MEP then amended proposed air quality standards, which initially failed to include standards for $PM_{2.5}$ for political reasons, fearing potential backlash if emission numbers were too high (Hsu 2011; MEP 2012).

To address data availability gaps and to counter citizen claims of data secrecy and manipulation, the Chinese government made available real-time $PM_{2.5}$ monitoring data for 74 cities in China at the beginning of 2013, with the goal of expansion to 113 key environmental cities by 2016 (Xinhua 2013a). The example of the government's prompt reaction to Chinese citizen demands demonstrates the impact of active, engaged citizens in spite of an authoritarian governance structure that is led from the top down. It remains to be seen whether the success of the Chinese citizenry to effectively 'lobby' the government for better air quality will trickle into other aspects of Chinese society and politics.

What is evident, however, is that to sustain political legitimacy based on environmental performance, public participation has already proven critical. Political legitimacy, according to Haas (2004), 'rests on a process of knowledge development and diffusion that is scrupulously free of political interference,' implying the necessity of public participation for information provision. The inclusion of multiple actors for data can bolster the quality and legitimacy of policies, while building the capacity of those involved to engage in the policy process (NRC 2008). However, incorporating meaningful public participation in China's environmental governance, particularly for the purposes of citizen engagement in data and information provision, will require institutional and legal reform. Laws in China only allow for designated government bodies, such as the National Bureau of Statistics or the Ministry of Environmental Protection, to release environmental information or to issue statistics (SEPA 2007). Citizens,

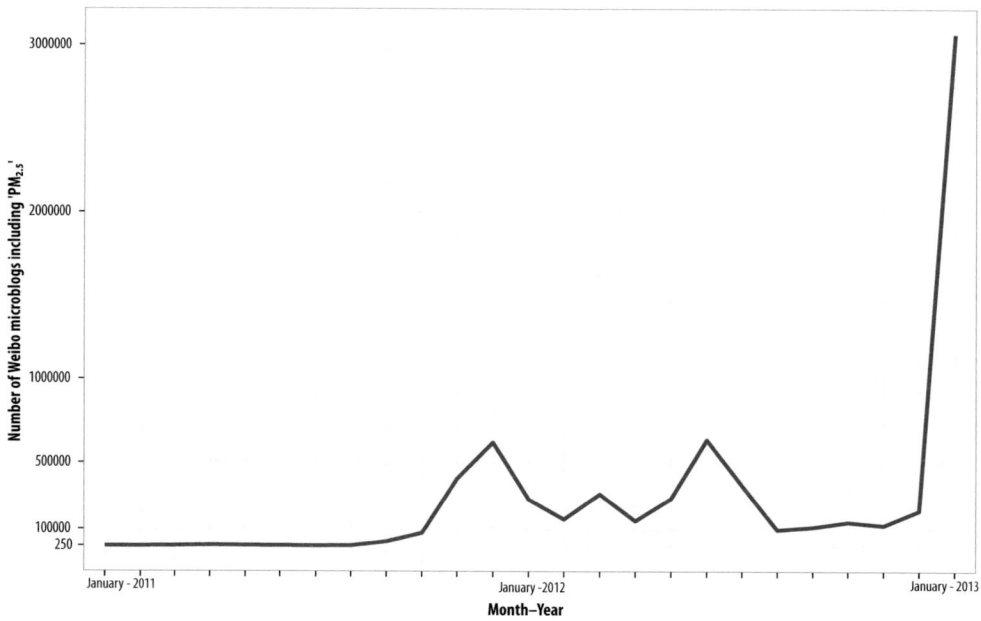

Figure 10.3 Mentions of '$PM_{2.5}$' on Sina Weibo, January 2011–2013

while able to collect environmental data, are forbidden from publicly releasing or reporting the information. Public engagement can lead to improved results for environmental quality and other social objectives, as well as to enhance trust and understanding among parties (NRC 2008).

There are examples that suggest citizens within China are already taking individual action and contributing to an increasingly vocal and active society. With more than 500 million internet users and at least half using social networking sites, the netizen outcry over misleading air pollution statistics demonstrates the power of social media and the internet – an increasingly menacing threat that a new Chinese leadership will need to address if it is to contain potentially explosive environmental conflicts (Kan 2012). Chinese real estate mogul Pan Shiyi conducted a poll in early 2013 on the ubiquitous microblogging platform Sina Weibo to gauge citizen support for a Clean Air Act. Within less than 10 hours of voting, Pan garnered nearly 32,000 votes in support of his call for government action (Chin 2013). He was also one of the driving forces behind an online campaign demanding the release of $PM_{2.5}$ data by the Beijing municipal government (Page 2011). The role of mobile applications, backpack air monitors, and data-crowdsourcing could usher in a form of citizen science to improve air pollution monitoring and governance within China.

Regional consequences within East Asia

Due to the transboundary nature of air pollution, China's air quality has effects beyond its immediate borders. China's sulfur dioxide emissions have been estimated to contribute to one-fourth of global emissions and up to 90 percent of East Asia's since the 1990s (Lu et al. 2010). Nowhere is this more visible than through satellite imagery of the 'Asian Brown Cloud' – thick plumes of aerosols and particulate pollution that regularly sweep over East Asia and can migrate into the Western United States (Jacobs 2008). While primarily originating from the sources of pollution discussed earlier in this chapter from China and parts of South and Southeast Asia, seasonal weather patterns mean that these pollution clouds permeate the Korean peninsula and Japan.

What fraction of air pollution in other East Asian countries originates from China is a subject of debate within scientific communities. For example, scientists within China and Japan vary widely on the amount that China contributes to Japan's sulfur deposition. China claims only 3.5 percent (Huang et al. 1995), while some studies conducted by Japanese scientists suggest that the sulfur dioxide contribution of China to Japan is 'conservatively, above 40 percent' throughout Japan (Aikawa et al. 2010). China has even disputed the results of scientific models, citing methodological issues of models such as the RAINS-Asia (Regional Air Pollution Information Simulation in Asia) – a research effort jointly funded by the Netherlands, Norway and Sweden through the World Bank and Asian Development Bank (Kim 2007; Nam & Lee 2013).

The lack of consensus within the scientific communities in East Asia could partly explain why to date there is no legally binding regional agreement to address air pollution, given the success of efforts such as the Convention on Long-Range Transboundary Air Pollution (LRTAP) in Europe. Severe air pollution events like the 1952 London Fog, in which 3,000 people died from air pollution-related mortality, catalyzed efforts to understand the link between air quality and epidemiology, and underlined the need to effectively control harmful anthropogenic emissions (Bell & Davis 2001). Since that time, and following other large pollution events, an evolving international movement has sought to monitor the global spread and impact of air pollutants, including LRTAP, which began a successful effort to track and reduce sulfur dioxide emissions across Europe. LRTAP was the first internationally, legally binding instrument to address regional issues and is widely lauded as a success.

It is still puzzling, then, why regional cooperation within East Asia with respect to air pollution has not followed the success of the European model. While there are examples of inter-governmental cooperation, including the Acid Deposition Monitoring Network in East Asia (EANET), Northeast Asian Subregional Programme on Environmental Cooperation (NEASPEC) and the U.S.–China Regional Air Quality Management Program (RAQM), such efforts are primarily focused on capacity building, information sharing, and technology transfer. Efforts stop short of specifying targets or goals for regional air pollution reduction or a legally-binding agreement similar to that of LRTAP. In fact, the case could be made that the need for regional cooperation within East Asia is even more urgent than three decades ago when LRTAP was formed, given the pace and scale of industrialization and urbanization in East Asian countries. Kim (2007) suggests the lack of regional cooperation comparable to Europe's successful regulatory regime could be due to two explanations. First, the high economic costs of reducing emissions and debate as to the ecological vulnerability to transboundary pollution could be to blame (Levy 1993). Second, as described, the lack of consensus within epistemic communities as to the origination of transboundary pollution could hinder progress toward greater regional cooperation (Kim 2007). Moreover, political sensitivities surrounding responsibility and attribution, as demonstrated through other international treaties such as the Kyoto Protocol of the UN Framework Convention on Climate Change, could be another factor inhibiting similar binding agreements for air pollution in East Asia.

Regional cooperation on air pollution is necessary for long-term emissions reductions, given the anticipated economic and urban growth in Asia. Seto et al. (2012) predict a 185 percent increase in global urban land area extent by 2030, with nearly half of the growth forecasted to occur in Asia and predominantly in China and India. Data suggest that other countries, particularly those in South Asia, may experience even worse air quality than China, highlighting the urgency for China to develop solutions that could have salience for these other countries (YCELP & CIESIN 2012). Many of these countries lag behind China in terms of air pollution control policies, including pollution monitoring and air quality standards (CAI-Asia 2010).

China's air pollution could also have further reaching economic and political impacts, both regionally and globally. For example, border-adjustment taxes on energy and carbon-intensive products have been discussed within the United States and Europe as a way of leveling the playing field between domestic producers, who may face costly environmental regulation policies, and those abroad, who may be able to produce goods more cheaply because of lax environmental regulation (Houser et al. 2008). One could imagine similar trade measures imposed by major competitors within East Asia as a means of incentivizing policymakers within China to take even more ambitious measures to reduce air pollution. While such border-adjustment tax policies are controversial in terms of the uncertain effectiveness in driving environmental results and feasibility, the full consequences of China's air pollution regionally are yet to be seen.

Conclusion

This chapter has provided an introduction to an understanding of China's air pollution – its composition, sources, and consequences both domestically and regionally within East Asia. Understanding what comprises China's air pollution, where it is derived, and what is being done to address it provides insight as to why predictions for its improvement project several decades into the future. What is encouraging are policy developments over the last few years that suggest the Chinese government is actively working to address severe levels of air pollution that have not only affected its citizens, but have damaged its international reputation through near continuous media attention criticizing the deleterious environmental effects of its rapid industrialization. The State Council's approval of a costly air pollution reduction and control plan at least signals high-level commitment and financial backing. Nonetheless, implementation of environmental policies, despite the inclusion of a growing number of hard or binding environmental targets for local government officials, is still a challenge in most of China (Hsu 2013).

In the short term, Chinese citizens are unfortunately experiencing a 'new normal' of spikes in extreme air pollution and levels of emissions that regularly exceed safe recommended levels. The term '$PM_{2.5}$' has almost become ubiquitous throughout China, a term which prior to 2013 did not even have official Chinese recognition (Hsu & Miao 2013). Rock music concerts in China have even adopted '$PM_{2.5}$' for headlines and advertisements, complete with a silhouette of a person donning a pollution mask (Zhu 2012). Face masks, indoor air purifiers, and retreats to the less polluted countryside are coping mechanisms urban dwellers in China's cities have had to adopt in recent years. School closings and emergency alert systems to stay indoors during high pollution events are becoming everyday realities.

Regardless, air quality in China and East Asia will not see significant improvement unless there are regional mechanisms for cooperation that have more teeth and can move beyond intellectual disputes and political sensitivities. Regional coordination efforts focus almost exclusively on data monitoring and information sharing. While these initiatives are critical to establish baseline parameters, the adaptation of frameworks that have proved successful elsewhere or the creation of new models that can move beyond isolated efforts are needed. As air pollution is by nature transboundary, so should cooperation extend beyond borders to improve air quality not just in China, but the whole of East Asia.

Bibliography

AFP (2011) *Beijing air pollution 'hazardous'*, US embassy: Associated Foreign Press. Available: http://www.google.com/hostednews/afp/article/ALeqM5iJTkt3-cVITDVI6xipFX4aAyYjpw?docId=CNG.d957d0999e1088b0ce61729ec5b6c9f1.5f1. Last accessed: November 1, 2012.

Aikawa, M., Ohara, T., Hiraki, T., Oishi, O., Tsuji, A., Yamagami, M. et al. (2010) 'Significant Geographic Gradients in Particulate Sulfate over Japan Determined from Multiple-site Measurements and a Chemical Transport Model: Impacts of Transboundary Pollution from the Asian Continent', *Atmospheric Environment* 44(3), 381–391.

Bailey, A., Chase, T.N., Cassano, J.J. & Noone, D. (2011) 'Changing Temperature Inversion Characteristics in the US Southwest and Relationships to Large-scale Atmospheric Circulation', *Journal of Applied Meteorology and Climatology* 50(6), 1307–1323.

Beard, J.D., Beck, C., Graham, R., Packham, S.C., Traphagan, M., Giles, R,T. et al. (2012) 'Winter Temperature Inversions and Emergency Department Visits for Asthma in Salt Lake County, Utah, 2003–2008', *Environmental Health Perspectives* 120, 1385–1390.

Beijing Environmental Protection Bureau (Beijing EPB) (2013) 'Status of Yellow-Label Vehicles'. Available at http://www.bjepb.gov.cn/portal0/tab189/info10474.htm.

Bell, M.L. & Davis, D. (2001) 'Reassessment of the Lethal London Fog of 1952: Novel Indicators of Acute and Chronic Consequences of Acute Exposure to Air Pollution', *Environmental Health Perspectives* 109(Suppl 3), 389–394.

Bloomberg (2012) China Detains 2 for Wuhan Pollution Rumor, Daily Reports. Bloomberg News Agency. June 12. Available: http://www.bloomberg.com/news/2012-06-12/china-detains-2-for-wuhan-pollution-rumor-daily-reports.html. Last accessed: January 8, 2013.

Bradsher, K. (2012a) 'Chinese City Suspends Factory Construction After Protests', *New York Times*, 3 July. Available at http://www.nytimes.com/2012/07/04/world/asia/chinese-city-suspends-factory-construction-following-protests.html?_r=0.

Bradsher, K. (2012b) "'Social Risk" Test Ordered by China for Big Projects', *New York Times*, 12 November. Available at http://www.nytimes.com/2012/11/13/world/asia/china-mandates-social-risk-reviews-for-big-projects.html.

Brauer, M., Amann, M., Burnett, R.T., Cohen, A., Dentener, F., Ezzati, M. et al. (2012) 'Exposure assessment for Estimation of the Global Burden of Disease Attributable to Outdoor Air Pollution', *Environmental Science & Technology* 46(2), 652–660.

Chen, Y., Jin, G.Z., Kumar, N. & Shi, G. (2011) The Promise of Beijing: Evaluating the Impact of the 2008 Olympic Games on Air Quality (No. w16907), National Bureau of Economic Research.

Chen, Y., Ebenstein, A., Greenstone, M. & Li, H. (2013) 'Evidence on the Impact of Sustained Exposure to Air Pollution on Life Expectancy from China's Huai River Policy', *Proceedings of the National Academy of Sciences* 110(32), 12936–12941.

Chin, J. (2013) 'China Internet Users Scream for Clean Air Act', *Wall Street Journal*, 29 January. Available at http://blogs.wsj.com/chinarealtime/2013/01/29/chinese-internet-users-scream-for-clean-air-act/.

China Daily (2010) 'China: World's biggest auto producer, consumer', *China Daily*. Available: http://www.chinadaily.com.cn/bizchina/2010-01/12/content_9309129.htm. Last accessed: October 21, 2013.

China Daily (2013) 'Beijing to shut coal-fired boilers to clean up air', *China Daily*. May 27. http://www.chinadaily.com.cn/china/2013-05/27/content_16533835.htm. Last accessed: October 10, 2013.

China National Environmental Monitoring Centre (CNEMC) (2013) Available at http://www.cnemc.cn.

China National Statistical Yearbook (2012) China Data Online. Available at http://chinadataonline.org.

Chinese Academy for Environmental Planning (CAEP) (2010) Emission Trading Scheme made big progress in the 11th Five-year Plan. Briefing of Environmental Economic Policy Pilot Project, December, 167 (in Chinese).

Chinese Academy of Social Sciences (CASS) (2012) *The Urban Blue Book: China City Development Report* (Social Sciences Academic Press).

Clean Air Initiative – Asia (CAI-Asia) (2010) 'Air Quality in Asia: Status and Trends'. Available at cleanairinitiative.org/portal/system/files/documents/AQ_in_Asia.pdf_.

Clean Air Initiative – Asia (CAI-Asia) (2012) 'Accessing Asia: Air Pollution and Greenhouse Gas Emissions from Road Transport and Electricity. Clean Air Asia, Pasig City, Philippines'. Available at http://cleanairinitiative.org/portal/node/11573.

Cohen, A.J., Ross Anderson, H., Ostro, B., Pandey, K.D., Krzyzanowski, M., Künzli, N. et al. (2005) 'The Global Burden of Disease Due to Outdoor Air Pollution', *Journal of Toxicology and Environmental Health, Part A* 68(13–14), 1301–1307.

Deng, C. (2013) 'Pollution Halves Visitors to Beijing', *Wall Street Journal*, 31 October. Available at http://blogs.wsj.com/chinarealtime/2013/10/31/beijing-air-pollution-drives-50-drop-in-visitors/.

Duan, F., Liu, X., Yu, T. & Cachier, H. (2004) 'Identification and Estimate of Biomass Burning Contribution to the Urban Aerosol Organic Carbon Concentrations in Beijing', *Atmospheric Environment* 38(9), 1275–1282.

Energy Information Agency (EIA) (2012a) 'Countries Analysis: China', 4 September. Available at http://www.eia.gov/countries/cab.cfm?fips=CH.

Energy Information Agency (EIA) (2012b) 'Countries Analysis: Japan', 4 September. Available at http://www.eia.gov/countries/country-data.cfm?fips=JA.

Energy Information Agency (EIA) (2012c) 'Countries Analysis: Korea – South', 4 September. Available at http://www.eia.gov/countries/country-data.cfm?fips=KS.

Ewing, R. & Rong, F. (2008) 'The Impact of Urban Form on US Residential Energy Use', *House Policy Debate* 19, 1–30.

Fang, M., Chan, C.K. & Yao, X. (2009) 'Managing Air Quality in a Rapidly Developing Nation: China', *Atmospheric Environment* 48(1), 79–86.

Global Burden of Disease (GBD) (2012) 'Global Burden of Disease Study 2010', *Lancet*. Available at http://www.thelancet.com/themed/global-burden-of-disease.

Haas, P.M. (2004) 'When Does Power Listen to Truth? A Constructivist Approach to the Policy Process', *Journal of European Public Policy* 11(4), 569–592.

He, G., Lu, Y., Mol, A.P.J. & Beckers, T. (2012) 'Changes and Challenges: China's Environmental Management in Transition', *Environmental Development* 3, 25–38.

Hopke P., Cohen, D., Begum, B., Biswas, S., Ni, B., Pandit, G., et al. (2008) 'Urban Air Quality in the Asian Region', *Science of the Total Environment* 404, 1103–1112.

Houser, T., Bradley, R., Childs, B., Werksman, J. & Heilmayr, R. (2009) 'Leveling the Carbon Playing Field: International Competition and US Climate Policy Design', *World Trade Review* 8(4), 611.

Hsu, A. (2011) 'China Amends Air Quality Measures but Misses Key Pollutant – PM 2.5'. Personal blog, 22 March. Available at http://hsu.me/2011/03/china-amends-air-quality-measures-but-misses-key-pollutant-pm-2–5/.

Hsu, A. (2012) 'Wuhan's Decade's Worst Air Pollution'. Personal blog, 17 August. Available at http://hsu.me/2012/08/wuhans-decades-worst-air-pollution/.

Hsu, A. (2013) 'Limitations and Challenges of Provincial Environmental Protection Bureaus in China's Environmental Monitoring, Reporting, and Verification', *Environmental Practice* 15(3), 280–292.

Hsu, A. & Miao, W. (2013) 'What's In A Name? That Which We Call "PM$_{2.5}$", China Doesn't'. *The Metric*, Yale Center for Environmental Law and Policy. Available at http://environment.yale.edu/envirocenter/post/whats-in-a-name-that-which-we-call-pm2.5-china-doesnt/.

Huang, M., Wang, Z., He, D., Xu, H., and Zhou, L. (1995) 'Modeling studies on sulfur deposition and transport in East Asia', *Water, Air, Soil Pollution* 85, 1921–1926.

Huntington, S. (1991) *The Third Wave: Democratization in the Late 20th Century. Vol. 4* (University of Oklahoma Press).

Jacobs, A. (2008) 'UN Reports Pollution Threat in Asia', *New York Times*, 13 November. Available at http://www.nytimes.com/2008/11/14/world/14cloud.html.

Jerrett, M., Su, J., Apte, J., and Beckerman, B. (2010) Estimates of Population Exposure to Traffic-related Air Pollution in Beijing, China and New Delhi, India: Extending exposure analyses reported in HEI Special Report 17: Traffic-Related Air Pollution: Critical Review of the Literature on Emissions, Exposure, and Health Effects. Health Effects Institute: Boston. Available: http://www.healtheffects.org/International/Jerrett_Asia_Traffic_Exposure.pdf. Last accessed: June 27, 2014.

Jiang, Y.C. (2013) 'Shanghai EPB Will Eliminate Yellow Cars' (in Chinese), *New People's Daily*, 27 November. Available at http://news.sina.com.cn/c/2013–11–27/150928826461.shtml.

Jie, F. & Tao, W. (2012) 'Officials Struggling to Respond to China's Year of Environment Protests, *China Dialogue*, 12 June. Available at http://www.chinadialogue.net/article/show/single/en/5438-Officials-struggling-to-respond-to-China-s-year-of-environment-protests-.

Kan, M. (2012) 'China's Internet Users Cross 500 Million', PC World. Available at http://www.pcworld.com/article/248229/chinas_internet_users_cross_500_million.html.

Kim, I. (2007) 'Environmental Cooperation of Northeast Asia: Transboundary Air Pollution', *International Relations of the Asia-Pacific* 7(3), 439–462.

Larson, C. (2013) 'China's Autos Need to Emit Less Pollution', *Bloomberg Businessweek*, 4 February. Available at http://www.businessweek.com/articles/2013-02-04/chinas-autos-need-to-emit-less-pollution.

Levy, M.A. (1993) 'European Acid Rain: the Power of Tote-board Diplomacy', *Institutions for the Earth: Sources of Effective International Environmental Protection*, 75–132.

Lim, L. (2013) 'Beijing's "Airpocalypse" Spurs Pollution Controls, Public Pressure', *NPR*, 14 January. Available at http://www.npr.org/2013/01/14/169305324/beijings-air-quality-reaches-hazardous-levels.

Lim, S.S., et al. (2012) 'A Comparative Risk Assessment of Burden of Disease and Injury Attributable to 67 Risk Factors and Risk Factor Clusters in 21 Regions, 1990–2010: A Systematic Analysis for the Global Burden of Disease Study 2010', *Lancet* 380.9859: 2224–2260.

Lu, Z., Streets, D.G., Zhang, Q., Wang, S., Carmichael, G.R., Cheng, Y.F., et al. (2010) 'Sulfur Dioxide Emissions in China and Sulfur Trends in East Asia since 2000', *Atmospheric Chemistry and Physics* 10(13), 6311–6331.

Ma, C., Tohno, S., Kasahara, M. & Hayakawa, S. (2004) 'Properties of Individual Asian Dust Storm Particles collected at Kosan, Korea during ACE-Asia', *Atmospheric Environment* 38, 1133–1143.

Macur, J. (2007) 'Beijing Air Raises Questions for Olympics', *New York Times*, 26 August. Available at http://www.nytimes.com/2007/08/26/sports/othersports/26runners.html?pagewanted=all.

Matus, K., Nam, K.M., Selin, N.E., Lamsal, L.N., Reilly, J.M. & Paltsev, S. (2012) 'Health Damages from Air Pollution in China', *Global Environmental Change* 22(1), 55–66.

Ministry of Environmental Protection (MEP) (2011) 'Chinese Ministry of Environmental Protection's Motor Vehicle Pollution Prevention and Control Situation in 2012' (in Chinese). Available at http://www.mep.gov.cn/gkml/hbb/qt/201212/t20121227_244340.htm.

Ministry of Environmental Protection (MEP) (2012) 'Ambient Air Quality Standards' (in Chinese). Available at http://kjs.mep.gov.cn/hjbhbz/bzwb/dqhjbh/dqhjzlbz/201203/W020120410330232398521.pdf.

Ministry of Environmental Protection (MEP) (2013) 'The Ministry of Environmental Protection issues Status of Urban Air Quality in the 74 Key Areas in the First Half of 2013' (in Chinese), June. Available at http://www.mep.gov.cn/gkml/hbb/qt/201307/t20130731_256638.htm.

Nam, S. & Lee, H. (2013) 'Reverberating Beyond the Region in Addressing Air Pollution in North-East Asia'. Available at www.reg-observatory.org/docs/EEREG_NamLee.doc_.

National Research Council (NRC) (2008) *Public Participation in Environmental Assessment and Decision Making, Panel on Public Participation in Environmental Assessment and Decision Making* (National Academies Press).

Page, J. (2011) 'Microbloggers Pressure Beijing to Improve Air Pollution Monitoring', *Wall Street Journal*, 8 November. Available at http://blogs.wsj.com/chinarealtime/2011/11/08/internet-puts-pressure-on-beijing-to-improve-air-pollution-monitoring/.

Price, L., Levine, M.D., Zhou, N., Fridley, D., Aden, N., Lu, H., et al. (2011) 'Assessment of China's Energy-saving and Emission-reduction Accomplishments and Opportunities during the 11th Five Year Plan', *Energy Policy* 39(4), 2165–2178.

Reuters (2013) 'China smog emergency shuts city of 11 million people', *Reuters*. October 21. Available: http://www.reuters.com/article/2013/10/21/us-china-smog-idUSBRE99K02Z20131021. Last accessed: October 30, 2013.

Seligsohn, D. (2013) 'China's New Regional Air Quality Regulations: A Win-Win for Local Air Quality and the Climate', *ChinaFAQs*. Available at http://www.chinafaqs.org/blog-posts/china%E2%80%99s-new-regional-air-quality-regulations-win-win-local-air-quality-and-climate#sthash.qwladSqv.dpuf.

Seligsohn, D. & Hsu, A. (2011) 'How Does China's 12th Five-Year Plan Address Energy and the Environment?', *World Resources Institute*. Available at http://www.wri.org/blog/how-does-china%E2%80%99s-12th-five-year-plan-address-energy-and-environment.

Seto, K.C., Guneralp, B.& Hutyrac, L.R. (2012) *Global Forecasts of Urban Expansion to 2030 and Direct Impacts on Biodiversity and Carbon Pools* (PNAS).

Shan, S. & Wang, X. (2007) 'Why it is So Difficult to Adjust the Industrial Structure of Steel Industry in China?', *China Steel* 5, 9–11.

State Council (2012) 'Notice on the Implementation of the Ambient Air Quality Standards' (in Chinese). Available at http://www.gov.cn/zwgk/2012–03/02/content_2081004.htm.

State Council (2013) 'State Council Notice on the Issuance of Air Pollution Control Plan of Action' (in Chinese). Available at http://www.gov.cn/zwgk/2013–09/12/content_2486773.htm.

State Environmental Protection Agency (SEPA) (2007) 'Regulations for Environmental Monitoring and Management' (in Chinese). Available at http://www.zhb.gov.cn/info/gw/juling/200708/t20070807_107652.htm.

Streets, D.G., Fu, J.S., Jang, C.J., Hao, J., He, K., Tang, X., et al. (2007) 'Air Quality During the 2008 Beijing Olympic Games', *Atmospheric Environment* 41(3), 480–492.

U.S. Energy Information Agency (EIA) (2013) 'China Profile'. Available at http://www.eia.gov/COUNTRIES/cab.cfm?fips=CH.

U.S. Environmental Protection Agency (USEPA) (2013) 'Benefits and Costs of the Clean Air Act: Second Prospective Study – 1990 to 2020'. Available at http://www.epa.gov/air/sect812/prospective2.html.

Wagner, D.V. & He, H. (2013) 'China Announces Breakthrough Timeline for Implementation of Ultra-low Sulfur Fuel Standards' International Council on Clean Transportation, Policy Update, March. Available at http://theicct.org/sites/default/files/publications/ICCTupdate_CH_fuelsulfur_mar2013.pdf.

Wang, T., Nie, W., Gao, J., Xue, L.K., Gao, X.M., Wang, X., et al. (2010) 'Air Quality during the 2008 Beijing Olympics: Secondary Pollutants and Regional Impact', *Atmospheric Chemistry and Physics* 10(16), 7603–7615.

Wilkinson, L. (2013) 'China's Asthma Problem is Bad – and it's Getting Worse', *The Atlantic*, 26 June. Available at http://www.theatlantic.com/china/archive/2013/06/chinas-asthma-problem-is-bad-and-growing-worse/277250/.

Woetzel, J., Mendonca, L., Devan, J., Negri, S., Hu, Y., Jordan, L., et al. (2009) 'Preparing for China's Urban Billion', McKinsey Global Institute. Available at http://www.mckinsey.com/insights/mgi/research/urbanization/preparing_for_urban_billion_in_china.

World Bank and State Environmental Protection Agency (SEPA) (2007) 'Cost of Pollution in China: Economist Estimates and Physical Damages', World Bank, Washington, D.C. Available at http://documents.worldbank.org/ curated/en/2007/02/7503894/cost-pollution-china-economic-estimates-physical-damages.

World Health Organization (WHO) (2005) 'Air Quality Guidelines Global Update 2005'. Available at http://whqlibdoc.who.int/hq/2006/WHO_SDE_PHE_ OEH_06.02_eng.pdf.

World Health Organization (WHO) (2006) Air Quality Guidelines. Global Update 2005. Particulate Matter, Ozone, Nitrogen Dioxide and Sulfur Dioxide, World Health Organization, Geneva.

Xinhua (2009) 'Chinese Official Urge Early Exit of High-emission Vehicles to Curb Air Pollution', Xinhua News Agency. Available at http://english.gov.cn/2009–08/13/content_1391467.htm.

Xinhua (2010) 'China: World's Biggest Auto Producer, Consumer', *China Daily*. Available at http://www.chinadaily.com.cn/bizchina/2010–01/12/content_9309129.htm.

Xinhua (2011) 'Beijing Air Pollution "Hazardous"', US Embassy, Associated Foreign Press. Available at http://www.google.com/hostednews/afp/article/ALeqM5iJTkt3-cVITDVI6xipFX4aAyYjpw?docId= CNG. d957d0999e1088b0ce61729ec5b6c9f1.5f1.

Xinhua (2012) 'Beijing Releases PM 2.5 Air Quality Readings', *Caixin*, 21 January. Available at http://english.caixin.com/2012–01–21/100350762.html.

Xinhua (2013a) '74 Cities Release Real-time PM 2.5 Data', Xinhua News Agency. Available at: http://news.xinhuanet.com/english/china/2013–01/01/c_132075595.htm.

Xinhua (2013b) 'Beijing Releases Heavy Air Contingency Plan', Xinhua News Agency. Available at http://news.xinhuanet.com/english/video/2013–10/22/c_132821147.htm.

Yale Center for Environmental Law and Policy (YCELP) and the Center for International Earth Science Information Network (CIESIN) (2012) 'The 2012 Environmental Performance Index'. Available at http://epi.yale.edu.

Yang, A. & Cui, Y. (2012) 'Global Coal Risk Assessment: Data Analysis and Market Research', WRI Working Paper. World Resources Institute, Washington, D.C. Available at http://www.wri.org/publication/global-coal-risk-assessment.

Yang, A. & Cui, R.Y. (2013) 'Can China's Action Plan to Combat Air Pollution Slow Down New Coal Power Development?', *ChinaFAQs*. Available at http://www.chinafaqs.org/blog-posts/can-china%E2%80%99s-action-plan-combat-air-pollution-slow-down-new-coal-power-development.

Yangzong, Y., Shi, Z., Nafstad, P., Håheim, L.L., Luobu, O. & Bjertness, E. (2012) 'The Prevalence of Childhood Asthma in China: A Systematic Review', *BMC Public Health* 12(1), 860.

Yee, A. (2013) 'India Increases Effort to Harness Biomass Energy', *New York Times*, 8 October. Available at http://www.nytimes.com/2013/10/09/business/energy-environment/india-increases-effort-to-harness-biomass-energy.html.

Zeng, H. (2013) 'On the Move: Reducing Car Usage and Ownership in China, Latin America, and other Developing Economies', *The City Fix*, EMBARQ Center for Sustainable Transport. Available at http://thecityfix.com/blog/on-the-move-reducing-car-usage-ownership-china-latin-america-developing-economies-heshuang-zeng/.

Zhang, Y., Zhu, X., Slanina, S., Shao, M., Zeng, L., Hu, M., et al. (2004) 'Aerosol Pollution in Some Chinese Cities', *Pure Applied Chemistry* 76(6), 1227–1239.

Zhu, W. (2012) '"PM2.5" Theme of the Midi Music Festival Beijing'. Available at http://totobobo.com/blog/2012/06/pm2–5-theme-of-the-midi-music-festival-in-beijing/.

11

Municipal solid waste

The burgeoning environmental threat

Dickella G.J. Premakumara and Toshizo Maeda

Introduction

Municipal solid waste (MSW) is one of the major public health and environmental concerns in urban areas of many developed and developing countries in Southeast and East Asia. The growing economy associated with the rapid population growth in many parts of the region has resulted in a remarkable increase in waste volume in recent years. It was estimated that the urban areas of the region produce about 33% of the world's total waste quantities. Individually, China has surpassed the United States (US) in 2004 to become the world's largest waste generator and would likely produce twice as much MSW as the US in 2030 (World Bank 2012).

In general, MSW management is a responsibility of local governments. It is the largest single budget allocation, estimated at about 40% of the municipality's operating budget in low- and middle-income countries and also one of the largest employers (ADB 2004). However, many municipalities in the region are struggling to provide at least the most basic level of waste services to its citizens. Throughout the region, the urban poor suffer most from the life-threatening conditions deriving from the deficient MSW management (Kungskulniti 1990). Typically, one-to two-thirds of the MSW generated in low- and middle-income countries are not properly collected (World Resource Institute et al. 1996). The uncollected waste, which is often mixed with human and animal excreta, is dumped indiscriminately in the streets and in drains, so contributing to flooding, breeding of insects and rodent vectors and the spread of diseases (UNEP-IETC & HIID 1996).

Further, the collected MSW is dumped on land in a more or less uncontrolled manner. Such inadequate waste disposal is leading to the emission of harmful methane gas, which not only adds to global warming and associated climate change, but also leads to increased public and environmental health risks (ADB 2011). The events of July 2000 which happened at the Payatas dump in Quezon City on the outskirts of Manila, Philippines, where hundreds of people were killed by the collapse of a seven-storey high open dump, shows the direct potential consequences of uncontrolled dumping in the region.

This chapter therefore reviews the broad trends related to MSW management in Southeast and East Asian countries and gives some recommendations for regional collective action for achieving a sound material-cycle society, based on the literature review and some case study analysis. It is organised into four sections covering the key aspects of MSW management. Section one gives a brief introduction to the study. Section two highlights the current practices of MSW management in the region, particularly the identification and review of the status quo and issues of waste generation, collection and disposal. Section three provides some policies and regulations as well as good practices in promoting 3R (reduction, reuse and recycling) in establishing a sound material-cycle society in the region. Finally, section four concludes by identifying some recommendations.

Overview of municipal solid waste management in Southeast and East Asian countries

Municipal solid waste generation: status and trends

As Table 11.1 shows there is no commonly agreed definition for MSW among the countries in the region. It includes waste generated from households, offices, hotels, shops, schools, other institutions, and from municipal services such as street cleaning and maintenance of recreational areas. Often only residential waste is referred to as MSW in high-income countries. However, some low- and middle-income countries include industrial waste and demolition and construction debris also in their definitions. Thus, comparing generation rates for various countries is problematic.

MSW generation rates vary from country to country, and have a strong correlation with levels of economic development and activities. A clear understanding of the quantities and characteristics of the MSW generated is important in planning and development of cost-effective solid waste management strategies (Eawag & Sandec 2008). Although waste quantification and characterisation makes a basis for MSW management planning and development interventions in high-income countries of the region, including Japan, Singapore and the Republic of Korea, most low- and middle-income countries have given a low priority to the systematic surveying of waste generation, its quantities, characteristics, seasonal variations and the future trends.

In general, the greater the economic prosperity and the higher percentage of urban population, the greater the amount of solid waste produced. Table 11.2 shows current waste generation per capita by income level in selected countries of the region. High-income countries (such as Japan, Hong Kong, the Republic of Korea, and Singapore) produce between 1.1 and 4.5 kg/capita/day. Middle-income countries (such as the People's Republic of China, Indonesia, Malaysia, Philippines, and Thailand) generate between 0.8 and 1.5 kg/capita/day, while low-income countries (such as Cambodia) have generation rates between 0.6 and 1.0 kg/capita/day (World Bank 2012). Taken as a whole, East Asia produces some 270 million tonnes of MSW per year and this is expected to more than double by 2025. This quantity is mainly influenced by waste generation in China, which makes up 70% of the regional total. An average per capita waste generation ranges from 0.44 to 4.3 kg/capita/day, with an average of 0.95 kg/capita/day (World Bank 2012).

MSW composition also varies significantly across the region. As countries advance to higher stages of economic development, the composition of their waste also changes. Figure 11.1 presents the current average MSW composition in selected low-, middle- and high-income countries. Generally, low- and middle-income countries have a high percentage of organic materials, averaging 50% of the total. However, the organic fraction in high-income countries, about 28% on average, is significantly lower. In contrast, the proportion of non-organic material is increased relative to the population's degree of wealth and urbanisation. The presence of paper, plastic, glass and metal becomes more prevalent in the waste stream of middle- and high-income countries.

Dickella G.J. Premakumara and Toshizo Maeda

Table 11.1 Definition of municipal solid waste (MSW) in selected East Asian countries

Country	Definition of MSW
Brunei Darussalam	MSW includes waste from residential, commercial, institutional, industrial as well as soil waste, construction waste, toxic and hazardous waste and used oil
Cambodia	MSW includes household waste that does not contain toxins or hazardous substances and is discarded from dwellings, public buildings, factories, markets, hotels, business buildings, restaurants, transport facilities and recreational sites
PR China	MSW, which is commonly known as urban house refuse in China, includes all solid waste discharged from urban daily life activities and also includes household hazardous waste such as fluorescent tubes, mercurial thermometers, paints, cleansers etc. that are generated from urban daily life activities
Indonesia	MSW includes household waste derived from household daily activities, excluding faeces and specific hazardous waste, household-type waste derived from commercial areas, industrial areas, special areas, social facilities, public facilities and other facilities
Japan	MSW is defined as waste other than industrial waste (the Waste Management and Public Cleansing Law defined 21 types of waste generated through industrial and other business activities as industrial waste) generated from households, business establishments such as offices, restaurants, shops etc.
Republic of Korea	MSW is household, office, industrial, construction and hazardous waste
Malaysia	MSW includes public solid waste, imported solid waste, household solid waste, institutional solid waste and special solid waste such as waste from commercial, construction, industrial and controlled activities
Mongolia	MSW is domestic and industrial solid waste from materials produced during the process of consumption, production and services, including unwanted waste
Myanmar	MSW is that which comes from human and animal activities and is normally solid, discarded as useless and unwanted. It is all inclusive encompassing the heterogeneous mass disposed from urban community as well as the more homogeneous accumulation of agricultural, industrial and mineral wastes
Philippines	MSW is produced from activities within the local government units including a combination of domestic, commercial, institutional and industrial waste and street litters. It refers to all discarded household, commercial waste, non-hazardous institutional waste, street sweepings, construction debris, agriculture waste and other non-hazardous/non-toxic solid wastes
Singapore	MSW includes household waste and waste from offices, hotels, shopping complexes, schools, institutions, trade premises, hawker centres, markets and municipal services such as street cleaning and maintenance of recreational facilities
Vietnam	MSW includes waste discharged from production, service, daily life and other activities. This can include both domestic and industrial waste as well as hazardous and non-hazardous waste

Source: AIT & UNEP-RRC.AP (2010), compiled by the author.

Table 11.2 Average municipal solid waste (MSW) generation rates by income

Country	GNI per capita (2011 US$)	Current urban population (% of total)	Current urban MSW generation (kg/capita/day)
Low income	Less than 876	29	0.6–1.0
Cambodia	820	20	0.7
Middle income	876–10,725	37	0.8–1.5
PR China	4,940	51	0.75
Indonesia	2,940	51	0.52
Philippines	2,210	49	0.5
Thailand	4,420	34	1.7
Malaysia	8,420	73	1.52
High income	Over 10,725	80	1.1–4.5
Republic of Korea	20,870	83	1.24
Hong Kong, China	35,160	100	1.49
Singapore	42,930	100	1.49
Japan	45,180	91	1.71

Source: World Bank (2012), compiled by the author.

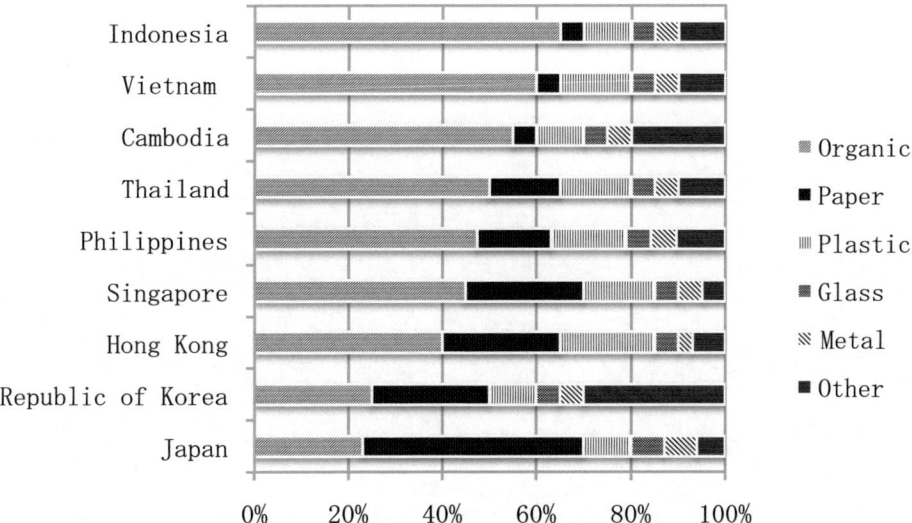

Figure 11.1 Types of waste composition in selected countries
Source: World Bank (2012), compiled by the author.

MSW storage, collection and transfer

Collection of MSW is by far the largest cost allocation in many cities in the region, accounting for less than 10% of costs in high-income countries, but 70–90% of costs in middle- and low-income countries (World Bank 2012). A wide variety of waste collection methods are used by cities including house-to-house collection (waste collectors visit each individual house to collect garbage), communal collection (users bring their garbage to community bins that are placed at fixed points in the neighbourhood) and kerbside collection (users leave their garbage directly outside their houses according to a garbage pick-up schedule).

In the high-income countries of Japan, the Republic of Korea, and Singapore, collection and transfer services are capital intensive and highly mechanised employing standardised collection vehicles, compactors and containers and provide higher collection rates exceeding 98% collection services to most urban and even rural areas (World Bank 2012). Source separation and subsequent collection of recyclables is governed by regulation and is facilitated by the provision of colour-coded bins or bags or by the establishment of area recycling centres.

However, in the low- and middle-income countries of the region, waste collection and transfer tend to be labour intensive. It is undertaken by persons directly employed by the municipal authorities. A low level of mechanisation with handcarts and tractor-trailers is commonly used to collect waste from communal bins and neighbourhood transfer areas. The collection systems are relatively inefficient as the collection vehicles and containers are not fitted with compactors, necessitating the transportation of loose waste and hence, imposing a constraint on the capacity of the collection system. The average waste collection rates in these countries range from 40% to 80% (World Bank 2012). There are also large disparities in collection services between the rich and poor residential areas and often, a city's waste collection service is not extended to the poor, especially those who are living in illegal or slum and shanty settlements. Figure 11.2 shows uncollected waste dumped in a low-income neighbourhood in Cebu, Philippines.

The financial constraints, lack of technical expertise, shortages of storage bins, collection vehicles, non-existent or inadequate transfer stations and traffic congestion severely limit the

Figure 11.2 Uncollected waste dumped in the low-income neighbourhood of Cebu, Philippines

effectiveness of MSW collection service in the cities and towns of the low- and middle-income countries. The lack of coordination and overlapping of responsibilities among various government agencies and different levels of local government also contribute to the problem (Eawag & Sandec 2008).

There is a trend, however, driven by failing municipal systems or by pressure from national governments and international agencies, to outsource the provision of waste collection services to the private sector. This practice is gaining momentum, especially in Hong Kong, PR China, Malaysia, the Republic of Korea, Singapore and Thailand. However, it was evident that private sector involvement in the provision of MSW management services should not be seen as a miracle. Building appropriate partnership with the private sector requires a careful scheme including deliberate design of service, distinctive responsibility assumed by governments and private sector, appropriate selection of private providers, ensuring fair competition and a feasible monitoring system (JICA 2009).

By contrast, community-based, decentralised primary collection has proved effective in achieving increased collection rates in low- and middle-income countries. The *kampongs* (villages) of Indonesian cities have formal responsibility for primary collection. The waste collected from each *kampong* is delivered to a transfer station or temporary storage point for collection by the city service. The Philippines also uses a similar system at barangay (a lowest level political and administrative body) and has achieved reasonably good results. However, a lack of efficient transfer facilities represents a weak link in the primary MSW collection and transportation system. The transfer stations often serve as material recovery facilities (MRF) where recyclables are separated for reuse and recycling.

Privatisation of SWM in Malaysia

MSW in Malaysia was traditionally under the jurisdiction of local governments following the Local Government Act of 1976. The local governments (including city halls (seven), municipal councils (33) and district councils (104)) were responsible for handling the waste collection, transportation, treatment and final disposal of MSW. The state governments, the second tier of administration in Malaysian government, were responsible for allocating lands for landfill sites and other treatment facilities, while the federal government was responsible for developing policies and financing for infrastructure development. However, the privatisation of solid waste management (SWM) commenced in 1994 with a privatisation policy made as part of Vision 2020 focusing on reducing the government's financial and administrative burden, promoting competition, increasing the role of the private sector in nation building and providing opportunities for meeting the targeted new economic policy. Privatisation has also resulted in the growth of private companies specialising in the waste business and these often complement the services that are mainly provided by the local authorities. The Solid Waste and Public Cleansing Management Corporation (PPSPPA) was newly established under the Ministry of Housing and Local Governments (MHLG) to enforce and monitor the privatisation activities, taking over the local authority's responsibilities. The waste management system, however, has not attained full privatisation. The system is still in an interim period and is not running as expected due to some problems arising from the lack of funds, the length of the interim period, the unavailability of financial resources, and confusion among the stakeholders.

Source: Sakawi 2011.

Community-based waste collection system in Indonesian cities

In late 1970s, a neighbourhood primary collection system was piloted in Surabaya and Jakarta, Indonesia involving local neighbourhood leaders and workers hired from the community. Primary collection is now common in most Indonesian cities, reaching about 70% of the country's urban population. In these cities, primary collection involves handcarts or tricycle carts collecting the waste from door to door. In some neighbourhoods, primary collection is done by individuals hired at the neighbourhood level and paid a salary by the neighbourhood leader. In other neighbourhoods, pre-collection is part of a zonal service contract with the city for solid waste collection. In both cases, residents pay through direct user charges to cover the cost of pre-collection and part of the cost of secondary collection by the municipality.

Source: Tience 2005.

Disposal methods of MSW

Various disposal methods of MSW in selected countries in the region are given in Table 11.3.

Open dumping is still the most widespread method of solid waste disposal in the region and typically involves the uncontrolled disposal of waste without measures to control leachate, dust, odour, landfill gas or vermin (see Figure 11.3). In some cities, open burning of waste is practised at both neighbourhood sites and dumpsites. In many coastal cities, waste is dumped along the shoreline and into the sea or dumped in coastal and inland wetlands and rivers.

The indiscriminate dumping of organic wastes into landfills or open dump sites results in the generation of large quantities of methane, a greenhouses gas (GHG) with 21 times greater global warming impact than that of carbon dioxide. Methane generation produced at landfill sites contributes approximately 3–4% to the global anthropogenic GHG emissions (IPCC 2006). The generation of methane in solid waste landfill sites is dependent on: the total amount of solid waste, which is determined by population size and affluence, composition of the waste, and the characteristics of the solid waste disposal sites, such climate, size/depth, pH level, and moisture (IGES 2008). As shown by Table 11.4, growing populations, urbanisation and economic growth in the region are expected to lead to increasing amounts of solid waste and potentially escalating methane emissions from solid waste disposal sites.

Table 11.3 Disposal methods for municipal solid waste (MSW) in selected countries of the region

Country	Disposal methods (%)				
	Open dumping	Landfill	Composting	Incineration	Others (animal feed, dumping in water and open burning)
PR China	50	30	10	2	8
Hong Kong, China	20	60	—	5	15
Indonesia	60	10	15	2	13
Japan	—	15	10	75	—
Republic of Korea	20	60	5	5	10
Malaysia	50	30	10	5	5
Myanmar	80	10	5	—	5
Singapore	—	30	—	70	—
Philippines	75	10	10	—	5
Thailand	65	5	10	5	5
Vietnam	70	—	10	—	20

Source: World Bank (2012), compiled by the author.

Figure 11.3 Open dump site in Talisay City, Philippines

Table 11.4 Waste generation rates and methane emissions from solid waste disposal sites, 1995–2025

Country	1995			2025		
	Urban MSW generation rate (kg/cap/day)	Methane emissions (kt/year)	Methane emissions (kg/cap/year)	Urban MSW generation rate (kg/cap/day)	Estimated methane emissions (kt/year)	Estimated methane emissions (kg/cap/year)
PR China	0.79	898.52	2.35	0.90	4,075.12	4.93
India	0.46	474.55	1.92	0.70	2,774.92	5.37
Indonesia	0.76	457.49	6.52	1.00	1,581.74	0.05
Thailand	1.10	165.33	9.44	1.50	424.39	13.58
Philippines	0.52	127.83	3.46	0.80	451.11	5.61
Malaysia	0.81	68.91	6.08	1.40	281.11	11.09
Bangladesh	0.49	38.66	1.46	0.60	243.69	3.29
Vietnam	0.55	31.76	1.96	0.70	189.87	4.60
Myanmar	0.45	18.46	1.61	0.60	106.41	3.94
Cambodia	0.69	2.67	1.64	0.80	25.50	3.50
Lao PDR	0.69	1.33	1.64	0.80	10.41	3.50

Notes: Data on urban MSW rate are cited from Hoornweg et al. (1999) and emissions of methane were calculated using the mass balance methodology of IPCC (1997).

Source: IGES (2008).

The disposal of solid waste at a semi-engineered or full sanitary landfill has gained most attractive alternative disposal options in the region. Bandung, Singapore, Hong Kong, Seoul, and Fukuoka do have some well-designed and reasonably operated sanitary landfills. Figure 11.4 shows the sanitary landfill in Fukuoka City, Japan.

The generation of landfill gas has been turned to advantageous use at a number of landfills in the region through the development of electricity generation facilities. However, recovery rates are much lower in low- and middle-income countries, with up to 60% leakage expected (World Bank 1999). The guidelines for clean development mechanism (CDM) projects also recommend that a recovery rate of 50% be used in project proposals. Hence, from a climate protection perspective, landfills with gas recovery systems are not entirely suitable options for treating organic waste (IGES 2008).

In the densely populated cities and towns of the region, finding the suitable land for landfill siting is a major constraint. For example, in Hong Kong and Singapore severe land constraints have led to complex engineering infrastructure solutions being developed to ensure high standards of operational and maintenance control and have enabled the development of acceptable landfill solutions in coastal areas, offshore islands and mountainous terrain.

Considering its significant benefits in reducing the waste weight (up to 75%) and volume (up to 90%) (Eawag & Sandec 2008) effectively eliminating the hygienic hazards of organic waste, and generating electricity and heat which can replace energy from fossil fuels (IGES 2008), incineration of MSW is widespread in high- and middle-income countries, such as PR China, Japan, Singapore and the Republic of Korea.

Figure 11.4 Sanitary landfill in Fukuoka City, Japan

Although incineration as a strategy for dealing with MSW is increasing in many cities, these planned new incinerators have provoked large-scale not in my back yard (NIMBY) protests in a number of cities. The local residents often fear the dioxin emissions and other threats, leading them to oppose the construction of incineration facilities in their local areas, even while accepting that a need for such facilities exists. For example, a Tokyo waste war happened in 1971, when local government planned to establish a new incineration plant in Suginami Ward. It took three years to negotiate between the local authority and citizens. The Liulitun anti-incinerator campaign in Beijing is also famous for its 'successful' outcome (cancellation of the project at that location, and a later decision made to locate an incinerator at a more remote location).

To overcome the NIMBY situation, it is important to make sure that citizens have a correct picture of the pros and contras of waste incineration through a public participatory consultation process. It is important to communicate information on waste incineration technology, as well as the global and local environmental impacts, in a trustworthy and detailed manner. The community should be encouraged to express its concerns at an early stage – for example, during public information meetings and hearings. Here, the local government can present the potential risks and impacts as well as the environmental protection measures that will be introduced. If necessary, additional environmental protection measures or community nuisance control measures can be planned and announced after the public meetings. Further, the plant can be opened to the local residents, after starting its operation. This gives local residents an opportunity to monitor its operation.

Although incineration offers the advantage of high volume reduction and helps to conserve landfill space, it is not adequate if more waste is generated each year. This would then lead to additional demand for more incineration plants and landfills. Thus, some cities began to realise the necessity of reducing and recycling wastes.

Waste reduction by promoting incineration in Singapore

Solid waste generation in Singapore has increased from 1,200 tonnes/day in 1970 to 7,000 tonnes/day in 2005. To overcome the constraint of limited land, Singapore has adopted waste-to-energy incineration as a waste treatment method. Currently, Singapore has four incinerators to handle more than 90% of MSW that is collected each day. These incineration plants are designed with advanced pollution control measures and energy is recovered to generate electricity. The total electricity generated by the existing plants is about 980 million kWh per year (2–3% of electricity demand for Singapore), a portion of which is used to run incinerator operations, and the balance is sold to the national electricity grid. There is only one landfill in Singapore (Semakau Landfill), constructed at a cost of US$610 million by joining two small islands with earth/rock bunds lined with an impermeable membrane.

Source: AIT/UNEP-RRC.AP 2010, NEA 2006.

Moving from incineration to promoting waste reduction in Yokohama City, Japan

Being one of the largest cities in Japan, with a population of 3.67 million people, Yokohama had to continue building incineration plants to face the growth of MSW in the city. In 2001, seven incineration plants were operated and a reduction of the amount of combustible waste was the high priority in the city's waste management measures. In January 2003, therefore, the Yokohama G30 Plan was formed to aim at achieving a 30% waste reduction by 2010 compared with 2001 by increasing the separating and the recycling of MSW. The number of categories for separating household waste was increased from five to 10. Since 2005, this system has been in place in the entire city. To encourage the citizens to separate their waste at home, some 12,000 meetings have been held to explain separating, and a variety of awareness raising events have been organised including 8,300 early morning explaining events at collection stations. Waste that has not been separated properly is left behind by refuse collectors with a warning sticker. A non-penal fine of US$25 is imposed if someone is repeatedly warned and still fails to separate waste properly. As a result, MSW was successfully reduced by 42% compared with 2001. Due to the reduction of combustible waste, two incineration plants were closed permanently and one temporarily deactivated. This brings a US$6 million waste management cost saving for the city.

Source: IGES 2011.

Integrated solid waste management (ISWM) and 3R policies and practices

Waste management, like many other urban environmental issues, is multi-sectorial in nature and encompasses policy making, strategic thinking, and establishment of legal-institutional-financial and administrative frameworks, as well as functional design, implementation, operation and management of waste-handling facilities. Thus, the UNEP International Environmental Technology Centre (1996) describes the importance of viewing solid waste management from an integrated approach. ISWM is defined as the selection and application of appropriate techniques, technologies, and management programmes to achieve specific waste management objectives and goals. Understanding the interrelationships among various waste activities makes it possible to create an ISWM plan where individual components complement one another (World Bank 1999).

A waste management hierarchy is usually established to identify the key elements of an ISWM plan. The general waste hierarchy accepted by industrialised countries is comprised of the following order: source reduction, material recovery, reuse and recycling, recovered waste transformation through physical, biological and chemical processes and land filling at the end. Even though there are some best practices of ISWM and 3R in Japan, Singapore, the Republic of Korea, Malaysia, Philippines, Indonesia and Vietnam, some countries in the region still do not have national 3R strategies, action plans and framework for solid waste management and necessary regulation, institutional setup and economic instruments for the implementation of these action plans. Waste management is often hampered by a lack of national policy direction with no clear allocation of responsibilities and little or no national level planning to develop integrated waste management policies and strategies.

Best practice policy and instruments to promote 3R in the region

Formulation of national plans and target setting for waste minimisation and recycling

Some countries in the region have begun to set national targets in 3R-related laws and regulations. For example, Malaysia has targeted its recycling rate to reach 22% by 2020. Vietnam's 2025 targets cover waste separation at source (80% of cities), collection (100%) and recycling (90%) of solid waste from urban households (UNCRD 2010). However, Japan displays the best example in the region in formulating national plans and target setting for establishing a sound material-cycle society (SMC) based on 3R principles.

Establishing a sound material-cycle society in Japan

The 3R approach in Japan was designed to overcome two interrelated problems, including limited land available for landfills and an increasing reliance on imported raw materials (METI

2010). The government made some early efforts dating back to the plague epidemic in 1887 and the subsequent passage of the Unsanitary Substance Cleaning Law in 1900. In 1963, the government set up the first Five-Year Plan for Development of Living Environment Facilities, which marked a transition to incineration of urban waste, with the considerably reduced residues sent to landfills. In 1976, Hiroshima City introduced waste segregation, sorting waste into combustible, non-combustible, recyclable, large-sized waste and hazardous waste (MOEJ 2008). The Recycling Promotion Association (this was renamed the Reduce, Reuse and Recycle Promotion Association in 2002) was established in 1991.

In 2000, the following laws were enacted in supporting 3R: Basic Law for Promotion of a Recycling Oriented Society, Waste Management Law, Law for Promotion of Effective Utilisation of Resources, Food Recycling Law and Green Purchasing Law. Additional laws were passed in subsequent years, including the Home Appliance Recycling Law (2001), End of Life In Vehicles Recycling Law (2005) and Containers and Packaging Recycling Law (amended in 2006) and in 2001, the Environment Agency was upgraded to the Ministry of Environment (MOEJ).

These successful institutional innovations and the ever increasing waste volumes convinced the government to introduce the Basic Plan for Establishing a Sound Material-Cycle Society in 2003, to be reviewed every five years (METI 2010). The second plan (2008) established quantitative targets for 2015 (based on 1990) including: material productivity from JPY210,000/ tonne (US$2,100/tonne) to JPY420,000/tonne (US$4,200/tonne), usage rate of recycled goods from 8% to 14–15%, and final disposal from 110 million/tonnes to 23 million/tonnes.

A variety of indicators have been added to measure the level of efforts put into implementation of the 3R initiative (such as reduction of municipal solid waste and industrial waste, change in awareness and behaviour, promotion of SMC business, and strict enforcement of recycling laws) and to set up the baselines for future policies (e.g., resource productivity of fossil fuels, biomass resource input rate, hidden flows and total material requirements, industry specific resource productivity and sales of disposal products). The progress shows that the 2015 targets are likely to be met. MSW peaked in 2000 at 1.2 kg/cap/day and has declined to 1.0 kg/cap/day. The recycling rate increased from 4.5% in 1989 to 20.3% in 2007.

Source: IGES 2011.

Promotion of source segregation

Waste separation at source is an important element in promoting 3R and proper MSW management. Separation of organic waste, which is the largest portion of household wastes, can be effectively utilised for composting or feed as animal food. Waste generated from high-income households tends to contain more recyclable waste and thus the ratio of waste collected by collectors and collection dealers rises if the waste is separated into glass, cans, plastic, etc., at the point of generation. Figure 11.5 shows the community waste collection station in Minamata City, Japan.

Although there are some good practices to promote source segregation in the region, it is still not a common practice throughout the region and represents a critical area for improvement. Implementation of source segregation requires establishing a clear definition and classification rules for waste and recyclables, establishing a proper sorting, collection, and treatment mechanism, raising citizens' awareness on waste classification, establishing a recycling system involving citizens and the private sector (both formal and informal) and establishing recycling stations or MRF to convert waste into resources after it is collected (UNCRD 2010).

According to Section 10 of the Republic Act (RA) 9003, known as the Ecological Solid Waste Management Act of 2000 in the Philippines, the segregation and collection of solid waste is the responsibility of the local governments. To strengthen the national policy framework,

Figure 11.5 A community-separated waste collection station in Minamata City, Japan

Cebu City enacted the City Ordinance No. 2031 (enforcing the No Segregation No Collection) and No. 1361 (detailing the penalties for violators of the waste segregation policy), requiring waste segregation at source into biodegradable, recyclable, special (hazardous) and residual. To promote the participation of communities in waste segregation at source, the Cebu Environmental Sanitation Enforcement Team (CESET), comprising members from the Barangay Solid Waste Management Committee and the team of Barangay Environmental Officers (recruiting five staff for each barangay based on the community leadership elements) organised a series of about 300 information, education and communication campaigners at barangay level to explain in detail the implementation procedures. Further, a two-week trial period was introduced so that the residents and barangay officials could familiarise themselves with the types of waste to be segregated and the schedule of collection.

Since the end of the trial period the city government with the assistance of barangay officials strictly enforced the No Segregation No Collection policy. As an enforcement team of the city, the CESET carries out monitoring and issuing citation tickets to ordinance violators. Any person who is found guilty of violating Ordinance 2031 shall be punished by a fine (not less than 1,000 Peso (US$20) but not more than 5,000 Peso (US$100)) or by imprisonment (not less than one month to not more than six months), or both fine and imprisonment at the discretion of the court. If the violator cannot pay the compromise fee, the person must render community service of one day to 15 days at any barangay as determined by the Monitoring/Enforcement Unit of the CSWMB. Segregation of waste increases the rate of recycling of both biodegradable and non-biodegradable waste at barangay level and has helped to reduce waste to be land filled by 30% during the last three years (Ancog et al. 2012; Premakumara et al. 2013).

Material recovery, reuse and recycling

In many countries of the region, including Japan, the Republic of Korea, and Singapore, the rate of recovery of recyclable materials from MSW has improved significantly in recent years. In the high-income countries, recycling is undertaken at source (i.e., at household, business and industry level) and is actively promoted by governments, NGOs and the private sector. Elsewhere, such as in Vietnam, Philippines, Indonesia and Thailand, informal recycling networks have flourished despite the lack of formal promotion or support of the government (see Figure 11.6).

It has been estimated that up to 2% of the population survives by recovering materials from waste to sell for reuse or recycling or for their own consumption. In some cities, these waste scavengers constitute large communities. Approximately 15,000 squatters make their living by scavenging in rubbish dumps in the Philippines and Jakarta is served by between 15,000 to 20,000 waste pickers (UN–Habitat 2008). Some of these communities have high levels of organisation and the creation of scavenger co-ops has gained momentum in some countries of the region including the Philippines, Indonesia and Vietnam.

In the second half of the 1980s, Vietnam enjoyed rapid economic growth. Rural communities became industrialised, especially in the Northern region and craft villages where handicrafts have become the main industry. At the same time, waste created by industrial production and waste from urban cities were discarded in great quantities. As a result, recycling businesses that collect and separate such waste developed. In the suburbs of Hanoi, several communities called recycling villages have made the recycling business their economic core. Currently, there are

Figure 11.6 A family engage in waste picking at the landfill in Mandaue City, Philippines

around 90 recycling villages that collect and separate waste such as paper, metals and plastics. The waste scavengers of Hanoi operate at no cost to the city and provide both financial benefits to the society in the form of avoided costs (such as landfill space, collection and transportation, energy, employment generation, protection of public health) as well as ecological benefits in the form of resource conservation and environmental protection. A network of scavengers and junk buyers (estimated to comprise some 6,000 people during the August peak season) collect discarded goods for onward sale to junk dealers, who in turn resell the materials in bulk to factories and exporters.

However, the business of waste scavenging is not without its human health costs and the rewards for some engaged in extracting materials from waste are inadequate to alleviate their poverty. In many cases, the scavengers picking over the mixed waste of the dumping grounds do not wear protective clothing, neither do they have access to washing facilities. The majority of dumpsite scavengers are women and children, who live in overcrowded, poorly ventilated temporary huts, often on the peripheries of the waste dump. The scavengers seldom to have access to public or private latrines, are often malnourished, and suffer from a range of illnesses including worm infections, scabies, respiratory tract infection, abdominal pain, fever and other unspecified diseases.

Source: UNCRD 2010.

Composting

The composting process ferments the organic constituent of waste by microorganisms functioning under aerobic conditions. The compost thus generated can be used as soil conditioner, agricultural fertilizer for organic farming, or as cover material on final disposal sites. Composting is also effective in reduction and reuse of organic waste including kitchen waste, which helps to decrease the amount of waste to be finally disposed, leading to longer life of disposal sites.

While small-scale composting of organic waste is widespread in the region, attempts to introduce large-scale composting as a means of reducing the quantities of municipal solid waste requiring disposal or with the intention of creating a revenue stream from the sale of compost have been met with limited success. Most of the larger composting plants in the region are neither functioning at full capacity nor producing compost of marketable value (see Figure 11.7). The high operating and maintenance costs result in compost costs that are higher than commercially available fertilisers, while the lack of material separation produces compost contaminated with plastic, glass and toxic residues. Under such circumstances, little of the compost produced is suitable for agriculture. The forced-air composting plant in Hanoi is a typical example. The plant is currently operating at 20% of its design capacity, while the municipal authorities have been unable to persuade local farmers to take the product free as it is too contaminated with plastic and glass.

Surabaya, the provincial capital of East Java, is the second largest city in Indonesia with a population of 3 million. In 2000, a university-based NGO called Pusdakota was running campaigns trying to increase people's awareness around waste issues, at a time when the system of waste collection and treatment in Surabaya City was in poor condition and many citizens were concerned about the situation. In 2004 the Japanese organisation Kitakyushu International Techno-Cooperative Association (KITA) helped improve waste management in Surabaya and provided technical assistance in composting. The severe waste-related problems seem to have motivated people to take actions to improve the situation and over the years small-scale household composting gradually became common.

The city government became interested at an early stage and introduced some supportive policy measures to replicate the composting practices at city-wide scale, building partnerships with the women's network, local NGOs, informal waste pickers, academic institutions, private

Figure 11.7 Community composting centre in Surabaya City, Indonesia

ventures and media. A system of community facilitators and environmental volunteers was established to share information and assist new families in starting household composting. Once the families were willing to use household compost, a free compost bin was provided. In addition, a municipal budget was allocated for establishing composting facilities in the area, for the households that cannot carry out the composting themselves or that prefer door-to-door collection. Currently, the city has distributed more than 17,000 composting baskets and established 16 composting facilities. A total of 5,760 tonnes of compost annually is used for the maintenance of city parks and landscaping. As a result, a 30% waste reduction was achieved, saving the municipality solid waste management costs estimated at US$4 million annually.

Source: Premakumara et al 2011.

Private sector participation

Business and industries have made our lives more comfortable and convenient by producing more goods that meet consumer needs. With the increase in manufacturing activities, the quantities of waste and pollution from hazardous chemicals have also increased. Thus, some countries have begun to enforce regulations to avert these environmental problems. Business and industries need to comply with these regulations while fulfilling their social obligation as manufacturers. In response, many progressive companies are working as equal partners with government in developing comprehensive waste management programmes.

Extended producer responsibility (EPR) is a voluntary measure, that taken by the manufacturers to reduce the environmental impacts of their product at each stage of the product's lifecycle, from the time the raw materials are extracted, produced and distributed, through the end use and final disposal phases (World Bank 1999). EPR does not consider only the manufacturers to be accountable for environmental impacts; it also deems this responsibility to extend to all those involved in the product chain, from manufacturers, suppliers, retailers, to consumers and disposers of the products.

Some high-income countries have introduced environmental labelling of consumer products, which has helped raise environmental consciousness among low-income countries. Under this type of scheme, businesses voluntarily label their products to inform consumers and promote products that are more environmentally friendly than other functionally and competitively similar products. It helps in achieving a number of goals, including improving the sales or image of labelled products, raising consumers' environmental awareness, providing accurate, complete information regarding products and making manufacturers more accountable.

Further, waste minimisation by waste exchange programmes is receiving attention in some countries, including Japan, Thailand and Philippines. The Waste Exchange of the Philippines (IWEP) serves as a link between companies that mutually benefit from waste-to-waste exchange. At least 600 industrial waste products are advertised for exchange with other industries and the IWEP catalogue lists over 130 further waste products that are sought for exchange. Each product is assigned a code to ensure that the producing company's identity and location remain confidential and technical information, such as pH and industrial, the presence of any contaminants present is indicated. When two companies come to an agreement, the IWEP withdraws and leaves the producers and users to negotiate directly. In addition to achieving reductions in the quantities of materials disposed of as waste, the waste exchange scheme has provided substantial benefits to a variety of companies through providing savings in disposal and raw materials costs and in improving the company's public image. The waste exchange programmes were established with the aim of exchanging of valuable information between generators and potential users of industrial and commercial wastes, whereby a beneficial use rather than disposal is the end result. It identifies both the producers and potential markets for by-products, surpluses, unspent materials and other forms of solid waste that is no longer needed for operation. If companies cannot find onsite recycling options, the best alternative may be to offer the waste to another company through the waste exchange programme. The waste lists are published regularly and most exchanges have a website on the internet with links to other exchanges. Through waste exchanges, companies can save thousands of dollars by avoiding disposal costs or obtaining raw materials at reduced prices (World Bank 1999).

Economic and financial mechanisms

With the rapid growth of waste generation rates, more investments are now required to bring improvements in the MSW management. In some countries of the region, including Japan, the Philippines, the Republic of Korea, Indonesia, Thailand, Malaysia, China and Singapore, a number of different economic tools have been integrated into their strategic waste management plan to ensure that waste in all its forms is minimised, that revenues for waste management are raised and that wherever possible, the polluter/user pays. Some of the widely practised economic instruments include charging for collection services, charging fees for garbage bags (see Figure 11.8), and charging for waste inputs to final disposal site. The method of charging fees for waste collection services is commonly practised by the cities where primary collection is done by the private collectors or community collectors. In this case, the collectors charge each household at a fixed rate to collect the waste at each household and carry it to the transfer stations.

Figure 11.8 Designated bags used in Kitakyushu City, Japan

There are some cities in Japan and the Republic of Korea that have successfully implemented a kerbside charging system. Residents are charged per bag which appears to be more effective than the charging collection fees. These bags are provided for non-separated waste, and aluminium cans, glass, cardboard, and newspapers are collected separately. Bag users can save money by putting out fewer bags in a given week. Households have to use garbage bins with differentiated charges according to the size and number of bins that are used. Residents are not charged for the removal of various types of separated waste. Another system is a vehicle or weight-based charging system at the final disposal site. Some countries such as Australia, Japan and Singapore impose fees for disposing waste into the designated waste disposal facilities.

Volume-based collection fee system for MSW in the Republic of Korea

The volume-based collection fee system for MSW was launched in the Republic of Korea to minimise the generation of waste and encourage households to separate their waste for recycling. The system was put into effect nationwide in 1995. Under the system, household waste was to be discarded in the officially designated plastic trash bags, which were manufactured and sold by city, county and district governments. The programme had a far reaching effect on the reduction of waste generation and recycling in the MSW sector. The results show that the system led to a 17% reduction in MSW generation and 21% increase in recyclable wastes in the first year. It has also served as an opportunity to heighten the general public's awareness of the environment in addition to producing visible benefits such as a meaningful reduction of the volume of the waste generated, an increase in recycling, and an improvement in the waste administration service.

The entire process of production, distribution and consumption of goods was shifted to a more environmentally friendly paradigm. The launching of the system served as an opportunity for local governments to become business minded regarding waste management administration. With the supply of recyclable goods increasing, the recycling industry is also beginning to flourish, although the system still needs some improvements, especially in preventing illegal dumping and burning in rural areas and increased recycling costs (MOE Korea 2003).

Conclusions and recommendations

Recently, the area has been witnessing some of the world's fastest economic growth and rapid rates of urbanisation in the region, which are much higher than those the industrialised countries has experienced. This is most clearly illustrated by China, Indonesia, Philippines, Malaysia and Thailand. Even as the financial crises threatened most industrialised countries, these emerging countries in the region still registered high economic growth rates of about 5–10%. Along with this economic growth, these countries have seen the rise of a new consumer class, with lifestyles that largely emulate those seen in more industrialised countries. As ADB (2010) argues Asia's emerging consumers are likely to assume the traditional role of the US and European middle classes and global consumers. The economic dynamism of the region has not only enabled a growth in new middle classes with consumption behaviour, but also increased demand for natural resources. Unsustainable production and consumption are becoming an issue that both governments and international partners are starting to address.

This chapter has identified MSW management as one of the major concerns in both developed and developing countries in the region. A typical MSW management system displays an array of problems including increase in the volume of waste generation, increase and emergence of new types of waste, increase in the number of urban poor families who make their living collecting waste at landfills, low collection coverage, lack of suitable land for landfilling, lack of basic data and information, lack of long-term planning, lack of proper policy formulation, insufficient government priority, lack of finance, lack of skilled personnel, lack of public awareness and participation, inadequate legislation and institutions, and poor monitoring and enforcement. These have led to escalating environmental pollution, public health problems and emissions of GHG in the region.

However, evidence from the region indicates that the promotion of 3R policies and practices is taking root, especially in high- and middle-income countries. Japan, Singapore, and the Republic of Korea are leading in establishing a sound material-cycle society in the region by not only actively pursuing environmental protection through proper institutions and legislation but also developing new and innovative measures to minimise the amount of waste which has to be landfilled. Other countries such as PR China, Indonesia, Vietnam, Thailand, Malaysia, Cambodia and Philippines have also made some progress in implementing the 3R measures. The lessons from good practices in these countries show how cities can improve their waste management and achieve a sound material-cycle society with very limited financial resources, if the politicians, the city administration, citizens and private sector identify solid waste as a priority, showing clear vision and strong determination, and establishing supportive institutional structures, financial mechanisms and organisational capacities. The modernisation of MSW management is a process based on specific local situations, starting from existing strengths of the city and building on them, adopting the best solutions that will work in particular situations.

Technical transfer and investments are essential for MSW management, especially for middle- and low-income countries. The multilateral and bilateral development agencies (including various United Nations agencies (UNDP, UNEP, UNIDO and ESCAP), World

Bank, ADB and some donor countries, especially Japan) are actively offering both technical and financial assistance for improving the MSW management system in countries of the region. However, most of these international projects aim at creating waste management plans and investment projects to improve the waste collection and treatment facilities based on the internationally recognisable standards. Most often these plans do not convert into actions on the ground and investments in facilities that cities cannot afford to operate. As a result, a major priority still remains capacity and knowledge building in order to develop inclusive approaches, sound institutions and proactive policies that are acceptable, appropriate and affordable to the local circumstances.

Thus, international efforts to develop and strengthen inter-regional networks and platforms (including Regional 3R Forum in Asia, 3R Knowledge Hub, ASEAN-ESC Model Cities Programme) to facilitate high-level policy dialogue on regular basis, facilitate good practices among countries, conduct international collaborative research and capacity building on establishing a regional sound material-cycle society in Southeast and East Asia are essential.

Bibliography

AIT and UNEP-RRC.AP (2010) *Municipal Waste Management Report: Status-quo and Issues in Southeast and East Asian Countries* (AIT/UNEP-RRC.AP).

Ancog, R.C., Archival, N.D & Rebancos, C.M. (2012) 'Institutional Arrangements for Solid Waste Management in Cebu City, Philippines', *Journal of Environmental Science and Management* 15(2), 74–82.

Anthonio, L.C. (2008) 'Study on 3R Policy and Waste Exchange in the Philippines', in M. Kojima & E. Damanhuri (Eds.), *3R Policies for Southeast and East Asia* (ERIA Research Project Report).

Asian Development Bank (ADB) (2004) *The Garbage Book: Solid Waste Management in Metro Manila* (ADB).

Asian Development Bank (ADB) (2010) 'Key Indicators for Asia and the Pacific 2010'. Available at http://www.adb.org/Documents/Books/Key_Indicators/2010/pdf/Key-Indicators-2010.pdf.

Asian Development Bank (2011) *Toward Sustainable Municipal Organic Waste Management in South Asia: A Guide Book for Policy Makers and Practitioners* (ADB).

Clarke, M.J. (1993) 'Integrated Municipal Solid Waste Planning and Decision-making in New York City: The Citizens' Alternative Plan', *Journal Air and Waste Management* 43(4), 453–462.

Eawag & Sandec (2008) *Sandec Training Tool 1.0-Module 6: Solid Waste Management* (Eawag/Sandec).

IGES (2008) White Paper 2, Institute for Global Environmental Strategies, Kitakyushu Urban Centre, Japan.

IGES (2011) *Local-level Innovations towards an Environmentally Sustainable City: Case Studies from Japanese Cities* (Institute for Global Environmental Strategies).

IPCC (2006) *IPCC Guidelines for National Greenhouse Gas Inventories* (IPCC).

JICA (2009) *Thematic Guidelines on Solid Waste Management* (Global Environmental Department).

Kungskulniti, N. (1990) 'Report: Public Health Aspects of a Solid Waste Scavenger Community in Thailand', *Waste Management & Research* 8(2), 167–170.

METI (2010) Towards a 3R Oriented Sustainable Society: Legislations and Trends 2010, Ministry of Economy, Trade and Industry, Tokyo.

Ministry of Environment-Korea and Korea Environment Institute (2003) 'Volume Based Waste Fee System', *Korea Environmental Policy Bulletin* 1(1), 1–23.

MOEJ (2008) The World in Transition and Japan's Efforts to Establish a Sound Material-Cycle Society, Ministry of Environment, Tokyo.

National Environment Agency and Ministry of the Environment & Water Resources (2006) 'Integrated Solid Waste Management in Singapore', Asia 3R Conference, 30 October–1 November 2006. Available at http://www.env.go.jp/recycle/3r/en/asia/02_03–3/05.pdf.

Premakumara, D.G.J. (2012) 'Establishment of the Community-Based Solid Waste Management System in Metro Cebu', in *Report for the Establishment of the Waste Management System in Metro Cebu, Philippines* (KITA and IGES).

Premakumara, D.G.J., Abe, M. & Maeda, T. (2011) 'Reducing Municipal Waste Through Promoting Integrated Sustainable Waste Management (ISWM) Practices in Surabaya City, Indonesia', in Y. Villacampa & C.A. Brebbia (Eds.), *Eco System and Sustainable Development VIII* (WIT Press).

Premakumara, D.G.J., Canete, A.M.L. & Nagaishi, M. (2013) Policy Implementation of the Republic Act (RA) 9003 in the Philippines: A Case Study of Cebu City, the 1st IWWG-ARB Symposium, 18–21 March 2013, Hokkaido University, Japan.

Sakawi, Z. (2011) 'Municipal Solid Waste Management in Malaysia: Solution for Sustainable Waste Management', *Journal of Applied Science in Environmental Sanitation* 6(1), 29–38.

Tience, D. (2005) 'Maximising the Potential for Community-Based Solid Waste Management in Indonesia', Unpublished PhD thesis, the University of Queensland, Australia.

UNCRD (2010) *3R Source Book* (United Nations Centre for Regional Development).

UNEP-IETC & HIID (1996) *International Source Book on Environmentally Sound Technologies for Municipal Solid Waste Management* (UNEP, International Environmental Technology Centre).

UN-Habitat (2008) *State of the World's Cities 2008/2009: Harmonious Cities* (Earthscan).

World Bank (1999) *What a Waste: Solid Waste Management in Asia* (World Bank).

World Bank (2012) *What a Waste: A Global Review of Solid Waste Management* (World Bank).

World Resource Institute, United Nations Environment Programme, United Nations Development Programme and the World Bank (1996) *World Resources 1996–97: The Urban Environment* (Oxford University Press).

12

Water scarcity and pollution in South and Southeast Asia

Problems and challenges

M. Dinesh Kumar, P.K. Viswanathan and Nitin Bassi[1]

Introduction

South Asia accounts for nearly 4.5 per cent of the global freshwater resources, but one-quarter of its population (Babel & Wahid 2008: x). Some of the oldest civilizations of the world flourished and perished on the banks of the rivers of this region. But the region displays high heterogeneity with respect to hydrology, climate, topography, culture and socio–economic milieu. It has some of the most productive farming systems of the world and also regions of precariously low productivity. The economic conditions vary drastically with the eastern Gangetic Plains having more poor people than in sub-Saharan Africa.

With a predominantly agrarian population and tropical and sub-tropical climate, irrigation development was necessary for sustaining agricultural production. Irrigation is through surface and groundwater systems. Today, the region is the world's largest user of groundwater. However, with excessively high demand for water for agriculture and limited freshwater resources, some agriculturally prosperous regions face physical scarcity of water. In contrast, some other regions, in spite of being water rich, face economic scarcity of water owing to the extreme poverty of farming communities. The poor peasants pay prohibitive costs to well owners for accessing water for protective irrigation.

With rapid economic growth, demand for water for urban and industrial uses is also growing rapidly, resulting in not only increased withdrawal of water from rivers and aquifers, but also pollution of rivers and lakes from indiscriminate disposal of untreated and partially treated effluents. This seriously threatens the ecosystem health and water security of millions of rural and urban households that have no access to formal water supply systems to meet their basic survival needs. The ecosystem deterioration is not confined to naturally water-scarce basins alone, but also exists in relatively water-rich basins. The hydrological and ecological changes occurring in South Asia's river basins as a result of the rapid socio-economic drivers and the way they ultimately impact on the society can offer important lessons for certain countries of Southeast Asia, which, although having abundant water resources, lack institutional capability to deal with problems caused by poor access to water for human health and livelihoods, water pollution and water-related disasters such as floods and droughts.

The second section discusses the hydrological regime of South Asia, focusing on the inter-regional and inter-annual variability. The water resources availability in the region are analysed from an anthropogenic perspective, showing how the per capita annual renewable water resources vary across countries. The socio–economic factors driving the demand for water, such as the structure of the economy, the proportion of people dependent on agriculture, population distribution between rural and urban areas and access to arable land are analysed in the third section. The water supply systems for agriculture and other sectors, and water withdrawal for agriculture are analysed and discussed in the fourth section. In the fifth section, we discuss the water security challenges of South Asia through the lens of the three remarkable river basins of South Asia, viz., Ganges-Brahmaputra-Meghna (GBM), Indus and the Helmand, which constitute nearly 45 per cent of the region's geographical area and more than 50 per cent of its population, and represent the region's diversity. Emerging water management challenges are covered in the sixth section. In the final section, we discuss water security challenges in Southeast Asia and compare them with those of South Asia.

Hydrological regime of South Asia

South Asia is one of the most water-rich regions of the world, but it also displays the highest spatial variability (FAOSTAT data). Monsoon being the largest source of water, the region is also characterized by very high seasonal variation in water resource availability. The role of institutions and policies in managing water resources for human wellbeing is very important in this region, where poor management of water costs millions of lives in floods, water-borne diseases and malnutrition. It is also one of the most water- and food-insecure regions of the world, second only to Sub-Saharan Africa.

The rainfall in South Asia varies from nearly 50mm in the Southwest of Afghanistan to around 11,000mm in Chirapunji in Northeast of India. The Indian monsoon is characterized by a high degree of spatial and temporal heterogeneity. The mean annual rainfall varies from less than 100mm in Western Rajasthan to around 11,700mm in Meghalaya. The regions that receive low mean annual rainfall experience high year to year variation, whereas the regions that receive high mean annual rainfall experience low year to year variation. Further, the number of rainy days is large in high rainfall regions, and small in low rainfall regions. The coefficient of variation in rainy days shows more or less the same spatial pattern as the mean annual rainfall (Pisharoty 1990: 1–3). India has diverse climates, and varies from hyper-arid to arid to semi-arid to sub-humid to humid. It has mountainous regions, middle mountains, plateaus, plains, deserts, and coastal plains and deltas (Kumar 2011: 346).

Annual rainfall in Sri Lanka is 2,540mm to over 5,080mm in the southwest of the island. The rainfall is less than 1,250mm in the northwest and southeast of the island. Hence, there are two rainfall zones in the country, viz., the dry zone and the wet zone. The southwest monsoon is from May to August, whereas the northeast monsoon is from November to February. Rainfall pattern in Sri Lanka is influenced by differences in terrain and elevation across the island. It has a coastal climate in many parts. It also has mountains in the central and south central regions with a cold climate, and the remaining low-lying regions have a humid tropical climate (Kumar 2011: 347).

In Pakistan, there is significant spatial variation in the rainfall, though most of the country receives very low rainfall. The whole of Sindh, most parts of Baluchistan, major parts of Punjab and central parts of northern areas receive less than 250mm of rainfall in a year. Northern Sindh, southern Punjab and northwestern Baluchistan receive less than 125mm of rainfall (Kumar 2011).

Point rainfall recorded in different stations in the upper Indus basin show a lowest mean annual rainfall of 131.2mm at Gilgit located at an elevation of 1,460m to a highest of 1,429mm at Shahpur located at an elevation of 2,012m (Archer & Fowler 2004: 50). True humid conditions occur after the rainfall increases to 750mm in the plains and 625mm in the highlands. There are two sources of rainfall: the monsoon from July to September, and the western depression from December to March (Kumar 2011: 347).

Rainfall in Nepal ranges from more than 6,000mm along the southern slopes of the Annapurna range in central Nepal to less than 250mm in the north central portion near the Tibetan plateau. On average, about 80 per cent of the precipitation is confined to the monsoon period. High mountains cover nearly 35 per cent of the geographical area, followed by the middle mountains covering nearly 42 per cent and the Terai region covering nearly 23 per cent. The climate varies from alpine to sub-alpine in the higher Himalayas to temperate in the lesser Himalayan region to sub-tropical in the Terai and Siwalik regions of the south within a distance of 200km (Kumar 2011: 347).

Precipitation in Afghanistan has a very pronounced annual cycle with a dry period in summer, generally from June to September, except for the western region where the dry season starts in May, lasting until October. Annual precipitation ranges from 50mm in the southwest to 700mm in the region of Salang. In the eastern part of the country, the total annual precipitation decreases to about 100mm. It has the central highlands, which are part of the Hindukhush Himalayan range, northern plains and southwestern plateau, which consists of sandy desert and semi-desert (Kumar 2011: 348).

Bangladesh has tropical monsoon climate characterized by high seasonal rainfall, high temperature and high humidity. The average annual rainfall varies from a maximum of 5,690mm in the northeast to 1,110mm in the west. The groundwater, however, provides adequate storage to compensate for annual variations in rainfall and stream flow. Most of Bangladesh is alluvial plains formed by the delta of the Ganges, Brahmaputra and Meghna river systems, except the southeastern and northeastern hills (Kumar 2011: 348).

The high spatial variability in water resource endowment in South Asia is evident from the climate characteristics of Indus, one of the most important river basins of the sub-continent, feeding three countries (Laghari et al. 2012: 1066). Within the same basin, there are regions that receive more than 1,000mm of water, and with humid and sub-tropical climates. There are also regions that receive less than 100mm of rainfall, with a hot desert climate. But the population density also varies remarkably from region to region. The regions that have poor natural water endowment such as Western Rajasthan in India have extremely low population density. Mountainous regions such as in the northeastern part of India and the sub-Himalayan region (in Nepal, Pakistan and Afghanistan) also have very low population density, whereas those that are naturally water rich but with a warm climate, such as the Eastern Gangetic basin in India and Bangladesh, have very high population densities. Therefore, it is important to look at freshwater resources in per capita terms. The per capita renewable water resource varies drastically from region to region and from country to country (Figure 12.1). It varies from 89m^3 per capita per annum in Maldives to 39,700m^3 per capita per annum the mountain kingdom of Bhutan. The per capita renewable water resource per annum is around 1,700m^3 in India, around 1,400m^3 in Pakistan and 8,000m^3 in Nepal. Except for Bhutan, renewable water resource availability in South Asia is below the global average, which is around 20,000m^3 per capita per annum (FAOSTAT data). But these figures are not indicative of how much water the population currently can access, which is determined by how much of this water is actually utilizable, access to financial resources and available technologies.

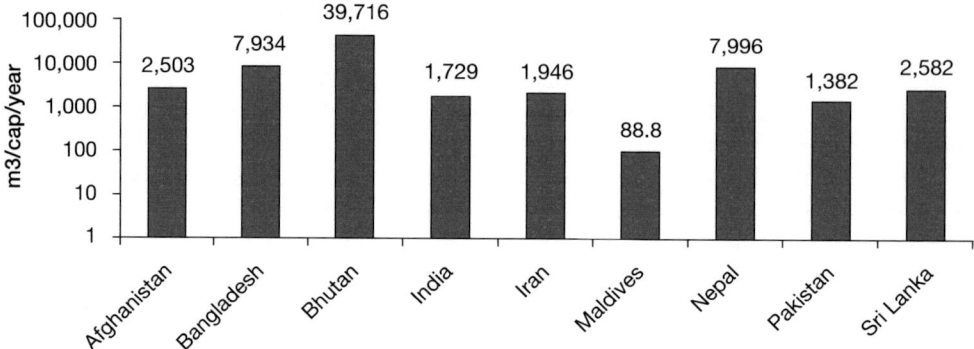

Figure 12.1 Annual per capita renewable water resources in selected South Asian countries, 2003–2007

Source: Author analysis using FAOSTAT data.

Drivers of water demand in South Asia

In South Asia, the agricultural component of GDP ranges from 17.1 per cent in India to 33.9 per cent in Nepal (Figure 12.2). The region is experiencing the fastest growth in population of all regions of the world. The region is also experiencing considerable economic growth, most of which is driven by India, which accounts for a major chunk of the sub-continent's population. Both population growth and rising per capita income resulting from economic growth act as major drivers of growth in water demand. As regards the impact of population growth, more water will be required to produce food for the growing population, and more water will be required to meet the water supply needs in rural and urban areas. Rising income levels, by way of contrast, will change the consumption patterns towards food items that are higher up in the food chain such as meat and dairy products (FAO 2009: 5). But for food demand to get translated into water demand, arable land availability is a pre-requisite. Since South Asia has

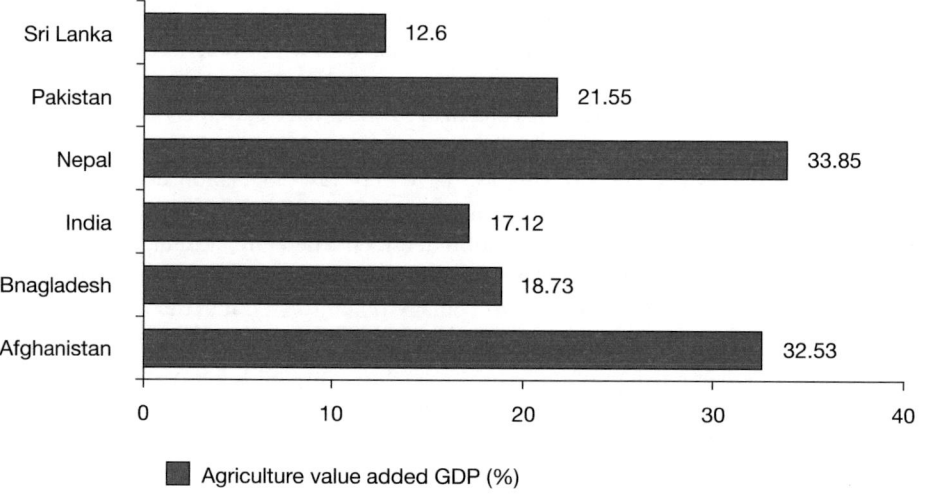

Figure 12.2 Contribution of agriculture to GDP in selected South Asian countries, 2008–2012

Source: Author analysis using FAOSTAT data.

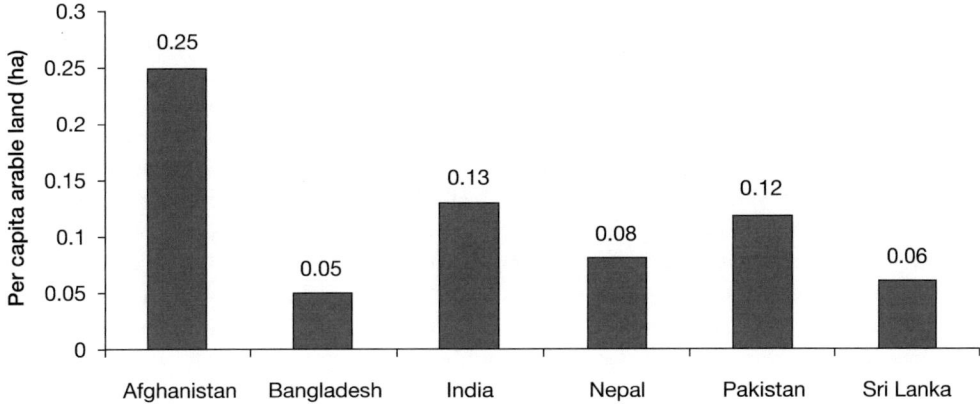

Figure 12.3 Per capita arable land in selected South Asian countries, 2004

Source: Author analysis using FAOSTAT data.

Table 12.1 Land use intensity in South Asia, 2008–2012

Sr. no.	Country	Total geographical area ('000 ha)	Total arable land ('000 ha)	Gross cropped area ('000 ha)
1	Afghanistan	65,223	7,793	7,910
2	Bangladesh	14,400	7,509	8,549
3	India	328,726	157,923	169,623
4	Nepal	14,718	2,400	2,520
5	Pakistan	79,610	20,430	21,280
6	Sri Lanka	6,561	1,200	2,170

Source: Author analysis using FAOSTAT data.

one of the highest land use intensities in the world in terms of the proportion of arable land under cultivation (Table 12.1), the additional food production can come from intensifying land use through expansion of the irrigated area, which is quite low in countries such as India. Only approximately 40 per cent of the total arable land of 157m. ha is irrigated in India. The per capita agricultural water demand is determined by the per capita arable land and climate. It is considerably high in countries that experience hot and arid climatic conditions (Figure 12.3). While the per capita arable land is 0.13ha and 0.12ha for India and Pakistan, it is as high as 0.25ha for Afghanistan, whereas it is very low in water-rich Bangladesh.

Urbanization is yet another factor that drives the demand for water in the region. As average per capita water requirement for domestic uses in urban areas is more than that of rural areas (Kumar 2010: 26), a demographic shift would mean a higher average per capita water requirement for domestic uses. In India, rapid industrialization is posing new challenges of water quality deterioration along with very rapid and continuing increase in the demand of water for manufacturing.

Water resource systems in South Asia

South Asia has a wide range of water supply systems that are both modern and traditional to cater to different water needs of the society, and that utilize surface water resources, groundwater resources, water from springs and even soil moisture. South Asia has some of the largest man-made reservoirs and water diversion systems created through the construction of dams and barrages, for irrigation and water supplies. It also has large hydropower dams and reservoirs, particularly in India, Pakistan, Bhutan and Nepal. Countries such as India and Bangladesh have a variety of traditional water harnessing systems such as tanks, ponds and lakes, which are used for multiple purposes such as irrigation, domestic water supply, livestock drinking, fisheries and recreation. For water supplies in rural areas, all these countries depend on many modern water systems such as hand-pumps, piped water supply schemes based on wells, bore wells and tube wells, tanks and lakes, traditional open wells and Persian wheels, although the extent of contribution of different types of source depends on the hydrological and geo-hydrological regime.

Large urban areas in India now depend on water imported from distant reservoirs, and their percentage contribution increases with increase in the size of the city. Regional water supply schemes based on surface water sources are also now replacing small, decentralized rural water supply schemes, based on wells (Mukherjee et al. 2010: 10–14). In Pakistan, about 10 per cent of the total pumped groundwater (around 4 BCM) is used to meet domestic and industrial requirements. In the most populous province of Punjab, about 90 per cent of the population depend on groundwater for their daily domestic needs. In Baluchistan province, about 4 per cent of the population depend on groundwater (Qureshi et al. 2010: 3).

Irrigation accounts for a major share of the water demand in South Asia (FAOSTAT data), and some of the world's largest irrigated areas are in this sub-continent. India alone has nearly 97 m. ha of gross area under irrigation. This is followed by Pakistan, which has a total of 19.02 m. ha of irrigation, mostly in the Indus irrigation system (Government of Pakistan 2007). Agricultural water withdrawal data for selected South Asian countries are given in Figure 12.4. The per capita water withdrawal is highest for Pakistan (1,012m³/annum), followed by Afghanistan and then India.

The irrigation water demand is met through many large, medium and minor irrigation schemes based on surface reservoirs and gravity flow. The total reservoir storage systems created by surface water systems, which cater not only for irrigation but also domestic water supplies and hydropower generation in different countries, and also the storage in per capita terms for these countries, are given in Figure 12.5. The per capita reservoir storage is an important indicator of water security of nations, as argued by Kumar et al. (2008). Sri Lanka has the highest per capita storage among all the six countries, followed by India and Pakistan.

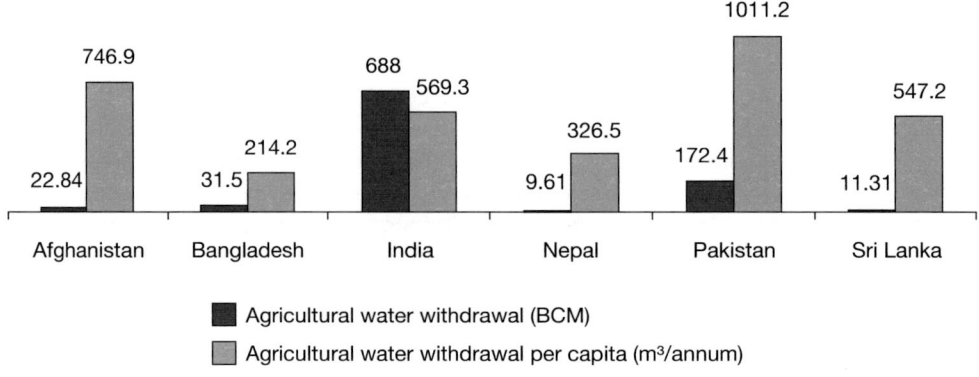

Figure 12.4 Agricultural water withdrawal in selected South Asian countries, 2008–2012
Source: Author analysis using FAOSTAT data.

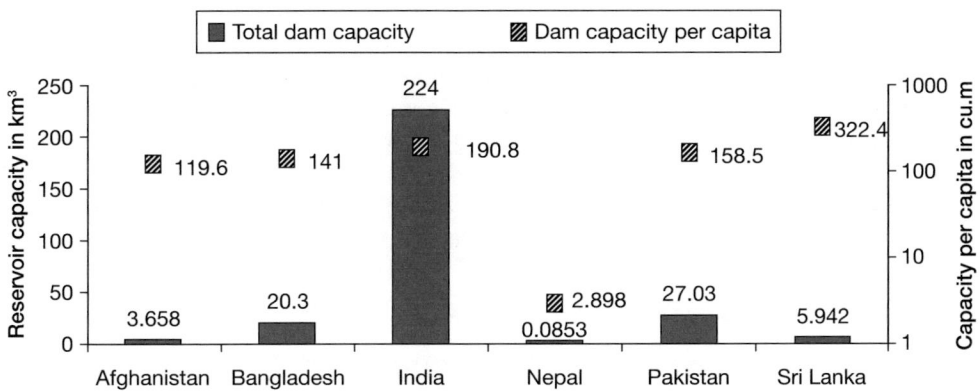

Figure 12.5 Reservoir water storage capacity in selected South Asian countries, 2003–2007
Source: Author analysis using FAOSTAT data.

Another most fascinating feature of irrigation in South Asia is the well-developed groundwater irrigation systems. Many millions of wells, which irrigate small plots of land, dot the rural landscape of South Asia. Groundwater is the major source of irrigation for semi-arid and arid regions of the sub-continent, where surface water resources are extremely limited and public irrigation systems are inefficient. India alone has nearly 25 million groundwater irrigation structures. Groundwater accounts for nearly 64 per cent of the net irrigated area in India, whereas it is as high as 76 per cent in Bangladesh. In Afghanistan, it is only 15 per cent and Sri Lanka has virtually no well irrigation (Figure 12.6). Tanks and large surface systems are the main sources of irrigation in this island country.

However, interestingly, inefficiency of public irrigation systems, caused by heavy losses in conveyance, low on-farm water use efficiency under gravity irrigation and unreliable supplies,

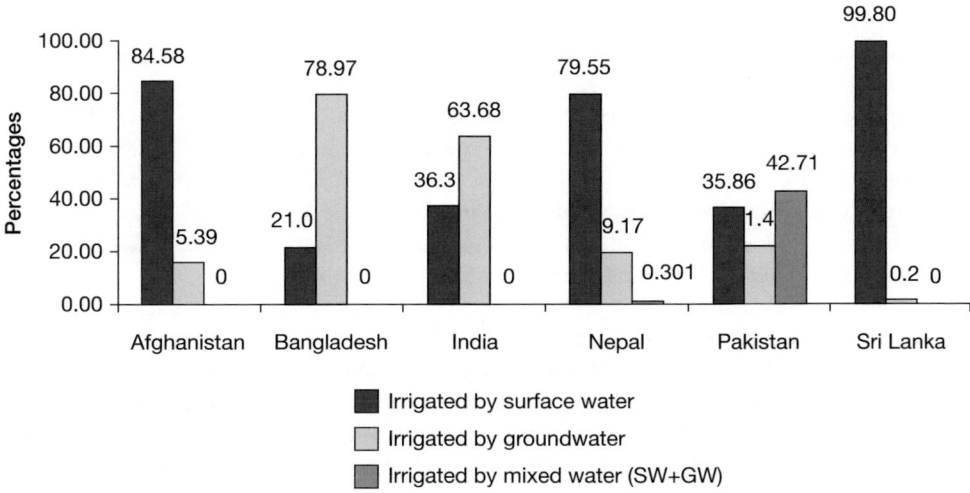

Figure 12.6 Proportion of area irrigated by different sources in selected South Asian countries, 2008–2012

Source: Author analysis using FAOSTAT data.

enriches groundwater significantly in the command areas in large gravity irrigation systems such as the Indus Basin Irrigation System (IBIS) in Pakistan (Qureshi et al. 2010: 2). The same groundwater is pumped out by farmers, including those who access canal water. IBIS also has vast areas irrigated by both groundwater and surface water conjunctively, accounting for nearly 43 per cent of the net irrigated area in Pakistan. The total groundwater pumping in Pakistan is estimated to be around 49.9 BCM, from around 0.13 million electric tube wells and 0.83 million diesel tube wells. There is extensive tube well irrigation in Indian Punjab also, which to a great extent is sustained by recharge from irrigation return flows, and canal seepage.

Water security challenges of the South Asian region

South Asia encompasses some of the most water-scarce regions of the world. The whole country of Pakistan, whose agriculture is heavily dependent on the Indus Basin Irrigation System, is highly water stressed. Many regions in India have renewable water resources falling below the threshold of 1,000m³ per capita. Here, we analyse the water security problems of the Ganges, Indus and Helmand basins using five major components, namely: i) water resource stress, which is reflective of the natural water resource endowment and its variability across the basin; ii) water development pressure; iii) the water access poverty; iv) ecological health of the river basins; and v) the management capacity that exists within these river basins to tackle these problems.

There is a clear pattern in spatial variation of rainfall and climate in South Asia. The regions of low rainfall have high aridity and regions of high rainfall have low aridity or high humidity (as in the Indus river basin). Such unique patterns vis-à-vis spatial variations in rainfall and climate create regions of extreme water stress in terms of renewable water availability due to poor runoff rates and low groundwater recharge rates, and fast depletion of soil moisture (such as Western Rajasthan, Punjab, Sindh and Baluchistan), and regions of water abundance (such as the Eastern Gangetic plains encompassing Bihar, Eastern Uttar Pradesh, most parts of West Bengal, and Bangladesh). Similarly, the regions of low rainfall also experience monsoon failure more frequently, causing extreme water stress resulting from hydrological droughts (Eriyagama et al. 2009: 16).

A river basin experiences a water resource stress when the available freshwater fails to support socio-economic development and maintain healthy ecosystems. The availability of freshwater is expressed in terms of per capita water resources (1,700m³ per person per day is considered as the threshold for water-stressed condition), and by variation of precipitation (a coefficient of variation of 0.3 is taken as the critical level, beyond which a water resources system is considered most vulnerable) in a basin (Babel & Wahid 2008: x). The Indus basin experiences highest water stress owing to a renewable water availability falling below the 1,700m³ per capita mark; but, at the same time, the Helmand basin experiences highest stress in terms of the water variability due to sharp spatial variation in rainfall across the basin. On both fronts, the Ganges experiences the least water resources stress.

The level of exploitation of water is not uniform across regions in the sub-continent. For instance, in India, it is high in regions of poor water resource endowment, and vice versa. The level of exploitation of groundwater is very low in regions of water abundance, owing to poor availability of arable land, high rainfall and low aridity. Excessive withdrawal of groundwater results in drying up of wells in the hard rock areas where sufficient groundwater stock is not available (Kumar 2007: 55). Surface water is also over-appropriated in the semi-arid and arid regions of India, leaving much less water as 'environmental flow' for meeting the ecological functions of the river, downstream. Such areas experience 'water development stress' (Kumar 2010: 32).

The quality of groundwater in the Indus plains varies widely (Qureshi et al. 2010: 5). Groundwater in areas receiving high rainfall in the upper parts of Punjab has low salinity. Similarly areas underlying the Indus River and its tributaries and canals have wide and deep belts of relatively fresh groundwater. The salinity of the groundwater generally increases away from the rivers and also with depth. In Punjab province, 23 per cent of the area has poor quality groundwater, while it is 78 per cent in Sindh (Haider 2000, cited in Qureshi et al. 2010: 8). In the lower parts of the Indus plain, the area of fresh groundwater is confined to a narrow strip along the river. In central areas of Punjab province, a layer of fresh groundwater floats over the saline water. Due to excessive pumping of this fresh groundwater lens, saltwater intrudes into fresh groundwater areas. Groundwater resources are over-exploited for irrigation in areas where rainfall is low, climate is more arid, and groundwater quality is good (Qureshi et al. 2010: 8).

In Nepal, groundwater utilization is mostly confined to the lower Terai region, with farmers using shallow tube wells (Kansakar 2003: 95–96). The Indus basin area of Pakistan also experiences extreme environmental water stress resulting from excessive diversion for irrigation in the IBIS.

The water development pressure was assessed considering: total water use against the renewable water resources, which reflects the level of water exploitation; and the proportion of the population having access to a safe drinking water source, which reflects the level of water infrastructure development (Babel & Wahid 2008: 5). Water exploitation is highest in the Indus basin (89 per cent), the basin is having the lowest renewable water resources in per capita terms. The water exploitation pressure is lowest for the Ganges, which has the highest renewable water resources and lowest level of use. But the proportion of population without access to safe drinking water is highest for the Helmand (57 per cent), followed by the Ganges (17 per cent). Although water development pressure is the highest in the Indus basin, a large majority of the people (from India and Pakistan) enjoy access to drinking water supplies. Contrariwise, in spite of having plenty of water resources, nearly 17 per cent of the people do not have access to safe drinking water sources in the Ganges basin, the reason for which can be attributed to the socio–economic backwardness of the region and the poor public investment in water supply sources.

Although richly endowed with water, South Asia experiences great difficulties in accessing it from the natural systems, for meeting irrigation needs. Some of the factors contributing to this 'water access poverty' are the high degree of poverty among farmers in the region, which reduces their financial capacity to invest in wells; the poor landholding size, which reduces the overall economic viability of having independent sources of water such as wells and pump sets for small and marginal holders; poor rural electrification, which increases the fuel cost of accessing well water; and high monopoly power of well owners, which increases the price at which irrigation services are offered to farmers. These regions (comprising the Eastern Gangetic basin in India and Bangladesh, and the Terai region of Nepal) therefore face economic water scarcity (Kumar 2007: 68).

As regarding ecological health, in India, with rapid industrialization and fast growing urban areas, even water-rich river basins are heavily polluted due to indiscriminate disposal of effluents from industries located on the banks through sewers, and direct disposal of human and animal waste from rural areas. The Ganges, which provides water to around 40 per cent of India's population, receives around 2,900 million litres of untreated sewage every day, while the existing treatment plants have capacity to treat only 1,100 million litres per day (*Hindustan Times* 2012). In semi-arid and arid regions, the intensity of this pollution is exacerbated by excessive diversion of water from the rivers and streams, reducing the ability of the rivers to assimilate pollution, while causing ecosystem deterioration. In Pakistan, pollution of the Indus River is caused by excessive salinity resulting from runoff from irrigated fields. Also, groundwater in the Indus

Basin Irrigation System is increasingly becoming saline. Salinization in IBIS is due to two significantly different processes: i) by shallow saline water tables; and ii) due to irrigation with marginal quality groundwater (Aslam & Prathapar 2006: 1).

The ecosystem health of the basins was assessed considering: water quality deterioration, which is evaluated as the volume of wastewater discharged into the basin's rivers; and ecosystem insecurity, which is evaluated as the extent of (natural) vegetation cover as a percentage of the basin area (Babel & Wahid 2008: 5). Very high levels of pollution were observed in both the Helmand and Indus river basins. The water quality deterioration is low in GBM, just because of the huge amount of renewable water resources these basins have. Further, a major share of the renewable water resources are in the Brahmaputra basin, which does not have any major polluting sources, although certain sub-basins of the Ganges such as the Yamuna, which have much less renewable water resources, are highly polluted. On ecosystem insecurity, GBM is far more seriously deteriorated than the other two basins with only 20 per cent of the land area under natural vegetation.

Obviously the capacity of the existing institutions to tackle water scarcity and pollution problems in these countries is precariously low, as is evident from low water use efficiency in the agriculture sector, which consumes by far the largest share of the water, poor access to sanitation in rural and urban areas, and poor capacities for resolving conflicts which emerge out of the growing competition for water at times of scarcity. The institutional capabilities to charge for water in the agriculture and domestic sectors, in a manner that reflects the scarcity value of the resource, or introduce charges or taxes for producing wastewater or causing pollution of water bodies, are largely lacking in the water bureaucracies of South Asian countries. The regulatory framework for pollution control, which uses norms vis-à-vis quality for effluents being discharged into natural water bodies, is not effectively implemented. The 'polluter pays' principle is not practised.

While competition among different uses of water such as between irrigation and urban water supply, and industrial use and rural water supply is growing in many water-scarce regions in India (Kumar 2010: 37–39), there is mounting tension between India and Pakistan over sharing of water from the Indus, as the flows in the river from upstream glacier-laden catchments reduce and India moves ahead with its decision to build a few more hydropower dams in its territory in Jammu and Kashmir. The countries have already been embroiled in two legal fights over water. In 2005, Pakistan challenged India's 450-megawatt Baglihar dam before a World Bank-appointed neutral expert and lost. In 2011, the countries went head to head at the International Court of Arbitration over India's 330-megawatt Kishanganga project in Jammu and Kashmir. The court ordered India to temporarily stop the dam construction while assessments are being made (*Time* 2012). There exists an institutional vacuum to resolve such conflicts within these countries.

Thus, in addition to the components (such as water resource endowment and ecological health) that focus on natural processes, management efficiency also plays an important role in sustainable development and use of water resources. The current management capacity to cope with the mismatch between water demand and supply is assessed in relation to: water use efficiency, assessed as the ratio of total annual GDP per capita from activities using water against the total annual water use per capita; and improved sanitation inaccessibility, assessed as an inverse function of the percentage of people having access to improved sanitation facilities (Babel & Wahid 2008: 5).While there is but a slight difference in water use efficiency, the inaccessibility to improved sanitation is high in both GBM and Helmand (about 60 per cent in each basin), and comparatively better in Indus (50 per cent).

Emerging water management challenges

Water resource development schemes are administered in a highly sectoral and segmented way in South Asian countries. Multiple agencies are engaged in water resource development in these countries for different sectors such as the irrigation department, rural water supply and sewerage department, urban water utilities etc. There is hardly any coordination among them in terms of policies and actions. Often, there is more than one agency to look after the same sector, be it irrigation or rural water supply, and they work in a segmented way. For instance, there are separate departments dealing with well irrigation and surface irrigation in these countries. Although groundwater and surface water systems are interconnected, these agencies work at cross purposes, and their actions are not coordinated at the hydrological system level.

Often, the same line agency looks after multiple functions – water resource planning, water development, water resource management and water allocation, reducing their effectiveness in managing the resource (Kumar 2010: 233). For instance, there is no separate agency concerned with catchment (basin) assessment in most Indian provinces. The same agency, which is engaged in irrigation development, is responsible for assessing the dependable yield of the catchments or groundwater resource assessment. This leads to a situation of over-assessment of the utilization potential and consequent over-development of water resources in the basin. Likewise, the agency that monitors water quality is also responsible for protection of water quality. This situation motivates them to downplay the magnitude of pollution problems.

These agencies neither pay for drawing water from the basins, nor are constrained by any limits on water diversion, in the absence of any agency for basin-wide allocation of water to different sectors and users within the sectors. They also do not have to pay for pollution of water, on the basis of the volume of effluent or the intensity of pollution, except being mandated to adhere to effluent standards. Hence, they have no special incentive in reducing their withdrawal or improving the efficiency of its use or reducing the intensity of pollution. At the same time, there is an absence of any legitimate agency to ensure sufficient flows in rivers and streams to meet the environmental flow requirements.

Another important issue is the lack of well-defined rights in groundwater. Although it is one of the most important sources of water to meet all human needs, the right to use this resource in the South Asian countries is attached to land ownership rights, and there is no restriction on the volume of groundwater that the landowners can pump. Again, most of the irrigation wells in India, Pakistan and Nepal (the countries in which there is significant groundwater irrigation) are in the private domain and very few wells are owned by government agencies.

Although groundwater is a major source of irrigation in India (Kumar 2007: 1) and electric wells account for a major share of the total number of groundwater abstraction structures (Scott & Sharma 2009), electricity used for groundwater pumping is not metered in most Indian states, with the exception of West Bengal in Eastern India (Kumar et al. 2011: 382–383). In Pakistan, although farmers pay for electricity on the basis of consumption (pro rata pricing), farmers who use electric wells are a small fraction of the total well owners, owing to high electricity costs, and the rest use diesel engines for their tube wells (Qureshi et al. 2010: 11). Because of this, the well irrigator farmers in Pakistan are already confronted with the 'positive marginal cost' (i.e., if farmers pump more water, it will cost them more) of using groundwater. Hence, one can presume that they are already using groundwater efficiently, particularly when we consider the fact that groundwater irrigation costs are very high there.

Many major river basins in South Asia are transboundary in nature. In spite of the vital role water resource management plays in the region's economic development, very little regional cooperation on water resources development and management exists among the co-basin countries (Brichieri-Colombi & Bradnock 2003: 47). This disables each region from making use of its comparative advantage over the other regions (in terms of climatic conditions, topography, water flows, presence of arable land and unique socio-economic features, finance, technology and human resources) in harnessing the resource for socio-economic development and sharing the benefits with its neighbours. The important transboundary basins are the Indus, Ganges, Brahmaputra, Meghna and Kosi.

But, the regional cooperation on water resource development is limited to only hydro-power development, and that too is very recent. For instance, in April 2013, Nepal, India and Bangladesh (NIB) decided to cooperate and exploit the hydropower sector and use water resources management for mutual advantage, including jointly developing and financing projects in the Ganges river basin (*The Hindu* 2013). Instead, only limited legal agreements on water-sharing were signed in the aftermath of longstanding disputes over the use of water from these basins.

The only two agreements that exist are the Indus Water Treaty, signed in 1960 between India and Pakistan over sharing of water from the Indus river basin and its five tributaries, viz., Jhelum, Chenab, Sutlej, Rabi and the Beas, and the Farakka Treaty of December 1996, signed between India and Bangladesh over sharing of water from the Ganges at Farakka barrage in West Bengal. Since the construction of Farakka barrage by India in 1975, the Ganges' waters have been a key source of conflict between India and Bangladesh. But the Farakka Treaty has not been successful in resolving all pending issues between the two countries (Rahman 2005: 200). India has recently raised serious concerns over the building of three hydropower dams by China in the Tibetan part of the Brahmaputra basin due to their potential negative impact on flows in the river in Arunachal Pradesh (*Economic Times* 2010).

There are no basin-wide initiatives among the co-basin countries over utilization of water in these transboundary river basins for the region's economic development, which makes full use of the comparative advantage each country has with respect to climate, topography, availability of arable land, finance, technology and human resources. This is in spite of the fact that the ability of the lower riparian states (countries) to fully utilize the benefits of the water from the basin or manage water in the basin to protect their interests, depends on the actions by upper riparian states.

The emerging water management challenges are: creating river basin organizations for basin-wide water allocation across and within sectors and also water resources management including WQM, and institutional platforms for resolving water-related conflicts between countries in transboundary (international) river basins; coordinated development and management of water resources within countries at the level of hydrological systems; integrated planning for development of surface water and groundwater resources; establishment of well-defined water rights for groundwater; removing electricity subsidies for pumping groundwater in the farm sector; and taking care of the hydraulic interdependence between groundwater and surface water in transboundary water allocation decisions.

Water security challenges of Southeast Asia

In this section, we explore the major factors influencing water security in the Southeast Asian region. Table 12.2 provides some interesting comparisons and contrasts between countries in the South and Southeast Asian regions.

The SEA countries have relatively lower population density, and have abundant renewable water resources on a per capita basis and the extent of withdrawal of annual renewable water resources is much lower as compared to South Asian countries, barring Bangladesh and Bhutan.

Table 12.2 Major development indicators of countries in South and Southeast Asia

Region/country	Population, 2011–12			Poverty (%) PPP $1.25 a day (2002–11) a	Agri. land (% of land area, 2009)	Available renewable water (m³/capita/annum, 2011)	Per capita arable land (ha)	Freshwater withdrawal (% of available, 2003–12)
	Total (million)	Urban (%)	Density (pop./km²)					
South Asia								
Afghanistan	32.4	23.8	50	20.0	58.1	2,069	0.27	35.6
Bangladesh	150.5	28.9	1,045	43.3	70.3	8,252	0.05	2.9
Bhutan	0.7	36.4	19	10.2	13.2	107,438	—	0.4
India	1,241.5	31.6	378	32.7	60.5	1,560	0.13	39.8
Pakistan	176.8	36.5	222	21.0	34.1	1,422	0.12	79.5
Sri Lanka	21.1	15.2	321	7.0	41.6	2,531	0.06	24.5
Southeast Asia								
Cambodia	14.3	20.1	79	22.8	31.5	33,675	0.27	0.5
Indonesia	242.3	51.5	127	18.1	29.6	8,417	0.10	5.6
Malaysia	28.9	73.5	88	3.8	24	20,422	—	2.3
Myanmar	48.3	33.2	71	—	19	24,352	0.21	2.8
Lao PDR	6.3	35.4	27	33.9	10.2	53,782	0.21	1.3
Philippines	94.9	49.1	317	18.4	40.1	5,136	0.06	17
Thailand	69.5	34.4	136	0.4	38.7	6,345	0.24	13.1
Vietnam	88.8	31.7	268	40.1	33.1	10,064	0.07	9.3

Source: Author compilation from ADB (2013); World Bank (WDI); FAOSTAT.

Table 12.3 Status of access to water, improved sanitation and living conditions, South and Southeast Asia

Region/country	Population lacking access to improved				Natural disasters (2011)		Natural resources depletion (% GNI, 2010)	People living on degraded land (%, 2010)
	water (% of pop.)		sanitation (% of pop.)		Death (no./annum)	% population affected		
	Rural	Urban	Rural	Urban				
South Asia								
Afghanistan (AF)	58	22	70	40	83	5.42	2.6	11.0
Bangladesh (BD)	20	15	45	43	102	1.11	2.3	11.0
Bhutan (BT)	6	0	71	27	1	2.70	3.6	—
India (IN)	10	3	77	42	1,038	1.01	4.4	10.0
Pakistan (PK)	11	4	66	28	511	3.06	2.8	4.0
Sri Lanka (SL)	10	1	7	12	254	1.40	0.3	21.0
Southeast Asia								
Cambodia (KH)	42	13	80	27	247	11.47	0.1	39.0
Indonesia (ID)	26	8	61	27	129	0.01	6.6	3.0
Malaysia (MY)	1	0	5	4	18	0.07	6.9	1.0
Myanmar (MM)	22	7	27	17	225	0.11	—	19.0
Laos (LA)	38	23	50	11	48	7.41	8.3	4.0
Philippines (PH)	8	7	31	21	1,989	12.35	2.1	2.0
Thailand (TH)	5	3	4	5	923	16.19	2.4	17.0
Vietnam (VN)	7	1	32	6	138	1.53	9.4	8.0

Source: Authors' compilation from ADB (2013); World Bank (WDI); FAOSTAT.

This creates a major difference in water management challenge between the two regions. But, some of the Southeast Asian countries are land scarce. The per capita arable land availability is extremely poor in countries such as Vietnam and Philippines (0.07ha and 0.06ha, respectively), reducing their ability to utilize the available water for crop production, whereas countries such as Thailand, Laos, Cambodia and Indonesia have a sufficient amount of arable land and their agricultural water demands could increase in future, and water demand could exceed supplies in a few countries such as Thailand and Indonesia.

Water security scenario in the Southeast Asian countries needs to be examined in terms of the sheer lack of access to good-quality water and improved sanitation facilities both in the rural and urban areas. As is evident from Table 12.3, the proportion of population lacking improved water facilities in the rural and urban areas is significant in many Southeast Asian countries. But, in terms of improved sanitation facilities, they are better placed in general, although Cambodia, Indonesia and Lao PDR remain as exceptions.

But unlike in the case of South Asia, the SEA countries report increasing numbers of human casualties due to natural disasters. Some of the most common natural disasters hitting Southeast Asia are typhoons, coastal cyclones and floods. The percentage of population affected by natural disasters is as high as 16.2 in the case of Thailand, 12.4 in the case of Philippines and 11.5 in the case of Cambodia. The extent of natural resources depletion and the proportion of people living on degraded lands are high in both the regions, however.

7.1 Water security status: comparing Southeast Asia and South Asia

A recent report (UN Water 2013) provides a comprehensive and widely accepted working definition of water security and claims that such a single description of the problem will help evolve global collaboration around water. Accordingly, *Water security* was defined as: 'the capacity of a population to safeguard sustainable access to adequate quantities of acceptable quality water for sustaining livelihoods, human well-being, and socio-economic development, for ensuring protection against water-borne pollution and water-related disasters and for preserving ecosystems in a climate of peace and political stability' (UN Water 2013: 1).

The Asian Water Development Outlook further explored the concept of water security at the national level and decomposed it into five key dimensions, viz., (a) household water security, (b) urban water security, (c) environmental water security, (d) economic water security, and (e) resilience to water-related disasters. Based on these five key dimensions and their integral sub-components (sub-indices), the Report developed a composite index, called the 'water security index' (WSI) (ADB 2013: 7).

At a regional scale, Southeast Asia has a WSI of 2.4 as against 1.6 for South Asia, indicating that the water security scenario is not that precarious as compared to South Asia. But, the scenario is much worse when we compare it with the advanced economies in the region, especially with respect to the key dimensions of urban water security and resilience related to water-related disasters. The advanced countries have an average water security index of 3.3.

7.2 Water security challenges of Southeast Asia

The national water security index and its key sub-indices for major South and Southeast Asian countries are presented in Table 12.4, which shows that access to piped water supply in rural areas and urban water security are some of the lowest in Vietnam, Cambodia, Lao PDR, Myanmar and Indonesia, and are comparable with the poorly performing South Asian countries. The data also suggest that access to water supply has got very little to do with the renewable water resources.

Table 12.4 National Water Security Index and its related components, major countries of South and Southeast Asia

Indicators		South Asia					Southeast Asia							
		AF	BD	IN	PK	SL	KH	ID	MY	MM	LA	PH	TH	VN
National Water Security Index (NWSI)		1.4	1.4	1.6	1.6	2.2	1.6	2.6	3.4	2	2.6	2.2	2.2	2.2
Household Water Security Index components	PWA (%)	4	6	23	36	29	17	20	97	8	20	43	48	23
	SANACC (%)	37	56	34	48	92	31	54	96	76	63	74	96	76
	DALY	5,289	1,217	1,246	1,072	153	2,170	483	181	1,551	1,078	528	504	296
Eco. Water Security Index (EWSI)		2.0	3.0	3.0	4.0	4.0	3.0	4.0	4.0	3.0	4.0	4.0	3.0	3.0
Urban Water Security Index (UWSI)		1.0	1.0	1.0	1.0	1.0	1.0	2.0	3.0	1.0	1.0	1.0	2.0	1.0
Environmental Water Security Index (EWSI)		2.0	1.0	1.0	1.0	1.0	2.0	3.0	3.0	3.0	3.0	2.0	1.0	2.0
Hazard and Vulnerability Index (HVI)		1.0	2.0	2.0	2.0	2.0	1.0	2.0	2.0	1.0	3.0	2.0	2.0	2.0

Notes: PWA – piped water access (%); SANACC – sanitation access (%); DALY (age-standardised disability-adjusted life years) is a measure of the diarrhoeal incidence per 100,000 people (hygienic measure).

Source: Author compilation from Statistical Yearbook for Asia and the Pacific (2012).

In spite of having abundant water resources, access to piped water supply in rural areas is very limited in countries such as Myanmar, Laos, Cambodia and Vietnam and is comparable with the South Asian countries such as Afghanistan, Bangladesh and India.

The growing urban population would imply concentrated demand for freshwater resources for drinking, food production as well as domestic uses. The share of urban population might reach the level of 40 per cent in most SEA countries with a significant increase in the withdrawal of water for domestic uses. There would be future problems of groundwater mining as cities grow rapidly. Such problems are already encountered in Thailand, and the challenges would be greater for coastal cities of this region. As regards agricultural water demand, in countries such as Vietnam and Philippines it would remain quite low due to the low per capita land availability and very high rainfall.

Nevertheless, widespread problems of groundwater mining and environmental water stress in rivers due to excessive diversion are least likely to be faced by most of the poor Southeast Asian countries. In Thailand, Malaysia, Lao PDR and Cambodia, the shrinkage of arable land and degradation of existing farm lands could pose new challenges. Simultaneously, water demand for irrigation has been growing in these countries. The increasing competition for the available water from domestic, municipal and manufacturing sectors and ecosystem services will reduce the effective availability of water for crop and livestock production in many basins/countries where renewable water availability is comparatively less (Thailand and Indonesia) and economic conditions are better. The growing water scarcity may have serious repercussions on future food production in these countries in the region (Viswanathan et al. 2013: 184–185).

At the same time, the deterioration of rivers in Southeast Asia occurring due to indiscriminate development pathways potentially threatens the livelihoods of tens of millions of people who depend on rivers. Fisheries, which support the livelihoods of about 1.6 million people in the Lower Mekong basin, may be seriously damaged if the migration routes of fish are blocked by dams on the Mekong River (MRC 2010).

The future of the rivers in the region is also at stake when we consider the potential pollution threat posed by the transport of dangerous goods and cargo by the major riparian countries, viz., Vietnam, Cambodia and Thailand through the waters of the Mekong river basin (MRC 2012). Much of the waters in Southeast Asia are polluted because of lack of wastewater treatment and disposal, adequate sanitation and proper management of sewage. As a consequence, the problem of pathogenic pollution is widely spread. Pathogens generally come from domestic sewage discharged untreated into watercourses.

The challenges facing provision of safe drinking water and improved sanitation facilities is immense in the poor SE Asian countries such as Laos, Cambodia and Myanmar with very high incidence of diarrhoea, a strong indicator of poor water-related hygiene. This is in spite of the people in these countries having relatively better access to improved sanitation in the rural areas. Environmental sanitation is also very poor in cities of many SEA countries, with a small fraction of the urban wastewater being treated to safe standards.

A much more serious issue facing water security in the Southeast Asian region is the increasing frequency of natural disasters. From the the mid-1990s to 2006, and in recent years, Southeast Asia has experienced a sharp increase in water-related disasters, particularly floods. Cambodia is known to be the least resilient due to its very low coping capacities, characterized by poor disaster preparedness and large proportion of poor inhabitants. Furthermore, as the floods in Thailand in 2011 demonstrated, even slow-onset events can disrupt the lives of millions and impact economic activities beyond the immediate flooded area (ADB 2013: 68). Evidence from Bangladesh suggests that natural disasters also affect food security of countries adversely (Herrmann & Svarin 2009: 5). Such impacts of large magnitude emphasize the need for long-term plans and investment strategies to enable SEA countries to cope with water-related disasters. Investments might include efforts to improve early warning systems, formulate improved flood and drought management plans, and implement upgraded data collection and surveillance systems. Although some countries have already made substantial advances in their disaster preparedness, risk reduction strategies are not uniformly or widely implemented in many countries.

8 Conclusions

Water security challenges of South Asia are distinctly different from those of Southeast Asia. While South Asia will have to deal with problems of extreme variations in climate, groundwater mining and environmental water scarcity due to excessive diversion of water from rivers and lakes along with increasing pollution from urban areas, the Southeast Asian countries will largely face environmental challenges of water pollution and vulnerability to natural disasters. While these challenges will magnify in future, countries such as Thailand and Indonesia are also likely to face problems of excessive withdrawal of groundwater and surface water for competing uses, including agriculture. In terms of managing its international basins, South Asian countries could learn a few lessons from Southeast Asia taking cues from the recent joint initiatives between ASEAN and Mekong River Commission in developing plans for development and management of the river basin and helping the member countries, which are also located along the Mekong river basin. Some of the poor countries of Southeast Asia such as Cambodia and Laos, where agricultural water withdrawal is likely to increase, could draw lessons from India and Pakistan on how uncontrolled exploitation of freshwater resources could pose challenges for the rural economy and the environment. On the other hand, the countries that are experiencing fast growth with industrialization and urbanization such as Philippines, Thailand and Indonesia should be prepared to face the challenges of managing water quality in rivers and aquifers.

Bibliography

ADB (2013) *Asian Water Development Outlook 2013: Measuring Water Security in Asia and the Pacific*, Asia Pacific Water Forum, Asian Development Bank, Philippines, 128p. Available at http://www.adb.org/sites/default/files/pub/2013/asian-water-development-outlook-2013.pdf.

Archer, D.R. & Fowler, H.J. (2004) 'Spatial and Temporal Variations in Precipitation in Upper Indus Basins, Global Teleconnections and Hydrological Implications', *Hydrology and Earthsystems Sciences* 8(1), 47–61.

Aslam, M. & Prathapar, S.A. (2006) Strategies to Mitigate Secondary Salinization in the Indus Basin of Pakistan: A Selective Review. Research Report 97, Colombo, Sri Lanka, International Water Management Institute (IWMI).

Babel, M.S. & Wahid, S.M. (2008) Freshwater Under Threat: Vulnerability Assessment of Freshwater Resources to Environmental Change – Ganges-Brahmaputra-Meghna River Basin Helmand River Basin Indus River Basin, United Nations Environment Programme (UNEP), Nairobi, Kenya. Available at http://www.unep.org/pdf/southasia_report.pdf.

Brichieri-Colombi, S. & Bradnock, R.W. (2003) 'Geopolitics, Water and Development in South Asia: Cooperative Development in the Ganges-Brahmaputra Delta', *Geographical Journal* 169(1), 43–64.

Economic Times (2010) 'China dams Brahmaputra River in Tibet', 16 November.

Eriyagama, N., Smakhtin, V. & Gamage, N. (2009) Mapping Drought Patterns and Impacts: A Global Perspective, Colombo, Sri Lanka, International Water Management Institute (IWMI Research Report 133).

Food and Agriculture Organization of the United Nations (FAO) (2009) Report of the FAO Expert Meeting on How to Feed the World in 2050, Rome, Italy, Food and Agriculture Organization.

Government of Pakistan (2007) *Agricultural Statistics of Pakistan 2006–07*, Ministery of Food, Agriculture and Livestock, Islamabad.

Haider, G. (2000) Proceedings of the International Conference on Regional Groundwater Management, 9–11 October, Islamabad, Pakistan.

Herrmann, M. & Svarin, D. (2009) 'Environmental Pressures and Rural–Urban Migration: The Case of Bangladesh', *MPRA Paper No. 12879.* Available at http://mpra.ub.uni-muenchen.de/12879/.

The Hindu (2013) 'Nepal, India & Bangladesh to Make Most of Ganga Water, Hydropower', 15 April.

Hindustan Times (2012) 'Ganga Receives 2900 Million Ltrs of Sewage Daily', *Hindustan Times*, 17 April.

Human Development Report HDR (2013) Human Development Report 2013: The Rise of the South: Human Progress in a Diverse World, New York, United Nations Development Programme.

Kansakar, D.R. (2003) 'Understanding Groundwater for Proper Utilization and Management in Nepal'. Available at publications.iwmi.org/pdf/H039311.pdf.pp95–104.

Kumar, M. Dinesh (2007) *Groundwater Management in India: Physical, Institutional and Policy* (Sage Publications).

Kumar, M. Dinesh (2010) *Managing Water in River Basins: Hydrology, Economics and Institutions* (Oxford University Press).

Kumar, M. Dinesh (2011) 'Potentials of Rainwater Harvesting in Different Hydrological Regimes: Practical and Policy Implications for South Asian Countries', in P.K. Mishra, M. Osman & B. Venkateswarlu (Eds.), *Techniques of Water Conservation and Rainwater Harvesting for Drought Management*, SAARC training programme, CRIDA and SAARC Disaster Management Centre, Hyderabad.

Kumar, M. Dinesh, Shah, Z., Mukherjee, S. & Mudgerikar, A. (2008) Water, Human Development and Economic Growth: Some International Perspectives, Paper presented at the IWMI-Tata water policy research program's seventh annual partners' meet, ICRISAT, Hyderabad, 2–4 April 2008.

Kumar, M. Dinesh, Scott, C.A. & Singh, O.P. (2011) 'Inducing the Shift from Flat Rate or Free Agricultural Power to Metered Supply: Implications for Groundwater Depletion and Power Sector Viability in India', *Journal of Hydrology* 409(1–2), 382–394.

Laghari, A.N., Vanham, D. & Rauch, W. (2012) 'The Indus Basin in the Framework of Current and Future Water Resources Management', *Hydrology and Earth System Sciences* 16, 1063–1083.

MRC (2010) Mekong River Commission. Strategic Environmental Assessment of Mainstream Dams. Available at http://www.mrcmekong.org/about-the-mrc/programmes/initiative-on-sustainable-hydropower/strategic-environmental-assessment-of-mainstream-dams/.

MRC (2012) Carriage, Handling and Storage of Dangerous Goods along the Mekong River, Mekong River Commission Navigation Programme, Volume I – Risk Analysis, Navigation Programme, April. Available at http://www.mrcmekong.org/assets/Publications/basin-reports/NAP-Risk-Analysis-Vol-I-Full-report.pdf.

Mukherjee, S., Shah, Z. & Dinesh Kumar, M. (2010) 'Sustaining Urban Water Supplies in India: Increasing Role of Large Reservoirs', *Water Resources Management* 24(10), 2035–2055.

Pisharoty, P.R. (1990) Characteristics of Indian Rainfall, Monograph, Ahmedabad, Physical Research Laboratories.

Qureshi, A.S., McCornick, P.G., Sarwar, A. & Sharma, B.R. (2010) 'Challenges and Prospects of Sustainable Groundwater Management in the Indus Basin, Pakistan', *Water Resources Management* 24(8), 1551–1569.

Rahman, M.M. (2005) Integrated Water Resources Management in the Ganges Basin: Constraints and Opportunities, Department of Civil and Environment Engineering, Laboratory of Water Resources, Helsinki University of Technology, Finland.

Scott, C.A. & Sharma, B. (2009) 'Energy Supply and the Expansion of Groundwater Irrigation in the Indus-Ganges Basin', *International Journal of River Basin Management* 7(1), 1–6.

Time (2012) 'Water Wars: Why India and Pakistan Are Squaring Off Over Their Rivers', 16 April.

UN Water (2013) 'Water Security and the Global Agenda, A UN-Water Analytical Brief', United Nations University, Institute for Water, Environment & Health (UNU-INWEH), Hamilton, Ontario, Canada. Available at http://www.unwater.org/downloads/watersecurity_analyticalbrief.pdf.

Viswanathan, P.K., Dinesh Kumar, M. & Sivamohan, M.V.K. (2013) 'Investment Strategies and Technology Options for Sustainable Agricultural Development in Asia: Challenges in Emerging Context', in M. Dinesh Kumar, M.V.K. Sivamohan & N. Bassi (Eds.), *Water Management, Food Security and Sustainable Agriculture in Developing Economies* (Earthscan/Routledge).

Note

1 Executive Director, Institute for Resource Analysis and Policy (IRAP), Associate Professor, Gujarat Institute of Development Research and Senior Researcher, IRAP, respectively.

13

Dams

Controlling water but creating problems

Darrin Magee

Introduction

Dams have been a key component of development scenarios around the world, and East Asia is no exception. The International Commission on Large Dams (ICOLD) defines as 'large' any dam 15 m or taller, and estimates that some 40,000 large dams punctuate the world's rivers. China alone may hold half of those (Gleick 2009). Dams come in all sizes, serve numerous purposes, utilize a range of construction techniques, and are funded and built by diverse actors including governments, development banks, and private companies. They bring varied socioeconomic, biophysical, and geopolitical impacts, both positive and negative, and have been the flashpoints for some of East Asia's most heated moments of social unrest and political change, ranging from the protests and violent suppressions that led to the cancellation of the Chico Dam in the Philippines in 1982 (Goldsmith & Hildyard 1986), to the more measured vote against the Three Gorges Project by roughly one-third of China's National People's Congress in 1992 (Gleick 2009), to the international outcry and eventual halting of the proposed dam cascade on the Nu River in China in 2004 (Magee 2006). During the last half-century, thousands of dams have been built in East Asia, a herculean effort whose social and ecological imprint across the region is significant.

For the purposes of this chapter, dams are built structures that impede the flow of a river in order to derive social benefits such as irrigation or power generation. Many of the dams covered here exceed ICOLD's definition of 'large' by an order of magnitude, and thus might more appropriately be labelled 'mega-dams' based on a common parlance in the dam industry (height greater than 150 m). Not surprisingly, many dam-related impacts scale directly with dam size; yet the cumulative impacts of many small dams may well outweigh one large dam. The ability of a dam and its reservoir to regulate a river is defined by the reservoir's volume as a fraction of the river's mean annual discharge. Regulating capacity is sometimes expressed in temporal terms such as daily, monthly, or seasonal, depending on the size of the fraction. By design, dams are disruptive; even the most carefully designed dams alter the natural flow regime of the rivers in which they are located. Swift-flowing water slows down in the reservoir area upstream of a dam, consequently dropping some of the silt it was carrying and undergoing a number of biochemical and physical changes. Alterations in temperature, erosion patterns, dissolved gases and other factors lead to changes to aquatic and terrestrial ecosystems, which in turn have implications for social systems that depend on those ecosystems.

Dams can and do provide important social benefits such as flood control, irrigation, power generation, transportation, drinking water, and recreation. In many parts of East Asia such as Indonesia and China, 'micro-hydro' dams provide relatively low-cost and low-technology electricity to rural communities where centralized power plants and electrical grids are nonexistent. Similarly, dams that store water for irrigation or drinking water supply may provide vital services to populations not connected to centralized infrastructure. At the same time, reservoirs can inundate valuable farmland that is fertile and flat, a phenomenon particularly problematic in a region where population densities are high and arable land relatively scarce. Moreover, populations displaced by dam projects often find themselves unwelcome in the areas into which they are resettled due to cultural differences and pressure on land resources (Brown & Xu 2010; Heggelund 2006). Documented cases of corruption, at times involving funds intended as compensation for resettled migrants, and the World Commission on Dams (2000) study highlighting the shortcomings of large dams around the world, do little to improve the public image of dams.

This chapter briefly tells the story of dams as a widespread and pervasive feature of the environment in East Asia. With more than a dozen countries and perhaps as many as 50,000 dams large and small, East Asia's dam-scape is immense and changing rapidly, and one chapter cannot hope to be exhaustive. The aim, then, is to showcase the scale of dam building across East Asia and the role dams have played in altering ecological and social systems in the region, in the hope of providing a starting point for understanding dams as complex socioecological components of the hybrid landscape humans inhabit. With lifespans of 50 to 100 years or more, dams are an enduring feature of many environments, so much so that only now, after nearly a century of fast-paced dam construction around the world, are we seeing the first attempts at systematically removing dams that have reached the end of their useful lifespan, as in the United States Northwest. After first explaining the purposes for which dams are constructed, I contextualize East Asia's dams, then profile several case studies and countries that exemplify the complicated nature of dams and the complex mix of costs and benefits they bring. The conclusion summarizes the story of dams in East Asia and offers directions for further research.

Purposes of dams

Governments and private developers build dams for hydropower, flood control, irrigation and navigation. Some dams also provide other benefits such as drinking water, aquaculture, and recreation. All provide some degree of sediment control by impeding the transport of suspended sediment. Many modern dams are multi-purpose dams capable of providing two or more of these services, with irrigation and hydropower being the most prominent (International Commission on Large Dams n.d.). Generating electricity from the potential energy in a mass of water stored at a height behind a dam is relatively simple and relies on two physical characteristics of the system: flow volume and hydraulic head, or the vertical distance between the reservoir surface and the power generation machinery. The machinery – in simplest form, turbines coupled with generators – extracts a fraction of that energy based on factors such as design, materials, operating parameters, and age. It is possible for a low-head, high-flow dam to have a similar or identical power rating as a high-head, low-flow dam even when the two appear very different to the casual observer.

The suitability of a country (or other geographic unit) for hydropower is often given by three measures: theoretical potential, technically feasible potential, and economically feasible potential. These indicators are usually measured in terms of installed capacity in units of megawatts (MW) or gigawatts (GW), or in actual energy output in units of gigawatt hours per year (GWh/yr). Taken together, the terms are relatively straightforward. Theoretical potential is derived from elevation drops and flow volumes. Technically feasible potential and economically feasible potential are fractions of the theoretical potential that vary based on availability of expertise, technology, and capital. Although clearly related – with enough capital, even the most technically challenging hydroelectric dam becomes more feasible – the two feasibility measures are usually given separately. Installed capacity is a function of the size and design of the turbine and generator components and captures a dam's technological capability to generate power, regardless of how that dam is actually operated (i.e., its capacity factor). Since installed capacity does not measure actual use, the unit has no time component. Figure 13.1 shows each country's contribution to East Asia's total theoretical potential.

Conventional hydroelectric dams convert energy in falling water in rivers and streams to electrical energy, preventing that energy from being 'wasted' as the river makes its way to the sea. A second type, pumped storage dams, actually consume electricity to pump water uphill to a storage reservoir, from which it is later released to generate electricity in the same fashion as a conventional hydropower dam. From an energy perspective, this is a 'lossy' process; the energy required to pump the water uphill is greater than the energy captured when the water is released, since some energy is converted to heat through friction in pipes and mechanical systems. Yet the reality of electricity demand, supply, and pricing means that pumping water uphill at night, when demand and prices are lower, allows the upper reservoir to serve as a battery whose energy can be released to meet daytime peak electrical demands. In areas in which conventional hydropower resources have already been exploited, developers may continue to build pumped storage units.

Irrigation and drinking water dams impound water for direct or indirect consumption by humans, with or without treatment. Ditches, pumps, and other technologies transfer water from the reservoir to the place of use; since pumps and treatment plants require energy, irrigation and

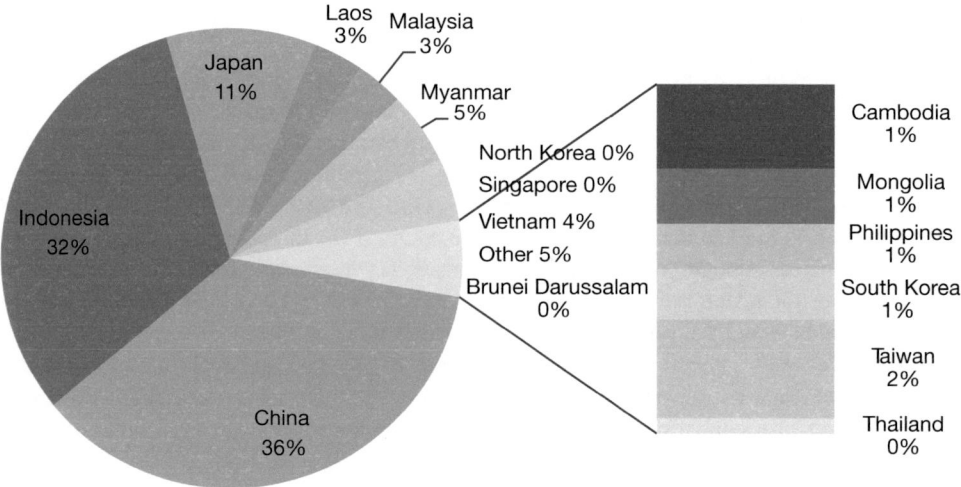

Gross theoretical hydro potential (% of estimated total 6,811,431 GWh/yr)

Figure 13.1 Country comparisons of theoretical hydropower potential

drinking water dams without a hydroelectric component are net energy consumers. Once removed from the reservoir, the fate of water applied to farmland or put into domestic drinking water systems varies. Some will be taken up by plants and metabolized into sugars and oxygen, or lost to evapotranspiration. Some will evaporate directly from the soil, while some will percolate into the soil and help recharge groundwater resources, perhaps eventually rejoining the river. The most important measure of a drinking water or irrigation dam's size is its reservoir volume, usually given in cubic metres. Figure 13.2 shows each country's irrigated area as a fraction of total area.

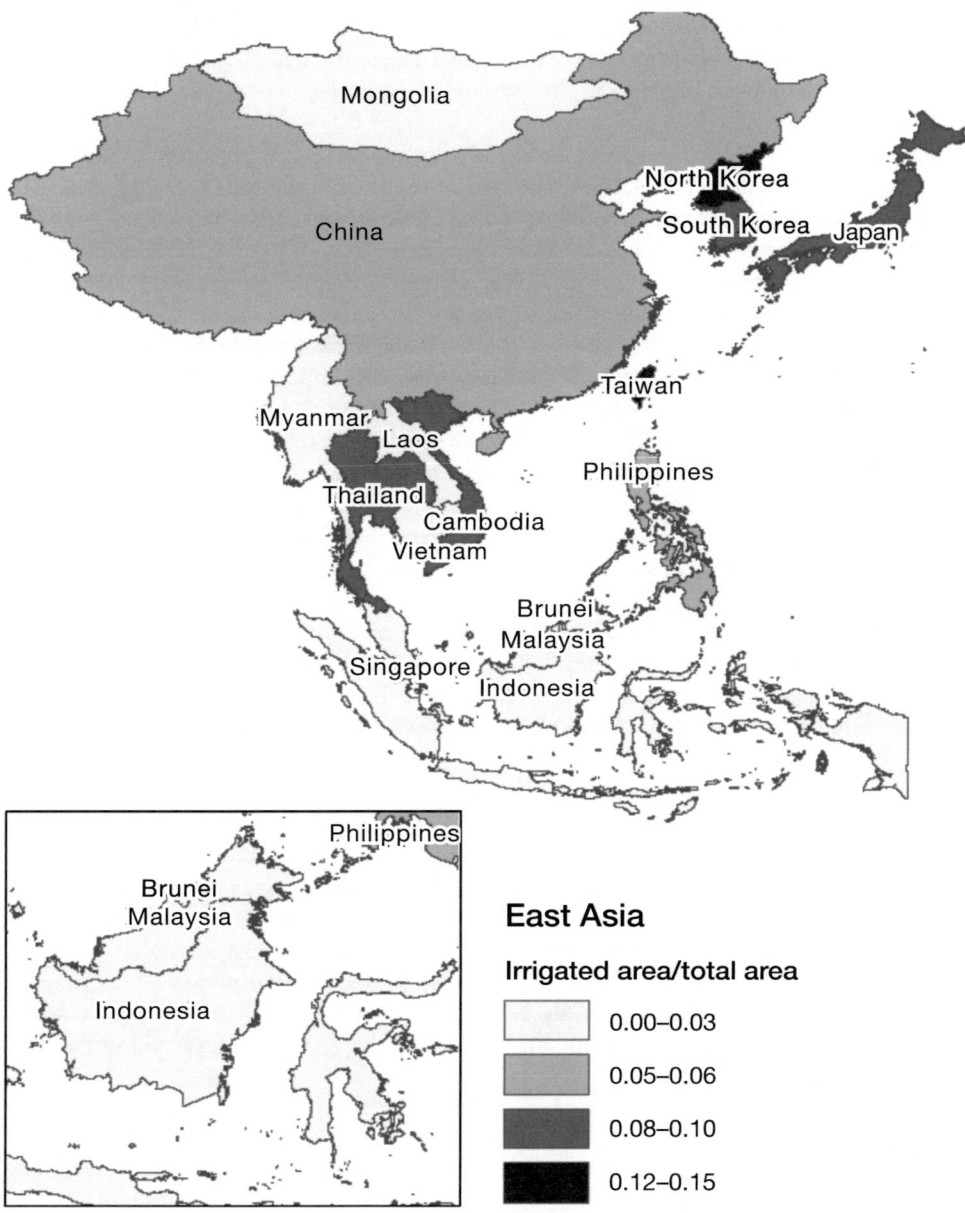

Figure 13.2 Irrigated area relative to total area

Dams whose purpose is to improve navigation do so by flattening the gradient of a river over a certain reach, effectively turning a flowing river into a lake. Transporting people and goods by water is highly energy efficient due to water's buoyancy, yet shallow or fast-moving rivers with treacherous currents, underwater hazards, or dramatic seasonal fluctuations in water level present significant challenges to all but the smallest vessels. Shallow, rocky stretches of the middle Mekong, for instance, have been dynamited to facilitate navigation, a controversial technique that destroys aquatic habitats and brings uncertainty to fishing peoples who depend on them. Locks, integrated into the dam or in an adjacent river channel, allow vessels to pass from upstream to downstream or vice versa. The Three Gorges Dam in China also includes a ship lift that will allow vessels up to 3,000 tons to traverse the dam faster than via the locks. Challenging to design and implement, this portion of the project is expected to be completed in 2014.

Although multi-purpose dams offer technical solutions to several problems at once, many fail to perform optimally when demands for two or more services conflict. In East Asia, such conflicts arise in summer months, when the need for capturing and controlling monsoon-swollen floodwaters coincides with the need for boosting electrical output to cool factories and homes in tropical heat. Operating a dam for maximum electricity production requires releasing large volumes of water over short periods of time. Conversely, operating the same dam for maximum flood control capacity requires first draining the reservoir to low levels in the weeks preceding the rains, in order to then capture and retain large runoff volumes resulting from heavy rains over short periods. Thus the same dam cannot be simultaneously operated for maximum power output and maximum flood control. Indeed, many hydroelectric dams operate at roughly half their full capacity in terms of electricity output since they must balance other objectives and respond to the seasonal hydrology.

All dams are subject to phenomena that may reduce their effectiveness on any number of measures. Siltation is perhaps the most obvious and most vexing. Fast-moving water transports more silt than slow-moving water, so when water enters a reservoir and slows down, it naturally dumps much of its sediment load, which then settles to the bottom of the reservoir, gradually reducing the amount of water the reservoir can hold. This reduction has clear consequences for flood control and power generation, even though those consequences may be relatively small on an individual dam basis. Siltation rates vary with local hydrology and geology; dams in the USA lose capacity due to siltation at a rate of around 0.5% per year, whereas rates in China are more than four times that (McCully 2001). Silt also abrades turbines and other components, reducing their longevity and lowering their efficiency. Downstream of the dam, water that has lost silt in the reservoir is now 'hungrier', that is, capable of transporting more suspended sediments, and therefore more prone to erode banks and sand bars that may provide habitat for humans, plants and animals. Valley lands tend to be fertile, thanks to the periodic flooding and deposition of sediment over geologic time. In areas where pressures on arable land are great, marginal lands in floodplains are often farmed by peasants. Unannounced releases from dams upstream, as has occurred repeatedly on transboundary rivers in mainland Southeast Asia such as the Se San, can cause loss of life, wipe out crops, and increase vulnerability of downstream populations.

Impacts of dams

In an early authoritative volume on the negative impacts of large dams, Edward Goldsmith and Nicholas Hildyard (1986: 331) flatly contest the utility of large dams: 'There is clear evidence . . . that building large dams is not an appropriate means of feeding the world's hungry, of providing energy, or of reducing flood damage.' Writing more recently, Thayer Scudder, professor emeritus of anthropology at California Institute of Technology and one of 12 commissioners on the World Commission on Dams (WCD), calls large dams 'a flawed yet still necessary development option' (Scudder 2005: 1). Drawing on the work of the WCD and his own long-term studies, he argues large dams are flawed simply because, in most cases, their 'benefits are overstated while costs are understated' (Scudder 2005: 1). Conversely, he asserts they are necessary only because of their potential role in meeting the 'needs of a human population that is expanding beyond the carrying capacity of the world's life support systems' (Scudder 2005: 2). With half a century of involvement in dams as a consultant and researcher, Scudder pulls no punches in his account of the failures of large dams to restore, much less improve, the livelihoods of millions of people they displace: 'Within the major dam-building countries, including the US, China, and India, I am aware of none that can document that they have been able even to restore the incomes of the majority of resettlers' (Scudder 2005: 20). He cites as one pernicious example of this failure the fact that very few reliable data on people displaced by dams are maintained by most countries or financing institutions such as the World Bank, even when details on all other project components are carefully documented. In my own research on dam impacts in China, I have had to rely on population grids intersected with reservoir areas in a geographic information system (GIS) to estimate resettlement figures.

Goldsmith and Hildyard (1986: 127) assert that humans 'live in an age of technological euphoria, in which man's ability to control the forces of nature is taken for granted'. Large dams, they argue, are monuments to the hubris of engineers, planners, and financiers, at once technological wonders and socioecological disasters. They document numerous problems with large dam projects, including the displacement of millions of rural inhabitants; failed flood control promises; changes to flow and sedimentation patterns; destruction of fisheries and agricultural land; and salinization of reservoir water and irrigated lands. The problem of salinization is compounded when water is drawn from reservoirs to irrigate cropland, and can lead to land that is unsuitable for growing anything but the most salt-tolerant crops. The large surface area of reservoirs, especially in warm regions such as obtain in much of East Asia, yields significant evaporative losses, which, in turn, raises the concentration of natural salts within the remaining water. When that water is used for irrigation, it changes the water–salt balance in the soil and inevitably leads to greater salt concentrations over time as excess water evaporates from the surface of waterlogged land, leaving salts behind. Given the widespread dependence on irrigated agriculture in East Asia, the challenges presented by waterlogging and salinization and their relation to food security are daunting.

One schema for understanding dam impacts is integrative dam assessment modelling (IDAM) (Brown et al. 2009; Tullos et al. 2010). The schema separates dam impacts based on three pillars of sustainability – biophysical, socioeconomic, and geopolitical – using a range of indicators to assess 21 impacts (seven per pillar). Each impact is assumed to have a positive or negative magnitude that is measurable or at least predictable, and a salience that captures the importance of that impact for various stakeholders evaluating the sustainability of an existing or proposed dam. Thus while two stakeholders might agree on the magnitude of a particular impact, they might disagree on the salience of that impact.

Tables 13.1, 13.2, and 13.3 list the impacts identified in the IDAM schema, along with positive and negative examples of each impact, and the indicators used to assess each. While some impacts may be ascertained through data aggregated at provincial, state or national scale, many require careful micro-scale data collection and analysis. Here, the IDAM schema serves simply as a guide to understanding in broad strokes the positive and negative impacts – some of which are more easily quantified as benefits or costs in impact assessments – of East Asia's dams. Scudder's own survey of 50 dams, focused primarily on socioeconomic impacts, found living standards improved for communities resettlement as a result of dams in only 7% of cases; living standards were restored (essentially maintained) in another 11% of cases (Scudder 2005). Among the greatest contributors to the decline in living standards were landlessness, joblessness, food insecurity, social marginalization, and loss of access to common property resources.

Dams in East Asia

Dams have a long history in East Asia. If weirs (which do not impound reservoirs) are included, one could argue that the Dujiangyan Irrigation Project on the Min River in Western China, built in the third century BCE to prevent floods that regularly devastated nearby farmlands, is

Table 13.1 Biophysical impacts and indicators

Impact name	Positive and negative scope of impact; indicator used
BP1: Water quality	(+) Pollutants stored in reservoir, prevented from migrating downstream
	(−) Nutrient cycling, dissolved oxygen or temperatures altered
	(i) Change in residence time of water flowing through reservoir reach
BP2: Biodiversity	(+) New habitats created, predation on rare/endemic species reduced
	(−) Disruption/destruction of aquatic or terrestrial habitat, migration routes
	(i) Index of habitat quality
BP3: Impact area	(+) New habitats created, predation on rare/endemic species reduced
	(−) Disruption/destruction of aquatic or terrestrial habitat
	(i) Index of habitat quantity (area, length of river reach)
BP4: Sediment	(+) Sediment trapping may decrease turbidity and aggradation downstream
	(−) Sediment trapping may increase downstream bank instability or negatively impact habitat or water infrastructure
	(i) Trap efficiency of dam, percentage of basin contributing sediment to dam
BP5: Natural flow regime	(+) Dam may re-regulate flows altered by upstream dams
	(−) Alteration of historic flow regimes may degrade ecosystems, infrastructure or river morphology, spawning patterns, etc.
	(i) Changes compared to flow and flood baselines
BP6: Climate change/air quality	(+) Hydropower may reduce greenhouse gases and particulate emissions
	(−) Methane emissions from reservoirs may offset other GHG savings
	(i) GHG emissions for equivalent coal-fired power plant
BP7: Landscape stability	(+) N/A
	(−) Reservoir construction/filling may induce landsides, increase seismicity
	(i) Weight and depth of reservoir, distance to faults, slope, soils

Source: Adapted from Tullos et al. (2010).

Table 13.2 Socioeconomic impacts and indicators

Impact name	Positive and negative scope of impact; indicator used
SE1: Social capital	(+) Dams may facilitate cross-river transportation and community links
	(−) Resettlement disrupts communities and weakens social cohesion
	(i) Buckner Scale, surveys, interviews
SE2: Cultural change	(+) Dams may instill national pride
	(−) Dams may inundate sites of cultural importance, lead to loss of traditions (including traditional ecological knowledge)
	(i) Index of impacts on material culture, sense of place, surveys, interviews
SE3: Local hydro access	(+) Dams may facilitate connection to power grid
	(−) Prices of electricity may rise
	(i) Index of frequency of power losses and prices of electricity from surveys
SE4: Health impacts	(+) Water treatment facilities with dam may improve drinking water
	(−) Reservoirs create still water breeding grounds for parasites
	(i) Index of water quality, water-borne illness, toxicity from surveys
SE5: Income	(+) Off-farm work may bring rise in incomes
	(−) Farmers may lose income due to loss of farmland
	(i) Income share of watershed average
SE6: Wealth	(+) Housing and/or land in resettlement villages may be better than in villagers' original location prior to resettlement
	(−) Evacuees may deplete resources (e.g., funds) while re-establishing themselves in new communities
	(i) Housing and land values as share of watershed average
SE7: Macro impacts	(+) New infrastructure and activities related to dam may have positive spillovers for local economy
	(−) Resettlement of villagers is costly
	(i) Index of resettlement, infrastructure and hydropower costs

Source: Adapted from Tullos et al. (2010).

the world's best example of sustainable hydraulic engineering since it survives and functions in essentially its original design today. Yet the real boom in dam building in East Asia began with industrial modernization in the late 1800s, led in large part by Meiji Japan's push into the Korean peninsula and Manchuria in search of mineral and energy resources. The mountainous terrain and steep rivers of the peninsula provided attractive sites for hydropower, and Japanese engineers harnessed several rivers in order to provide electricity for smelting metals from their ores.

One of the focal points for dam building in East Asia in recent years has been the Mekong River Basin. The 4,800–kilometre-long river traverses six countries (China, Myanmar, Thailand, Laos, Cambodia, and Vietnam), the most of any river in East Asia. The first half lies within China, where the river is known as the Lancang, and since the late 1980s has been the site of a contentious series of eight to 14 dams known broadly as the Lancang cascade. Although the dams in the cascade are referred to as multi-purpose, they are primarily hydroelectric dams designed to connect to eastern load centres via the China Southern Grid and, eventually, to a

Table 13.3 Geopolitical impacts and indicators

Impact name	Positive and negative scope of impact; indicator used
GP1: Basin population affected	(+) Dam provides benefits to basin residents (hydro, navigation, etc.)
	(–) Dam brings costs to basin residents (loss of farmland, resettlement, etc.)
	(i) Share of population affected as percentage of entire basin population
GP2: Political complexity	(+) Basin-wide management may increase dialogue and cooperation
	(–) Basin-wide management may increase tensions and reduce cooperation
	(i) Number and type/level of political boundaries crossed
GP3: Legal framework	(+) Strong laws and institutions help mitigate vulnerability
	(–) Weak or non-existent laws and institutions may increase vulnerability
	(i) Administrative level of highest legal framework governing dam
GP4: Domestic governance	(+) Decision processes are open and transparent, management is robust
	(–) Decision processes are opaque, management is constrained/limited
	(i) Democracy Index
GP5: Political stability (domestic)	(+) Cooperation strengthens ties among administrative areas/units
	(–) Lack of cooperation increases tension among administrative areas/units
	(i) Internal Basins at Risk (BAR) Scale
GP6: Political stability (international)	(+) Cooperation strengthens international ties, builds institutions
	(–) Lack of cooperation increases international tensions
	(i) International Basins at Risk (BAR) Scale
GP7: Impacts on non-constituents	(+) Dam provides benefits for communities outside immediate area
	(–) Dam brings negative impacts for communities outside immediate area
	(i) Index of spatial extent and magnitude of dam impacts

Source: Adapted from Tullos et al. (2010).

'subregional' power grid that would provide electricity to countries throughout the Mekong watershed (Magee 2006). The completion of the Manwan Dam in 1995 marked the first time a dam disrupted the flow of the main stem of the Mekong, although by that time dozens if not hundreds of smaller dams already existed on tributaries throughout the basin. Since then, roughly half a dozen more dams, with a total installed hydroelectric capacity of nearly 15 GW, have been completed, with several others in design and planning stages (Magee 2012).

Development along the Lower Mekong is ostensibly subject to cooperation among the four countries of the Mekong River Commission (MRC), Cambodia, Laos, Thailand, and Vietnam. The MRC was reorganized from its predecessor, the Mekong Committee, in 1995 and is funded by member countries, international development agencies, and countries outside the region. Its mission is 'to promote and coordinate sustainable management and development of water and related resources for the countries' mutual benefit and the people's well-being' (Mekong River Commission n.d.). Despite early interest in the hydroelectric potential of the Lower Mekong, the MRC remained officially sceptical of mainstream hydropower development through the early 2000s. According to the first version of its hydropower development strategy (Mekong River Commission 2001: 39), the MRC's hydropower plan for the four lower basin countries did *not* include mainstream Mekong dams because of regional geopolitical concerns, project costs and scale, and environmental and human resettlement costs. The richness of Mekong-related fisheries such as Cambodia's Tonle Sap lake, and the number of people who depend on the river for sustenance and livelihoods, made large-scale hydropower on the main stem of the river politically untenable and, in the wake of the WCD report, unattractive to potential funders.

Official unease over mainstream Mekong hydropower development attenuated by the time the Commission released its 2006–2010 strategic plan (Mekong River Commission 2006). Since then, Chinese and Thai developers have been actively engaged in survey and planning work for hydropower on the Mekong and its tributaries. Among lower Mekong countries, Laos holds the greatest hydropower potential, and the proposed 1,285-MW Xayaburi Dam, recently given the green light by Laotian authorities even as Vietnam and Cambodia call for a moratorium (Hunt 2013), symbolizes the ensemble of concerns about damming the main stem of the Mekong (International Rivers n.d.(b)). If completed – preliminary work is already underway – the dam will displace some 2,100 Laotians and export 95% of its electricity to Thailand. Scholars and activists have warned of negative impacts in the biophysical, socioeconomic, and geopolitical spheres: interruption of migration and alteration of riparian habitat for dozens of fish species; resultant increases in vulnerability of fishing communities dependent on the Mekong; disruption of sediment transport downstream to the Tonle Sap and delta areas; and potential destabilization of regional political dynamics. International Rivers, a California-based non-profit, argues that the Thai developers and the Laotian government are in violation of the 1995 MRC agreement, a claim the latter flatly denies as it moves forward with the project. If construction of the Xayaburi Dam proceeds, other mainstream dams will likely follow.

Concerns about sediment may actually be used to justify more dams. In China, the rate at which the sediment-rich Yangtze has discharged sediment into the Three Gorges reservoir has been cited as one reason more dams needed to be built upstream and on upper Yangtze tributaries. As noted earlier, sediment decreases a reservoir's flood control capacity and can cause premature wear on power generation equipment, problems developers and government proponents did not want. Among their other benefits (primarily power generation), dams on the upper Yangtze would trap silt before it made its way to the Three Gorges. The upstream projects are themselves behemoths, more than a dozen mega-dams whose combined power generation capacity exceeds that of Three Gorges and the Lancang cascade, and whose reservoirs will submerge some of the most biologically and culturally diverse regions of China.

A second transboundary river on which hydropower plans have proved even more contentious than the Lancang–Mekong system is the Salween. The 2,800-kilometre Salween has its source in China at the eastern edge of the Tibetan plateau, in the same region as the sources of the Mekong, Yangtze, and Yellow Rivers. Plans to build a 13-dam cascade with over 21 GW of installed capacity on the Chinese portion of the river were halted in 2004 by Premier Wen Jiabao, in response to domestic and international outcry and ostensibly on grounds that the developer had failed to follow appropriate environmental impact assessment protocols (Magee & McDonald 2009). Preliminary work, including site testing and even the construction of resettlement villages, has crept along since then, and project proponents in Beijing recently expressed confidence that the cascade will go forward under China's new leadership (Zhu 2013). Meanwhile several projects downstream on the portion of the river that lies in Myanmar and on the Myanmar–Thailand border have raised additional concerns about environmental and social impacts among local and international groups. Such concerns are exacerbated by the fact that much of the work is being conducted by Chinese companies such as Sinohydro (Magee & Kelley 2009).

These and other highly controversial large dams in the Greater Mekong subregion, including Nam Theun 2 in Laos and Pak Mun in Thailand, are canonical examples of top-down decision making, opaque financing, and negative social and ecological impacts invariably felt more acutely in the immediate dam and reservoir area. They are also only three examples of thousands of other projects already built, under construction, or in the planning phase across East Asia, a snapshot of which follows.

Country profiles

In light of the geographic diversity of the countries of East Asia, it is hardly surprising that the dam profiles of different countries should be similarly diverse. Figure 13.3 shows the number of dams per thousand square kilometres for countries throughout the region. Brunei Darussalam, with its area of 5,765 sq km, is roughly 1/2,000th the size of China and has no hydroelectric dams and only 10 sq km of irrigated agriculture. China, meanwhile, has an area of roughly 9.6 million sq km, with some 640,000 sq km of that land occupied by irrigated agriculture. China is traversed by numerous great rivers whose volumes and gradients combine to give the country the world's greatest hydroelectric potential and set it apart in terms of its leaders' ability to indelibly shape the landscape through dam-building activities. The following profiles briefly sketch the importance of large dams in several countries in which they are prevalent. Profiles are omitted for countries where large dams are nonexistent, have a minimal role, or where reliable data are unavailable. Unless otherwise indicated, demographic and geographic data are drawn from World Bank (2013) country profiles, while dam data are drawn from the World Register of Dams (WRD) and the World Atlas and Industry Guide (WAIG). Irrigated area and total (internal) renewable water resources data are from the Food and Agriculture Organization's Aquastat (2013) database. Data for displaced populations, the most difficult to ascertain, are based on media, government, academic, and company sources and corroborated as carefully as possible using publicly available data. Where I was unable to obtain reliable figures for displaced populations, I have indicated with 'N/A' (not available). Table 13.4 summarizes basic data for each country. Unless otherwise indicated, socioeconomic and demographic data are 2011 figures.

China

Ever since the decision to build the massive and controversial Three Gorges (Sanxia) Project in 1992, China has been the focus of scrutiny of East Asia's dam-building efforts. Current plans for hydropower development on single stretches of China's major rivers such as the Jinsha (upper Yangtze), Lancang (upper Mekong), or Nu (upper Salween) rival Japan's entire existing installed hydropower capacity, and developers plan to nearly double the country's 2010 installed hydropower capacity by 2020. The WRD lists nearly 5,200 entries for large dams in China; as noted earlier, one estimate put that number as high as 20,000 or more. Regardless of which is more accurate, the scale and pace of China's existing dam-building efforts are huge, and pale in comparison only to the plans that remain on the drawing boards of the country's water engineers, macroeconomic planners, and politicians. Storage capacity of China's five largest reservoirs, all of which are behind multi-purpose dams, is 125,532,000,000 m³. Key data on

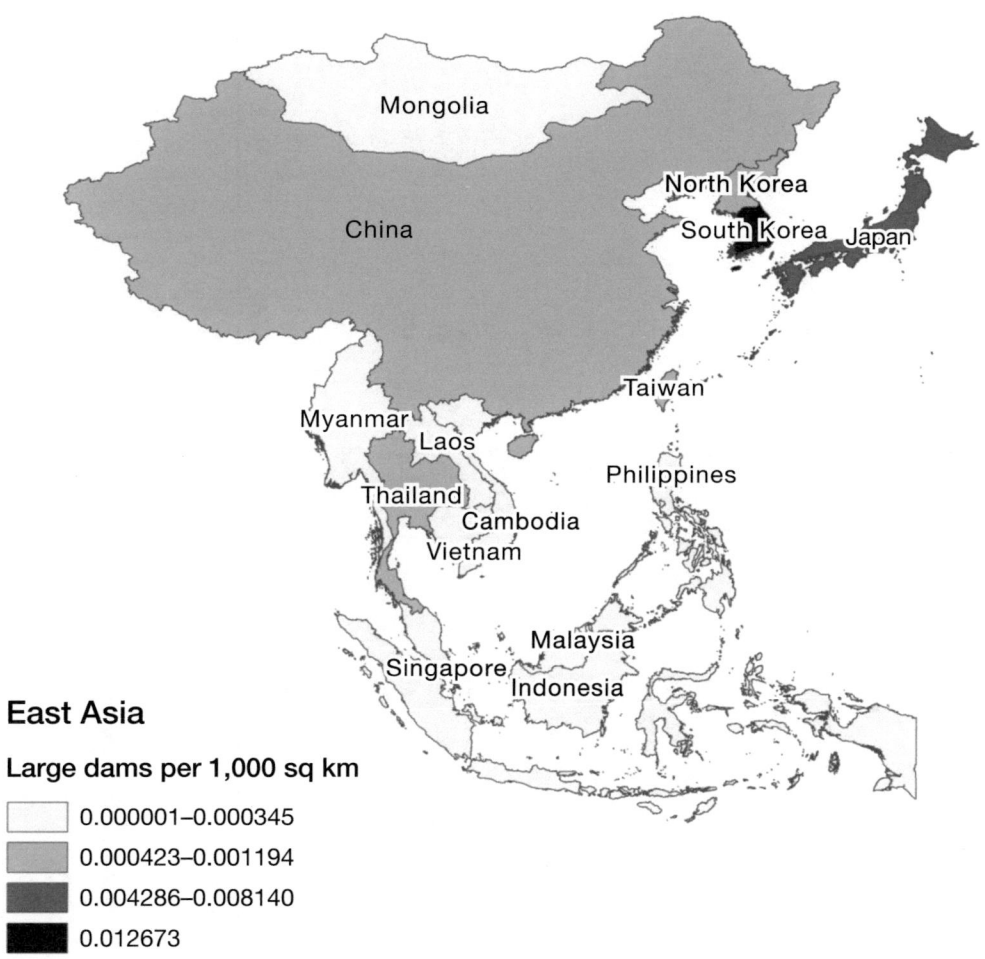

Figure 13.3 Large dams (>15 m high) per thousand square kilometres

the five largest dams in China based on reservoir volume are presented in Table 13.5. These five dams are in the 99th percentile of dams worldwide by a number of measures, including reservoir volume.

Much has been written about the impacts of the Three Gorges Dam, with numerous scholars focusing on the challenges of resettling more than a million dam-displaced residents (Heggelund 2006; Padovani 2004; Qing et al. 1998; Tan et al. 2005). Yet as impressive as it is for its size, flood control and power generation capacities, and havoc caused to social, physical, and ecological systems, the Three Gorges Dam is but one giant Chinese dam among many. Current dam-building efforts are focused on the great rivers of the Southwest, including the upper Yangtze, upper Mekong, and upper Salween, projects whose combined capacities will be equivalent to several Three Gorges. The upper Yangtze projects have caused concern among scientists and others who fear such dams will mean the demise of habitat vital to Yangtze fisheries already under threat. Construction of major cascades on the upper Mekong and upper Salween, both transboundary rivers, has the potential to create geopolitical uncertainty, as downstream

Table 13.4 Basic condition and water indicators

Country	Area (10^3 km^2)	Population (millions)	GNI per capita (Atlas method) (US$/person)	Electricity use (kWh per capita)	Irrigated area (ha)	Renewable water resources (km^3/yr)	Large dams (units)
Brunei Darussalam	5.8	0.4	31,800	8,759	1,000	8.5	2
Cambodia	181	14.14	760	146	284,172	120.6	2
China	9,600	1,337.83	4240	2,944	53,821,000	2,813	5,188
Indonesia	1,904.6	239.87	2500	641	4,459,000	2,019	128
Japan	377.9	127.45	42,050	8,394	3,129,000	430	3,076
Laos (a)	236.8	6.2	1,010	No data	295,535	190.4	5
Malaysia	330.8	28.4	8,090	4,117	362,600	580	52
Mongolia (b)	1,564.1	2.76	1,870	1,530	57,300	34.8	2
Myanmar (c)	676.6	47.96	790	131	1,841,320	1,003	23
North Korea	120.5	24.35	No data	749	1,460,000	67	70
Philippines	300	93.26	2,060	643	1,550,000	479	14
Singapore	0.7	5.08	39,410	8,307	0	0.6	3
South Korea	99.9	49.41	19,720	9,744	880,365	64.85	1,266
Taiwan (d)	36	23.16	21,000	10,257	525,528	67	43
Thailand	513.1	69.12	4150	2,243	4,985,708	224.5	217
Vietnam	331.1	86.93	1160	1,035	3,000,000	359.4	51

Notes:
(a) Laos dam data from International Rivers (www.internationalrivers.org).
(b) Mongolia dam data from Ministry of Environment and Green Development (2013).
(c) Myanmar GNI per capita calculated from ADB data (www.adb.org).
(d) Taiwan data (except irrigation, renewable water resources and large dams) from ADB (www.adb.org). Large dams data from National Taiwan Ocean University (n.d.).

Table 13.5 Characteristics of largest dams in China

Dam	Reservoir volume (m³)	Reservoir area (10³ m²)	Height (m)	Installed capacity (MW)	Completion date	Displaced population (no. of people)
Sanxia/Three Gorges	39,300,000	1,084,000	181	22,500	2010	1,300,000
Longyangxia	27,630,000	380,000	178	1,280	1989	3,300
Longtan	27,270,000	360,000	216	6,300	2010	80,000
Sanmenxia	16,200,000	2,350,000	106	400	1960	400,000
Xiaowan	15,132,000	190,000	292	4,200	2012	35,000

Notes: Original generation capacity at Three Gorges was 18,200 MW, but additions completed in 2012 raised the total to 22,500 MW.

Source: Chinese National Committee on Large Dams (n.d.); Heggelund (2004); Longtan Hydropower Begins Water Storage (Longtan Hydropower Begins 2006); Qinghai Bureau of Land Resources (2010).

governments and peoples anticipate and experience disturbances to the flow regimes of rivers on which millions depend for their livelihoods. As noted already, the main stem of the Lancang–Mekong ceased to flow freely in 1995 when the 1,500-MW Manwan, with its reservoir of just over a billion cubic metres, was completed. Since then, six more dams on the Lancang have been completed, are under construction, or are in advanced planning stages, while at least another six are at various stages of planning for stretches of the river even further upstream in Northern Yunnan and Tibet (Magee 2012). Figure 13.4 shows the early stages of construction of the Xiaowan Dam on the Lancang which, now complete, is the tallest arch dam in the world.

Figure 13.4 Xiaowan dam construction site

Indonesia

Indonesia is rich in water resources, but like most Southeast Asian countries, is also subject to temporal variations in precipitation due to the monsoons. Its physical geography as an archipelago comprising more than 15,000 islands contributes to uneven geographic distribution of fresh water resources as well. The WRD lists roughly 130 large dams throughout the country. The majority of those dams, especially those completed before 2000, are for water storage rather than power generation. WAIG estimates that only 6% of Indonesia's technically feasible hydropower has been developed, and there is great interest within the government and state-owned power sector to promote large-scale hydropower development. While it is technically feasible to add hydropower-generating capacity to existing dams, it is unclear at this point to what extent, if any, that will play a role in Indonesia's future dam projects. Storage capacity of Indonesia's five largest reservoirs is some 7,256,000,000 m³. Key data on the five largest dams in Indonesia based on reservoir volume are presented in Table 13.6.

Given the hydrology of the Indonesian archipelago, there are relatively few major rivers suitable for large-scale hydropower. As in other parts of East Asia where national power grids are sparse or nonexistent, small hydropower may have a role in meeting rural electricity demand. The authors of one recent study argue strongly for small and micro hydro (300 kW to 5 MW), across Indonesia, estimating some 6,000 are economically feasible and attractive for private sector involvement and rural energy provision (Hasan et al. 2012).

Table 13.6 Characteristics of largest dams in Indonesia

Dam	Reservoir volume (10³ m³)	Reservoir area (10³ m²)	Height (m)	Installed capacity (MW)	Completion date	Displaced population (no. of people)
Jatiluhur	2,556,000	77,800	105	175	1967	N/A
Kotopanjang	1,545,000	12,400	58	114	1997	20,000
Riam Kanan	1,200,000	3,200	57	30	1974	5,000
Saguling	982,000	8,340	99	700	1986	13,737
Cirata	973,000	6,200	125	1,008	1988	27,978

Source: Hasan et al. (2012); Suutari (2003); Yuliansyah (2002); Scudder (n.d.).

Japan

The WRD lists more than 3,000 large dams in Japan, making Japan one of the most densely dammed countries in East Asia. This can be explained by both geographic and financial reasons. First, Japan is a densely populated mountainous archipelago, with the majority of that population concentrated in relatively compact floodplains. Japan has limited fossil fuel reserves, yet its per capita incomes and electricity consumption levels are among the highest in the world, creating strong demand for hydropower. Thus most of Japan's dams are multi-purpose and more than half of those have hydroelectricity generation as one of their functions (Bartle 2012: 129). In recent years, roughly 30% of Japan's electrical energy has come from nuclear power, but the March 2011 tsunami and resultant Fukushima Daiichi nuclear accident prompted a public conversation about the future of nuclear power in Japan, which could increase the country's demand for hydropower. Most of the country's technically and economically feasible sites for large dams have already been utilized, which means that future hydropower dams in Japan will primarily be a mix of smaller units and pumped-storage facilities (Bartle 2012: 130). Storage capacity of Japan's five largest reservoirs is 2,463,000,000 m³. Key data on the five largest dams in Japan based on reservoir volume are presented in Table 13.7.

Table 13.7 Characteristics of largest dams in Japan

Dam	Reservoir volume (10^3 m^3)	Reservoir area (10^3 m^2)	Height (m)	Installed capacity (MW)	Completion date	Displaced population (no. of people)
Tokuyama	660,000	13,000	161	153	2008	1,500
Okutadami	601,000	11,500	157	560	1961	N/A
Tagokura	494,000	9,950	145	395	1960	N/A
Miboro	370,000	8,800	131	215	1960	1,000
Ikehara	338,000	8,400	111	350	1964	N/A

Notes: Displacement estimates on dams built in the 1960s are based on available data on households displaced.

Source: Japan Commission on Large Dams (2012); Japan Dam Foundation (n.d.); Suiryoku (n.d.).

The March 2011 earthquake also caused damage to a number of Japan's large dams. According to a report on inspections of over 400 dams in the weeks following the earthquake, 'several dams suffered minor or moderate damage', with only one 18.5-metre-high water retention structure at Fujinuma-ike experiencing catastrophic failure, resulting in eight deaths (Japan Commission on Large Dams 2011). Clearly, the news could have been much worse, and some of the effects of the quake may not yet be evident. Large dams seem to have proven similarly resilient in the 2008 earthquake in Wenchuan, China. Ironically, many scientists attribute the weight of the reservoir behind the Zipingpu Dam with triggering the 2008 Wenshan quake. Large reservoirs have been implicated in earthquakes elsewhere ('reservoir-induced seismicity') in Japan and around the world, a phenomenon captured by IDAM Indicator BP7.

Malaysia

While Malaysia currently has just over 50 large dams listed in the WRD, many more are on the drawing board for the island states of Sarawak and Sabah. The geography of the island states yields a technically feasible hydropower potential of 87 GW, over five times that of peninsular Malaysia, where one-quarter of all technically feasible potential has already been developed (Bartle 2012: 150). The primary purpose of most of Malaysia's existing dams is storage, but hydropower will grow in importance if the country pursues a supply-side approach to meeting projected growth in electricity demand (Sovacool & Bulan 2011). Malaysia derives approximately 10% of its total generated electricity from hydropower; if development plans for Sarawak proceed, power exports to peninsular Malaysia may occur in the near future, assuming prices for generation and transmission (also a 'lossy' process) are competitive with those from coal-fired sources. Storage capacity of Malaysia's five largest reservoirs is 23,709,800,000 m^3. Key data on the five largest dams in Malaysia based on reservoir volume are presented in Table 13.8.

The scale and potential impact of the pending hydropower boom in Sarawak cannot be overstated. Sarawak has been estimated to hold some 20 GW of hydroelectric potential (Sarawak Energy n.d.). Large-scale dam construction plans as part of the so-called Sarawak Core of Renewable Energy (SCORE), with up to 12 new dams planned for 2020 and dozens more in early planning stages, have raised concerns about corruption by project controllers and the inundation of biologically diverse rainforests. Transparency International labelled the 2,400-MW Bakun Dam, begun in 2002, a 'monument of corruption' that ignores ecological and social concerns and rewards corrupt Sarawak state officials with lucrative development contracts (International Rivers n.d.(a)). Built in partnership with Chinese developers in the Malaysia China Hydro Joint Venture, the project has inundated roughly 700 square kilometres and displaced an estimated 10,000 people (Keong 2005). Despite the fact that methane emissions from

Table 13.8 Characteristics of largest dams in Malaysia

Dam	Reservoir volume (10³ m³)	Reservoir area (10³ m²)	Height (m)	Installed capacity (MW)	Completion date	Displaced population (no. of people)
Kenyir	13,600,000	369,000	155	400	1985	N/A
Temengor	6,050,000	152,000	127	348	1978	N/A
Batang-Ai	2,650,000	90,000	83	108	1985	N/A
Pedu	1,047,800	N/A	61	10	1969	N/A
Kenering	352,000	40,500	48	120	1983	N/A

Source: Sovacool & Bulan (2011); World Bank (n.d.).

submerged biomass will be substantial in a project this size situated in the tropics, the project website claims Bakun is 'emission-free and has a 0% impact on global warming' (Bakun National Hydroelectric Project 2011), implausible even if methane emissions were ignored, since activities such as forest clearing, concrete production and excavation with diesel-powered heavy equipment are hardly emissions free.

The 2013 International Hydropower Association World Congress was held in Kuching, Sarawak, from 21 to 23 May, a clear indication of the importance accorded large-scale hydropower development plans for Sarawak. As a sign of just how contentious the SCORE hydropower projects have become, IHA officials barred local indigenous leader Peter Kallang, a vocal opponent of the projects, from attending the Congress, in spite of his having registered and paid the congress fee (Environment News Service 2013).

South Korea

The WRD lists nearly 1,300 large dams in South Korea, making it the most densely dammed country in East Asia, with roughly 50% more dams per 1,000 square kilometres than Japan. Yet South Korea remains almost entirely dependent on thermal power generation from fossil and nuclear sources, with hydropower contributing less than 1% of total electricity generation. Additionally, the country holds more pumped storage than conventional hydroelectric capacity, and a leading industry publication cites a decrease in suitable dam sites, lack of local government cooperation, and public opposition as among the reasons 'conditions for the development of new water resources are deteriorating' in South Korea (Bartle 2012: 143). Storage capacity of South Korea's five largest reservoirs, all of which are impounded by multi-purpose dams, is 6,141,000,000 m³. Key data on the five largest dams in South Korea based on reservoir volume are presented in Table 13.9.

Table 13.9 Characteristics of largest dams in South Korea

Dam	Reservoir volume (10³ m³)	Reservoir area (10³ m²)	Height (m)	Installed capacity (MW)	Completion date	Displaced population (no. of people)
Soyanggang	1,900,000	70,000	123	200	1973	18,546
Chungju	1,789,000	97,000	98	412	1986	38,663
Andong	1,000,000	51,500	83	90	1977	19,657
Daechong	790,000	72,800	72	90	1981	26,178
Yongdam	672,000	36,240	70	24	2001	12,616

Source: Dooho et al. (2006); Korea Water Resources Corporation (n.d.).

Implications

Since the publication of the WCD Final Report in 2000, large dams have come under increased scrutiny for their negative impacts on resettled communities, cost overruns, disruption and degradation of ecosystems, and other reasons. A review of scholarship on large dams in East Asia invariably yields articles focused on the direct and indirect costs of resettling dam area residents, the negative impacts on fisheries, the increased incidence of landslides and erosion, and similar concerns. Much rarer are scientific studies touting the benefits large dams have brought to the region, despite the fact those benefits do exist and can be very materially relevant for some individuals and communities. Industry groups such as the International Hydropower Association and the International Commission on Large Dams, along with government proponents, continue to advance large dams as a sustainable alternative to meeting society's energy and water needs.

Some of the backlash against large dams has led to a renewed interest in the promise of small dams, including small-scale impoundments for water storage, as well as so-called micro-hydropower projects on tributaries aimed at providing power to villages not connected to power grids or reliant on expensive, polluting sources such as diesel generators. It is clear that the capital, technology, and expertise requirements for micro-hydropower projects can be far lower than those for mega-dams such as China's Sanxia or Thailand's Pak Mun. Less certain are the cumulative impacts of dozens to thousands of small dams whose combined storage capacities or electricity-generating potential might equal that of one very large dam. This is not, however, a fair comparison, since small hydropower generally targets local electricity provision whereas

Figure 13.5 **A small dam in Asia**
Source: Photo by Thomas Hennig.

mega-projects more often export electricity to distant urban and industrial load centres. For indicators such as biodiversity and impact area (BP2 and BP3), where habitat fragmentation and/or loss weigh heavily, the cumulative impacts of many small dams on different tributaries could be comparable or greater than for one large dam and reservoir (Kibler & Tullos 2013). Figure 13.5, for instance, shows a section of stream completely dewatered by a small hydropower station, a phenomenon with clear negative habitat implications for that section of stream. In truth, it makes little sense to consider many small dams as a direct alternative to one large dam; small dams may be a viable option for rural areas and poor communities where 'demand is low and costs of connecting villages to national distribution systems high' (McCully 2001: 227).

Conclusion

The dominant narrative surrounding large dams in East Asia is one where strong states, in conjunction with international development banks or domestic capital infusions, support major infrastructure projects designed to meet national or regional needs for flood control, power provision, irrigation, or drinking water. While diverse but marginal benefits from large dams may accrue to a large swathe of the population, the negative socioeconomic impacts of displacement, flooded lands, and disrupted ecosystems are acutely felt by those closest to the dam. Those communities, by virtue of geography, ethnic minority status, poverty, or some combination of the three, are often already marginalized politically and therefore have little say in decision making about the projects. Compensation, where it is paid, is almost universally perceived as insufficient by recipients, and the track record of hundreds of closely studied large dams around the world suggests that most people resettled by dams see declines, rather than improvements, in their living standards. The handful of dams profiled in this chapter collectively represent roughly 3,000 square kilometres of submerged land, perhaps as many as 2 million displaced residents, and complex ecological and social impacts. While some benefits of large dams can be easily quantified in terms of kilowatt-hours produced, hectares irrigated, or floods avoided, many of the negative impacts on biodiversity or rural social fabric, for example, are more difficult to measure and therefore often ignored in cost–benefit analyses of dam projects.

Finally, most new dams underway or planned for East Asia are designed as multi-purpose dams with hydropower as a key function. It is important to recall that most hydroelectric dams operate at about 50% of their capacity due to water level fluctuations and flood control demands. Thus while dams with gigawatt-scale hydroelectric capacity may seem at first glance to be important components of a low-carbon, renewable energy future for the countries of East Asia, care must be taken to scrutinize operating parameters when assessing the actual contribution of dams to the region's energy needs and pollution control efforts, and when weighing those contributions against the negative socioeconomic, geopolitical, and biophysical impacts of the projects.

Acknowledgements

I am grateful to Anna Hertlein, Katherine Marino, and Congjing Zhong for their data collection assistance, to Rob Beutner for masterful cartographic assistance, and to the Henry Luce Foundation for its support of Asian environmental studies at Hobart and William Smith Colleges. All errors are, of course, my own.

Bibliography

Bakun National Hydroelectric Project (2011) 'Bakun Dam'. Available at http://www.bakundam.com/home.html.

Bartle, A. (Ed.) (2012) *International Journal on Hydropower and Dams: World Atlas and Industry Guide* (Aqua-Media International).

Brown, P. & Xu, K. (2010) 'Hydropower Development and Resettlement Policy on China's Nu River', *Journal of Contemporary China* 19, 777–797.

Brown, P., Tullos, D., Tilt, B., Magee, D. & Wolf, A. (2009) 'Modeling the Costs and Benefits of Dam Construction from a Multidisciplinary Perspective', *Journal of Environmental Management* 90, S303–311.

Chinese National Committee on Large Dams (n.d.) Three Gorges Project. Available at http://www.chincold.org.cn/dams/rootfiles/2010/07/20/1279253974143251-1279253974145520.pdf.

Dooho, P., Yangsoo, Y. & Youngdu, S. (2006) 'Dam Construction and Sustainable Livelihood Support for Displaced People', in L.E.A. Berga (Ed.), *Dams and Reservoirs, Societies and Environment in the 21st Century* (Taylor & Francis).

Environment News Service (2013) 'Sarawak Native Leader Barred from Hydropower World Congress', Kuching, Sarawak: Environment News Service Newswire. Available at http://ens-newswire.com/2013/05/20/sarawak-native-leader-barred-from-hydropower-world-congress/.

Food and Agriculture Organization (2013) Aquastat. Available at http://www.fao.org/nr/water/aquastat/main/index.stm.

Gleick, P.H. (2009) *The World's Water 2008–2009* (Island Press).

Goldsmith, E. & Hildyard, N. (1986) *The Social and Environmental Effects of Large Dams* (Sierra Club Books).

Hasan, M.H., Mahlia, T.M.I. & Nur, H. (2012) 'A Review on Energy Scenario and Sustainable Energy in Indonesia', *Renewable and Sustainable Energy Reviews* 16, 2316–2328.

Heggelund, G. (2004) *Environment and Resettlement Politics in China: The Three Gorges Project* (Ashgate).

Heggelund, G. (2006) 'Resettlement Programmes and Environmental Capacity in the Three Gorges Dam Project', *Development and Change* 37, 179.

Hunt, L. (2013) 'Laos Finally Called Out Over Xayaburi Dam', *The Diplomat*. Available at http://thediplomat.com/asean-beat/2013/01/23/laos-finally-called-out-over-xayaburi-dam/.

International Commission on Large Dams (n.d.) 'Role of Dams'. Available at http://www.icold-cigb.org/GB/Dams/role_of_dams.asp.

International Rivers (n.d. (a)) 'Bakun Dam'. Available at http://www.internationalrivers.org/campaigns/bakun-dam.

International Rivers (n.d. (b)) 'Xayaburi Dam'. Berkeley, International Rivers. Available at http://www.internationalrivers.org/campaigns/xayaburi-dam.

Japan Commission on Large Dams (2011) 'Quick Report on Dams'. Available at http://www.jcold.or.jp/e/Pdf/Quick_report_of_the_Earthquake.pdf.

Japan Commission on Large Dams (2012) 'Overview'. Available at http://www.jcold.or.jp/e/outline/Pdf/JCOLDOutline.pdf.

Japan Dam Foundation (2005) 'Tokuyama Dam' (in Japanese). Available at http://damnet.or.jp/cgi-bin/binranA/All.cgi?db4=1130.

Japan Dam Foundation (n.d.) 'Dams in Japan'. Available at http://damnet.or.jp/Dambinran/binran/TopIndex_en.html.

Keong, C.Y. (2005) 'Energy Demand, Economic Growth, and Energy Efficiency: The Bakun Dam-induced Sustainable Energy Policy Revisited', *Energy Policy* 33, 679–689.

Kibler, K.M. & Tullos, D.D. (2013) 'Cumulative Biophysical Impact of Small and Large Hydropower Development, Nu River, China', *Water Resources Research*, 49, 1–15.

Korea Water Resources Corporation (n.d.) 'Multi-Purpose Dams'. Korea Water Resources Corporation. Available at http://english.kwater.or.kr/.

McCully, P. (2001) *Silenced Rivers: The Ecology and Politics of Large Dams* (Zed Books).

Magee, D. (2006) 'Powershed Politics: Hydropower and Interprovincial Relations under Great Western Development', *China Quarterly*, 23–41.

Magee, D. (2012) 'The Dragon Upstream: China's Role in Lancang-Mekong Development', in J. Öjendal, S. Hansson & S. Hellberg (Eds.), *Politics and Development in a Transboundary Watershed: The Case of the Lower Mekong Basin* (Springer).

Magee, D. & Kelley, S. (2009) 'Damming the Salween River', in F. Molle, P. Sokhem, T. Foran & M. Käkönen (Eds.), *Contested Waterscapes in the Mekong Region* (Earthscan).

Magee, D. & McDonald, K. (2009) 'Beyond Three Gorges: Nu River Hydropower and Energy Decision Politics in China', *Asian Geographer* 25, 39–60.

Mekong River Commission (2001) MRC Hydropower Development Strategy, Phnom Penh, MRC Water Resources and Hydrology Programme.

Mekong River Commission (2006) 'MRC Strategic Plan 2006–2010'. Phnom Penh: Mekong River Commission. Available at http://www.mrcmekong.org/annual_report/2006/strategic-plan.htm.

Mekong River Commission (n.d.) 'Vision & Mission'. Available at http://www.mrcmekong.org/about-the-mrc/vision-and-mission/.

Ministry of Environment and Green Development (2013) 'Technology Needs Assessment for Climate Change Mitigation in Mongolia'. Available at http://tech-action.org/Mongolia/TechnologyNeeds AssessmentMitigation_Mongolia.pdf.

National Taiwan Ocean University (n.d.) 'Taiwan Dam and Reservoir Information' (in Chinese). Available at http://wrm.hre.ntou.edu.tw/wrm/dss/resr/wk.htm.

Padovani, F. (2004) 'Les Effets sociopolitiques des migrations forcées en Chine liées aux grands travaux hydrauliques: l'exemple du barrage des Trois-Gorges', *Les Etudes du CERI* 103.

Qing, D., Thibodeau, J., Williams, P.B., Probe International & International Rivers Network (1998) *The River Dragon Has Come! The Three Gorges Dam and the Fate of China's Yangtze River and its People* (M.E. Sharpe).

Qinghai Bureau of Land Resources (2010) 'Qinghai Province Actively Resolving Problems for Reservoir Migrants' (in Chinese). Available at http://sym.mwr.gov.cn/mtgz/201007/t20100720_6914.htm.

Sarawak Energy (n.d.) 'About Hydropower'. Sarawak Energy. Available at http://www.sarawakenergy.com.my/index.php/hydroelectric-projects/about-hydropower.

Scudder, T. (2005) *The Future of Large Dams: Dealing with Social, Environmental and Political Costs* (Earthscan).

Scudder, T. (n.d.) 'A Comparative Survey of Dam-induced Resettlement in 50 Cases'. Available at http://www.hss.caltech.edu/~tzs/50%20Dam%20Survey.pdf.

Sovacool, B.K. & Bulan, L.C. (2011) 'Meeting Targets, Missing People: The Energy Security Implications of the Sarawak Corridor of Renewable Energy (SCORE)', *Contemporary Southeast Asia* 33, 56–82.

Suiryoku (Hydropower) (n.d.) 'Power Development: Ikehara Power Plant' (in Japanese). Available at http://www.suiryoku.com/gallery/nara/ikehara/ikehara.html.

Suutari, A. (2003) 'ODA in the Dock over a Dam', *Japan Times* 14 August.

Tan, Y., Hugo, G. & Potter, L. (2005) 'Rural Women, Displacement and the Three Gorges Project', *Development and Change* 36, 711.

Tullos, D., Brown, P., Kibler, K., Magee, D., Tilt, B. & Wolf, A. (2010) 'Perspectives on the Salience and Magnitude of Dam Impacts for Hydro Development Scenarios in China', *Water Alternatives* 3, 71–90.

World Bank (2013) 'Country Profiles as of January 2013'. Available at http://data.worldbank.org/data-catalog/country-profiles%20as%20of%20January%202013.

World Bank (n.d.) 'Projects & Operations: Power Project (09) [Project P004260]'. Available at http://www.worldbank.org/projects/P004260/power-project-09?lang=en.

World Commission on Dams (2000) *Dams and Development: A New Framework for Decision-making* (Earthscan).

Zhu, J. (2013) 'Nujiang Hydro Project Back on Agenda', *China Daily*. Available at http://www.chinadaily.com.cn/china/2011–02/01/content_11949587.htm.

14

Food and agriculture

Security, globalization and technology

Amy Zader

Introduction

Agricultural production has certainly been a source of livelihood for many in East Asia over time. The recent shift in East Asian societies from agrarian to industrial has indeed made dramatic impacts on agricultural production, distribution, and consumption. Just as it has in many Western nations, agriculture in Asian countries has expanded beyond local production methods as globalization and multinational corporations have reached these countries. Asian diets are changing to reflect changing social, economic, and environmental conditions in the region. Throughout this chapter, I will discuss the various changes in agricultural practices and diets throughout East Asia over time and space. I define East Asia broadly in the sense that many nations in the region have similar – yet quite different – experiences in the relationship to food and agriculture. While most of my examples will come from the 'core' region of East Asia – including China, Korea, Taiwan, and Japan – I sometimes draw on the experiences of Southeast Asian nations. As I explore where the region has come over the past few decades (sometimes centuries) in terms of food and agricultural production, I pay special attention to the ways that nations are attempting to create a future of sustainable agriculture in the region.

This chapter explores the multiplicity of changes facing East Asian food and agriculture. It examines how recent changes from agrarianism to industry have impacted agricultural production. I begin by exploring the ways that the challenge of maintaining food security has framed agricultural production efforts and goals by East Asian countries over time and into the twenty-first century. East Asia is unique in that each nation has an intense commitment to maintaining food security in the country. Governments maintain a strong presence in agricultural production systems because they believe that food security is a matter of national security and that if grain supply dwindles far below demand, the government has a crisis on its hands. I then move on to an explanation of how globalization has impacted agricultural production methods as well as distribution and consumption of food in East Asian societies. Like many regions of the world, East Asia has experienced increased interaction with Western powers as well as other parts of the world. These interactions have both shaped the ways that agriculture is produced by bringing new knowledge and technology and has also opened doors for cash crop production and trade around the world. Third, this chapter looks at the role that technology has played in agricultural production as societies are challenged to not only produce more food, but also to produce more food when fewer people are farming and less land is available for production purposes. Finally, I conclude this chapter by looking ahead at future environmental and social challenges that East Asian nations will likely face in the coming decades as they attempt to create and maintain a path towards sustainable agriculture.

Security

Fears of famine have long dominated the rulers of Asian empires. Indeed, to many Asian leaders – both in the past and contemporary – food security is viewed as an issue of national security. Many Asian empires throughout history were primarily concerned with what was happening in their domestic spheres first; they wanted to make sure that their people were taken care of, content and calm before addressing other external issues. To many Asian nations, notions of security are based on inward-looking policies that see food security as a national security issue; if the country knows it has a larger demand than supply for grain products, it will do what it can to look outside the nation in order to secure grain supply. During the nineteenth century, many East Asian nations – with perhaps the exception of Japan – pursued isolationist foreign policies to prevent their nations from opening up to Western trade. Leaders adopted nationalistic policies to begin to define their territory and peoples as a 'nation.' To justify their legitimacy as an empire, ensuring food supplies was a high priority for leaders because of Asia's long history with famine. These inward-looking policies continued as empires transformed into nation-states. Today, as I demonstrate throughout the chapter, political leaders continue to pay special attention to grain production and food security policies for fear of losing power. Although today's challenges to food security exist in a much different political and physical environment, they continue to remain a priority for East Asian leaders.

According to the Food and Agricultural Organization (FAO), food security exists 'when all people, at all times, have physical and economic access to sufficient, safe and nutritious food preferences' in order to sustain an active lifestyle (FAO 1996). Food security in Asia has revolved around one grain: rice. Rice production and distribution was a tool that emperors and landlords, and later, political leaders, used to ensure that populations were well-fed. Moreover, rice was – and continues to be – embedded in nationalism and state-building efforts throughout East Asia (Anderson 1988; Bray 1986). In Japan, rice developed as the staple food of society at the beginning of the Yayoi agricultural period (350 BC) and laid the foundation for rice to become a symbol of life for the Japanese nation-state. Over time, rice became the symbol of the settled agrarian population (Ohnuki-Tierney 1993). When the Guomingdang took control of modern China in the early twentieth century, rice sat at the heart of China's 'food problem' which the Nationalists approached with modern, Western science and technology (Lee 2010). Similarly, in Vietnam, while other staples such as corn, potatoes, and yams were widely eaten at the time, the emperor Minh Mang, who adamantly opposed French involvement in Vietnam in the nineteenth century, refused to acknowledge these as 'staples' because they were seen as a last resort when rice harvest was poor (Peters 2012). This section attempts to draw a picture of the importance of food security in East Asian societies by looking at 1) how ideas and notions of food security transitioned from empire to the modern East Asian nation-state; 2) the ways in which globalization and Western environmentalism in the late twentieth century affected national food security ideas when Lester Brown of the Worldwatch Institute asked '*Who Will Feed China?*' (Brown 1995); and 3) food security in a changing climate. Throughout the past century, Asian nations have approached food security as a scientific challenge that needs to be overcome in order to benefit the nation. As natural disasters associated with climate change intensify and as world markets for grain fluctuate, East Asian nation-states will continue to employ modern scientific approaches to solve the problem.

Food security in the modern Asian nation-state

As mentioned already, the period of modern nation building in East Asia saw grain production and food security policies take a priority as leaders wanted to base their authority and legitimacy on their ability to feed their people. In China, it is customary to store surplus grain to guard against famine. Ancient records describe how the Chinese emperor's 'Ever Normal Granary' not only prevented famine, but also allowed the state to stabilize prices for the benefit of both farmers and consumers. In Republican China, Seung-Joon Lee illustrates how Canton's constant rice shortage led the Guomindang to create a propaganda campaign around 'national' rice because they were suspicious of foreign rice imports. The Guomindang were busy structuring and organizing the modern Chinese nation. Central to this process was the recognition that science and technology provided answers to rice deficiency problems while they also helped establish a modern nation. The party devised an integral set of national grain policies, massive railroad construction, and scientific development designed to ensure the transport of rice from Hunan, the center of grain production in China, to Canton. Even though rice from Hunan was abundant, Cantonese considered it inferior in quality. In their haste to develop the economy, technology, and culture of 'national' rice, the issue of quality and taste of rice had slipped past the Guomindang. By 1936 Canton experienced a famine, and the outbreak of war in 1937 shifted leaders' attention to external political matters. Ultimately, China's national rice policy was revamped and redesigned to include all measures to ensure grain supplies at reasonable prices, rather than the previous policy designed to centralize power (Lee 2010).

Meanwhile, in Japan, agriculture during the Meiji period went through significant structural changes, even though the farmers' lifestyles were fairly consistent through broader political changes. Agriculture was based on a tenant system; landowners played a significant role in the development of agriculture and asked not for cash or capital but rice. As cash crops began to develop around the turn of the century, however, capitalist ideas began to take over. Agricultural cooperatives were formed to help bring farmers' tools and knowledge together. They hoped to modernize Japanese agriculture and adapt it to a cash crop economy. Regional cooperation began to grow, and agricultural cooperatives served as credit unions and assisted in the marketing and sales of new products. At this time, government subsidies, loans, and education were provided to farmers as well. Following the Meiji, the Taisho period brought central organization of agricultural cooperatives. The Imperial Agricultural Organization was formed to assist smaller, individual cooperatives in agricultural research and facilitating the sale of farm products. During the summer of 1918, prices of rice increased dramatically and resulted in what is known today as the 'rice riots.' The wives of fishermen in Toyoma Prefecture came together to protest the high prices of rice. These protests spread in scale and scope, ultimately forcing the Prime Minister to resign (Ohnuki-Tierney 1993). From then on, the Japanese leadership paid very close attention to the quantities and prices of rice to ensure there would be no further political unrest surrounding rice. As we will see later in this chapter, these riots instigated Japan's need to secure agricultural trade with nations external to Japan (Howe 1999).

Once the Chinese Communist Party (CCP) took over in 1949, self-sufficiency, a core ideal of Communist policies, prevailed in agricultural policies. In order to accomplish their grain self-sufficiency goal in the Chinese countryside, the Communist leaders created large, mass campaigns in the late 1950s and early 1960s. Kicking off the Great Leap Forward (GLF) were Mao's campaigns to 'Take Grain as the Key Link' (*Yiliang Weigang*) and to 'Take Steel as the Key Link'

(*Yigeng Weigeng*). Grain and steel production formed the basis of agricultural and industrial goals; while grain would pull agricultural production, steel would pull industry. However, the emphasis on village steel production, plus a combination of local officials reporting higher production levels than they were actually producing and bad weather, led to famine in the early 1960s. Following the disaster of the GLF, the CCP leadership began to loosen its tight control in the countryside. Grain became the focus – if not obsession – of agricultural policies (Shapiro 2001). Although farmers were now permitted to grow their own crops (mostly vegetables) on the side in individual plots, collectively they were required to produce grain. The grain as the key link campaign continued after the GLF. The campaign emphasized grain production; grain was seen as the source of development for China's countryside and producing grain was a moral duty of the farmers. Areas and regions that had been barren wastelands (particularly the Northwest) were turned over to grain cultivation. However, as Peter Ho points out, aside from grain production, this campaign also emphasized the diversification of crops and not just grain (2003).

Who will feed China?

The techniques and practices that have allowed Asian agriculture to sustain the land and feed its large populations over a long history have long been at the center of Westerners' understanding of Asian agriculture (King 1911; Perkins 1969). With regard to India and China, researchers have looked at how intense agricultural production can survive in these largely populated areas. In China, Mao's massive collectivization experiment in the 1950s, combined with the state's determination to be self-sufficient in food production in the 1960s, garnered attention to the politics of agricultural production (Buck et al.1966). Throughout the 1980s, the Chinese economy grew as rural and then urban reforms took off, and while production levels remained important, there was little focus directly on food security; the fear of food security lingered in the background. However, it re-emerged in the 1990s. This time, the fear of food security came from abroad and was mixed in with fears of a global food shortage and global environmental crisis.

In 1995, Lester Brown of the Worldwatch Institute in Washington, DC, published a book entitled *Who Will Feed China?*, which emphasized the stress of a growing population and increasing environmental issues on China's food production. Brown's proclamation was not made out of feelings of concern or being impressed at innovativeness of the Chinese peasants, but it was made with fear for the global market for grain and the global environment. Whereas both King and Perkins sought to understand how China could accomplish its tasks, Brown brought a negative fear-generating approach to his study. From this neo-Malthusian view, Brown argued that China's increasing population and growing economic development are placing increased demands on grain. At the same time, land resources are decreasing due to increased urbanization and industrialization. This drop in cropland area comes at the same time as awareness grows of environmental issues stemming from increased development such as decreased water shortages, soil erosion, and land pollution. This combination of environmental degradation and economic development, Brown argues, will lead to China purchasing more grain on the world market, thus raising grain prices worldwide. This price increase will then affect the poor in developing countries the most because they have the least to begin with and will lead them into famine.

Brown's book created a heated debate that delved into the depths of Chinese nationalism and notions of environmental security (Boland 1998). China's response to Brown's infamous question was that China would feed China. Within China Brown's claims generated a plethora of rigorous studies addressing China's food security issue. These projections, most coming from within China and using different methods, vary greatly (Huang, Rozelle & Rosegrant 1997).

In the short term, most scientists and agricultural economists project that China can continue to produce enough grain and food to feed its population for the first half of the twenty-first century. According to Huang, Rozelle, and Rosegrant (1997), while many Chinese economists, following up on Brown's predictions, argue that China's capacity to produce grain will outpace grain demand in the next few decades, other studies report that China could start importing grain within the next decade. Different models using various factors as well as the uncertainty of the direction of China's developing economy likely cause the disparities of these studies. However, the Chinese Ministry of Agriculture (MOA) maintains that China is and can remain 95 percent self-sufficient in grain production (including soybeans) with heavy export restrictions and a goal to be 100 percent self-sufficient in rice, wheat, and corn (USDA FAS 2012).

Food security in a changing climate

The latest challenge Asian nations are facing with regard to food security is the uncertainty and the risk that changing climates have. According to the Intergovernmental Panel on Climate Change (IPCC) Assessment Report, some of the most adverse effects of climate change will hit the environment, economies, and agriculture and food systems of developing countries. These changes and risks will be a result of increasing CO_2 fertilization affecting the growth of crops, increasing variability of precipitation potentially leading to flooding and drought conditions, and increased variability and declining runoff in rivers leading to decreased crop production. In East Asia, specifically, the effects of climate change are likely to impact meteorological conditions, affect coastal areas (including fishing industries), and cause short-term production disappointments and overall crop yield declines (Su, Weng & Chiu 2009).

Because of the amount of coastal areas that encompass Southeast Asian nations, this group has been particularly adept at taking action. In order to adapt strategies for climate change adaptation, the member states of the Association of Southeast Asian Nations (ASEAN) adopted a five-year Strategic Plan of Action on Food Security from 2008–2013. The plan identified climate change as an impact factor on food security in the region, and they hope this plan will ensure long-term food security goals and protect and improve the livelihood of farmers. Individual nations have brought this international agreement back to adapt to their own national policies. For example, apart from a thriving fisheries industry, the city-state of Singapore grows very little to none of its own food. As part of its national efforts to ensure food security in the coming years, Singapore has adopted a plan for urban agriculture and food security, which involves developing strategic agri-business connections with other East and Southeast Asian nations centered in Singapore, the development of rooftop and urban vegetable gardens, and emphasis on the local production of eggs, leafy green vegetables, and fish (Lee & Tan 2011).

In China, the world's most populated country and a large emitter of carbon dioxide, the impacts of climate change are both diverse and vast. Indeed, water resources and agriculture are expected to experience dramatic impacts from climate change. With & percent of the world's available arable land to feed 22 percent of the world's population (Piao et al. 2010), China is in a vulnerable position to experience widespread effects of climate change. Chinese scientists are working to understand, highlight areas of uncertainty, and potentially mitigate the environmental impacts of climate change. Piao et al. (2010) have found that around the country, 'the consequences of recent climate change on water resources and agriculture have been limited because climate trends remained moderate due to natural variability' and the intervention of technology (49). However, due to China's vast size and natural diversity, there are a number of regional differences of potential impacts of climate change. With technological progress and sustainable economic development practices intended to counteract the impacts of climate change,

Chinese scientists are proactively working to prevent drastic changes, but the vast territory and diversity of China complicate the impacts of climate change on food security.

In addition to climate itself, large-scale natural disasters, which are expected to increase and/or intensify with global climate change, can have a significant impact on a nation's food security. This was illustrated in March 2011 when Japan was struck with a massive earthquake and tsunami off the coast, setting off a nuclear disaster. This crisis at the Fukushima Daiichi nuclear plant left radioactive contamination around the area and led to major worries about the safety of Japanese food with radiation in the soil (Frid 2012). Prior to the disaster, Japan imported nearly 50 percent of its food supplies, but that number has increased since the spring of 2011. Additionally, all food exported from Japan now undergoes significant radiation testing by importing countries. The time and cost – along with consumer fears, despite the testing – of these procedures have dramatically impacted Japan's agricultural sector.

Globalization

Just as efforts to preserve food security have shifted over the past few decades, so too have the interconnections between different countries affected food and agriculture in the region. From colonialism in the region to the GATT and WTO, international trade of agricultural products has certainly changed the historical agrarian settlement patterns of the past few centuries. Indeed the twentieth century saw the introduction of cash crops, regional specialization, government institutions for agriculture, and the development of organizations promoting international trade. With each of these developments, Asian nations had to restructure their economies so they were compatible with international standards.

Food and agriculture in the nineteenth century

The nineteenth century was a time when East Asia began to interact with the West and new ideas, knowledge, and technology were introduced. What is striking about this period – as many others have noted (Brown 1986; Kelly 2006; Mulgan 2000) – is the contrast between how China reacted to new Western powers and how Japan did. While Japan was quick to consolidate itself into a modern nation-state in the Meiji Restoration Period, China, by its own design, was closed off and tried to resist incursions by traders, missionaries, and armed expeditions. These two differing approaches were evident in the changes agriculture and food production experienced in these two nations at the time. While China reluctantly changed its production methods to accommodate the tea trade to Europe, Japan under the Meiji eagerly embraced new concepts to change agriculture from a practice to a science.

China's tea trade and interactions with the West have been well documented since the 1700s (Yong 2007) to show how trading with the West impacted China's own domestic situation by creating a solid trade and commodity system. In the early 1800s, China was largely cut off from the rest of the world. Canton, known today as Guangdong, was the main trade port with Europe. All Chinese products went through an extensive process to leave the port. While Chinese products such as porcelain and silk were in high demand in Europe, tea was certainly the most popular trade item. Although China's emperors resisted opening China up to the rest of the world, Britain's insatiable taste for tea led it to pursue it at any cost. Determined to defend its borders from outsiders, the Qing Dynasty in China could only do so much before falling in 1911. Landlords controlled much of the land and, thus, agricultural production during this tumultuous time in China.

Meanwhile, Japan's mid-nineteenth century political transition from the Tokugawa Shogate to the Meiji emperor in 1868 brought Western ideas of agronomy and agricultural science to

Japan. One of the first goals of the Meiji agricultural regime was to change the 'backward' Japanese agriculture to more advanced and superior Western methods. By the late nineteenth century, the Meiji had set up and established a series of agricultural universities, educational organizations, agricultural textbooks, and experiment stations that were modeled after Western agricultural systems (Kelly 2006). At the same time as the Japanese were creating a system of agricultural science based on Western models, they were also in the process of re-evaluating their own traditional agricultural systems. The Meiji paid special attention to 'experienced farmers' from different regions of Japan who were able to gather together and give lectures on farming techniques in their region. The combination of the introduction of Western ideas and the Meiji consolidation of agriculture led to the 'Meiji agrarian system,' which introduced techniques such as dry rice fields, the use of cattle and horses, the introduction of high-yielding crops, and heavy use of soybean crops for manure (Kelly 2006). By combining Western scientific ideas with traditional Japanese techniques, the Meiji period had established what it saw as its own unique form of agriculture.

Food and agriculture during the twentieth century

While the nineteenth century was a time of transition from closed-off Asian societies to opening to Western powers, the twentieth century witnessed a number of changes related to agriculture and food from widespread Japanese agricultural colonization in East Asia to wartime rations to post-war industrialization. Japanese colonization dominated East Asia in the early part of the nineteenth century. While the Japanese were busy building a military empire, they were also concerned about maintaining the ability to feed their nation. Much of their colonial efforts in other countries was devoted to agriculture when they were not too busy building their military base.

Following the collapse of the Qing Dynasty in China in 1911, the government of Northeast China, headed by warlord Zhang, organized Han Chinese migration to the region for cultivation activities to strengthen the Northeast frontier and defense at the borders. The central and Northern regions of Heilongjiang and Jilin provinces were cleared for cultivation. Soybeans dominated the agricultural landscape and provided greater interest for the Japanese. In his commodity history of soybeans in Northeast Asia, David Wolff documents the role that soybean production in Northeast China played as the source of Japanese interest in the region. Once a crop for the bourgeoisie, widespread soybean cultivation across Manchuria enabled the Japanese to import soybeans in mass quantities. This allowed for the soybean to become a staple of the Japanese diet. From the early 1900s onward, the Japanese established a strong base in producing soybeans in Northeast China and exporting them to Japan. Access to Manchuria enabled the Japanese to import large quantities of soybeans, thus making soy products available to the middle classes in Japan. In turn, demand for soybeans intensified Japanese determination to maintain control in the region (Wolff 2000).

Although the Japanese had great interest in agricultural production in Manchuria, production decreased due to increased military and political upheaval when they gained control of the region in the early 1930s. However, the Japanese had long-term, ambitious plans to move farmers from Japan to Manchuria and to further develop the region agriculturally. Prior to 1931, the Japanese in Manchuria consisted primarily of an urban, transient population. Following Japanese takeover of the region in 1932, the Japanese planned a massive emigration of Japanese farmers to settle and farm the land. In 1936, the project had turned from trial colonization to a project of mass colonization, and the Japanese had created a 20-year plan for full agricultural colonization of Manchuria. The idea behind the plan was that the Japanese could strengthen their military defense by furthering economic development in the region through agricultural production

(Guelcher 1999). The 20-year plan envisioned the relocation of over 1 million Japanese farming households from Japan to Manchuria by 1956. By 1937, over 270,000 Japanese had settled in Manchuria, the largest emigration in modern Japanese history (Guelcher 1999).

Although the mass colonization of Manchuria by the Japanese was never realized, the impact that the Japanese had on the region remains. During their time of colonization, the Japanese brought to Manchuria the technology necessary to form the basis of mass agricultural production in the region. Drawing on the Han Chinese, the Korean minority already in the region, and the Japanese farming households that came to Manchuria with the promise of land, the Japanese had ample labor opportunities. With the technology and labor in place, the Japanese colonialists were able to create agricultural production opportunities that made up the base of their Manchurian empire. Although most of the Japanese left at the end of the colonization experiment, the Korean minority and Han settlers remained in the region. The land that had been cleared and set up for agricultural production remained, along with most of the technology brought over by the Japanese initially.

Food and agriculture and the introduction of the WTO

The establishment of the GATT (General Agreement on Trade Tariffs) following World War II allowed for nation-states from around the world to be part of this global trade initiative. The road leading to the accession of Japan into the GATT in 1955 was long after the nation had been cut off from the rest of the world in 1945. Despite US opposition at the time, Japan insisted it get most favored nation (MFN) status as well as tariff reductions abroad (Forsberg 1998). Once it entered the global trade scene, Japan's internationalization had profound impacts on its economic restructuring, especially in the agricultural sector. This sector had been tradition bound, but needed to quickly meet the international economic standards of other nations at the time (Kihl 1990). Although tensions from this transition emerged, Japan's transition to international agricultural trade was relatively smooth.

When China announced it wanted to join the World Trade Organization, it faced much international attention and pressure. Global agricultural trade has had a large impact on China's agricultural production and distribution methods. Indeed, the publication of Brown (1995) raised a lot of eyebrows about the role that other countries might have to play in Asia's food security (Anderson et al. 1997). The book was not just about China's own future, but also those in Asia and around the world as China was expected to open its trade doors to the WTO in 2001. Agricultural economists both within China and outside of it focused their attention on the ways China's WTO accession impacted its agricultural production yields and techniques, land reform, farmers' incomes, and food security and self-sufficiency. While China is the world's fifth largest exporter of agricultural products, it is also a net importer. It has seemingly found a balance (albeit weighted to imports) between imports and exports. Farmers' incomes have increased and are reportedly more linked with global economic trends than in the past, and land reform continues to improve (Carter et al. 2009). Overall, it appears as though China's WTO accession has brought positive changes to the agricultural sector (Carter et al. 2009), even though it has also shifted its priorities of producing for export markets and importing other goods, thereby affecting China's goals for self-sufficiency.

Technology

As East Asian nations have sought to find a balance between meeting their own internal domestic food security goals and interacting with external countries, they have each experimented and worked with different forms of technology. Technology enables more rapid and efficient agricultural growth. In modern Asian nation-states, the promises technological advancement have brought include alleviation of the fear of food security through higher yielding seeds. This was most visible with the Green Revolution. Although the Green Revolution is considered a success in terms of producing more grain for Asia's growing populations, it did not benefit the livelihoods of marginal farmers in developing countries who had little access to technology. These farmers were those that were vulnerable to begin with (Pengali & Raney 2005). Moreover, Green Revolution technology, especially the use of chemical fertilizers and pesticides, created many environmental problems that vulnerable regions were not equipped to deal with. Fears of the way that Green Revolution technologies were implemented continue to haunt officials and researchers as the newest form of agricultural technologies – genetically modified seeds – enters Asian societies.

Green Revolution technologies

The Green Revolution was driven by the assumption that higher yields and the production of more food will lead to a well-fed society. While India is the most cited example of where Green Revolution technology was incorporated into agricultural production, it took place throughout East Asia as well, though at different scopes and scales. It has certainly been acknowledged that 'hundreds of millions of rural people in less-favored environments of this region still live in poverty and received limited benefit from the Green Revolution' (Pender 2007). Furthermore, ecosystems of South and East Asia are so diverse that no single approach to technology will work for each different circumstance (Pender 2007). Because of the contrast in the ways high-yielding rice was developed and implemented in Southeast Asia and China, these cases are outlined in the following.

In Southeast Asia, most of the Green Revolution technological development was centered in the International Rice Research Institute (IRRI) in the Philippines. With rice as the staple food throughout Southeast Asia, IRRI developed and provided farmers with high-yielding rice varieties, irrigation or controlled water patterns, improved moisture utilization, chemical fertilizers and pesticides, and management skills (Pengali & Raney 2005). Southeast Asian governments helped distribute these technology packages to farmers. However, in order for farmers to access the technologies, governments had to improve transportation and communication infrastructure. This allowed the famers closer to transportation hubs more access and resulted in uneven distribution and uneven gains for farmers. Even though socio-economic gains were not evenly distributed, the Green Revolution in Southeast Asia is credited as averting a major food crisis during the 1980s as food prices crept up and shortages were feared (Pengali & Raney 2005). Throughout the 1980s and 1990s, Southeast Asian governments shifted to a market approach to rural development in the agricultural sectors. The state no longer provided necessary technologies, but expected farmers to get them through the market. This has allowed more farmers to access the technologies and grow seeds with an increased amount of chemical inputs such as fertilizers and pesticides.

With China being the world's largest producer of rice, rice breeding sits at the core of Chinese agricultural technological development. Since famous scientist Yuan Longping developed Chinese hybrid rice in the 1970s (Deng & Deng 2007) and Deng Xiaoping promoted developing agricultural technology as part of the Four Modernizations, a great deal of China's agricultural research has gone into rice research. As the world has learned from the Green Revolution, however, advanced agricultural technology does not solve the problem of feeding the poor. Although the Green Revolution achieved its initial goal of producing higher crop yields, it has left a legacy of many unintended social and environmental consequences. The technological optimism that accompanied Green Revolution thinkers is alive and prevalent throughout many discussions of agriculture. When hybrid rice was introduced on a large scale to farmers in the 1970s and 80s, the Chinese countryside had already experienced massive organizational changes such as rural collectivization and the Great Leap Forward campaign of 1958–1962. The introduction of a new kind of rice that required a change in production techniques simply added to the complex changing nature of the countryside. Determining the best use for technology requires significant collaboration and negotiation between the state, society, and scientists. The Chinese state has invested significantly in rice breeding technology; since the Green Revolution, the state has always been on the global frontiers of rice technology. The Chinese state has carefully prepared and planned how it would best utilize the technology that it has in its possession to ensure that the technology will assimilate with larger state goals.

China holds a unique story in its hybrid rice breeding. While most of the hybrid rice varieties developed as part of Green Revolution technology at IRRI, Chinese scientists developed hybrid rice on their own before international technological collaboration started. They patented the rice, the first time a Chinese scientist had obtained a patent since the PRC was formed. With the introduction of hybrid rice China had not only solved a practical problem of producing enough rice for its large population, but it had conquered this problem without external help or assistance. The story of hybrid rice is important in understanding recent trends and developments in Chinese rice breeding technology.

Hybrid rice breeding came to China in three stages. The first stage, 1950 to the early 1960s, saw the rise in rice breeding of local varieties based on selection, evaluation, and use. During this time, the dwarf plant was introduced. Many Southern rice farmers quickly adopted this high-yielding variety. While this first stage was important, it was not as significant as the development of F1 hybrid rice varieties in the mid-1970s. The second stage of hybrid rice breeding was during the large-scale disruption and change that took place from the early 1960s to the early 1970s. With Mao's 'mass line of technology development,' there was mass participation in science and technology. Programs for rice breeding technology began in 1964, primarily in Hunan province. At this time Yuan Longping ('the father of hybrid rice' – see later), after finding natural growing male sterile rice, developed his theory of heterosis in rice. This theory claimed that the progeny of two distinctly different parents grows faster, yields more, and is more stress resistant than either parent. Once scientific research, headed by Yuan Longping, began to show promising results, the Hunan provincial government elevated the status of hybrid rice breeding. Finally, the third stage of rice-breeding technology came in the early 1970s when China made its most significant technological breakthroughs of the 'super-hybrid' varieties of conventional strains. In 1971, Hunan province initiated a cooperative research program in hybrid rice breeding. By the following year, this research program had gained national attention and thus national funding. The Chinese Academy of Agricultural Sciences teamed with the Hunan Academy of Agricultural Sciences in 1972 when scientists had successfully isolated the male sterile species necessary to produce hybrid rice. By 1974, scientists began a small demonstration field. In 1976, China became the first country in the world to commercialize hybrid rice. After successful results, hybrid rice cultivation enjoyed a rapid increase each year in the 1970s to reach 5 million ha by the end of the decade.

Genetically modified crops

Throughout Asia, there is a big push to use and implement GM technologies. According to Jonathan Crouch, the Director of the Genetic Resources and Enhancement Unit of the

International Maize and Wheat Improvement Center: 'A Green Revolution pace of progress is once again needed . . . All possible innovations will be required to achieve this . . . GM technologies are one of many important contributions' (quoted in Healy 2008). Because of Asia's increased population in the future and because new cropland is scarce, Asian countries will need to intensify rice production greatly. Scientists at IRRI in Manila doubt that conventional methods of production can meet the need for rice in the near future. To meet Asia's rice needs, IRRI is planning to use GM technologies to boost yields by more than 50 percent. They also plan to make sure traditional rice varieties stay pure for cultural and scientific purposes (Healy 2008). Preliminary research results from both IRRI and the Chinese government indicate that new strains of genetically modified rice produce higher yields than traditional varieties, require fewer chemical pesticides, and have more nutritional content than non-GM rice (Healy 2008).

China has made significant leaps in using biotechnology to develop transgenic rice. While China sees this research as another step in its rice research trajectory, many international groups, particularly environmental NGOs opposed to genetic modification, have opposed commercial production of GM rice, causing a controversy in China. The fragmented structures of GMO regulation, discussed later, indicate that the concept of biosafety regulation emerges as China becomes closely integrated in the global capitalist economy. The Chinese state's investment in biotechnology research and development is high. In fact, the national research budget increased from $8 million to $112 million between 1986 and 1999, and, in 2001, the state announced plans to raise its research budget in plant technology by 400 percent by the year 2005 (Huang et al. 2002). Currently, over 100 public sector institutions are involved in plant and agricultural biotechnology research. Jikun Huang, an agricultural economist and a leading non-scientific advocate of biotechnology, has published two articles in *Science* advocating the use of GM products (Huang et al. 2002, 2005). China's researchers are eager not only to see China progress and advance in biotechnology, but also to boast its capabilities to the world. Currently China falls behind the United States, Canada and Argentina as being the fourth-largest producer of GM crops. However, most of the GM crops currently grown in China are cotton, not food crops. The state's biosafety regulation thus far has prevented widespread use of GM food crops, especially rice. For the state, rice is a symbol of science and progress that has dominated state discourses in the reform era. Science politics have guided reform-era China (Simon & Goldman 1989; Suttmeier 1980; Yu 1999), and Yuan Longping is an icon in China for developing the technology to grow high yields of rice that have dramatically raised China's grain production. Today, China remains one of the only Asian countries with the capacity to distribute and use hybrid rice extensively, but it has not yet done so due to concerns about its safety. Hybrid rice, which contributed greatly to the success of grain production in reform-era China, serves as an emblem of national pride based on the technological success of a culturally symbolic food crop. Genetically modified rice is the next step in the state's march towards technological progress through rice.

Towards a sustainable future?

As East Asian nations have overcome a number of economic and environmental challenges in recent years, they have proven their resilience and power to overcome obstacles. However, as we move further into the so-called 'Asian Century,' a number of social, political, economic and environmental challenges face agricultural production in East Asian states. These challenges include: incorporating more ecologically efficient forms of agriculture such as organic and ecological production, meeting the food demands of growing urban middle classes, maintaining rural economic development to lift subsistence farmers out of poverty, dealing with contaminated environments and changing climates, and finding a balance for safety mechanisms and regulation of new agricultural technologies. Indeed, these are all large changes that Asian nations must address individually and collectively.

Many Asian nations have successfully adopted organic and ecological agricultural techniques. After the idea of organic production emerged and was advertised in Japan in the 1990s, there was a lot of confusion about what organic actually meant or who could claim certain products to be organic (FAO 2006). The Japanese government then decided to implement laws outlining the standards and, in 2006, required these standards to be compatible with international regulations for organic production. Many East Asian nations, such as Taiwan, Korea, and arguably China, followed Japan in implementing such domestic and international regulations. However, programs sponsoring the widespread adoption of these techniques have been limited. Many family farmers continue to use traditional organic techniques to grow their own food in their yards, but recognize that they can gain more income from using chemical inputs on food they sell in the market because the yields are higher and the losses due to pests are greatly reduced. According to Ong Kung Wai (2009), in a report put out by the International Federation of Organic Agriculture Movements, organic trends started in Asia as a result of some farmers wanting to get back to nature and others finding a niche in export agriculture trends. However, over the past decade, organic production has grown dramatically and domestic markets have developed. Asian consumers are certainly aware of the health benefits of eating organically, but they still lack information about what organic really means and cost is a barrier, as organic produce is more expensive than what is produced conventionally. Food safety, especially with a number of food scandals exposed in China and with the radiation contamination from the Japanese earthquake have contributed to consumer demand for organic and higher quality produce as well.

Aside from the production of organic or ecological agriculture, many scientists predict that the biggest threat of food insecurity to Asia in the near future will be the loss of grain production due to climate change. According to the United Nations Economic and Social Commission for Asia and the Pacific (UNESCAP), 'Climate change holds the potential to radically alter agroecosystems in the coming decades and there is already evidence of devastating crop failures' (2009: 104). In 2008 the Asia region experienced short-term food security fears as grain prices increased dramatically. While these fears have subsided in the short term, the overall factors that affected the rise in price have not been fixed. Moreover, when grain production is projected along with long-term issues caused by climate change, the region is facing a number of challenges. The anticipated short-term impacts of climate change on agriculture in the region include the possible effects of heat on rice production and counter-balancing the increased carbon dioxide in the atmosphere. The long-term impacts of climate change in the region center around the Tibetan Plateau. Many of China and Southeast Asia's major rivers originate in this region. Ice-pack melting will reduce water on the plateau, but also will cause flooding in downstream rivers, and the rise of sea levels will impact the land productivity near water. Additionally, concerns arise from 'black carbon' deposits in Tibetan glaciers (Qiu 2010). In order to combat these impacts, UNESCAP recommends that governments take climate change projections seriously and do what they can to reduce their own emission of climate-altering emissions while also maintaining a system of equitable and sustainable agricultural production in the region (UNESCAP 2009).

Indeed, environmental threats in East Asia from industrial pollution, synthetic pesticides and fertilizers, and climate change present major challenges that East Asian nations will face in the future. The governments of major East Asian countries are proactively attempting to mitigate these challenges through industrial clean-up efforts, the promotion of organic and natural agriculture, and the use of technology to slow and alleviate the problems caused by climate change. Climate change certainly presents the most uncertain challenges in the future, yet with many regions in Asia vulnerable to potential climate change impacts, it will be the biggest challenge. In the short term, governments must work closely with experts to understand projections and develop technological and sustainable practices and plans for the future. In the long term, countries must be prepared for the challenges of food security as a result of climate change by developing safe, sustainable agricultural practices.

Bibliography

Anderson, E.N. (1988) *The Food of China* (Yale University Press).

Anderson, K., Dimaranan, B., Hertel, T. & Martin, W. (1997) 'Asia Pacific Food Markets and Trade in 2005: A Global, Economy-Wide Perspective', *Australian Journal of Agriculture and Resource Economics* 41(1), 19–44.

Boland, A. (2000) 'Feeding Fears: Competing Discourses of Interdependency, Sovereignty, and China's Food Security', *Political Geography* 19, 55–76.

Bray, F. (1986) *The Rice Economies: Technology and Development in Asian Societies* (Blackwell).

Brown, L. (1995) *Who Will Feed China? Wake-up Call for a Small Planet* (W.W. Norton & Co.).

Brown, L.K. (1986) 'Agriculture in Japan: The Crisis of Success', *Japan Society Newsletter* 34(1), 2–5.

Carter, C.A., Zhong, F. & Zhu, J. (2009) 'Development of Chinese Agriculture Since WTO Accession', *EuroChoices* 8(8), 10–16.

Deng, X. & Y. Deng. (2007) *The Man Who Puts an End to Hunger: Yuan Longping, 'Father of Hybrid Rice'* (Foreign Languages Press).

FAO (Food and Agricultural Organization) (1996) 'The State of Food and Agriculture, 1996' (Food and Agriculture Organization).

FAO (2006) Japan Country Profile for Organic Agriculture. Available at http://www.fao.org/organicag/display/work/display_2.asp?country=JPN&lang=en&disp=summaries.

Forsberg, A. (1998) 'The Politics of GATT Expansion: Japanese Accession and the Domestic Political Context in Japan and the United States, 1948–1955', *Business and Economic History* 27(1), 185–195.

Frid, M.J. (2012) 'Food Safety in Japan: One Year After the Nuclear Disaster', *Asia-Pacific Journal: Japan Focus* 10(12), March. Available at http://www.japanfocus.org/-Martin_J_-Frid/3722.

Guelcher, P. (1999) Dreams of Empire: The Japanese Agricultural Colonialization of Manchuria (1931–1945) in History and Memory, PhD dissertation, University of Illinois.

Healy, E. (2008) 'Genetically Modified Foods Gain Ground', *Development Asia* June. Available at http://development.asia/issue01/feature-03.asp.

Ho, P. & Zhao, J.H. (2005) 'A Developmental Risk Society? The Politics of GMOs in China', *International Journal of Environment and Sustainable Development* 4(4), 370–394.

Howe, C. (1999) *The Origins of Japanese Trade Supremacy: Development and Technology in Asia from 1540 to the Pacific War* (University of Chicago Press).

Huang J. & Rozelle, S. (1996) 'Technological Change: Rediscovering the Engine of Productivity Growth in China's Rural Economy', *Journal of Development Economics* 49, 337–369.

Huang, J., Rozelle, S. & Rosegrant, M.W. (1997) 'China's Food Economy to the Twenty-first Century: Supply, Demand, and Trade', *Economic Development and Social Change* 47(4), 737–766.

Huang, J., Rozelle, S., Prey, C. & Wang, Q. (2002) 'Plant Biotechnology in China', *Science* 295, 674–676.

Huang, J., Hu, R., Rozelle, S. & Prey, C. (2005) 'Insect-Resistant GM Rice in Farmers' Fields: Assessing Productivity and Health Effects in China', *Science* 308, 688.

Kelly, W.W. (2006) 'Rice Revolutions and Farm Families in Tohoku: Why is Farming Culturally Central and Economically Marginal?', in C.S. Thompson & J.W. Traphagan (Eds.), *Wearing Cultural Styles in Japan: Concepts of Tradition and Modernity* (SUNY Press).

Kihl, Y. Whan (1990) 'Japan: The Political Economy of Trade Policy Reform', in G. Skogstad & A. Fenton Cooper (Eds.), *Agricultural Trade: Domestic Pressures and International Tensions* (Institute for Research on Public Policy, Toronto).

Lee, M.Y.E. & Tan, H.H.W. (2011) 'Growing Your Own Food: The Need for Urban Agriculture in Singapore', *Innovation Magazine*, National University of Singapore. Available at http://rmbr.nus.edu.sg/news2/2010/ innovgrow.pdf.

Mulgan, A. George (2000) *The Politics of Agriculture in Japan* (Routledge).

Ohnuki-Tierney, E. (1993) *Rice as Self: Japanese Identities Through Time* (Princeton University Press).

Pender, J. (2007) 'Agricultural Technology Choices of Poor Farmers in Less-Favored Areas of South and East Asia', *International Food Policy Research Institute Discussion Paper*, 709. Available at http://www.ifpri.org/publication/agricultural-technology-choices-poor-farmers-less-favored-areas-south-and-east-asia.

Pengali, P. and Raney, T. (2005) From the Green Revolution to the Gene Revolution: How Will the Poor Fare? ESA Working Paper 05–09, Agriculture and Development Economics Division of the United Nations FAO. Available at ftp://ftp.fao.org/DOCREP/fao/008/af276e/af276e00.pdf.

Piao, S., Ciais, P., Huang, Y., Shen, Z., Peng, S., Li, J., et al. (2010) 'The Impacts of Climate Change on Water Resources and Agriculture in China', *Nature* 467(2), 43–51.

Qiu, J. (2010) 'Measuring the Meltdown', *Nature* 468(7321), 141–142.

Sanders, R. (2006) 'A Market Road to Sustainable Agriculture? Ecological Agriculture, Green Food and Organic Agriculture in China', *Development and Change* 37(1), 201–226.

Su, Y.Y, Weng, Y.H. & Chiu, Y.W. (2009) 'Climate Change and Food Security in East Asia', *Asia Pacific Journal of Clinical Nutrition* 18(4), 674–678.

United Nations Economic and Social Commission for Asia and the Pacific (UNESCAP) (2009) 'Sustainable Agriculture and Food Security in Asia and the Pacific'. Available at http://www.unescap.org/65/documents/Theme-Study/st-escap-2535.pdf.

USDA (United States Department of Agriculture) FAS (Foreign Agriculture Service) (2012) 'Global Agriculture Information Network: People's Republic of China: Grain and Feed Annual Report'. Available at http://gain.fas.usda.gov/Recent%20GAIN%20 Publications/Grain%20and%20Feed%20Annual_Beijing_China%20-%20Peoples%20Republic%20of_3-2-2012.pdf.

Wai, O. Kung (2009) 'Organic Asia – From Back to Nature Movement & Fringe Export to Domestic Market Trend', in Willer, H. & Kilcher, L. (Eds.), *The World of Organic Agriculture: Statistics and Emerging Trends 2009* (International Federation of Organic Agriculture Movements).

Wolff, D. (2000) 'Bean There: Toward a Soy-Based History of Northeast Asia', *South Atlantic Quarterly* Winter.

Yong, L. (2007) *The Dutch East India Company's Trade with China, 1757–1781* (Brill).

Part V
Fisheries, forests and wildlife

15

Fisheries in East Asia

Political, economic and security challenges

Tabitha Grace Mallory

The marine fisheries in East Asia are the most productive in the world. Fish are crucial to the economies and food security of East Asian countries. Asia has the most fishermen in the world, and the Asian diet depends more heavily on fish than does any in the rest of the world. At the same time, fish contribute to healthy ecosystems. But fisheries in East Asia are under heavy strain. China and Japan are already two of the largest fish-producing and consuming nations in the world, and demand for fish is projected to increase over the next few years as the middle classes in developing Asian countries expand.

This chapter provides an overview of fisheries in East Asia and discusses the political, economic, environmental, and security challenges to sustainable management of East Asia's fish stocks. This chapter covers the marine capture (wild catch) fisheries in East Asia; it does not address aquaculture or freshwater fisheries. This chapter covers the areas of East Asia corresponding to FAO major fishing areas 61 and 71 (see Figure 15.1).

Global fisheries crisis

Global marine fisheries are in crisis. In its 2012 report *State of World Fisheries and Aquaculture*, the UN Food and Agriculture Organization (FAO) states that as of 2009, an unprecedented 87 percent of the world's marine fish stocks were fully exploited, overexploited or depleted (FAO 2012a: 11).

World fisheries resources were decimated in only about five decades. In the early 1950s, less than 5 percent of global marine fisheries were fully exploited or overexploited. By the mid-1990s, marine production plateaued at about 80 million tons annually (see Figure 15.2).

Fish stock depletion is the result of four categories of changes that took place from the middle of the twentieth century. First, by the 1950s the dominant model for socioeconomic development in the industrialized world was modernization theory, premised on the idea that humans could control nature through science (Chuenpagdee 2005: 28). With its focus on large production volumes, fisheries modernization disregarded natural patterns of fish resources, which are not always predictable or quantifiable. Second, fishing shifted from small-scale, traditional fishing to state-led industrialized fishing. Once the UN Convention on the Law of the Sea

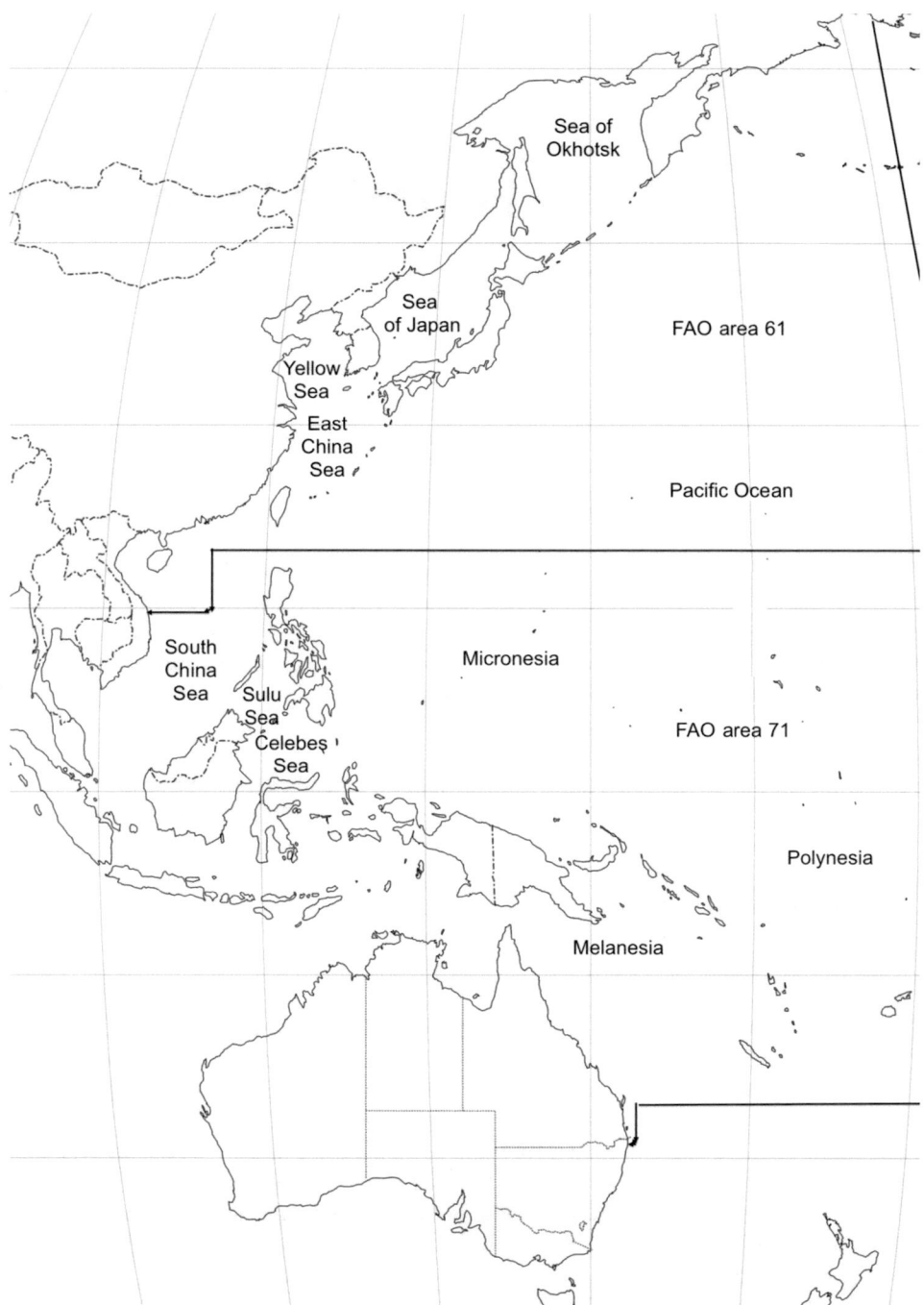

Figure 15.1 Map of FAO major fishing areas 61 and 71

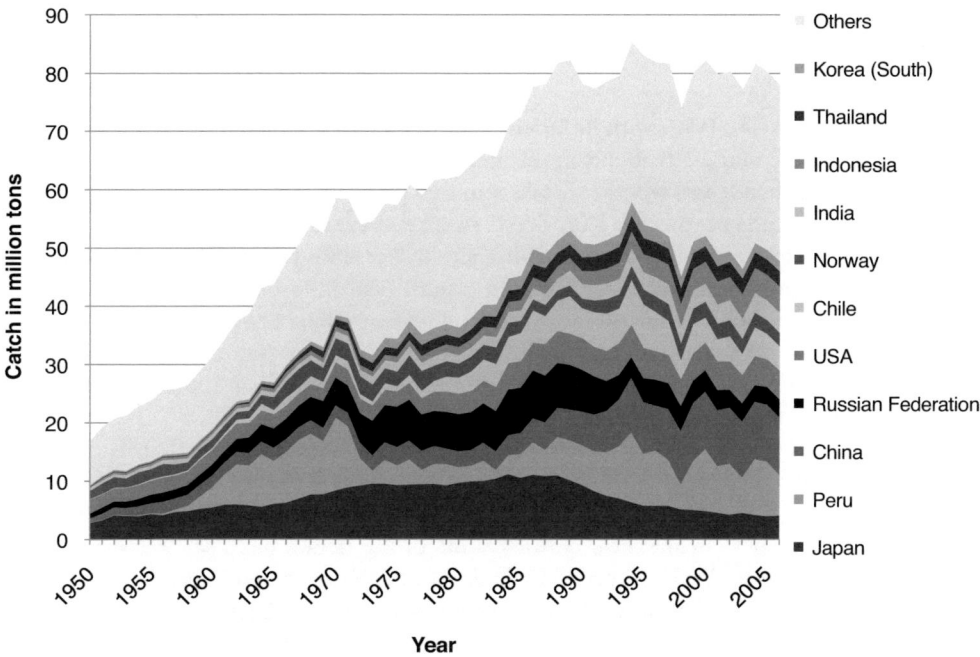

Figure 15.2 Landings by fishing country in the global ocean

(UNCLOS) was signed in 1982, giving coastal states exclusive jurisdiction over the marine resources up to 200 nautical miles off their shores, states developed domestic fishing industries through massive subsidies. The number of fishermen increased from 12.5 million to 36 million between 1970 and 1998 (FAO 1999a). Between 1970 and 1995, the number of non-decked vessels grew by about 55 percent, while the number of decked vessels more than doubled (FAO 1998). Industrial fishing has negatively affected the livelihoods and food security of coastal populations, especially in Asia, where populations are heavily dependent on fisheries resources. Third, technological advancements – such as sonar and satellite technology, better materials, and larger and more advanced ships – contributed to overfishing. Fourth, during the second half of the twentieth century, global demand for fish soared, especially in the United States, the European Union, and Japan as the incomes of those populations increased. Fish exports increased over five times between 1961 and 1999, even though production only doubled (FAO 2003). This increased demand drove fishing effort.

Why do we care about fisheries management? Having healthy fisheries is important for a number of reasons. Robust fish stocks add to biodiversity, which is beneficial for the environment and life on the planet. Having a healthy natural environment directly relates to human health. If fisheries are managed sustainably, they can be a nutritious food source for coastal populations and generate long-term economic benefits. But currently poor fisheries management threatens the long-term viability of capture fisheries as a protein source for Asia.

Fisheries in East Asia

Biological overview

The vast Western Pacific Rim, with its hundreds of thousands of miles of coastline and diverse ocean habitats, is teeming with abundant sea life. The North Equatorial Current runs east-to-west across the Pacific Ocean, where it breaks into two on hitting the Philippine landmass. The northern-bound strand becomes the Kuroshio Current and the southern-bound strand becomes the Mindanao Current. Off the coast of China, freshwater from the Yangzi River flows into the East China Sea, where it converges with the warm Kuroshio Current, creating a fecund habitat for sea life and productive fishing grounds. The East China Sea has over 800 species of fish, with about 40 to 50 species of significant commercial value (Ou & Tseng 2010). The Yellow Sea is one of the world's most heavily fished areas, with over 100 commercial fish, cephalopods, and crustaceans (Teng et al. 2005: 17–21).

The Kuroshio Current continues northward, becoming the North Pacific Current, looping around Alaska and returning to the Russian and Japanese coasts as the cold Kamchatka Current. The convergence of these warm and cold currents in the Northwest Pacific provides ideal conditions for fish stocks in the area, creating some of the richest biological resources in the world. Half of the North Pacific's marine biological resources are found in the Sea of Okhotsk (Alekseev et al. 2006: 9). The North Pacific is home to bountiful pollock stocks, whitefish that are often processed into frozen items such as fillets and fish sticks. Anadromous fish stocks such as salmon migrate up rivers on the Kamchatka Peninsula and in the Amur River basin in parts of the Russian Far East and Northern China to spawn.

In Southeast Asia, the Mindanao Current feeds surface water into the Celebes Sea, before joining the North Equatorial Countercurrent. The currents in the warm, tropical waters of Southeast Asia are affected by seasonal monsoons. Freshwater rivers feed many of the seas in Southeast Asia. The shallow South China Sea basin, fringed with mangroves and seagrass, is home to some 2,500 species of marine fish and 500 species of coral (Wilkinson et al. 2005: 18–19). The archipelagic waters of Indonesia possess 10 to 14 percent of the world's coral reefs; its coastlines account for two-thirds of Southeast Asia's mangroves (De Vantier et al. 2005a: 17–18). The Sulu and Celebes Seas provide habitat to about 2,500 fish species and 400 coral species (De Vantier et al. 2005b: 20).

Oceania's expansive spray of 23 Pacific Island Nations (PINs) comprises Melanesia, Polynesia, and Micronesia. Despite small land masses and populations, the combined total area of their exclusive economic zones (EEZs) makes up 30 percent of worldwide EEZs (Dalzell et al. 1996: 397). The most fertile ocean areas lie in Melanesia, particularly off the coast of Papua New Guinea (South et al. 2004: 16). The rest of Oceania is characterized by warm surface water, which, because it does not mix well with deeper waters, provides poor nutrient content for marine species. The only significant species in this kind of ocean water are highly migratory species, such as tuna and billfish. Thus, Oceania hosts the world's largest tuna industry (Wauthy 1986: 39) (see Table 15.1).

Threats to fisheries resources in East Asia

East Asian fisheries are directly threatened by unsustainable exploitation. Because of Asia's large population and the importance of fish in the Asian diet, the demand for fish outstrips supply. Demand for fish also comes from developed countries outside Asia. Overfishing has caused the depletion of many larger fish species, leaving only smaller, lower quality species.

Table 15.1 Catch of major fish species in tons

FAO area 61 – Northwest Pacific Ocean

	2004	2005	2006	2007	2008	2009	2010
Alaska/Walleye pollock	1,169,079	1,241,737	1,314,775	1,514,336	1,612,751	1,652,790	1,938,611
Large hairtail	1,282,512	1,175,114	1,293,833	1,244,460	1,290,042	1,278,154	1,262,862
Japanese anchovy	1,629,486	1,481,442	1,509,170	1,390,596	1,265,763	1,068,334	1,202,212
Large yellow croaker	75,804	64,157	66,792	69,469	56,088	62,955	63,550
Yellow croaker	284,936	293,674	316,866	373,249	368,018	407,081	438,837
Chum salmon	256,292	234,366	239,759	233,906	200,159	251,048	43,396
Sockeye salmon	11,818	10,589	16,892	15,686	15,471	17,790	14,124
Red king crab	2,101	1,317	1,882	4,729	2,084	2,206	1,606
Blue king crab	2,391	4,024	4,587	4,831	4,774	3,267	2,915

FAO area 71 – Western Central Pacific Ocean

	2004	2005	2006	2007	2008	2009	2010
Skipjack tuna	1,217,868	1,267,094	1,373,627	1,494,890	1,360,818	1,598,167	1,484,863
Yellowfin tuna	322,858	368,162	364,046	395,220	497,310	418,556	480,333
Bigeye tuna	93,350	95,686	106,529	109,546	119,654	95,468	93,756
Albacore tuna	55,906	58,020	59,575	58,256	69,138	72,039	79,730

Source: FAO (2012b).

Poor management and enforcement of existing regulations governing fisheries plague East Asian countries. In the past half-century, the number of fishing vessels in East Asia has increased dramatically. Destructive fishing gear or practices also deplete resources. Inappropriate mesh sizes for fishing nets results in the capture of juvenile species, which are removed before they can reproduce. Trawling, whereby a fishing net is dragged along the seabed, catches non-target species and destroys fragile marine habitats. In some areas, fishermen use cyanide or dynamite to kill fish, which obliterates entire ecosystems.

Illegal, unreported, and unregulated (IUU) fishing is an enormous global problem that also impacts Asia. IUU fishing accounts for somewhere between 14 percent and 30 percent of global catch, for a value between $10 billion and $23.5 billion annually (Agnew et al. 2009). In both the Northwest Pacific Ocean and the Western and Central Pacific Ocean, it was estimated that IUU fishing accounted for one-third of the total catch between 2000 and 2003 – one of the highest rates in the world (Agnew et al. 2009: 2). IUU fishing is greater in areas where countries score low on governance indicators; according to one study, poor governance was the most statistically significant predictor of IUU fishing (MRAG 2005). Poaching is a considerable problem in the Russian Far East, as demand from other Northeast Asian countries, especially China, is fed by an economically depressed population in Russia seeking an income source. Demand for live reef fish in China and elsewhere in East Asia contributes to fisheries depletion in Southeast Asia. As coastal waters become empty, desperate fishermen venture further in search of fish, and often are caught illegally fishing in areas under the jurisdiction of other countries.

East Asian fishing vessels are active in fisheries beyond their coastal waters too. East Asia is home to significant distant water fishing (DWF) fleets – Japan, Taiwan, South Korea, and China – which fish in the EEZs of other countries or on the high seas. Fisheries access agreements allow DWF fleets to operate in host countries in exchange for licensing fees. China's central government is implementing policies to expand China's DWF operations around the world (Mallory 2013). The Russian government also wants to revive Russia's DWF industry. DWF fleets tend to operate in developing countries such as the PINs, affecting the economic development and food security of those countries (Hanich & Tsamenyi 2009).

The health of East Asian fisheries is threatened by pollution. Land-based pollution such as toxic wastes from factories and agricultural runoff is a serious problem. In China, pollution originating from land is the largest source of marine pollution because half of China's wastewaters are discharged to the sea (Zou 2005a). Pollutants such as oils; eutrophical salts such as nitrogen and phosphoric acid (found in fertilizers); organic substances, such as human waste; and heavy metals spill into the ocean from rivers, poisoning marine species. Air pollution, shipping, and mariculture also contribute to marine pollution. Toxic chemicals, such as mercury, move up the food chain and accumulate in the bodies of larger sea life, a process called biomagnification, and are harmful to human health when these species – for example large carnivorous fish including salmon and tuna – are eaten. Eutrophical salt and organic substances cause harmful algal blooms that suffocate marine life. Pollution can pose a transnational problem: pollution originating in China affects waterways in the sparsely populated Russian Far East.

Habitat loss or degradation is another threat to fisheries in East Asia. The Three Gorges Dam in China has diminished the flow of the Yangzi River, which brings freshwater to marine life along the continental shelf in the East China Sea. The dam has also altered the sediment flows that traditionally fed wetland areas at the mouth of the Yangtze River, which is now threatened by saltwater. Throughout Southeast Asia, mangroves, sea grass, and coral reefs – important habitats for fish – are being lost because of changing land use, aquaculture, and pollution. Asia accounts for over half of the world's existing mangrove species and about one-third of the earth's mangrove forest coverage, yet Asia lost about 25 percent of its mangrove forest between 1980 and 2005, more than any other part of the world (FAO 2007: 24). About 40 percent of Southeast Asia's coral reefs were lost between 1969 and 2009 (Hoegh-Guldberg et al. 2009: 3).

Climate change may as yet pose the largest threat to fisheries of all (Doney et al. 2012; Hoegh-Guldberg & Bruno 2010). About 25 percent of anthropogenic carbon dioxide is absorbed by the ocean, where it reacts with water and increases acidity, creating a chain reaction that decreases the amount of carbonate available to marine organisms like corals and other shell-forming species (Hoegh-Guldberg et al. 2007). Many of these organisms provide habitats or food for other sea life. The earth's rising average temperatures alter ocean currents, which may have significant detrimental effects on the productivity of the ocean by making habitats unsuitable to species adapted to previous conditions. The warming of the surface of the ocean combined with the change in currents causes water to circulate less, decreasing both the availability of oxygen, a condition known as hypoxia, as well as nutrients. Scientists have already documented 'ocean deserts' that have developed because of this lack of ocean water mixing, or upwelling. Higher temperatures melt polar ice, destroying habitats for arctic marine organisms. Rising sea levels contribute to the destruction of coastal habitats such as mangroves.

Importance of fisheries to East Asia

Fisheries are important to East Asian food security and economies. Not only do fish provide a source of food and income, they also contribute to the cultures and cuisines for Asian coastal peoples. East Asia provides the greatest share of the world's catch of marine fish. In 2010, catch in the Pacific Northwest Ocean reached 20,945 million tons, making area 61 by far the most productive FAO fishing area in the world (FAO 2012b: 500). The Western Central Pacific Ocean, area 71, was the second most productive area in the world in 2010, producing 11,710 million tons (FAO 2012b: 507).

The Asian population relies on fish as a source of protein more than any other part of the world. The greatest increase in fish consumption in the past half-century occurred in Asia. While global per capita fish consumption was 18.4 kg in 2009, per capita fish consumption in Northeast and Southeast Asia was 34.5 and 32.0 kg, respectively. At 85.4 million tons out of 126 million tons in 2009, Asia accounts for two-thirds of global fish consumption (FAO 2012a: 84–86). An increasing amount of fish demand is met by aquaculture, accounting for 47 percent of fish consumption in 2009, up from 5 percent in 1960. China's rise has had the largest impact on global fish consumption and production. China alone accounts for about half of Asia's fish consumption. Asia's fish consumption is projected to rise further as the middle class population develops and its demand for fish grows.

Fishing is a key contributor to the economies of East Asia. Marine fisheries contribute significantly to annual GDP in East Asian countries, particularly in Oceania (see Table 15.2). Fisheries are also an important source of employment in Asia. In 2010, of the world's 38.3 million people engaged in capture fisheries, 31.8 million – 83 percent of them – were in Asia (FAO 2012a: 41–43). About 9 million capture fishermen – close to one-quarter of the world's fishermen – live in China alone. Indonesia has 2.62 million people working in capture fisheries.

Within East Asia, there are a few fishing 'giants.' Japan, an island nation in which fish have long been a culinary mainstay, is one of the largest, most advanced fishing nations in the world. Even though today fewer Japanese are employed as fishermen, especially in Japan's depleted coastal waters, domestic demand for fish in Japan drives fishing efforts elsewhere. In 2009, Japan imported over half of the fish it consumed (FAO 2012a: 71). Much of this comes from other Asian nations.

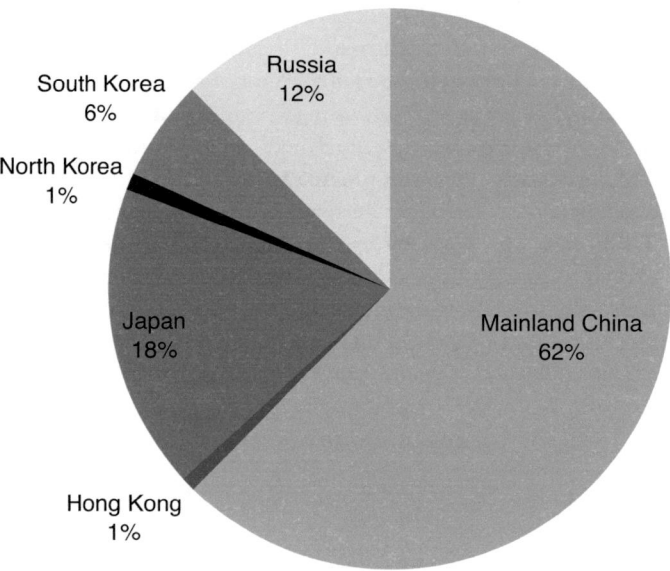

Figure 15.3 FAO fisheries area 61, percentage share of capture production by major fishing nations, 2010

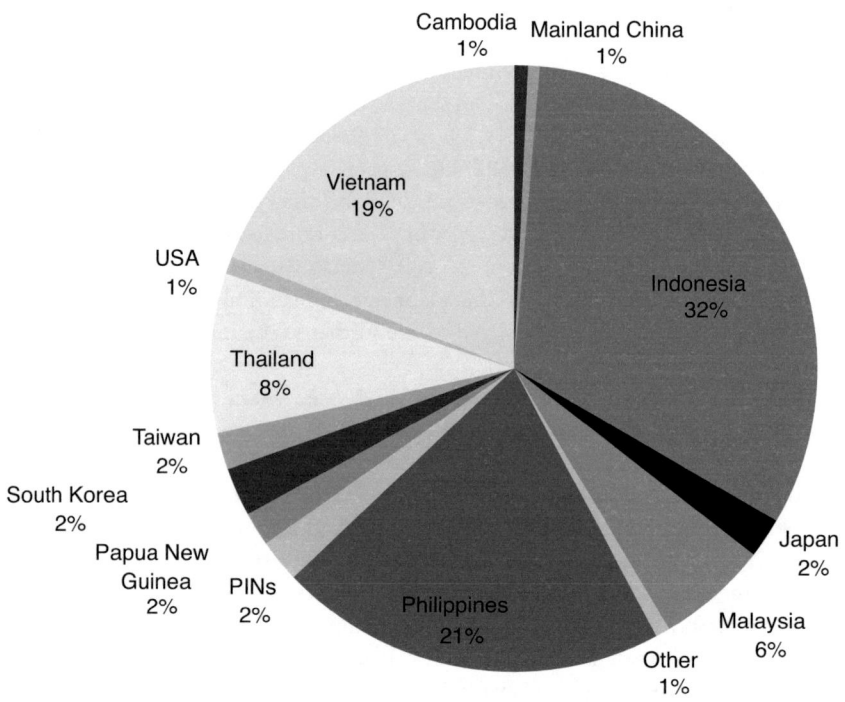

Figure 15.4 FAO fisheries area 71, percentage share of capture production by major fishing nations, 2010

Table 15.2 Key economic indicators on fisheries for East Asian countries

	Fish as % of animal protein (FAO 2012c)	Fish as % of total protein (FAO 2012c)	Fishing as % of GDP (FAO 2008)	Total catch in tons; * means estimated (FAO 2012b)	Total production (FAO 2012c)	Fish imports (FAO 2012c)	Fish exports (FAO 2012c)
NE Asia							
China	21.0	8.2	0.9	13,129,624	49,699,466	2,801,624	6,168,397
China HK	19.8	14.2	—	168,010	163,784	588,886	233,267
China Macau	25.2	16.8	—	1,500*	1,500	29,821	1,090
Japan	37.2	21.1	—	4,004,274*	4,903,173	3,607,320	592,918
North Korea	26.1	4.7	1.6	200,000*	269,050	47,358	58,724
South Korea	41.2	18.2	—	1,722,672	2,331,632	1,431,613	725,903
Russia	14.6	7.7	—	3,806,641	3,942,700	1,228,477	1,513,064
Chinese Taipei	18.5	8.8	—	851,307	1,056,380	337,792	727,767
SE Asia							
Brunei	13.6	17.1	11.4	2,272	2,863	8,398	306
Cambodia	59.0	14.6	1.9	85,094	514,200	8,037	49,361
Indonesia	53.7	20.6	1.1	5,035,294	6,837,036	208,220	930,304
Malaysia	39.1	18.2	3.0	1,428,882	1,563,942	245,422	514,614
Philippines	42.6	15.0	—	2,426,314	3,339,938	276,596	276,495
Singapore	22.8	12.4	—	1,732	5,688	302,708	78,260
Thailand	32.0	12.4	1.6	1,617,399	3,287,370	1,613,192	2,516,189
Timor-Leste	0.6	0.2	—	3,125*	350	—	—
Vietnam	35.3	13.7	9.5	2,226,600	4,836,580	116,009	1,438,751
Oceania							
Cook Islands	32.7	24.2	3.9	10,019	3,200	351	1,259
FSM	—	—	7.0	30,866*	16,990	9,000	2,370
Fiji	33.6	13.6	2.5	39,671*	48,857	35,197	40,889
Kiribati	54.9	27.6	57.7	44,599	21,603	1,743	2,648
Marshall Islands	—	22.9	62.8	59,730	60,409	948	22,959
Nauru	50.4	—	—	200*	220	0	0
Niue	47.9	24.0	1.1	113	200	—	—
Palau	—	—	—	1,000*	1,003	461	332
Papua New Guinea	13.4	7.2	4.6	211,007	263,960	28,355	143,207
Samoa	27.3	16.3	—	12,999	4,609	6,079	3,176
Solomon Islands	64.9	20.1	14.9	35,179*	31,272	2,744	17,282
Tonga	24.6	14.3	1.6	2,150*	2,549	2,380	1,321
Tuvalu	31.1	19.2	13.6	11,324*	2,201	52	100
Vanuatu	36.2	14.7	37.6	97,807*	85,387	3,035	73,565

China's fisheries production has grown substantially in the last half-century. According to official statistics, China is the world's largest producer of marine catch. However, some experts suspect that Chinese catch statistics may be artificially inflated for domestic political purposes (because local officials are promoted based on the economic performance of their respective jurisdictions) (Watson & Pauly 2001). Nonetheless, even with revised figures, China is either second behind or tied with Peru as the largest producer of wild catch, both maintaining a considerable lead over the next five producers (FAO 2012a). China has also become a global fish-processing hub.

Trade

Trade and the global economy are 'the single-most important factors determining international fisheries development and trends' but governance in these areas has been historically poor and remains weak (Hosch 2009: 31). Fish is a highly traded commodity globally; in value terms, fish is second only to fruits and vegetables. In 2010, trade in total fishery products reached about $109 billion (FAO 2012a: 68). China is the world's largest seafood exporter, with other developing countries such as Thailand and Vietnam providing significant exports. The world's largest seafood destination markets are developed regions: the European Union, the United States, and Japan.

Because trade is so important, many solutions to addressing the global fisheries crisis lie in this area, for example, through product certification, port-state measures that prohibit illegal landing of fish, fighting corruption, and regulating the capacity and economic performance of fishing fleets. Trade issues in fisheries management have gradually become a higher priority for governments, indicating that trade-related measures may serve as an effective fisheries management tool moving forward (Hosch et al. 2011: 192).

One key trade issue is fishing subsidies, which distort prices and incentives, creating false market signals and making fishing profitable where it may not have been otherwise. Fisheries subsidies fall into three categories: beneficial subsidies that contribute to the growth of fish stocks; harmful, capacity-increasing subsidies that have a deleterious effect on fisheries; and ambiguous subsidies, where the effect is unclear (Sumaila et al. 2010). Harmful subsidies include fuel subsidies; vessel construction and renovation programs; fishing port construction and renovation programs; price and marketing support, processing and storage infrastructure programs; fishery development projects and support services; tax exemptions; and foreign access agreements. Japan and China are the world's largest providers of subsidies to their fishing industries.

In the early 1990s, the negative effect of subsidies on sustainability of fisheries resources became clearer. By the end of the decade, the matter appeared on the World Trade Organization (WTO) agenda. While the declaration from the November 2001 Fourth Ministerial Conference of the WTO in Doha, Qatar, acknowledged that fisheries subsidies can be environmentally damaging, and ministers agreed to clarify and improve WTO measures on fisheries subsidies, few concrete measures have been taken to eliminate these subsidies (von Moltke 2011).

International fisheries law

The UN Convention on the Law of the Sea (UNCLOS) is the foundational international treaty that deals with ocean matters, including fisheries. The pinnacle of decades of negotiation, UNCLOS was signed in 1982 and entered into force globally in 1994. UNCLOS established exclusive economic zones (EEZs) extending 200 nautical miles from coastal baselines, which roughly correspond to continental shelves. Countries have sovereign jurisdiction over – but also a responsibility to conserve and protect – the marine resources found in their EEZs. The creation of EEZs placed the management of between 75 and 90 percent of the world's fisheries resources into the hands of national governments. The arrival of UNCLOS has proved challenging in East Asia because the convention imposes a contemporary legal framework over traditional fishing practices – fishing grounds that fishermen had used for centuries suddenly became off-limits as countries claimed their EEZs in Asia. This issue is at the root of some of the fisheries disputes seen particularly in the East China and South China Seas.

Several subsequent international agreements attend to fisheries issues that UNCLOS left unaddressed. The 1995 Fish Stocks Agreement deals with highly migratory and straddling fish stocks (United Nations 1995). While most of the world's fisheries lie within EEZs, the rest are found in the high seas, which are under the jurisdiction of no country. Some stocks straddle the EEZs of more than one country and/or the high seas. According to the agreement's guidelines, in such cases regional fisheries management organizations (RFMOs) shall be created to manage the fish stocks. The Fish Stocks Agreement also introduces the precautionary approach, which calls on states to be 'more cautious when information is uncertain, unreliable or inadequate' and create conservation and management measures that take into account these uncertainties.

The 1995 Compliance Agreement is most relevant to high seas fishing. The Compliance Agreement requires flag states to license high seas fishing vessels; monitor vessels fishing on the high seas so that they act in accordance with sustainable fishing practices; not allow authorization of vessels that act in violation of conservation measures; and share relevant vessel information with the FAO (FAO 1995a).

The 1995 Code of Conduct for Responsible Fisheries is a non-binding agreement that lays out guidelines for the sustainable and responsible use of fisheries (FAO 1995b). A 2009 study scores 53 of the top fishing countries according to Code of Conduct guidelines across nine indicators in six evaluation fields (Pitcher et al. 2009). Based on these measures, the study created three scores – good, passing, and failing – and found that no country in the world had a good score.

The Code of Conduct for Responsible Fisheries was followed by four non-binding international plans of action (IPOAs). The most well-known IPOA, the 2001 Plan of Action on IUU Fishing is a non-binding agreement that addresses IUU fishing (FAO 2001). The Plan was followed in 2009 by a binding Port State Measures Agreement, which would require port states to inspect fishing vessels and deny entry to those engaged in IUU fishing (FAO 2009). This agreement will take effect once 25 countries have ratified, accepted, approved or acceded to it. Three other IPOAs address seabirds, sharks, and fishing capacity (FAO 1999b) (see Table 15.3).

Regional fisheries bodies and other actors

Because East Asia is diverse geographically, biologically, economically, and culturally, countries are less regionally integrated than other parts of the world. East Asia has a surfeit of regional political, economic, and security organizations, many with overlapping mandates; no less is true with fisheries (see Table 15.4). Regional fisheries bodies (RFBs) include treaties, commissions or other supervisory or advisory bodies formed by at least three states to manage fisheries resources. According to the FAO's framework, RFBs can be categorized into three types: RFMOs, which are tasked with creating and implementing fisheries management measures; advisory bodies that provide management and scientific advice; and scientific organizations. In addition to RFBs, other environmental projects and arrangements as well as scientific networks exist (Lymer & Funge-Smith 2009). Organizations for economic cooperation also have an impact on fisheries.

Table 15.3 Countries/entities that have ratified international fisheries laws

	UNCLOS	Fish stocks	Compliance
NE Asia			
China	7 June 1996	—	—
Japan	20 June 1996	7 August 2006	24 April 2003
North Korea	—	—	—
South Korea	29 January 1996	1 February 2008	24 April 2003
Russia	12 March 1997	4 August 1997	—
SE Asia			
Brunei	5 November 1996	—	—
Cambodia	—	—	—
Indonesia	3 February 1986	28 September 2009	—
Malaysia	14 October 1996	—	—
Philippines	8 May 1984	—	—
Singapore	17 November 1994	—	—
Thailand	15 May 2011	—	—
Timor-Leste	8 January 2013	—	—
Vietnam	—	—	—
Oceania			
Cook Islands	15 February 1995	1 April 1999	10 October 2006
FSM	29 April 1991	23 May 1997	—
Fiji	10 December 1982	12 December 1996	—
Kiribati	24 February 2003	15 September 2005	—
Marshall Islands	9 August 1991	19 March 2003	—
Nauru	23 January 1996	10 January 1997	—
Niue	11 October 2006	11 October 2006	—
Palau	30 September 1996	26 March 2008	—
Papua New Guinea	14 January 1997	4 June 1999	—
Samoa	14 August 1995	25 October 1996	—
Solomon Islands	23 June 1997	13 February 1997	—
Tonga	2 August 1995	31 July 1996	—
Tuvalu	9 December 2002	2 February 2009	—
Vanuatu	10 August 1999	—	—

Civil society organizations (CSOs) such as research institutes, foundations, and non-profit organizations play a role in fisheries conservation as a source of information, expertise, funding, ideas, transparency, and pressure on other key actors, for example through bringing attention to issues for government officials. For example, the World Wildlife Fund helped launch the Marine Stewardship Council (MSC), a program that certifies fish caught sustainably. The Asian Fisheries Society is a CSO based in Asia that brings together scientists to discuss fisheries scientific and technical research.

East Asian organizations that work on fisheries suffer from fragmentation and poor funding. A summit that brought together the RFBs and other relevant organizations in the area to better integrate efforts would contribute to improving fisheries governance in East Asia (Chung 2010).

Table 15.4 Membership/participation by Asian countries in RFBs and fisheries-related arrangements

Column groups: **Regional fisheries bodies** — Regional fisheries management organisations (NPAFC, CCBSP, WCPFC, SPRFMO); Fisheries advisory bodies (APFIC, FFA, SEAFDEC); Scientific bodies (INFOFISH, SPC). **Regional arrangements/cooperation/networks/projects** — Fisheries/environmental arrangements (COBSEA, PEMSEA, NOWPAP, SPREP, YSLME, CTI, RPOA); Scientific networks (IOC/WESTPAC, PICES); Economic cooperation (APEC, ASEAN, PIF).

Country	NPAFC	CCBSP	WCPFC	SPRFMO	APFIC	FFA	SEAFDEC	INFOFISH	SPC	COBSEA	PEMSEA	NOWPAP	SPREP	YSLME	CTI	RPOA	IOC/WESTPAC	PICES	APEC	ASEAN	PIF
NE Asia																					
China		X	X	(S)	X					X	X	X		X		X	X	X	X		
Japan	X	X	X		X		X				X	X				X	X	X	X		
North Korea								X			X			X			X				
South Korea	X	X	X	X	X					X	X	X		X			X	X	X		
Russia	X	X		X								X						X	X		
Taipei			X	X															X		
SE Asia																					
Brunei DS							X				X					X			X	X	
Cambodia					X		X	X		X	X					X				X	
Indonesia			X		X		X	X		X	X				X	X	X		X	X	
Laos							X				X									X	
Malaysia					X		X	X		X	X				X	X	X		X	X	
Philippines			X		X		X	X		X	X				X	X	X		X	X	
Singapore							X			X	X					X	X		X	X	
Thailand					X		X	X		X	X					X	X		X	X	
Timor-Leste					X						X				X	X					
Vietnam					X		X			X	X					X	X		X	X	

continued . . .

Table 15.4 Continued

| Country | Regional fisheries bodies | | | | | | | | | Regional arrangements/cooperation/networks/projects | | | | | | | | | | | |
| | Regional fisheries management organisations | | | | Fisheries advisory bodies | | | Scientific bodies | | Fisheries/environmental arrangements | | | | | | | Scientific networks | | Economic cooperation | | |
	NPAFC	CCBSP	WCPFC	SPRFMO	APFIC	FFA	SEAFDEC	INFOFISH	SPC	COBSEA	PEMSEA	NOWPAP	SPREP	YSLME	CTI	RPOA	IOC/WESTPAC	PICES	APEC	ASEAN	PIF
Oceania																					
Australia			X	X	X	X			X	X			X			X	X		X		X
Cook Islands			X	X		X			X				X			X	X				X
Fiji			X			X			X				X				X				X
FSM			X			X			X				X								X
Kiribati			X			X			X				X								X
Marshall Islands			X			X			X				X								X
Nauru			X			X			X				X								X
New Zealand			X	X	X	X			X				X				X		X		X
Niue			X			X			X				X								X
Palau			X			X			X				X								X
Papua New Guinea			X			X		X	X				X		X	X			X		X
Samoa			X			X			X				X				X				X
Solomon Islands			X			X		X	X				X		X		X				X
Tonga			X			X			X				X				X				X
Tuvalu			X			X			X				X								X
Vanuatu			X			X			X				X				X				X

				Other		
France		X		X	X	
United Kingdom			X		X	
United States	X	X	(S)	X	X	X	X

Notes: X = member.

(S) = signed but not ratified.

North Pacific Anadromous Fish Commission (NPAFC); Convention on the Conservation and Management of Pollock Resources in the Central Bering Strait (CCBSP); Western and Central Pacific Fisheries Commission (WCPFC); South Pacific Regional Fisheries Management Organisation (SPRFMO).

Asia-Pacific Fisheries Commission (APFIC); Pacific Islands Forum Fisheries Agency (FFA); Southeast Asian Fisheries Development Centre (SEAFDEC).

INFOFISH; Secretariat of the Pacific Community (SPC).

Coordinating Body on the Seas of East Asia (COBSEA); Partnership in Environmental Management for the Seas of East Asia (PEMSEA); Northwest Pacific Action Plan (NOWPAP); Secretariat of the Pacific Regional Environment Programme (SPREP); Yellow Sea Large Marine Ecosystem (YSLME); Coral Triangle Initiative (CTI); Regional Plan of Action for Responsible Fishing (RPOA).

IOC Sub-Commission for the Western Pacific (WESTPAC); North Pacific Marine Science Organisation (PICES).

Asia Pacific Economic Cooperation (APEC); Association for Southeast Asian Nations (ASEAN); Pacific Islands Forum (PIF).

Source: Lymer & Funge-Smith (2009), updated and amended by the author.

The Northwest Pacific and Russian Far East

Fishing is a significant portion of the Russian Far East economy. About 70 percent of Russia's marine biological resources come from Russia's Far East seas (Shvarts & Simonov 2011). A natural resource in high demand by Japan, Korea, China, and the United States, fish contribute as much as half of export revenues in the Russian Far East. Along with oil and gas, the Russian fishing industry is the part of the economy with the most involvement from foreign companies (Allison 2001). Russia now exports approximately 90 percent of its pollock catch (Shvarts & Simonov 2011: 112)

The rise and subsequent collapse of the Soviet Union had an enormous impact on Russia's fisheries management, industry structure, and consumption patterns. Soviet fishermen caught less than 500,000 tons of fish in 1950. Yet, in the mid-1950s, the Soviet Union initiated a massive state program of shipbuilding and fishing around the world. By 1970, annual catch reached over 6 million tons. The Soviet Union's fishing fleet became the largest in the world, and its total catch just second to Japan's. Over 90 percent of the fish were sold domestically, and per capita consumption of seafood rose to over 20 kg annually (Allison 2001: 70).

Institutional changes as the Soviet Union transitioned from a planned economy caused upheaval in the fishing industry. Fisheries personnel shrank and other state agencies vied for control over the fishing industry because of its significant wealth-generating potential. Internal conflict over fisheries prevented the establishment of long-term, reliable and effective institutions for resource management. Corruption and crime flourished. Because domestic demand for fish collapsed and capital investment dried up, unemployment in the fisheries industry rose. DWF operations were almost completely terminated. Unemployment led to poaching as local populations worked to improve their economic conditions by satisfying the huge demand from other countries in East Asia. Catch in the Russian Far East dropped from 4.5 million tons in 1990 to 2.3 million tons in 1994.

As the global fisheries decline became evident, foreigners quickly seized opportunities in the Russian fishing industry in the mid-1990s. To avoid the costly and overly bureaucratic Russian ports, much of the catch was landed in foreign ports, benefiting overseas economies instead. Russians often imported fish that was caught in Russian EEZs but landed elsewhere (Jørgensen 2009: 90–91). With the advent of highly advanced foreign vessels in Russian waters has come the depletion of fisheries resources. Catch rebounded, peaking in the late 1990s at nearly 10 million tons annually, and then sank again as stocks dwindled.

The central problems in Russia's fisheries governance remain poor regulation and control over fisheries resources. Problems with quota allocation allowed too many actors into the market. In the early 2000s, Russia's 5,000 fishing companies owned 0.7 vessels per company on average, meaning that many had no vessels, but profited from selling their quotas illegally to other fishing companies (Anferova et al. 2005). Catches in Russia far exceed sustainable quotas. For example, even though the catch quota for crab was 1,500 tons, in 2004 Russian fishing enterprises sold more than 30,000 tons of crab to Japan (Lukin 2007). Between 2003 and 2005 actual catches of Russian sockeye salmon were 60 to 90 percent higher than reported catch, between 8,000 and 15,000 tons annually, for a value of $40 to $74 million (Clarke et al. 2009). In 2007, the total loss to the Russian economy because of IUU fishing was estimated at $3 billion annually (Lukin 2007: 28). Russian IUU fishing in most recent years is driven by Chinese demand, but measures that address illegal fishing are lacking or not enforced. Trade data reporting between Russia and China is poor, and problems with customs procedures exist.

Reforms beginning in 2007 attempt to reduce the role of foreign enterprise in Russian fisheries. As of January 2009, all catches from Russia's EEZ must be landed in Russia, and quotas are allocated for 10 years in order to encourage long-term investment in the fisheries sector and only to companies that actually own and operate fishing vessels themselves (Jørgensen 2009: 97–98). Issues remain, however, as true ownership can be difficult to determine (Interfax 2012). In September 2010, China and Russia signed an inter-departmental memorandum of understanding to combat illegal fishing. The MOU aims to monitor China's seafood imports from the Russian Far East, establishing a mechanism of regular bilateral cooperation meetings on the issue (Ministry of Agriculture Bureau of Fisheries 2012).

Yellow Sea and East China Sea

UNCLOS has affected how East Asian countries manage their fish stocks. Since UNCLOS established EEZs at 200 nautical miles (nm) from coastal baselines, in cases where less than 400 nm separates countries, agreements must be worked out between neighboring countries on delimiting EEZs and how to share resources. This situation has negatively impacted fisheries in waters shared with other nations because several countries are using these fisheries while these arrangements are being worked out. Some Northeast Asian countries have signed bilateral fisheries agreements covering these issues (see Table 15.5). However, even though these bilateral agreements have contributed to managing fish stocks, they are not comprehensive enough – a management structure that covered both the Yellow Sea and the East China Sea would be better.

China–Japan

Both Japan and China fished in the East China Sea before World War II (WWII), but Japan's fishing ability was far greater than that of China. Japan began fishing further into Chinese coastal waters in the 1920s. Fisheries resources in the East China Sea were showing early signs of depletion in the 1920s and 1930s, but they rebounded during WWII when normal fishing was interrupted (Muscolino 2009). After WWII ended, and particularly after Japan's peace treaty entered into force in 1952, Japanese vessels returned to fishing off the coast of China. At the same time, in the 1950s, China's new Communist government was eager to reconstruct its economy after years of devastating warfare, and encouraged Chinese fishing as well. Chinese fishermen complained about unfair competition from better equipped vessels in Chinese fishing grounds. The Chinese state responded by arresting 158 fishing vessels and 1,919 fishermen between December 1950 and July 1954 (Zou 2005b: 90). In response to this increasing friction, the Japan–China Fisheries Council and the China Fisheries Association reached an agreement in June 1955. Because China and Japan had no diplomatic relations at the time, this agreement and two more agreements that followed in 1963 and 1965 were nongovernmental in nature. These agreements established measures such as fishing zones, fishing quotas, seasonal moratoria, and limits on the numbers and horsepower of fishing vessels. Once China and Japan normalized

Table 15.5 List of bilateral fisheries agreements

Countries	Signed	Entered into force
China–Japan	11 November 1997	1 July 2000
Japan–South Korea	28 November 1998	22 January 1999
China–South Korea	3 August 2000	30 June 2001
China–Vietnam	25 December 2000	30 June 2004

relations in 1972, they worked out the first government-level fisheries agreement, which entered into force in December 1975. This agreement was similar to the previous nongovernmental agreements, although it was more rigorous in its specifications. The agreement was revised in 1978 and 1985.

Because UNCLOS made vast changes to international ocean law and governance, China and Japan needed a new agreement that reflected these changes. The Agreement on Fisheries between the Government of the People's Republic of China and the Government of Japan entered into force in 2000. This most recent agreement is provisional, given that China and Japan have not yet finalized their boundary delimitations. Most notably, China and Japan dispute sovereignty over the Senkaku/Diaoyu Islands, and the fisheries agreement does not cover these islands. However, in the agreed on areas, China and Japan both have jurisdiction over their EEZs, with reciprocal fishing access by the other country, which can apply for a fishing license and is subject to the laws and regulations of the host country. In between the agreed on zones lies a Provisional Measures Zone (PMZ) that covers the areas where each country's EEZ claims overlap. The PMZ is governed by a joint fisheries committee, which determines fishing quotas, number of vessels allowed access, the species that may be caught, and other conservation and management measures.

The Senkaku/Diaoyu Islands have become a flash point in Sino-Japanese relations. While the issue is really about a territorial dispute, confrontation is often sparked through incidents that involve fisheries. Although the islands have been in dispute for decades, the issue heated up most recently starting in September 2010, when a Chinese fishing vessel collided with two Japanese coast guard vessels. Since then, both sides have sent fishing vessels and patrol forces to the islands in an effort to assert sovereignty over them. Taiwan also claims the islands, which it calls Diaoyutai. In April 2013, Japan and Taiwan signed a fisheries agreement to manage fishing in the disputed area.

Japan–South Korea

Japan and South Korea have a long history of struggle to manage the fisheries resources found between their two landmasses. The first agreement between these two countries dates to 1442, when South Korea tried to regulate Japanese fishing activities in its waters (Pak 2000: 57). Japan relinquished control over Korea at the end of WWII, however Japanese fishing vessels still operated in Korean coastal waters. In 1952, South Korea issued a presidential proclamation claiming jurisdiction over a zone extending 200 nm from the South Korean coast throughout which Japanese vessels were forbidden. Like China, throughout the next two decades South Korea arrested numerous Japanese fishing vessels in its coastal waters. The fishing disputes between these two countries led to a government-level agreement between Japan and South Korea in 1965, after 13 years of negotiation. This agreement created exclusive fishery zones off the coasts of both countries, a joint regulation zone, a flag state responsibility system, and limits on the size and number of fishing vessels. A joint fisheries committee was also created to govern this system. As the South Korean fishing fleet grew and expanded into Japanese waters, a new measure, the Voluntary Fishing Regulation, was signed in 1980 to manage the fishing of both fleets in each other's waters (Ou & Tseng 2010: 283).

UNCLOS provided the basic principles for the current bilateral fisheries agreement. Eight rounds of bilateral negotiations took place beginning in 1996. The countries disagreed over EEZ delimitation, and the main point of contention in the negotiations was the territorial dispute over Dokdo/Takeshima, a small group of islets situated between them. Japan unilaterally gave notice of terminating the 1965 fishing agreement in one year's time, which provided impetus to finalize a new agreement. The Agreement on Fisheries between Japan and the Republic of Korea entered into force in January 1999.

The new agreement set up two intermediate zones in between the two countries called Provisional Waters Zones (PWZs). The PWZs are similar to the PMZs with China.[1] The Dokdo/Takeshima islets are situated within the northeastern PWZ, the sovereignty over which shall be decided at a later date. A joint fisheries agreement was also created to decide on the details of the fishing operations of the two countries, such as quotas, number of fishing vessels, and harvestable species.

China–South Korea

China and South Korea both fish in the Yellow Sea and in the East China Sea. China and South Korea did not normalize relations until 1992, which prevented them from having a government-level fisheries agreement before that time. Historically, both China and South Korea had weaker fishing fleets compared to Japan, and the concern of both these countries through the 1950s was limiting these advanced Japanese fleets (Xue 2005: 183). However, Chinese fishermen gradually moved into South Korean coastal waters in the second half of the twentieth century, causing friction with South Korean fishermen and overfishing of stocks (Ou & Tseng 2010: 282–283). Bilateral fisheries negotiations between China and South Korea commenced in December 1993, though the process was hindered by maritime boundary disputes.

The Agreement on Fisheries between the People's Republic of China and the Republic of Korea entered into force in June 2001. The agreement is similar to that of other agreements in the East China Sea in that China and Korea have also established a joint fisheries committee that manages fishing operations in a PMZ situated in the Yellow Sea between the two countries. However, the agreement also established two transitional zones on either side of the PMZ that became the EEZs of the respective countries four years into the agreement.

Because the PMZ is heavily fished and stocks are dwindling, Chinese vessels are often caught illegally fishing in the South Korean EEZ. While the number of detained vessels had been falling since 2005, the number was again increasing since 2010, with 370 vessels seized in 2010 and 534 in 2011 (Korea Herald Online 2012). The run-ins have turned violent and even fatal, with both Chinese fishermen and South Korean coast guard officials becoming casualties. Chinese fishermen have also been accused of destroying the fishing gear of South Korean fishermen.

The South China Sea

The fisheries issues in the South China Sea (SCS) are intertwined with the maritime territorial disputes in the area. Resource management of only a small portion of the SCS has been worked out. The SCS is bordered by China, Vietnam, Malaysia, Indonesia, Brunei, the Philippines, and Taiwan. These entities all have various sovereignty or jurisdictional claims to the islands and/or waters of the SCS. The larger, more notable and disputed island groups in the SCS include the Paracels and the Spratlys, but other formations include the Pratas Islands, Macclesfield Bank, and Scarborough Reef. In addition to the rich fisheries resources in the area, the SCS is estimated to have abundant hydrocarbon resources in the seabed; as a potential source of wealth for whichever country has jurisdiction over the area, these resources motivate the maritime claims. Countries also vie for control of the sea lanes of communication (SLOCs) as the area is traversed by numerous vessels essential to the economies and security of the Asian nations.

This situation has several implications for fisheries. First of all, the significance of the fisheries resources in the area has been overshadowed by the immediate hunger for the hydrocarbon resources. Unfortunately, this means that the importance of fisheries to food security in the region is being overlooked, and the potential long-term wealth that could be generated by managing fisheries better is not harnessed. The true volume of the hydrocarbon resources in the area is unknown and these resources are ultimately finite in any case, whereas fisheries are renewable if managed well. Second, fisheries serve as a proxy for the maritime disputes in the area. Because fisheries law enforcement forces such as coast guards are paramilitary in nature, they are less directly confrontational to other countries in the way naval forces would be. Countries can use their coast guard forces to assert their control over disputed areas. Moreover, nationalism motivates the behavior of some actors in the area. This is one of the reasons many of the flare-ups seen in the news involve fishing, for example because one country detains the fishing vessels of another country, or fishing vessels harass vessels of another country. Third, because the legal framework on which to manage the SCS has not been agreed, no RFMO has been created to manage fisheries resources more optimally.

China–Vietnam

The only agreement on fisheries in the SCS covers Chinese and Vietnamese fishing in the Gulf of Tonkin. Both countries long relied on the waters in the Gulf of Tonkin for its rich fisheries resources. China and Vietnam signed their first bilateral fisheries agreement in 1957, which arranged for shoreline fishing to be regulated by the coastal state, but allowed fishing by both sides in the middle of the gulf. The agreement was signed in order to limit the activities of Chinese fishermen in Vietnamese coastal waters. A second agreement was signed in 1963, which limited Vietnamese fishing in Chinese coastal waters as well (Xue 2005: 204–228). Relations between China and Vietnam deteriorated in the late 1970s, fishing conflicts in the Gulf of Tonkin erupted, and resources declined. Diplomatic relations were reestablished in 1991, but the countries still disagreed over the boundary delimitation in the Gulf of Tonkin, which affected fisheries management.

In the mid-1990s, both countries ratified UNCLOS and worked toward agreements on boundary delimitation and on fisheries. China and Vietnam signed an agreement on their land border dispute in 1999, which contributed toward agreement in the other two areas. On 25 December 2000, China and Vietnam signed an agreement on their maritime boundary delimitation, as well as the Agreement between the Government of the People's Republic of China and the Government of the Socialist Republic of Vietnam on Fisheries Cooperation for the Gulf of Tonkin.

The fisheries agreement between China and Vietnam set up a management mechanism for the fisheries in the Gulf of Tonkin. The agreement created a Joint Fisheries Committee (JFC), tasked with a number of responsibilities, including agreeing on measures to conserve and manage resources and handling some dispute resolution. The agreement set up three different fishing zones: a Common Fishery Zone (CFZ), a Transitional Fishery Zone (TFZ), and a Buffer Zone (BZ). The CFZ straddles the EEZ delimitation between China and Vietnam and allows vessels from both countries to fish throughout. The JFC assesses the fisheries resources through joint surveys, and agrees on the number of vessels allowed in the CFZ based on these assessments. The coastal country grants fishing licenses to its own vessels, and each country patrols its side of the CFZ, though the other country must be notified of any fishing violations by one of its vessels in the other country's CFZ. The TFZ is an area in which both countries reduce fishing in the other country's EEZ by 25 percent annually for four years until catch equals zero. The BZ is an area in which small vessels, which often lack positioning technology, are issued warnings for fishing on the wrong side but are not detained.

The Sino-Vietnamese fishing agreement is distinctive in that it is a permanent agreement based on agreed maritime boundary delimitation. The JFC has more authority than those of other bilateral agreements in that it has power to deal with all issues related to fisheries management, including dispute resolution. The agreement also provides for cooperation in the area of scientific research.

Oceania

When the Western and Central Pacific Fisheries Commission (WCPFC) was established in 2004, it was heralded as the most progressive RFMO yet, incorporating 'state of the art' principles from the Fish Stocks Agreement, such as the precautionary principle (Hanich 2011: 17). The WCPFC covers the straddling and highly migratory fish stocks in Oceania, one of the most complex fisheries regions in the world in terms of diversity of fish species, types of fishing gear used, and the kinds of stakeholder involved. The fishing parties involved in the WCPFC can be categorized into two main coalitions: the PINs, and the DWF nations of China, Japan, South Korea, Taiwan, and the United States. Unlike the PINs, the DWF nations have the capital-intensive technology to fish highly migratory stocks.

Skipjack, bigeye, and yellowfin tuna are the more commercially significant species covered by the WCPFC. Two kinds of commercial fishing method predominate in the tuna industry. Skipjack tuna, destined for the canning industry, are mainly caught with purse seining vessels, which use large nets to haul in relatively significant amounts of fish. Bigeye and yellowfin species, which command high prices on the sashimi market, are caught with longline vessels, which send fishing lines with hooks deep into the ocean. Purse seining tends to take place in the EEZs of the PINs whereas longlining tends to take place in both the EEZs and on the high seas (Parris 2010: 11).

Conflict exists between the purse seine and longline vessels because of the impact that purse seining has on particularly bigeye but also yellowfin tuna. Purse seine vessels often use fish aggregating devices (FADs) – natural or artificial floating items – an increasingly popular method of attracting fish because tuna tend to aggregate around floating objects. While purse seining targets skipjack, the more slowly maturing juvenile yellowfin and bigeye are often taken as bycatch before they are able to reproduce or reach a more economically profitable size.[2] This trend, combined with excess fishing capacity, has resulted in overfishing of bigeye and full exploitation of yellowfin. If use of FADs was reduced or eliminated, while the purse seining industry would suffer losses, it could result in net benefits to the Pacific Island countries of an estimated $150 to $400 million (Bailey et al. 2011). Even though the industry is currently profitable, it is estimated that in over 50 years net economic losses would amount to at least $3.4 billion in present value terms, if no changes are made in the industry to increase conservation (Gillet 2009: xxxii).

Domestic and local initiatives

As transnational resources, fisheries require an element of international management, however domestic and local-level initiatives are also necessary. East Asian governments have made efforts to protect or rebuild their fish stocks. Hong Kong, a prominent center of fish consumption and trade with badly depleted local fish stocks, implemented a trawling ban in local waters from January 2013 to allow fish stocks to recover (Meigs 2013). Palau is a thriving marine ecotourism destination.

Mainland China has put in place measures such as seasonal moratoria and vessel decommissioning. However, the combination of depleted resources and increased fishing restrictions drives illegal fishing elsewhere, as fishermen venture farther to find fish, exacerbating maritime territorial conflicts with neighboring countries. China unilaterally enforces a fishing ban to 12°N in the South China Sea, which has potential to improve stocks but alienates other countries that have historically fished in the same area (Goldstein 2013).

Conclusion

Fish are a symbol of abundance in many Asian countries. Getting fisheries right in East Asia is essential for the region's economic and food security. The effects of overfishing and ecosystem degradation even affect military security.

Despite the proliferation of agreements and organizations, gaps remain in the governance of Asia's shared fisheries resources. Bilateral agreements are not sufficient to sustainably manage fish stocks. Because fish stocks migrate between the Yellow Sea and East China Sea, an RFMO that covered the whole region and included China, North Korea, South Korea, Taiwan, and Japan would be more effective (Ou & Tseng 2010; Schofield & Townsend-Gault 2011). This is also true of the South China Sea, where an RFMO would include China, Vietnam, Malaysia, Indonesia, Brunei, Taiwan, and the Philippines. Agreements that shelved territorial disputes but addressed fisheries would contribute to improving relations between countries in the region (Nguyen 2012).

Better trade measures would improve fisheries management, not only in Asia but globally, since the United States and the European Union are large fish importers. Traceability and certification systems would make it easier to identify fish. Enacting port states measures legislation would make it illegal to import IUU fish or mislabel fish. Removing harmful subsidies would eliminate artificial incentives to fish. Such efforts are crucial to the long-term survival of Asia's fish stocks.

Bibliography

Agnew, D.J., Pearce, J., Pramod, G., Peatman, T., Watson, R., Beddington, J.R., et al. (2009) 'Estimating the Worldwide Extent of Illegal Fishing', *PLoS ONE* 4(2), e4570.

Alekseev, A.V., Khrapchenkov, F.F., Baklanov, P.J., Blinov, Y.G., Kachur, A.N. & Medvedeva, I.A., (2006) Oyashio Current, GIWA Regional Assessment 31, UNEP, University of Kalmar, Kalmar, Sweden.

Allison, A. (2001) 'Sources of Crisis in the Russian Far East Fishing Industry', *Comparative Economic Studies* 43(4), 67–93.

Anferova, E., Veternaa, M. & Hannesson, R. (2005) 'Fish Quota Auctions in the Russian Far East: A Failed Experiment', *Marine Policy* 29, 47–56.

Bailey, M., Sumaila, R. & Martell, S.J.D. (2011) Can Cooperative Management of Tuna Fisheries in the Pacific Solve the Growth Overfishing Problem? Working Paper #2011–01, Fisheries Centre, University of British Columbia.

Chuenpagdee, R. (2005) 'Challenges and Concerns in Capture Fisheries and Aquaculture', in J. Kooiman, S. Jentoft, R. Pullin & M. Bavinck (Eds.), *Fish for Life: Interactive Governance for Fisheries* (Amsterdam University Press).

Chung, S.-Y. (2010) 'Strengthening Regional Governance to Protect the Marine Environment in Northeast Asia: From a Fragmented to an Integrated Approach', *Marine Policy* 34, 549–556.

Clarke, S., McAllister, M.K. & Kirkpatrick, R.C. (2009) 'Estimating Legal and Illegal Catches of Russian Sockeye Salmon from Trade and Market Data', *ICES Journal of Marine Science* 66, 532–545.

Dalzell, P., Adams, T.J.H. & Polunin, N.V.C. (1996) 'Coastal Fisheries in the Pacific Islands', *Oceanography and Marine Biology: An Annual Review* 34, 397.

De Vantier, L., Wilkinson, C., Lawrence, D. & Souter, D. (Eds.) (2005a) Indonesian Seas, GIWA Regional Assessment 57, UNEP, University of Kalmar, Kalmar, Sweden, 17–18.

De Vantier, L., Wilkinson, C., Souter, D., South, R., Skelton, P. & Lawrence, D. (2005b) Sulu-Celebes (Sulawesi) Sea, GIWA Regional Assessment 56, UNEP, University of Kalmar, Kalmar, Sweden, 20.

Doney, S.C., Ruckelshaus, M., Duffy, J.E., Barry, J.P., Chan, F., English, C.A., et al. (2012) 'Climate Change Impacts on Marine Ecosystems', *Annual Review of Marine Science* 4, 11–37.

Food and Agriculture Organization (FAO) (1995a) The Agreement to Promote Compliance with International Conservations and Management Measures by Fishing Vessels on the High Seas, Rome, FAO.

FAO (1995b) The Code of Conduct for Responsible Fisheries (CCRF), Rome, FAO.

FAO (1998) The State of World Fisheries and Aquaculture 1998, Rome, FAO.

FAO (1999a) Numbers of Fishers 1970–1996, Fishery Information Data and Statistics Unit, Rome, FAO.

FAO (1999b) International Plan of Action for Reducing Incidental Catch of Seabirds in Longline Fisheries; International Plan of Action for the Conservation and Management of Sharks; International Plan of Action for the Management of Fishing Capacity, Rome, FAO.

FAO (2001) The International Plan of Action to Prevent, Deter and Eliminate Illegal, Unreported and Unregulated Fishing, Rome, FAO.

FAO (2003) FAOSTAT: FAO Statistical Database, Rome, FAO.

FAO (2007) The World's Mangroves: 1980–2005, FAO Forestry Paper 153, Rome, FAO.

FAO (2008) Status and Potential of Fisheries and Aquaculture in Asia and the Pacific 2008, Bangkok, FAO APFIC.

FAO (2009) Agreement on Port State Measures to Prevent, Deter and Eliminate Illegal, Unreported and Unregulated Fishing, Rome, FAO.

FAO (2012a) The State of World Fisheries and Aquaculture 2012, Rome, FAO.

FAO (2012b) 2010 FAO Fisheries Yearbook on Fishery and Aquaculture Statistics: Capture Production, Rome, FAO.

FAO (2012c) Section 2: Food Balance Sheets and Fish Contribution to Protein Supply by Country, 2010 FAO Fisheries Yearbook Fishery and Aquaculture Food Balance Sheets, Rome, FAO.

Gillet, R. (2009) *Fisheries in the Economies of the Pacific Island Countries and Territories* (Asian Development Bank).

Goldstein, L. (2013) 'China Fisheries Enforcement: Environmental and Strategic Implications', *Marine Policy* 40, 187–193.

Hanich, Q. & Tsamenyi, M. (2009) 'Managing Fisheries and Corruption in the Pacific Islands Region', *Marine Policy* 33, 386–392.

Hanich, Q.A. (2011) *Interest and Influence: A Snapshot of the Western and Central Pacific Tropical Tuna Fisheries* (University of Wollongong Press).

Hoegh-Guldberg, O. & Bruno, J.F. (2010) 'The Impact of Climate Change on the World's Marine Ecosystems', *Science* 328, 1523–1528.

Hoegh-Guldberg, O., Mumby, P.J., Hooten, A.J., Steneck, R.S., Greenfield, P., Gomez, E., et al. (2007) 'Coral Reefs under Rapid Climate Change and Ocean Acidification', *Science* 318, 1737–1742.

Hoegh-Guldberg, O., Hoegh-Guldberg, H., Veron, J.E.N., Green, A., Gomez, E.D., Lough, J., et al. (2009) *The Coral Triangle and Climate Change: Ecosystems, People and Societies at Risk* (World Wildlife Fund Australia).

Hosch, G. (2009) Analysis of the Implementation and Impact of the FAO Code of Conduct for Responsible Fisheries since 1995, FAO Fisheries and Aquaculture Circular, No. 1038, Rome, FAO.

Hosch, G., Ferraro, G. & Failler, P. (2011) 'The 1995 FAO Code of Conduct for Responsible Fisheries: Adopting, Implementing or Scoring Results?', *Marine Policy* 35.

Interfax (2012) 'Far East Fishermen Deny Affiliation with Chinese Business', *Russia and CIS Food and Agriculture Weekly*, 18 April.

Jørgensen, A.K. (2009) 'Recent Developments in the Russian Fisheries Sector', in E. Wilson Rowe (Ed.), *Russia and the North* (University of Ottawa Press).

Korea Herald Online (2012) 'Editorial: Blame for Sea Tragedy', *Korea Herald Online*, 19 October.

Lukin, A.L. (2007) 'Environmental Security of Northeast Asia: A Case of the Russian Far East', *Asian Affairs* 34, 1.

Lymer, D. & Funge-Smith, S. (Eds.) (2009) *Handbook on Regional Fishery Bodies and Arrangements in Asia and the Pacific* (RAP Publications).

Mallory, T.G. (2013) 'China's Distant Water Fishing Industry: Policies and Implications', *Marine Policy* 38, 99–108.

Marine Resources Assessment Group Ltd (MRAG) (2005) Review of Impacts of Illegal, Unreported and Unregulated Fishing on Developing Countries, Report prepared by MRAG for the UK's Department for International Development (DFID), with support from the Norwegian Agency for Development Cooperation (NORAD), July.

Meigs, D. (2013) 'Fishermen Evolve after Trawl Ban', *China Daily*, 20 May.

Ministry of Agriculture Bureau of Fisheries (2012) *2012 China Fisheries Yearbook* (China Agricultural Press).

Muscolino, M.S. (2009) *Fishing Wars and Environmental Change in Late Imperial and Modern China* (Harvard University Press).

Nguyen Dang, T. (2012) 'Fisheries Co-operation in the South China Sea and the (Ir)relevance of the Sovereignty Question', *Asian Journal of Law* 2, 59–88.

Ou, C.-H. & Tseng, H.-S. (2010) 'The Fishery Agreements and Management Systems in the East China Sea', *Ocean and Coastal Management* 53, 279.

Pak, H.-G. (2000) *The Law of the Sea and Northeast Asia: A Challenge for Cooperation* (Martinus Nijhoff Publishers).

Parris, H. (2010) 'Is the Western and Central Pacific Fisheries Commission Meeting its Conservation and Management Objectives?', *Ocean & Coastal Management* 53.

Pitcher, T., Pramod, G., Kalikoski, D. & Short, K. (2009) 'Safe Conduct? Twelve Years Fishing under the UN Code', *Nature* 457, 658–659

Schofield, C.H. & Townsend-Gault, I. (2011) 'Choppy Waters Ahead in "a Sea of Peace, Cooperation and Friendship"? Slow Progress towards the Application of Maritime Joint Development to the East China Sea', *Marine Policy* 35, 25–33.

Shvarts, E. & Simonov, L.P. (Eds.) (2011) *Environmental Risks to Sino-Russian Transboundary Cooperation: From Brown Plans to a Green Strategy* (World Wildlife Fund).

South, G.R., Skelton, P., Veitayaki, J., Resture, A., Carpenter, C., Pratt, C., et al. (2004) Pacific Islands, GIWA Regional Assessment 62, UNEP, University of Kalmar, Kalmar, Sweden.

Sumaila, U.R., Khan, A.S., Dyck, A.J., Watson, R., Munro, G., Tydemers, P., et al. (2010) 'A Bottom-up Re-estimation of Global Fisheries Subsidies', *Journal of Bioeconomics* 12, 201–225.

Teng, S.K., Yu, H., Tang, Y., Tong, L., Choi, C.I., Kang, D., et al. (2005) Yellow Sea, GIWA Regional Assessment 34, UNEP, University of Kalmar, Kalmar, Sweden.

United Nations (1995) The Agreement for the Implementation of the Provisions on UNCLOS Relating to the Conservation and Management of Straddling Stocks and Highly Migratory Fish Stocks, UNFSA.

von Moltke, A. (Ed.) (2011) *Fisheries Subsidies, Sustainable Development, and the WTO* (UNEP).

Watson, R. & Pauly, D. (2001) 'Systematic Distortions in World Fisheries Catch Trends', *Nature* 414(29), 534–536.

Wauthy, B. (1986) Physical Ocean Environment in the South Pacific Commission Area, UNEP Regional Seas Reports and Studies No. 83.

Wilkinson, C., De Vantier, L., Talaue-McManus, L., Lawrence, D. & Souter, D. (2005) South China Sea, GIWA Regional Assessment 54, UNEP, University of Kalmar, Kalmar, Sweden.

Xue G. (2005) *China and International Fisheries Law and Policy* (Martinus Nijhoff Publishers).

Zou, K. (2005a) *China's Marine Legal System and the Law of the Sea* (Brill).

Zou, K. (2005b) *Law of the Sea in East Asia: Issues and Prospects* (Routledge).

Notes

1   In the PWZs, the reporting state cannot directly notify vessels from the other state of any violation.
2   Both type of fishing take other bycatch such as sharks, sea turtles, and birds.

16

Coral reefs

Artisanal fisheries and community-based management

Dan A. Exton, Paul Simonin and David J. Smith

Introduction

The nations that make up East Asia incorporate an extensive coastline, and therefore the reliance on marine resources is traditionally high in many regions. While economic development has led to increased livelihood diversification and more commercial fishing, large areas remain heavily reliant on nearshore artisanal fisheries, often to satisfy subsistence lifestyles. The most widespread examples of this are nearshore fisheries surrounding coral reefs, which are found throughout Southeast Asia in particular. When the direct impacts of heavy reliance on these fisheries are combined with ecosystem deterioration and an ever expanding human population, reef-associated fisheries become a global conservation and management priority. This chapter will focus on the status of reef fisheries in Southeast Asia due to their well-recognised status as a priority for fisheries management throughout the region. The following will attempt to summarise the current status of reef fisheries in Southeast Asia as well as levels of reliance on them, before discussing the management options in effect there.

Southeast Asia is home to the highest concentration of coral reefs anywhere in the world, contributing 28 percent of the global total area of reefs, equating to 70,000 km² (Burke et al. 2011). These reefs stretch from the 17,500-island archipelago of Indonesia in the south, to the southern islands of Japan in the north, which are home to some of the most northerly reefs in existence thanks to the warm waters of the Kuroshido Current increasing average sea temperatures in the area to levels able to support coral growth (Yamano et al. 2001). Biodiversity is high on coral reefs throughout the tropics, providing a home to roughly one-quarter of all marine species and possessing a level of biological diversity similar to their terrestrial equivalents rainforests (Lewis 1977). However, of all the regions of the world containing coral reefs, those of Southeast Asia, better known as the Indo-Pacific after the two great oceans the region straddles, exhibit the greatest diversity of all. In particular, an area of the Indo-Pacific known as the Coral Triangle is a true biodiversity hotspot, being home to the highest diversity of all major coral reef taxonomic groups (corals, fish and invertebrates) anywhere on the planet (Briggs 2005; Carpenter & Springer 2005; Veron et al. 2009). The Coral Triangle represents a roughly triangular area bound by the nations of Papua New Guinea and the Solomon Islands in the east, Indonesia in the west, and the Philippines in the North, and is home to 76 percent of the world's 798 coral species and 37 percent of the world's roughly 6,000 coral reef fish species (TNC 2008). Southeast Asia is also home to some of the largest areas of mangrove forests and seagrass beds anywhere in the tropics (Spalding et al. 2003, 2011), which are intrinsically linked to coral reefs and their associated fisheries due, among other things, to their importance as nursery grounds for coral reef species, a process known as habitat connectivity (Unsworth & Cullen 2010; Unsworth et al. 2008).

The importance of coral reefs

Coral reefs and their associated coastal habitats are important for a number of reasons, and have been estimated as being one of the most economically valuable ecosystems on the planet. The total global value assigned to coral reefs is a staggering US$375 billion per year, which equates to US$607,500 per km^2 (Costanza et al. 1997). Of course, the valuation of a natural ecosystem is fraught with uncertainties, and incorporates both direct and indirect benefits provided by their continued existence, and so should be treated as a guide only. Obvious direct benefits include their role in food security and livelihoods throughout the tropics, as well as others, and these goods and services have been estimated to provide net revenue to the global economy of US$30 billion per year (Cesar et al. 2003), US$2.4 billion of which comes directly from Southeast Asian coral reef fisheries (Burke et al. 2002).

The importance of coral reefs stretches beyond their ability to support local and regional human populations, however, and includes a range of less obvious but equally important services. A particularly apt example in light of recent catastrophes is their role in coastal protection, reducing erosion but also limiting damage from intense storm events (Berg et al. 1998). In extreme cases, coral reefs even provide some protection from tsunamis, a phenomenon reported from observations after the 2004 event in Asia (Liu et al. 2005) and through model simulations (Kunkel et al. 2006). Regardless of the economic valuation of coral reefs, when these many benefits are coupled with the other services they provide, such as their potential as sources for bioactive compounds for the pharmaceutical industry in their search for new cures and treatments, the full extent of the importance of coral reefs becomes clear (Smith et al. 2007).

The extent of reliance on these fisheries is perhaps a more useful way of gauging their true importance. For example at the beginning of the century, of the estimated 27 million fishers worldwide, 22 million were found in Asia (FAO 2002), a figure that is likely to be a major underestimation when small-scale subsistence fishers are taken into consideration. Globally, approximately 850 million people lived within 100 kilometres of a coral reef in 2007 (Burke et al. 2011) and, by the close of the twentieth century, 1 billion people received their major protein source from them (ICRAN 2002). In Southeast Asia specifically, 138 million people live within 30 kilometres of one, more than all other regions combined (Burke et al. 2011). Due to a combination of such close proximity, high human populations and low economic development, these fisheries have, in the recent past, accounted for one-quarter of total global fish catches and roughly half of all fish directly consumed by human populations (Matthew 2001). Throughout Southeast Asia, fish is a major part of the human diet, even in inland and urban environments (Burke et al. 2011), and 36 percent of all dietary animal protein for the region comes from fish and seafood, although coral reef associated fisheries do not account for all of this. They have, however, provided between 30 and 40 percent of all protein requirements to the approximately 100 million people living in the Philippines (Bernascek 1994). In Indonesia, a country approaching a population of 250 million people, 60 percent of people lived within the coastal zone at the beginning of the twenty-first century (Elliott et al. 2001), and subsistence fishing provided 70 percent of all nationwide protein requirements (Cesar et al. 1997; Resosudarmo 2005).

Overexploitation of tropical coastal fisheries associated with coral reefs is now one of the main threats to these ecosystems, and the majority of reefs within access to human populations are now fished unsustainably (Wilkinson 2008). Worryingly, the heavy reliance on coral reefs and their associated fisheries in the 1990s is predicted to double by 2050 as a result of population increases throughout the developing world (McManus 1997).

Coral reef fisheries pose unique barriers to sustainable management

The more widely publicised commercial fisheries in temperate regions are generally characterised by a small number of target species, limited numbers of fishing techniques being used, and centralised landing sites for catches. These factors combine to make the monitoring and management of these fisheries relatively straightforward, in theory if not in practice. But tropical nearshore fisheries focused around coral reefs present a completely different set of problems that act to significantly complicate management effort. This is in addition to the sheer extent of the practice throughout Southeast Asia, which makes the move towards sustainability a daunting one.

Tropical nearshore fisheries supported by coral reefs and their associated habitats are typically subsistence in nature, broadly meaning that fishing activities are primarily to support the fishers themselves, their close family and friends. Commercial activity, whereby fishing is carried out with financial profit in mind, is present in these systems throughout the region, but apart from local fishers selling any surplus catch within their community, this tends to be more seasonal and targets specific catches, for example to supply the live reef fish trade (Pet-Soede & Erdmann 2003). Although these commercial activities can make use of more developed techniques, the majority of fish caught on and around coral reefs, particularly those for subsistence, are extracted using traditional methods known as artisanal techniques. Artisanal techniques in Southeast Asia make use of a wide range of gear types including various types of net, trap, line, speargunning and aggregation devices (Exton 2010; Pet-Soede & Erdmann 2003; Weis & Weis 2005). Reef fisheries are therefore referred to as multi-gear fisheries (Davies et al. 2009; McClanahan & Mangi 2004), as many if not all of these various artisanal gear types will be used in any one location simultaneously. Individual fishers may favour one particular technique, or they may use a combination to exploit different aspects of the fishery at different times. This multi-gear aspect to the fisheries poses difficulties for those trying to conserve stocks as gear restrictions (e.g., minimum mesh size for nets, reducing the use of more ecologically damaging techniques etc.) become more complex if they are to adequately address each technique in use.

In addition to the management implications of a multi-gear fishery, another consequence is that a large number of species are subsequently exploited. At the broadest level, both the fin (fish) and shell (invertebrate) fisheries are exploited through a combination of the artisanal techniques mentioned already, as well as simple gleaning of invertebrates from exposed reef flats and seagrass beds at low tide (Ashworth et al. 2004; Dalzell et al. 1996; Munro 1989), often by the women and children of communities in Southeast Asian countries (Exton 2010). However, it is the sheer number of species caught that poses the problem, with individual fisheries throughout Southeast Asia, particularly in the Coral Triangle, commonly catching many hundreds of fish species alone. Although certain species are more profitable at market because of their taste, there is little concept of bycatch in a subsistence-based fishery, and juveniles are considered as good a catch as any. When only a small number of species are exploited, it allows scientists to model population dynamics and better understand the sustainability of fishing effort, but this is simply impossible in the face of so many species coexisting in such a complex ecosystem.

In addition to being multi-gear and multi-species, Southeast Asian artisanal reef fisheries also utilise a large number of landing sites, with individual fishers returning their catches directly to their home community as opposed to the more centralised system in commercial fisheries. This makes accurate monitoring of catch size and composition a major challenge, as data would need to be collected in multiple locations simultaneously. This also makes the use of quotas, whereby maximum catch sizes are enforced, unsuitable in most cases due to the difficulties associated with policing such a scheme (Exton 2010).

Despite all these factors, perhaps the greatest barrier to sustainable reef fisheries in the region is their open access nature. The fact that anyone is able to fish these systems leads to a 'race to fish' mentality among fishers, as each stakeholder is keen to receive his fair share of the available resources (Exton & Smith 2011). Without a feeling of ownership over the resource among local communities, which is prevented by an open access scenario allowing fishing pressure from external fishers, the desire to protect the resource and restrict extraction to sustainable levels is generally absent. The introduction of restricted access is therefore widely regarded as a top priority for fisheries managers (Grafton et al. 2008; Hilborn 2007), and will be discussed in more detail later.

In light of these specific barriers to conservation, it is important to put subsistence fisheries into context, and one must consider the role of poverty in Southeast Asian artisanal subsistence fisheries. The persistence of poverty among local communities is a major barrier to fishers exiting a fishery in a state of decline (Adato et al. 2006; Carter & Barrett 2006; Cinner et al. 2008), making fishing more than simply an occupation. This leaves these resource users in what is known as a poverty trap, driven by a lack of alternative livelihoods and an inability among poorer families to enter more profitable income streams. Fishers are subsequently forced to remain in the fishery even through severe stock depletion and reductions in catch (Cinner et al. 2008; Dasgupta 1997). This has also led to a unique form of overfishing evident throughout the region regarding reef fisheries, Malthusian overfishing, which is defined as 'the situation caused by poor fishers when, faced with declining catches, [they] induce wholesale resource destruction in an attempt to maintain income' (Pauly 1994). The causes of Malthusian overfishing are numerous, but include population growth and migration increasing fisher numbers, as well as a lack of alternative livelihoods and food sources available to tropical coastal subsistence communities (McManus 1997).

The status of reef fisheries in Southeast Asia

Communities throughout the region have relied on reef-based resources for millennia, but human population growth has increased demand, while technological developments have increased access to reefs and boosted the efficiency of fishing effort (Exton 2010). If properly managed, coral reefs can yield, on average each year, around 15 tonnes of fish and other seafood per km^2 (Cesar 1996). Unfortunately, this amount is often not sufficient to meet demand, while examples of suitable and successful management are few and far between. As a result, fishing above the sustainable limit is now regarded as the greatest threat to coral reefs throughout Southeast Asia and affects almost all reefs, even those remote from human populations (Wilkinson 2008). Very high local human populations combined with demands from elsewhere in East Asia mean the pressure on tropical coastal fisheries in the region is extremely high, and is forcing fishers to exploit even the most remote reefs. It is estimated that 95 percent of Southeast Asian reefs are currently at risk from local threats, and that by 2050 all will be threatened, 95 percent in the highest category (Burke et al. 2011). Recent years have seen a significant decline in the biomass of fish found on and around coral reefs, as demonstrated by decreases in the catch per unit effort (CPUE) of local fisheries. In the Wakatobi Marine National Park, Indonesia, CPUE of widely used fish fences fell by almost 70 percent in some areas in under five years (Exton 2010). This pattern was repeated throughout most commonly used extraction techniques in the park, and is a clear demonstration of a fishery in crisis.

The special case of the live food trade

Equally important in the region is fishing pressure to supply the live food and aquarium trades. Both these industries require fish to be caught and transported alive to satisfy the particular demands of consumer markets, and are both supplied significantly from the Coral Triangle region (Hoeksema 2007). In the case of the live reef food fish trade, the region forms the centre of the global supply, with the most important trade routes originating in the Philippines and Indonesia and being imported via Hong Kong to mainland China (Erdmann & Pet-Soede 1996) in an industry estimated in 2002 to be worth US$810 million per year (Sadovy et al. 2003). In the 1990s, approximately 60 percent of total volume was shipped directly to Hong Kong, with around half originating in Indonesia (Johannes & Riepen 1995; Lau & Parry-Jones 1999). The economic incentives to local fishermen are significant, with a live individual selling for between four and eight times the price of an equivalent dead individual, and certain species of high value fetching up to US$180 per kilogramme in some areas (Erdmann & Pet-Soede 1996).

In addition to those individuals removed for the live reef food fish trade, many more millions are fished for a global aquarium trade worth approximately US$200–300 million per year (Wabnitz et al. 2003). This industry supplies over 2 million homes and public aquaria (Smith et al. 2008) with organisms representing over 150 species of Scleractinian coral, several hundred non-coral invertebrate species, and a minimum of 1,472 fish species representing 50 families (Bruckner 2005; Wood 2001). The USA is the world's largest importer of marine species, with over 1 billion fishes entering the country officially between 2000 and 2008 (Smith et al. 2008). Based on 2005 figures, 55 percent of imported individuals originated in the Philippines and 31 percent in Indonesia (Rhyne et al. 2012), reinforcing the importance of the Southeast Asian region in this global trade. Despite several major concerns surrounding these trades as they currently operate, there is also optimism among many experts that the aquarium trade could prove useful in future conservation efforts, at least more useful than the demise of the practice altogether (Rhyne et al. 2012). Due to the economic benefits of the live fish trades to fishers, there is significant potential for specimen collection to increase the value of reefs and associated habitats, and thus for the trade to act as an important incentive for conservation and sustainable exploitation (Tlusty 2002). The continuation of the trade, supporting the availability of coral reef organisms to private and public aquaria, also undoubtedly has the potential to raise awareness of coral reefs and the threats to their survival, and to educate on these important ecosystems (Nijman 2009).

Before these conservation benefits can be widely felt, however, there are several issues that need addressing in the way the industry operates both for suppliers and consumers. One concern is the high mortality rate among captured individuals on the often long journeys from source to market, with estimates suggesting 50 percent of individuals are lost before reaching their destination (Indrawan 1999; Johannes & Riepen 1995). This makes the practice highly inefficient, and doubles the strain put on an already overexploited resource, although it would undoubtedly be difficult to remedy without significant investment in infrastructure. A second concern surrounds species selectivity, and the implications to overall ecosystem health of removing species providing important services to the reefs and other habitats (Rhyne et al. 2012). A good example of this would be the removal of herbivores, which are vital on reefs in preventing the overgrowth of seaweed and subsequent phase shifts to algal dominated systems. A difficult balance will need to be reached to minimise the removal of key groups while simultaneously avoiding excessive exploitation of those species deemed less critical to ecosystem health and functioning.

The most pressing issue associated with the live fish trade, however, is the collection method required. Catching a fish in a way that enables the individual to survive intact is no easy feat, and the most commonly used technique used to achieve this is cyanide fishing. The use of hydrocyanic acid, made by mixing water with either potassium cyanide or sodium cyanide, allows fishers to stun fish in a way that allows them to recover when moved to clean water (Pet-Soede & Erdmann 2003). This is typically performed using small bottles of the poison underwater, allowing valuable species to be targeted (Exton & Smith 2011). However, the practice has a number of negative impacts on the resource base, namely the reef itself. First, the poison has been shown to cause bleaching of nearby corals (Jones & Steven 1997). Second, targeted fish often manage to retreat into the complex reef structure before the poison takes effect, making it necessary for fishers to destroy parts of the reef in order to collect them (Mous et al. 2000).

Along with cyanide fishing, the other primary destructive fishing technique is blast or dynamite fishing. The materials used for blast fishing typically involve glass bottles filled with a fertiliser mixture and a petrol-soaked fuse (Pet-Soede et al. 1999). When thrown into the water, the shock wave produced either kills the fish outright or stuns them, causing them to float to the surface where they can be easily collected. However, the use of explosives causes significant damage to the physical structure of the reef (Fox & Caldwell 2006), which is vital for ecosystem functioning by driving the extremely high levels of biodiversity. This leaves an area of coral rubble (Edinger et al. 1998), which is an unstable substrate, thus delaying successful coral recruitment and meaning that even five years after a blast event there is no significant recovery (Fox et al. 2003). Overall, destructive fishing is estimated to affect over 60 percent of Southeast Asian reefs (Burke et al. 2011), with 0.06 percent of total reef area in Indonesia destroyed each year by cyanide fishing (Mous et al. 2000), and 3.75 percent by blast fishing (Pet-Soede et al. 1999). This is in spite of the use of destructive techniques being illegal in many countries, and therefore makes it a key challenge for conservation managers throughout the region. Particularly worrying are reports of a re-emergence of blast fishing in areas in which the practice has previously been eradicated. Personal observations by the authors in the Wakatobi Marine National Park, Indonesia, indicate an increase in the use of blast fishing since 2011 after no reported incidents since 2006.

Fisheries at the ecosystem level

As well as the direct impacts of overfishing and destructive fishing techniques, it is also important to examine the ecosystem as a whole. The highly diverse and productive fisheries associated with tropical coastal environments are highly reliant on the habitats found there and impacts to these habitats must also be taken into consideration. Increased sedimentation creates a number of issues on reefs, and has come about through land use change in the form of extensive deforestation and coastal development, the latter also tending to increase nutrient loading in the coastal marine environment (Burke et al. 2011). Added to this is the widespread loss of natural filters of both sediment and pollution, namely mangrove and seagrass habitats. These ecosystems also play a vital role in the lifecycles of many fish species through habitat connectivity (Nagelkerken et al. 2000, 2002) making them crucial at determining the community assemblage of fisheries found in tropical coastal waters (Mumby 2006). Unfortunately, mangrove loss in

Southeast Asia has been greater than anywhere else on earth, particularly as a result of the expansion of aquaculture practices (FAO 2007), while seagrass beds are also under considerable threat throughout the region (Unsworth & Cullen 2010) Added to these localised impacts are the threats associated with global climate change, which are expected to significantly affect the tropical coastal marine environment in coming decades, including ocean acidification and increased sea surface temperature among others.

Fortunately, there are a number of aspects of the Southeast Asian environment that bring a small sense of optimism to an otherwise worrying problem. Whereas other regions of the tropics, for example the Caribbean, have been unable to maintain their coral communities at healthy levels, coral cover in Southeast Asia has remained high despite the extremely raised levels of local threat. This has at least in part been thanks to the lower incidence of more global threats in the region, for example there have been relatively low levels of lethal coral bleaching, and certainly Southeast Asia has been spared the mass bleaching events seen elsewhere in recent decades (Burke et al. 2011). The region's high biodiversity may also be of assistance, as it is believed to increase reef resilience through what is known as functional redundancy (Worm et al. 2006). Where multiple species exist that provide similar or identical ecosystem services, a system is said to have high functional redundancy, as if one species is lost, the ecosystem function it provides is maintained. Although it is highly unlikely that these positives will prove to be the saviour of reefs and their fisheries, it is certainly feasible that they will provide much needed natural support to conservation management attempts and could prove the difference between success and failure.

Management options to achieve sustainability

With the high levels of human impacts affecting tropical nearshore fisheries in Southeast Asia, a move towards sustainability is a priority for conservation organisations and governments alike. However, balancing the reliance on marine resources among local communities with the need to protect biodiversity makes this a particularly difficult challenge, and a variety of management methods are in use throughout the region. Although many threats are described as external, such as sea surface temperature rises and ocean acidification, and therefore require a more global approach, the ability of organisms to cope with these broader disturbances is significantly reduced by continued exploitation (Chape et al. 2005; Palumbi 2004; Sale et al. 2005), and so protecting tropical coastal ecosystems is expected to help buffer them against predicted global climate change and give them a much better chance of survival.

Top-down protection

Marine protected areas (MPAs) have long formed the basis of management approaches throughout the tropics, although their suitability as a standalone conservation tool has become increasingly questioned. In theory, they are the simplest way to manage overexploited fisheries (Roberts 1997), by enforcing top–down policies to limit the extent of fishing pressure and the modes by which fish are removed from the system. In Southeast Asia, there are a total of 599 MPAs with coral reefs within their boundaries, and their scope and performance was assessed by Burke et al. (2011). This network of MPAs covers 17 percent of the region's coral reefs, which is significantly lower than the global total of 27 percent. More worryingly, only 2 percent of these parks were considered to be effective at reducing the threat of overfishing, and 69 percent were considered ineffective. This made Southeast Asian MPAs the lowest performers of any major region, with a total of 16 km² of coral reefs managed effectively. The report also found that the MPAs are disproportionately situated in areas where fishing pressure is low, a sign that sites at least threat are often favoured for conservation efforts due to the increased chance of success. These ineffective MPAs are commonly termed 'paper parks', where cooperation is not rewarded or violation punished (Rudd et al. 2003).

The most successful component of MPA use is the incorporation of no-take areas (NTAs), whereby resource extraction is strictly controlled (Mumby & Harborne 2010). However, outside Australia where NTA cover is significant, only 1 percent of coral reefs around the world are situated within NTAs. When established successfully, NTAs facilitate stock recovery and help protect biodiversity within the coastal zone. They also provide important benefits to neighbouring fishing grounds through what is known as a 'spill-over' effect. This describes the movement of adults, larvae and eggs from recovered populations within NTAs to impoverished areas nearby (Gell & Roberts 2003). If fishing effort within this buffer zone can be maintained below maximum sustainable yield (MSY), the benefits to fishers can be greatly enhanced (Guidetti & Claudet 2010). Some MPAs will therefore involve a complex zonation plan incorporating areas of varying protection to maximise the benefits of carefully placed NTAs, although overly complex zonation plans risk losing community compliance if the rules are not easy to follow.

The importance of community involvement

Although success is possible through the use of MPAs alone, particularly if NTAs are incorporated, they instead tend to form the basis for more complex management strategies including community-led components. The use of MPAs in isolation runs the risk of fishing pressure simply being displaced elsewhere (Grafton et al. 2006; Halpern et al. 2004), or a scenario of illegal fishing and poaching as local communities struggle to maintain livelihoods in the face of fishing regulations. With social factors the key drivers of success in MPAs (Kelleher & Recchia 1998), a community component is strongly encouraged (Exton & Smith 2011). This bottom-up approach to management, otherwise known as adaptive community management or co-management, aims to shift authority for management decisions towards local communities (Fabricius & Collins 2007), thus addressing key barriers to successful management such as stakeholder compliance (Pretty & Smith 2004) and effort displacement. Co-management involves a participatory approach, with stakeholders at the community level involved in all stages of the decision-making process (Fabricius & Collins 2007; Ostrom 2009) and empowered through this involvement. Education programmes often form an integral component of participatory management, increasing the understanding of sustainability and ecosystem threats among local stakeholders (Exton & Smith 2011). In addition, government decentralisation is also important as it allows management strategies to be developed on a case-by-case basis more easily through the benefits of localised legislative powers (Pomeroy & Berkes 1997).

Of the various components of community management of small-scale fisheries, two often considered key to achieving success are a move towards restricted access and the development of alternative income streams. Reef-associated artisanal fisheries tend to be treated as an open access resource, which induces a race to fish mentality often with a disregard for ecological consequences (Exton & Smith 2011). A move to dedicated, or restricted, access promotes sustainability and conservation by creating a sense of ownership among local fishers through the removal of existing external fishing pressure (Hilborn 2007). Restricted access can be developed through a number of approaches, with suitability on a case-by-case basis. In some areas, for example in Raja Ampat, Indonesia, legal ownership of reefs has been given to local communities, who are able to lease the access to resources to fishers (Beck et al. 2004). Where multiple communities exist, making village level ownership difficult to manage, registration of fishing gear has proved successful at eliminating external fishing effort and developing more informal ownership, for example around Kaledupa Island, Wakatobi, Indonesia (Exton 2010). An additional benefit of ownership schemes is the self-policing they promote. The identification and punishment of illegal fishing is an important consideration (Beddington et al. 2007), but dedicated access allows communities to identify and deal with external fishing pressure without the need for expensive enforcement.

Alternative incomes and livelihood diversification

A major concern with fisheries management is that effort will simply be displaced elsewhere, and thus the reduction of fishing pressure at one site will result in an increase at another (Halpern et al. 2004). It is therefore crucial that management attempts include the development of alternative livelihoods, to provide fishers with the range of incentives required to permanently exit the fishery. Compensation payments are an alternative option, for example through a licensing scheme, but this will only provide a short-term reduction in fishing pressure unless financial benefits are felt regularly enough to remove the need to fish for food and income.

Mariculture is just one example of an alternative livelihood developed to increase income diversification in tropical coastal communities. A number of invertebrates may be farmed, such as giant clams and shrimp, as well as fish, although the ecological impacts of these activities are often considerable. In Southeast Asia, mariculture in the form of seaweed farming has become a widespread alternative to fishing (Sievanen et al. 2005). Farmed seaweed is used to supply the hydrocolloid industry, reported to be worth an impressive US$800 million per year and continuing to grow rapidly. However, the uneven distribution of profits within the industry is a major obstacle to realising the true potential of seaweed farming as a true alternative to fishing. The financial benefits are predominately felt at the processing stage, which is currently restricted to a handful of large-scale factories, which tend to require farmers to use middlemen to export their product. A current project in Indonesia run by the Operation Wallacea Trust aims to localise the processing stage and incorporate the extraction into a fishing licence buyout scheme, which would significantly boost incomes for all stakeholders. Recent years have also seen a move towards increased ecotourism to provide those exiting the fishery with a viable income stream. Businesses may provide accommodation and food, as well as scuba-diving and snorkelling activities (Patterson et al. 2004). Although promising, achieving sufficient income from ecotourism is difficult in remote areas, while the financial benefits are often not maintained locally.

For better or worse: management case studies

In the Philippines, government decentralisation took place in 1991 as part of the Local Government Code, meaning that responsibilities were transferred to local government and communities (Pomeroy et al. 2001). This allowed the development of existing MPAs to take place, as well as the establishment of many new protected areas. By 2008, over 985 MPAs had been established, 95 percent of which incorporated community management and an NTA component, covering a total of between 2.7 and 3.4 percent of coral reefs within the country, although this is considered inadequate to fulfil management objectives (Weeks et al. 2010). Of these numerous MPAs, Apo Island is widely regarded as an example of management success, and formed the basis for the proliferation of no-take reserves throughout the Philippines. Established in the 1970s, only traditional non-destructive fishing methods have been permitted on the island since 1986, while a 0.45-kilometre-long NTA forms part of the management plan, accounting for around

10 percent of total coral reef area around the island (Russ et al. 2003). This led to the biomass of large predatory fish increasing sevenfold after 11 years of protection (Russ & Alcala 1996), and a tripling of biomass of certain herbivorous species over an 18-year period (Russ et al. 2003). This success has been attributed to the inclusion of local communities, with the responsibility of managing the area shared between local government, national government and local stakeholders (Alcala & Russ 2006). The positivity surrounding Apo Island has tempered in recent years since a transition from co-management to centralised national government management in the late 1990s. Although initially supported by local communities, support appears to have been lost due to stakeholder exclusion and poor institutional performance (Hind et al. 2010), and it will be interesting to see the consequences of this shift in the fisheries over coming decades.

It is also important to realise that there is no guaranteed recipe for success, and management must address sites on a case-by-case basis to maximise the chance of success. Elsewhere in the Philippines, co-management schemes with NTA incorporation were established on Balicasag and Pamilacan Islands in 1985. Despite success around Balicasag Island in terms of hard coral recovery (119 percent increase in percentage cover within the NTA and 67 percent outside), there was a decrease in coral cover in and around Pamilacan Island and a significant long term decline in target fish species abundance at both sites (Christie et al. 2002). This has led to calls for reduced reliance on scattered small-scale reserves without adequate attempts to reduce overall fishing pressure in the area that allow the spillover benefits to be truly appreciated. It also further highlighted the importance of management strategies that work towards improving the resource base itself (Christie et al. 2002), a step that would ultimately increase the potential maximum fishing extraction level that can be achieved sustainably.

Similar projects to those established in the Philippines have also been developed elsewhere, for example in Indonesia. In 1996, increasing use of destructive fishing practices within an area then known as the Tukangbesi Archipelago led to the establishment of the Wakatobi Marine National Park. The Wakatobi covers a total area of 13,900 km^2, making it the second largest MPA in Indonesia, and incorporates roughly 600 km^2 of coral reef habitat. Home to over 80,000 inhabitants with a high reliance on marine resources, there is a strong emphasis on sustainable exploitation in the park, and decentralisation of government has provided a strong foundation for successful management. Despite this promising start, initial research highlighted a severe lack of communication between managers and the local resource users themselves. This research demonstrated that, although 75 percent of interviewees were aware of the MPA, only 30 percent could refer to any rules relating to the use of marine resources in the area, leading to the author of the study to call for increased community management in the Wakatobi (Clifton 2003). In more recent years, there has been increased emphasis put on the empowerment of local communities to become involved in the management process, in collaboration with park authorities, local government and non-governmental organisations (NGOs) such as the Nature Conservancy (TNC) and the World Wildlife Fund (WWF). Another more localised co-management project has been ongoing since the early 2000s, led by academics from the international NGO Operation Wallacea, in an attempt to establish a best practice example of successful reef fishery co-management on the island of Kaledupa. This effort has focused on raising awareness among local communities on fishery declines and environmental concerns, while establishing a series of fisheries forums attended by elected representatives of each village. The first island-wide fisheries forum in 2009 culminated in the agreement of 21 new by-laws that were passed on to local government for ratification, and formed a significant milestone in the co-management process.

Conclusion

Overfishing and the use of destructive fishing techniques is undoubtedly a pressing concern throughout Southeast Asia, and a major priority for conservation managers in the region. The high reliance on marine resources among fishers using artisanal techniques makes the problem a complex one, as it incorporates not only the issue of livelihoods, but of food security and even cultural tradition. The threats facing tropical coastal ecosystems, primarily coral reefs but also mangroves and seagrass beds, are exacerbating the problem, reducing the capacity of nearshore tropical fisheries and combining with resource extraction to severely reduce resilience at the ecosystem level. This poses a unique set of questions to conservationists and makes achieving sustainability a particularly difficult challenge. Traditional top-down management through marine protected areas and no-take zones is beneficial in creating a framework for conservation, but, in isolation, is unlikely to achieve regional management goals. Instead, successes achieved to date have typically come where they form the basis for community-based management to reduce fishing pressure. This can be achieved through the provision of incentives such as restricted access via the promotion of ownership, and the development and provision of livelihoods that are sufficiently viable to be alternative rather than additional. To truly succeed in reducing fishing pressure throughout Southeast Asia, while simultaneously working towards the alleviation of poverty and securing regional food security, novel approaches are needed to fill the current gaps in management success, but existing success stories suggest this is achievable if approached in the right way.

Bibliography

Adato, M., Carter, M.R. & May, J. (2006) 'Exploring Poverty Traps and Social Exclusion in South Africa Using Qualitative and Quantitative Data', *Journal of Development Studies* 42, 226–247.

Alcala, A.C. & Russ, G.R. (2006) 'No-take Marine Reserves and Reef Fisheries Management in the Philippines: A New People Power Revolution', *Ambio* 35(5), 245–254.

Ashworth, J.S., Ormond, R.F.G. & Sturrock, H.T. (2004) 'Effects of Reef-top Gathering and Fishing on Invertebrate Abundance across Take and No-take Zones', *Journal of Experimental Marine Biology and Ecology* 303, 221–242.

Beck, M.W., Marsh, T.D., Reisewitz, S.E. & Bortman, M.L. (2004) 'New Tools for Marine Conservation: The Leasing and Ownership of Submerged Lands', *Conservation Biology* 18, 1214–1223.

Beddington, J.R., Agnew, D.J. & Clark, C.W. (2007) 'Current Problems in the Management of Marine Fisheries', *Science* 316, 1713–1716.

Berg, H., Öhman, M.C., Troëng, S. & Lindén, O. (1998) 'Environmental Economics of Coral Reef Destruction in Sri Lanka', *Ambio* 27, 627–634.

Bernascek, G. (1994) 'The Role of Fisheries in Food Security in the Philippines: a Perspective Study for the Fisheries Sector to the Year 2010'. Unpublished.

Briggs, J.C. (2005) 'Coral Reefs: Conserving the Evolutionary Sources', *Biological Conservation* 126, 297–305.

Bruckner, A.W. (2005) 'The Importance of the Marine Ornamental Reef Fish Trade in the Wider Caribbean', *Revista de Biología Tropical* 53, 127–138.

Burke, L., Selig, L. & Spalding, M. (2002) *Reefs at Risk in Southeast Asia* (World Resources Institute).

Burke, L., Reytar, K., Spalding, M. & Perry, A. (2011) *Reefs at Risk Revisited* (World Resources Institute).

Carpenter, K.E. & Springer, V.G. (2005) 'The Center of the Center of Marine Shore Fish Biodiversity: The Philippine Islands', *Environmental Biology of Fishes* 72, 467–480.

Carter, M.R. & Barrett, C.B. (2006) 'The Economics of Poverty Traps and Persistent Poverty: An Asset-based pproach', *Journal of Development Studies* 42, 178–199.

Cesar, H. (1996) *Economic Analysis of Indonesian Coral Reefs* (World Bank).

Cesar, H.S.J., Lundin, C.G., Bettancourt, S. & Dixon, J. (1997) 'Indonesian Coral Reefs: an Economic Analysis of a Precious but Threatened Resource', *Ambio* 26, 345–350.

Cesar, H.S.J., Burke, L. & Pet-Soede, L. (2003) *The Economics of Worldwide Coral Reef Degradation* (Cesar Environmental Economics Consulting).

Chape, S., Harrison, J., Spalding, M. & Lysenko, I. (2005) 'Measuring the Extent and Effectiveness of Protected Areas as an Indicator for Meeting Global Biodiversity Targets', *Philosophical Transcriptions of the Royal Society B* 360, 443–455.

Christie, P., White, A. & Deguit, E. (2002) 'Starting Point of Solution? Community-based Marine Protected Areas in the Philippines', *Journal of Environmental Management* 66, 441–454.

Cinner, J.E., Daw, T. & McClanahan, T.R. (2008) 'Socioeconomic Factors that Affect Artisanal Fishers' Readiness to Exit a Declining Fishery', *Conservation Biology* 23, 124–130.

Clifton, J. (2003) 'Prospects for Co-management in Indonesia's Marine Protected Areas', *Marine Policy* 27(5), 389–395.

Costanza, R., D'Arge, R., De Groots, R., Farber, S., Grasso, M., Hannon, B., et al. (1997) 'The Value of the World's Ecosystem Services and Natural Capital', *Nature* 387, 253–260.

Dalzell, P., Adams, T.J.H. & Polunin, N.V.C. (1996) 'Coastal Fisheries in the South Pacific Islands', *Oceanography Marine Biology Annual Review* 34, 395–531.

Dasgupta, P. (1997) 'Nutritional Status, the Capacity for Work, and Poverty Traps', *Journal for Econometrics* 77, 5–37.

Davies, T.E., Beanjara, N. & Tregenza, T. (2009) 'A Socio-economic Perspective on Gear Based Management in an Artisanal Fishery in South-west Madagascar', *Fisheries Management and Ecology* 16, 279–289.

Edinger, E.N., Jompa, J., Limmon, G.V., Widjatmoko, W. & Risk, M.J. (1998) 'Reef Degradation and Coral Biodiversity in Indonesia: Effects of Land Based Pollution, Destructive Fishing Practices and Changes over Time', *Marine Pollution Bulletin* 36, 617–630.

Elliott, G., Mitchell, B., Wiltshire, B., Manan, I.A. & Wismer, S. (2001) 'Community Participation in Marine Protected Area Management: Wakatobi National Park, Sulawesi, Indonesia', *Coastal Management* 29, 295–316.

Erdmann, M.V. & Pet-Soede, L. (1996) 'How Fresh is too Fresh? The Live Reef Food Fish Trade in Eastern Indonesia', *NAGA, the ICLARM Quarterly* 19, 4–8.

Exton, D.A. (2010) 'Nearshore Fisheries of the Wakatobi', in J. Clifton, R.K.F. Unsworth & D.J. Smith (Eds.), *Marine Research and Conservation in the Coral Triangle: the Wakatobi National Park* (NOVA).

Exton D.A. & Smith D.J. (2011) 'Coral Reef Fisheries and the Role of Communities in their Management', in J.S. Intilli (Ed.), *Fishery Management* (NOVA).

Fabricius, C. & Collins, S. (2007) 'Community-based Natural Resource Management: Governing the Commons', *Water Policy* 9, 83–97.

FAO (2002) The State of World Fisheries and Aquaculture, Rome, Fisheries Department of the Food and Agriculture Organization.

FAO (2007) The World's Mangroves 1980–2005, Rome, Forestry Department, Food and Agriculture Organization of the United Nations.

Fox, H.E. & Caldwell, R.L. (2006) 'Recovery from Blast Fishing on Coral Reefs: A Tale of Two Scales', *Ecological Applications* 16, 1631–1635.

Fox, H.E., Pet, J.S., Dahuri, R. & Caldwell, R.L. (2003) 'Recovery in Rubble Fields: Longterm Impacts of Blast Fishing', *Marine Pollution Bulletin* 46, 1024–1031.

Gell, F.R. & Roberts, C.M. (2003) 'Benefits beyond Boundaries: The Fishery Effects of Marine Reserves', *Trends in Ecology and Evolution* 18, 448–455.

Grafton, R.Q., Arnason, R., Bjorndal, T., Campbell, D., Campbell, H.F., Clark, C.W., et al. (2006) 'Incentive-based Approaches to Sustainable Fisheries', *Canadian Journal of Fisheries and Aquatic Sciences* 63, 699–710.

Grafton, R.Q., Hilborn, R., Ridgeway, L., Squires, D., Williams, M., Garcia, S., et al. (2008) 'Positioning Fisheries in a Changing World', *Marine Policy* 32, 630–634.

Guidetti, P., & Claudet, J. (2010) 'Comanagement Practices Enhance Fisheries in Marine Protected Areas', *Conservation Biology* 24, 312–318.

Halpern, B.S., Gaines, S.D. & Warner, R.R. (2004) 'Confounding Effects of the Export of Production and the Displacement of Fishing Effort from Marine Reserves', *Ecological Applications* 14, 1248–1256.

Hilborn, R. (2007) 'Moving to Sustainability by Learning from Successful Fisheries', *Ambio* 36, 296–303.

Hind, E.J., Hiponia, M.C. & Gray, T.S. (2010) 'From Community-based to Centralised National Management – a Wrong Turning for the Governance of the Marine Protected Area in Apo Island, Philippines?', *Marine Policy* 34(1), 54–62.

Hoeksema, B. (2007) 'Delineation of the Indo-Malayan Centre of Maximum Marine Biodiversity: The Coral Triangle', in W. Renema (Ed.), *Biogeography, Time, and Place: Distributions, Barriers, and Islands* (Springer).

ICRAN (2002) *People and Reefs: A Partnership for Prosperity* (International Coral Reef Action Network).

Indrawan, M. (1999) 'Live Reef Food Fish Trade in the Banggai Islands (Sulawesi, Indonesia): A Case Study', *SPC Live Reef Fish Information Bulletin* 6, 7–14.

Johannes, R.E. &Riepen, M. (1995) *Environmental, Economic and Social Implications of the Live Reef Fish Trade in Asia and the Western Pacific* (Nature Conservancy).

Jones, R.J. & Steven, A.L. (1997) 'Effects of Cyanide on Corals in Relation to Cyanide Fishing on Reefs', *Marine and Freshwater Research* 48, 517–522.

Kelleher, G. & Recchia, C. (1998) 'Lessons from Marine Protected Areas around the World', *Parks* 8, 1–4.

Kunkel, C.M., Hallberg, R.W. & Oppenheimer, M. (2006) 'Coral Reefs Reduce Tsunami Impact in Model Simulations', *Geophysical Research Letters* 33, L23612.

Lau, P.P.F. & Parry-Jones, R. (1999) *The Hong Kong Trade in Live Reef Fish for Food* (TRAFFIC East Asia and World Wide Fund for Nature).

Lewis, J.B. (1977) 'Processes of Organic Production on Coral Reefs', *Biological Review* 52, 305–347.

Liu, P.L.F., Lynett, P., Fernando, H., Jaffe, B.E., Fritz, H., Higman, B., et al. (2005) 'Observations by the International Tsunami Survey Team in Sri Lanka', *Science* 308, 1595.

McClanahan, T.R. & Mangi, S.C. (2004) 'Gear-based Management of a Tropical Artisanal Fishery based on Species Selectivity and Capture Size', *Fisheries Management and Ecology* 11, 52–60.

McManus, J.W. (1997) 'Tropical Marine Fisheries and the Future of Coral Reefs: A Brief Review with Emphasis on Southeast Asia', *Coral Reefs* 16, S121–S127.

Matthew, S. (2001) 'Small-scale Fisheries Perspectives on an Ecosystem-based Approach to Fisheries Management', in FAO (Ed.), Responsible Fisheries in the Marine Ecosystem, Rome, Food and Agriculture Organization.

Mous, P.J., Pet-Soede, L., Erdmann, M., Cesar, H.S.J., Sadovy, Y. & Pet, J. (2000) 'Cyanide Fishing on Indonesian Coral Reefs for the Live Food Fish Market – What is the Problem?', *SPC Live Reef Fish Information Bulletin* 7, 20–27.

Mumby, P.J. (2006) 'Connectivity of Reef Fish between Mangroves and Coral Reefs: Algorithms for the Design of Marine Reserves at Seascape Scales', *Biological Conservation* 128: 215–222.

Mumby, P.J., & Harborne, A.R. (2010) 'Marine Reserves Enhance the Recovery of Corals on Caribbean Reefs', *PLoS ONE* 5, e8657.

Munro, J.L. (1989) 'Fisheries for Giant Clams (Tridacnidae: Bivalvia) and Prospects for Stock Enhancement', in J.F. Caddy (Ed.), *Marine Invertebrate Fisheries: Their Assessment and Management* (John Wiley & Sons).

Nagelkerken, I., Van Der Velde, G., Gorissen, M.W., Meijer, G.J., Van't Hof, T. & Den Hartog, C. (2000) 'Importance of Mangroves, Seagrass Beds and the Shallow Coral Reef as a Nursery for Important Coral Reef Fishes, Using a Visual Census Technique', *Estuarine Coastal and Shelf Science* 51, 31–44.

Nagelkerken, I., Roberts, C.M., Van Der Velde, G., Dorenbosch, M., Van Riel, M.C., De La Morinere, E.C., et al. (2002) 'How Important Are Mangroves and Seagrass Beds for Coral Reef Fish? The Nursery Hypothesis Tested on an Island Scale', *Marine Ecology Progress Series* 244, 299–305.

Nature Conservancy (2008) 'Coral Triangle Facts, Figures and Calculations Part II: Patterns of Biodiversity and Endemism'. Available at http://ctatlas.reefbase.org/pdf/part%20II%20%20biodiversity%20stats.pdf.

Nijman, V. (2009) 'An Overview of International Wildlife Trade from Southeast Asia', *Biodiversity Conservation* 19, 1101–1114.

Ostrom, E. (2009) 'A General Framework for Analyzing Sustainability of Social-ecological Systems', *Science* 325, 419–422.

Palumbi, S.R. (2004) 'Marine Reserves and Ocean Neighbourhoods: the Spatial Scale of Marine Populations and their Management', *Annual Review of Environment and Resources* 29, 31–68.

Patterson, T., Gulden, T., Cousins, K. & Kraev, E. (2004) 'Integrating Environmental, Social and Economic Systems: A Dynamic Model of Tourism in Dominica', *Ecological Modelling* 175, 121–136.

Pauly, D. (1994) 'From Growth to Malthusian Overfishing: Stages of Fisheries Resource Misuse', *Traditional Marine Resource Management Knowledge Bulletin* 3, 7–14.

Pet-Soede, C., Cesar, H.S.J. & Pet, J.S. (1999) 'An Economic Analysis of Blast Fishing on Indonesian Coral Reefs', *Environmental Conservation* 26, 83–93.

Pet-Soede, L. & Erdmann, M. (2003) *Rapid Ecological Assessment of the Wakatobi National Park* (World Wildlife Fund Indonesia).

Pomeroy, R., & Berkes, F. (1997) 'Two to Tango: the Role of Government in Fisheries Comanagement', *Marine Policy* 21, 465–480.

Pomeroy, R.S., Katon, B.M. & Harkes, I. (2001) 'Conditions Affecting the Success of Fisheries Co-management: Lessons from Asia', *Marine Policy* 25, 197–208.

Pretty, J. & Smith, D. (2004) 'Social Capital in Biodiversity Conservation and Management', *Conservation Biology* 18, 631–638.

Resosudarmo, B.P. (2005) *The Politics and Economics of Indonesia's Natural Resources* (RFF Press).

Rhyne, A.L., Tlusty, M.F., Schofield, P.J., Kaufman, L., Morris, J.A. Jr. & Bruckner, A.W. (2012) 'Revealing the Appetite of the Marine Aquarium Fish Trade: The Volume and Biodiversity of Fish Imported into the United States', *PLoS ONE* 7(5), e35808.

Roberts, C.M. (1997) 'Ecological Advice for the Global Fisheries Crisis', *Trends in Ecology and Evolution* 12, 35–38.

Rudd, M.A., Tupper, M.H., Folmer, H. & Van Kooten, G.C. (2003) 'Policy Analysis for Tropical Marine Reserves: Challenges and Directions', *Fish and Fisheries* 4, 65–85.

Russ, G.R. & Alcala, A.C. (1996) 'Marine Reserves: Rates and Patterns of Recovery and Decline of Large Predatory Fish', *Ecological Applications* 6, 947–961.

Russ, G.R., Alcala, A.C. & Maypa, A.P. (2003) 'Spillover from Marine Reserves: the Case of *Naso vlamingii* at Apo Island, Philippines', *Marine Ecology Progress Series* 264, 15–20.

Sadovy, Y.J., Donaldson, T.J., Graham, T.R., McGilvray, F., Muldoon, G.J., Phillips, M.J., et al. (2003) *While Stocks Last: The Live Reef Food Fish Trade* (Asian Development Bank).

Sale, P.F., Cowen, R.K., Danilowicz, B.S., Jones, G.P., Kritzer, J.P., Linderman, K.C., et al. (2005) 'Critical Science Gaps Impede Use of No-take Fishery Reserves', *Trends in Ecology and Evolution* 20, 74–80.

Sievanen, L., Crawford, B., Pollnac, R. & Lowe, C. (2005) 'Weeding Through the Assumptions of Livelihood Approaches in ICM: Seaweed Farming in the Philippines and Indonesia', *Ocean & Coastal Management* 48, 297–313.

Smith, D.J., Pilgrim, S.E. & Cullen, L.C. (2007) 'Coral Reefs and People', in J. Pretty, A. Ball, T. Benton, J. Guivant, D.R. Lee, D. Orr, et al. (Eds.), *Sage Handbook on Environment and Society* (Sage Publications).

Smith, K.F., Behrens, M.D., Maxm L.M. & Daszak, P. (2008) 'U.S. Drowning in Unidentified Fishes: Scope, Implications, and Regulation of Live Fish Import', *Conservation Letters* 1, 103–109.

Spalding, M.D., Kainuma, M. & Collins, L. (2011) 'World Atlas of Mangroves', *Human Ecology* 39(1), 107–109.

Spalding, M.D., Taylor, M.L., Ravilious, C., Short, F.T. & Green, E.P. (2003) 'Global Overview: the Distribution and Status of Seagrasses', in E.P. Green & F.T. Short (Eds.), *World Atlas of Seagrasses* (University of California Press).

Tlusty, M. (2002) 'The Benefits and Risks of Aquacultural Production for the Aquarium Trade', *Aquaculture* 205, 203–219.

Unsworth, R.K.F. & Cullen, L.C. (2010) 'Recognising the Necessity for Indo-Pacific Seagrass Conservation', *Conservation Letters* 3, 63–73.

Unsworth, R.K.F., De Leon, P.S., Garrard, S.L., Jompa, J., Smith, D.J. & Bell, J.J. (2008) 'High Connectivity of Indo-Pacific Seagrass Fish Assemblages with Mangrove and Coral Reef Habitats', *Marine Ecology Progress Series* 353, 213–224.

Veron, J.E.N., Devantier, L.M., Turak, E., Green, A.L., Kininmonth, S., Stafford-Smith, M., et al. (2009) 'Delineating the Coral Triangle', *Galaxea, Journal of Coral Reef Studies* 11, 91–100.

Wabnitz, C., Taylor, M., Green, E. & Razak, T. (2003) *From Ocean to Aquarium: the Global Trade in Marine Ornamental Species* (UNEP World Conservation Monitoring Centre).

Weeks, R., Russ, G.R., Alcala, A.C. & White, A.T. (2010) 'Effectiveness of Marine Protected Areas in the Philippines for Biodiversity Conservation', *Conservation Biology* 24(2), 531–540.

Weis, J.S., & Weis, P. (2005) 'Use of Intertidal Mangrove and Sea Wall Habitats by Coral Reef Fishes in the Wakatobi National Park, Indonesia', *Bulletin of Zoology* 53, 119–124.

Whittingham, E., Campbell, J. & Townsley, P. (2003) *Poverty and Reefs: A Global Overview* (UNESCO).

Wilkinson, C.R. (2008) *Status of Coral Reefs of the World: 2008* (Global Coral Reef Monitoring Network and Reef and Rainforest Research Center).

Wood, E. (2001) 'Global Advances in Conservation and Management of Marine Ornamental Resources', *Aquarium Sciences and Conservation* 3, 65–77.

Worm, B., Barbier, E.B., Beaumont, N., Duffy, E., Folke, C., Halpern, B.S., et al. (2006) 'Impacts of Biodiversity Loss on Ocean Ecosystem Services', *Science* 314, 787–790.

Yamano, H., Hori, K., Yamauchi, M., Yamagawa, O. & Ohmura, A. (2001) 'Highest-latitude Coral Reef at Iki Island, Japan', *Coral Reefs* 20, 9–12.

17

Protecting the marine environment

Controlling pollution in the Coral Triangle

Ann Marie Manhart

The Coral Triangle is an area lying in the West Pacific spanning the seas of the Philippines, Malaysia, Indonesia, Timor Leste, Papua New Guinea and the Solomon Islands. A recent study conducted by the University of Queensland in Australia in cooperation with WWF (World Wide Fund for Nature) has identified this region to be the 'birthplace of the seas,' where a significant amount of marine organisms spawn only to travel to different parts of the world. The Coral Triangle encompasses only 1% of the world's oceans, yet it contains 76% of reef-building species and 37% of coral reef fish (Hoegh-Guldberg et al. 2009).

The Coral Triangle is the epicenter of marine biodiversity in the world, yet there is no mechanism in place to ensure its survival. A partnership between the six countries in 2007 brought forth the Coral Triangle Initiative on Coral Reefs, Fisheries and Food Security (CTI-CFF 2012). Yet this partnership is non-legally binding and tackling pollution does not seem to be a main priority. Close to 150 million inhabitants live in the Coral Triangle and a significant portion is directly dependent on it for their livelihood and food needs (Hoegh-Guldberg et al. 2009). Rampant increase in population adds more to the stress in these coastal areas. A study conducted by University of Queensland and WWF emphasized the need for concrete action plans beyond the Kyoto Protocol to lower carbon emissions that have affected the Coral Triangle. In the study, it is mentioned that carbon dioxide emissions have increased the temperature, in turn bleaching the corals and acidifying the seas. It is also mentioned, however, that besides a concrete gesture from the global community, regional efforts should also be undertaken to avert the destruction of this biosphere (Hoegh-Guldberg et al. 2009).

Having been a stakeholder in the Coral Triangle and culling from experiences in the aquaculture industry, the author is interested in pollution control as a subject matter. The absence of water treatment and the disposal of wastes into the sea are the most common causes of pollution afflicting Southeast Asia. An expedition by the California Academy of Sciences to the Verde Island Passage, said to be one of the most biodiverse areas in the Coral Triangle, resulted in findings that deep-sea fish develop tumors due to untreated human sewage dumped in the sea (California Academy of Sciences 2012).

This chapter will discuss the pollution from the three ASEAN states in the Coral Triangle Large Marine Ecosystem (CTLME), namely the Philippines, Indonesia and Malaysia. This chapter also proposes four pillars for solving pollution in the Coral Triangle Large Marine Ecosystem: society, economics and business, environmental science, governance and policy.

Society

It is much easier to sell the conservation of a whale shark than it is to sell recycling of wastes or the establishment of sewage treatment. Without delving deeper, we fail to recognize that having an efficient waste management system leads to better water quality and the survival and regeneration of our marine species. How many marine animals have fallen victim to ingestion of plastic waste or to poisoning from foreign substances or increase in concentration of certain substances in the sea? The answer to properly engaging the public is by examining a country or region's social structure. The Southeast Asian States in the CTLME share similarities and differences. Religion plays a big role in all three societies: the Philippines being largely Catholic, Indonesia (with the exception of Bali) and Malaysia being largely Muslim. At times, engaging religious institutions in the debate about pollution and the environment can get the message across, as these religious institutions are influential actors in society. In the Philippines, the Catholic Church was instrumental in installing two presidents through popular revolutions. The debate on mining is also spearheaded by Catholic institutions, such as the Ateneo School of Government, a Jesuit-run institution in the Philippines.

Social class is also something very important in environmental advocacy. Usually in diffusion of innovations, the more powerful higher strata of society set the tone. It is thus important to convince the upper classes on the importance of pollution control, as they have the power and influence in their respective countries. Nonetheless, social mobility in these societies is possible. Hence grassroots movement and the education and engagement of stakeholders are also measures to be taken. Another phenomenon to be tackled is urbanization in these countries. With the advent of urbanization, more and more people leave the countryside to work and live in cities. Megacities such as Metro Manila, Jakarta, and Kuala Lumpur are all bustling and growing cities. Aside from these capitals, there are other growing urban areas in the Philippines, Indonesia, and Malaysia.

The middle and upper classes, having access to education, possess the advantage while the lower classes are often times relegated to the slums and man the informal waste sector. How does one reach and engage each sector of society? How should one implement pollution control tailored to different sectors of society? Should the upper classes be taxed for pollution control to subsidize the share of the lower classes? Although the lower classes living in shanties with no proper toilets and with no garbage bins are often blamed for polluting waterways, it is the upper class consuming more and thus generating more waste. In areas without a proper waste management system not financed by the populace, it is the upper class carrying the burden of pollution. In the Rio Declaration of 1992 of the United Nations Conference of Environment and Development, Principle 16 or the 'polluters pay principle' describes how those who pollute are responsible to finance remediation (UNEP 2012b). Hence in the debate to preserve the Coral Triangle, one must ask who finances interventions to prevent and reduce pollution. Such interventions may be the installation of wastewater treatment plants and the construction of proper secured landfills that treat leachate before it is disposed into waterways and the sea. Consumption is related to pollution. In waste management, the more resources are used, the more waste products result. Hence, if one compares a rich household versus a poor household, the consumption of a rich household then exceeds that of the poorer one. It is therefore ideal for the upper class of society to share a bigger part to finance interventions that will reduce pollution. Instead of simply relying on foreign aid or the Brentwood institutions to finance these pollution prevention infrastructures, stakeholders of these developing countries are encouraged to make an investment in the environment, and for future generations.

Another phenomenon to utilize is the growing network of netizens. Because of urbanization, more and more people have become connected to the worldwide web. Social media are fast becoming tools for awareness and advocacies. Social network pages and blogs have cropped up as 'watchdogs' for conservation. In the Philippines, environmental NGOs have taken to social media to shed light on issues. From stopping reclamation of Manila Bay to lobbying for the U.S. government to pay fines on the destruction wrought on a UNESCO World Heritage reef, people have been taking their grievances to the internet. Pollution control may also be given a boost when awareness is tapped through the new tools available. Nowadays, politicians review the pulse of the public by looking at trends in networking sites. The initiation of pollution control in the Coral Triangle may therefore become a grassroots movement through social media. In particular, someone has initiated a Facebook page on the Coral Triangle. The page highlights biodiversity issues besetting the countries in this region.

Economics and business

Pollution is often the result of wastes or excess materials and substances. Knowing the livelihood sectors driving the economy of the CTLME countries will determine how to reduce pollution. For example, in an economy dominated by the agricultural sector, most of the pollution will come from fertilizer substances such as nitrogen (N) and phosphorus (P). Therefore, a sound knowledge of the economics of these countries will also help one determine and finance pollution control. Are livelihood sectors polluting the marine environment willing to pay for clean-up and future technologies? What is the economic trade-off to a cleaner environment? Can one quantify a pristine environment? And will these major sectors earn a profit from a cleaner environment? These questions must be asked and answered by thoroughly studying the economic setup of the Coral Triangle countries.

It is also important to note that the Philippines, Indonesia, and Malaysia have been increasing their share of tourism. Their tourism highlights are their beaches and marine life. Sustaining a tourism industry makes it more paramount for these countries to preserve their marine environment. Are the countries willing to pull funds together to finance a region-wide control of pollution? By the same token, UNEP's 'green economy' has become a byword of the past few years. Is there really a green economy or green industry? Can industries crop up from cleaning the pollution caused by other sectors? Can industries crop up to develop technologies to reduce pollution? This area of business is not yet highly tapped and promises a better future as the Philippines, Indonesia and Malaysia enter more prosperous times. While Malaysia has been faster in its growth, the Philippines and Indonesia are now touted by Ruchir Sharma, head of emerging market equities at Morgan Stanley as 'break out nations.' The theory is that increase in per capita growth rate will exceed market expectations such that increased investments will result (Reuters 2012). This development entails two key issues. As businesses grow, there is a greater influx of activity, thereby creating more cases of pollution. In the scenario that infrastructure such as waste water treatment plants, secured landfills, and incinerators are not available, it is the environment that will suffer. Take the case of Manila, a metropolis in which the increase in garbage is not controlled, and where secured landfills are not the norm. Trash ends up clogging city canals and rivers and eventually pollutants reach the bay. By way of contrast, increase in economic activity entails an increase in the finances of the country thereby making it more viable to invest in interventions that will reduce or prevent pollution.

Jakarta faces the same scenario during the monsoon season, where a mix of floodwaters and solid waste creates an unhealthy environment for the populace. Aljazeera documents that the problem of dumping wastes is seen as the biggest hindrance to keeping Indonesia free from floods (Aljazeera 2013).

Environmental science

A knowledge-based body of work is necessary for decisions to save the Coral Triangle. This is achieved by delving into the science of the environment. Policies should therefore be based on scientific studies. In democratic States in which a strong system of policymaking is based on citizens' initiatives, there are provisions wherein the academe, NGOs, and other interest groups are encouraged to give their contributions before a policy is made into law. A good example of this is the present initiative of the European Union on food waste. Before coming up with a common Food Waste Policy of the EU, a significant part of the initiative is to encourage scientists to come up with studies that analyze food waste, thereby providing a knowledge base from which decisions can be taken logically.

Pollution can be defined as the introduction of a material or substance in a given area, such that its concentration affects plant, animal, and human life and the environment itself. Pollution may be seen from both the material and the substance level; material meaning on a bigger dimension such as human sewage, animal waste or plastic; while on a substance level, it means looking at it on an elemental or compound level, such as studying pollutions caused by mercury, nitrogen, phosphorus or PCB (Polychlorinated biphenyl).

Visible pollution

To some extent, the campaign on plastics covering an area of the Pacific called the 'plastic soup' has come to the attention of many people, albeit somewhat too late. The United Nations Environment Program (UNEP) estimates 13,000 pieces of plastic waste for every square kilometer of ocean. This environment breeds ocean insects that eat plankton and fish eggs. This 'plastic soup' has said to have grown a hundredfold over the past 40 years (Discovery News 2012). In the Coral Triangle, marine animals such as whale sharks, and fish were found dead because of plastic ingestion.

Indeed, the remediation process will take years to clear the sea of the garbage it is still continuously holding. Consumption patterns definitely affect the increase of plastic use. Furthermore, the lack of facilities to process plastic waste leads to this accumulation of plastic in the seas. If one looks at the history of waste in the Philippines, the earlier populations used banana leaves and coconut fronds to package their food. Instead of plastic crates, reed baskets and wooden crates were used to transport products. Hence, landfills were sufficient to disintegrate these biodegradable materials after a certain period. The problem is that many of the garbage dumps did not evolve as the population switched to more plastic products. Indonesia is beginning to see the value of solving its household waste problem, while Malaysia has gone ahead to develop incinerators converting waste to energy. In the Philippines, it is forbidden to incinerate waste, as provided in its Clean Air Act, and as the law mentions to reduce the emission of CO_2 (CRA Law 2012). This is a clear example of how the lack of exposure to new technologies results in archaic laws. Present-day incineration systems have flue gas cleaning that rids the air of toxins. An even newer incineration system includes recovery of precious metals such as copper. Incineration can be a viable mechanism to reduce this 'plastic soup' in the sea. It is also important to note that garbage dumps filled with household wastes emit methane, one of the strongest greenhouse gases. Methane gas from secured landfills can also be tapped as an energy source but the efficiency is less than the heat from incineration.

Of course, a better recycling system also helps. However, not everything can be recycled and not all recycling is efficient. Each material has a certain stage in its lifetime when it is more costly to recycle it than to create a new one. An extensive study on the sources and pathways of plastic used in the CTLME countries is of the utmost importance. By knowing which sector and industry use which type of plastic, alternative materials or more efficient ways of consumption can be determined.

There is also the problem of hospital wastes. When incineration is banned, as is the case in the Philippines, these infectious wastes are likely to find their way into waterways. This is especially so when they are disposed in garbage dumps located near the water, and also because of the fact that this region is no stranger to very strong typhoons and earthquakes. Certainly studies are needed to document where these hospital wastes end up. Nonetheless, because of their nature and the probability that they can spread disease and affect human and animal lives as well as pollute the environment, Principle 15 of the Rio Declaration or the 'precautionary approach' is a fitting way to deal with the issue. To quote the Rio Declaration: 'Where there are threats of serious or irreversible damage, lack of full scientific certainty shall not be used as a reason for postponing cost-effective measures to prevent environmental degradation' (UNEP 2012b). Therefore, interventions such as proper incineration and secured landfills are needed to make sure these hazardous wastes are properly treated and contained securely.

Invisible pollution

In the field of regional nutrient analysis, the author has done work on nitrogen (N) and phosphorus (P), based on the contributions of three ASEAN States that straddle the Coral Triangle (Manhart 2012). Nitrogen (N) and phosphorus (P) are substances that are precursors to eutrophication. Eutrophication is a condition wherein nutrients are plentiful in an artificial manner such that they result in an abnormal algal and plankton bloom. This is harmful because in time these algae and plankton decay and contribute to oxygen depletion in the sea. This will then affect reef systems and eventually fish, which most of the population rely on for protein needs. N and P are also indicators of pollution because an excess of these nutrients indicates human sewage, agricultural, aquacultural and other wastes infiltrating an otherwise pristine marine environment.

In recent developments John Gunn, CEO of the Australian Institute of Marine Science, gives figures of how over the span of 27 years to 2012, more than 50% of the Great Barrier Reef cover has been lost. Extensive surveys from his team list the following: 10% of the damage is due to coral bleaching, 48% due to storms, and 42% due to infestation of crown of thorns starfish. Coral bleaching is due to warmed oceans, a result of increased CO_2; storms are a part of nature and the reef has the capacity to recover from them; however, the 42% damage comes from a local pest, the crown of thorns that eat up reefs and which, according to Gunn, is a result of increased nutrients coming from fertilizers reaching the reef (ABC News 2012). This may be the same scenario that may happen in the Coral Triangle Large Marine Ecosystem, and any efforts of remediation might prove futile if done too late.

The methodology used in the author's study is a substance flow analysis (SFA). Particularly, subsystems in the areas of agriculture, aquaculture, and the activity 'to nourish' are explored by a mass balance approach. The contribution of industry was not tackled in the study. The base year used is 2005. While the total contributions of the three investigated countries are estimated to be 700 kilo tons/annum of N and 190 kilo tons/annum of P, the study also deals with the contribution of each country and of selected sectors to the CTLME. Nonetheless, additional work to collect better and improved data is needed to verify the conclusions about nutrient loads in the CTLME drawn in the study.

The main sources of N and P are excess nutrients from commercial fertilizers in agriculture that accrue as stocks in soils and run-off to rivers; untreated human and animal wastes dumped in waterways; and effluents from aquaculture. In recent months the waters off Sabah have been found to contain harmful algal blooms commonly called 'red tide.' In January 2013 a warning on shellfish consumption was instituted in Sabah due to some deaths from paralytic shellfish poisoning. Paralytic shellfish poisoning is caused by the human ingestion of shellfish that have previously eaten this harmful alga. This recent turn of events collaborates the author's finding of high N and P outflows into the sea by Sabah, a phenomenon largely cause by soil run-off from palm oil plantations (Manhart 2012).

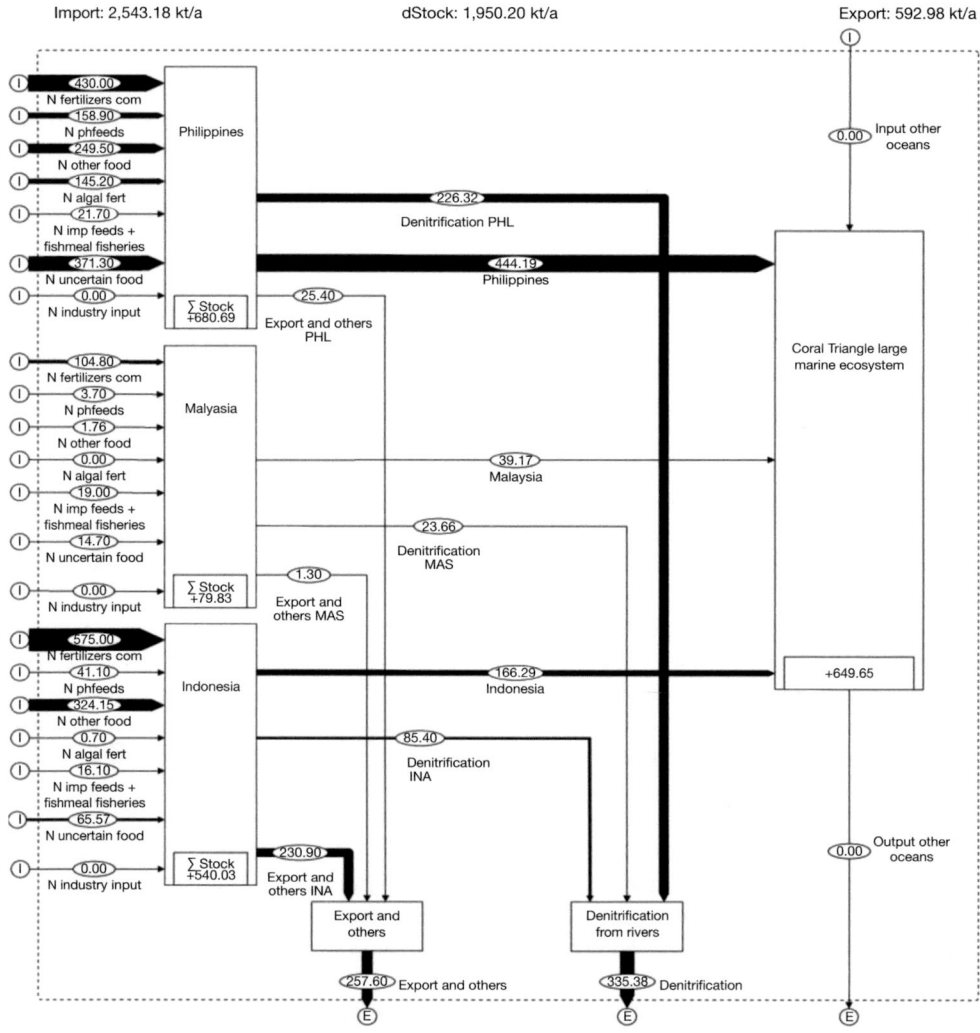

Figure 17.1 Nitrogen sources and sinks in the CTLME

Source: Manhart (2012).

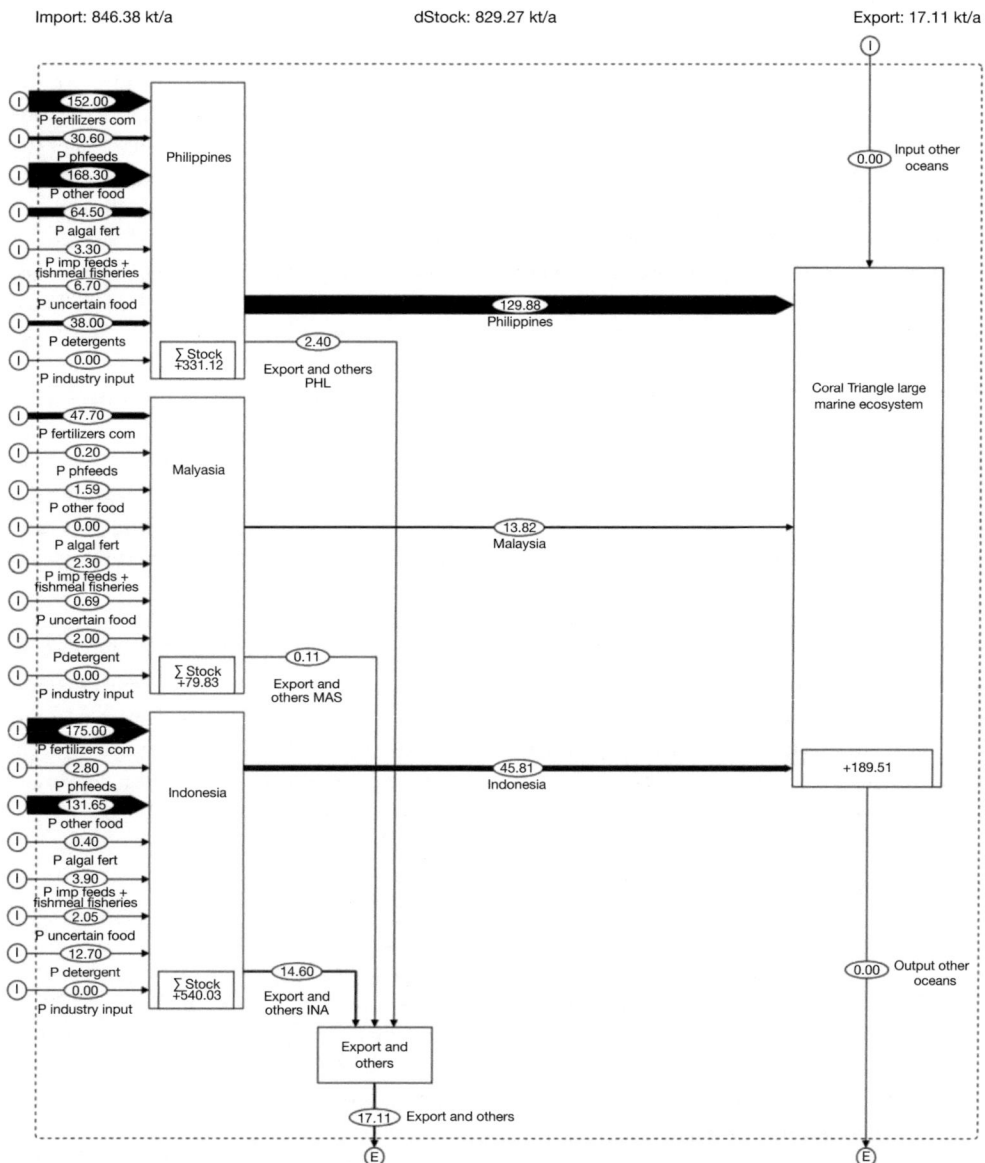

Figure 17.2 Phosphorus sources and sinks in the CTLME

Source: Manhart (2012).

The various countries have different scenarios on their sources and pathways for N and P. Nevertheless, the bottom line is that one day, without controlling N and P in the Coral Triangle, its water quality, marine life and the people who are directly relying on this large marine ecosystem will be greatly affected.

What is interesting to note is that nutrient flows into the country stocks are significantly higher than the flows to the CTLME. This is an environmental problem because even though reaction time is slow, these stocks will be future emissions into the CTLME. Although nutrient inflows in the future would have been reduced or even eliminated, the situation that these nutrients are already in stock creates a huge dilemma and will make remediation even more intensive and expensive.

It has also been proved that interventions are needed to prevent pollution. A more prudent use of commercial fertilizers and interventions such as landfills and wastewater treatment plants are recommended to decrease the nutrient load in the CTLME. There are other substances that are unseen and hence do not merit enough attention. A category of these harmful substances is called persistent organic polluters (POPs). POPs can lead to very serious health disorders in humans such as cancers, birth defects, dysfunctional immune and reproductive systems, an increased susceptibility to illnesses and even diminished intelligence (Stockholm Convention 2012). POPs' coverage is long range and transboundary such that they become even more lethal. They are also lipophilic, meaning they accumulate in fatty tissues of animals. Hence, increased concentration of POPs over the years make them even more dangerous.

The initial 12 POPs defined under the Stockholm Convention are further categorized: pesticides, industrial chemicals, and by-products. One of the industrial chemicals identified and also a by-product, PCB has been identified as the cause of death of fish and the culprit for decrease in fish spawning. Furthermore, it is linked to a disruption of both the reproductive and immune systems of seals. PCB is also used in industry as a heat exchange fluid and as an additive in paints and plastics (Stockholm Convention 2012). Indeed, much study still needs to be done on the sources, pathways, and sinks of POPs in the Coral Triangle.

Another substance needing further study in the region is mercury, often a result of mine tailings that find their way into rivers and seas. Like POPs, mercury is also lipophilic and can accumulate in the both animals and humans. Recently, states have come together to curb mercury pollution. Named after a city in Japan, Minamata, where the population suffered from mercury poisoning, the Minamata Convention aims to address the reduction of mercury in the environment by regulating the use of mercury in mining activities and the mercury emissions of coal power plants, incineration, and cement factories as well (UNEP 2013).

Governance and policy

'Governance is different from government.' Borrowing the words of the CEO of WWF Philippines, Jose Maria Lorenzo Tan, he opines that in order for a policy to work, it has to be formulated through consultation and implemented with participation as a key anchor (Tan 2012). Therefore it is also important to look at political structures that allow participation in the decision-making process. Are there legal mechanisms allowing stakeholders to decide on the future their locality will take? Are they equipped with the proper knowledge about pollution to enable them to make sound decisions?

In Europe, there is a European-wide convention called the Aarhus Convention. The Convention contains three themes: 'access to information, public participation and access to justice' (UNECE 2012). Basically, the Convention describes how ordinary citizens can participate in the decision-making process and the approval of a certain environmental policy, as well as having the means of redress for injustices. There is no similar convention in Southeast Asia at this point in time. But if someday the region were to have one, the question arises: How can the public initiate an environmental program or policy to address present needs?

Ipat Luna, an environmental lawyer and advocate, states that oftentimes legislation takes a long time to implement. Therefore, it is much faster to deal with problems by invoking existing laws using the right mix of strategies. She proposes that agreements can be forged among relevant sectors. Luna is the poster girl for this grassroots movement. As a founding member of a foundation she, along with professionals, established the Taal Lake Conservation Center through private funding from donors and an agreement with the local government. At present they are in a constant dialog with local stakeholders in the Philippines, as well as local and foreign experts on how to solve pollution in their area (Luna 2012).

As promising as this situation is, how do we multiply this several times such that other local governments and other like-minded individuals and groups can engage in similar efforts? The answer lies in a central body to encourage these efforts. Once again, the role of government should be to encourage and provide expertise in matters such as pollution control.

If local governments are not aware of problems and solutions, it is up to the central government to provide venues on how pollution can be dealt with. It is up to the departments involved such as the natural resources department or, in other cases, the water department to help initiate these movements. It is, nevertheless, understandable that developing countries would have to focus their efforts and resources in matters that they deem to be more urgent. And that is where it is the role of international organizations already having the expertise and experience in such issues to intervene. The United Nations Environment Programme (UNEP), United Nations Industrial Development Organization (UNIDO), United Nations Development Programme (UNDP), Global Environment Facility (GEF), World Bank and Asian Development Bank are active in facilitating knowledge, funding, and initiating cooperative efforts.

In fact, the Coral Triangle Initiative on Coral Reef Fisheries and Food Security (CTI-CFF) is in partnership with the Asian Development Bank among a host of other organizations such as USAID, Conservation International, WWF and the Australian government. In 2007, the six countries within the CTLME formed a multilateral partnership to address the threats facing coastal and marine resources. Although it is not a legally binding agreement the Coral Triangle Initiative is a very good start towards cooperation within the CTLME. The main themes of the Initiative are 'people-centered biodiversity conservation, sustainable development, poverty reduction and equitable benefit sharing' (Coral Triangle Initiative 2012). These are indeed urgent issues to deal with. Nevertheless, pollution could and should also be included as it is also a very urgent issue. In 2009, the CTI adopted a 10-year regional action plan. To quote the CTI, the action plan's goals are: 'priority seascapes designated and effectively managed, ecosystem approach to management of fisheries, marine protected areas established and effectively managed, climate change adaptation measures, and threatened species status improving' (Coral Triangle Initiative 2013). Again, a specific mention of pollution is nowhere to be found. It seems not to be a priority in this regional initiative, although pollution in this region is transboundary as these nations share one large marine ecosystem (LME). The need to raise awareness on this issue is therefore crucial, given that on an initial inventory and substance flow analysis by the author, the stock of nitrogen and phosphorus in the Coral Triangle has been accumulating dangerously over the years (Manhart 2012). It should also be a logical aspect towards maintaining marine protected areas (MPAs). MPAs will not be as effective if the waters hosting marine life are rife with pollutants.

Regional multilateral environmental agreements (MEAs) are indeed a tool for environmental protection through cooperation. Cooperation includes collaborative studies from experts within each country, a venue for those studies to be presented, and technical cooperation as in the case of Malaysia, already way ahead of the Philippines and Indonesia in the waste management field. MEAs also allow regional policies to flourish, ideally trickled down to the national level and then the local level. Biodiversity and pollution go beyond national borders as the Coral Triangle is a large marine ecosystem. Hence it is ideal for the stakeholder countries to come together. However, legally binding MEAs require money and financial contributions from the member states. If not all the countries are willing or capable to contribute, then usually the MEA is put on hold and an initiative like the Coral Triangle Initiative is put in place instead.

This non-legally binding agreement has also its advantages because it serves as a testing ground for countries to see if they can work with each other on a regional project before embarking on legally binding themselves to an obligation.

There are various marine regional MEAs across the world. A pioneering one is the Helsinki Convention. In 1974, the Baltic States, Denmark, Sweden, Finland, former East and West Germany, the former USSR, and Poland came together to tackle pollution in an entire sea for the first time. In 1980, the Convention entered into force, while in 1992 in the context of new political developments amendments were made and additional states including the European Community became party to the agreement. The Helsinki Commission is the governing body of the Convention. It is interesting to note that the Helsinki Convention deals largely with pollution control. The Convention differentiates between land-based pollution and pollution introduced to the sea by shipping vessels. The Convention is structured such that it provides for national laws within each country party to the agreement to come up with legislation individually or jointly to prevent and eliminate pollution in the Baltic Sea. This is done through 'best available practices' and 'best available technologies.' It also invokes the Rio Declaration's 'polluter pays principle' (Helsinki Convention 2014). What is interesting with this convention is that all parties are bound to calculate and measure emissions in a scientific matter. This determines the amount and concentration of the substances as well as the sources and pathways of the pollutant. By doing so, this convention is anchored on a very logical base, and decisions to halt pollution are carried out in a scientific manner.

In another part of the globe, decades after the Helsinki Convention, Angola, Benin, Cameroon, Congo, Côte d'Ivoire, Democratic Republic of the Congo, Equatorial Guinea, Gabon, Ghana, Guinea, Guinea-Bissau, Liberia, Nigeria, Sao Tome and Principe, Sierra Leone, and Togo came together to protect the Guinea Current Large Marine Ecosystem. Unlike the Helsinki Convention covering a sea, the Guinea Current LME Project by its very name covers a large marine ecosystem just like the Coral Triangle. The Project started in 1995 with a few countries tackling 'water pollution control and biodiversity conservation.' The GEF funded this Project through a US$6 million grant along with co-financing from other countries. In 1998, Benin, Cameroon, Côte d'Ivoire, Ghana, Nigeria, and Togo adopted the Accra Declaration (GCLME 2012).

The Accra Declaration deals with the sound management of the Gulf of Guinea region. Although it is soft law and not as concrete as the Helsinki Convention, the Declaration recognizes 'severe rates of coastal erosion, the threat of flooding, the seriousness of pollution, loss of biological diversity and depletion of fishery resources.' It recognizes that adoption of a 'standardized regional approach in a cooperative effort' is necessary to combat the environmental problems that this region faces (Accra Declaration 1998). Today there is an Interim Guinea Current Commission working in partnership with UNIDO (GCLME 2012).

There are several other marine MEAs. The Barcelona Convention (Mediterranean Sea) also deals with pollution and is in cooperation with UNEP's Regional Seas Program (UNEP 2012a). There is also the Convention on the Protection of the Black Sea against Pollution administered by the Black Sea Commission in partnership with GEF, UNDP, and the EU (Black Sea Commission 2012); and the Caspian Convention dealing with pollution in the Caspian Sea, a land-locked body of salt water in Central Asia (Caspian Environment Program 2012). There are more MEAs that apply the same regional cooperation principle to protect transboundary rivers and mountains.

Nonetheless, there is no one solution to regional pollution. The lessons from other regions can be taken as bases, however different the situations. But without a thorough knowledge of sociology, economics and business, and environmental science, all policy will fail miserably. It is simply not enough to transpose existing policy from other regions; a thorough understanding of how a region works is of the utmost importance.

Therefore there is a need for experts in the field to continue answering various questions posed in this chapter and contributing to the knowledge base. Contrariwise, since time is of the essence, governments should increase cooperation to bring their experts to work for a solution, as well as increase awareness so solutions in the grassroots level may be allowed to blossom.

Conclusion

The Coral Triangle holds a special place in the region and the whole world. The 'birthplace of the seas' must not be allowed to languish due to ignorance and negligence. Humankind cannot carry on making all the mistakes it possibly can. Let us learn from the previous generation and previous mistakes to help avert environmental degradation in this very special marine ecosystem.

This chapter proposes four pillars to aid in decisions to avert that destruction. First is the study of *society*: the dynamics of social classes within the three mentioned countries in this chapter; who holds the burden of pollution and who foot the bill; and the existing tools or media that actors in society use and have access to get the message across. Second is the study of *economics and business*: the livelihood sectors that drive the economy of the three countries; the sectors that pollute; financing remediation and prevention of pollution by said sectors; and the elaboration of 'green industries' as environmental service providers. Third is the study of *environmental science*: the science of pollution; visible and invisible pollution; sources, pathways and sinks of pollution-causing substances and materials; and the corresponding interventions to avert and remediate contaminated sites. Lastly, the study of *governance and policy* is crucial as the legal basis for the appropriate action plans and programs for pollution prevention. Legal frameworks tackling marine pollution control in other regions exist. But it is precisely this mix of variables defined by the first pillars that define the last pillar, as there is always a unique solution tailored to the needs of a region. In conclusion, decisions on saving the Coral Triangle require a holistic, multidisciplinary approach, taking into consideration the needs of its people whilst preserving this diverse marine environment.

We have only one Coral Triangle. It would be a tragedy to procrastinate, lose a sense of urgency, and eventually lose this large marine ecosystem.

Bibliography

ABC News (2012) 'Half of Barrier Reef Has Disappeared: Report'. Available at http://www.abc.net.au/news/2012-10-02/barrier-reef-coral-cover-halved-report/4290230.

Accra Declaration (1998) Available at http://gefgclme.chez.com/english/accra.htm.

Aljazeera (2013) 'Stemming the Tide'. Available at http://www.aljazeera.com/programmes/101east/2013/02/2013211837761773.html.

Black Sea Commission (2012) Convention on the Protection of the Black Sea against Pollution. Available at http://www.blacksea-commission.org/_convention.asp.

California Academy of Sciences (2011) 'Reefs to Rainforest the Great Expedition'. Available at http://abclocal.go.com/kgo/video?id=8404396&syndicate=syndicate&section.

Caspian Environment Program (2012) Caspian Convention. Available at http://www.caspianenvironment.org/newsite/Convention-FrameworkConventionText.htm.

CRA Law (2012) Clean Air Act of the Philippines. Available at http://www.chanrobles.com/philippinecleanairact.htm#.UHfi7BjDHu0.

Coral Triangle Initiative (2012) 'About the Coral Triangle Initiative'. Available at http://www.coraltriangleinitiative.org.

Coral Triangle Initiative (2013) 'Regional Plan of Action'. Available at http://www.coraltriangleinitiative.org/library/cti-regional-plan-action.

Discovery News (2012) 'Pacific Plastic Soup 100-fold Increase'. Available at http://news.discovery.com/earth/oceans/pacific-ocean-plastic-increase-120509.htm.

GCLME (Guinea Current Large Marine Ecosystem) (2012) 'Guinea Current Area'. Available at http://gclme.org/index.php?option=com_content&view=article&id=4&Itemid=5.

Hoegh-Guldberg, O., Hoegh-Guldberg, H., Veron, J.E.N., Green, A., Gomez, E.D., Lough, J., et al. (2009) *The Coral Triangle and Climate Change: Ecosystems, People and Societies at Risk* (WWF Australia).

Helsinki Convention (2014) 'Helsinki Convention' available at: http://helcom.fi.

Luna, I. (2012) Interview on Grassroots Environmental Movements, 23 September, 10 October.

Manhart, A.M. (2012) 'Assessing Nutrient Pathways and Sinks in the Coral Triangle', Master's thesis, Vienna University of Technology.

New Straits Times (2013) 'Warning of Shellfish Poisoning in Sabah', available at: http://www.nst.com.my/nation/general/warning-of-shellfish-poisoning-in-sabah-1.197232.

Reuters (2012) 'Breakout Nations'. Available at http://in.reuters.com/video/2012/04/24/forget-china-and-india-its-time-for-brea?videoId=233876298.

Stockholm Convention (2012) 'POPs'. Available at http://chm.pops.int/Home/tabid/2121/mctl/ViewDetails/EventModID/871/EventID/230/xmid/6921/Default.aspx.

Tan, J.-M.L. (2012) Interview on the CTLME, 23 September.

UNECE (2012) Aarhus Convention. Available at http://www.unece.org/environmental-policy/treaties/public-participation/aarhus-convention.html.

UNEP (2012a) Barcelona Convention. Regional Seas Program. Available at http://www.unep.ch/regionalseas/regions/med/t_barcel.htm.

UNEP (2012b) Rio Declaration. Available at http://www.unep.org/Documents.Multilingual/Default.asp?documentid=78&articleid=1163.

UNEP (2013) Minamata Convention. Available at http://www.unep.org/newscentre/default.aspx?DocumentID=2702&ArticleID=9373.

18

Governance of forests

Regional institutions in East Asia

Hang Ryeol Na, Jack P. Manno, David A. Sonnenfeld and Gordon M. Heisler

All forestry should be social in the sense that trees render valuable services to every form of society (Westoby 1989: 217). But given the global connectedness of human society and forests, forestry should also have a global or international perspective. One of the most important contemporary environmental challenges of our time is that forestry involves systems that are intrinsically global such as climate change or are tightly linked to global pressures, for example timber production for the world market, and that require governance at levels from the global to the local (Dietz et al. 2003).

Regional forest governance is important for East Asia.[1] The region has by far the most forest dwellers and the highest human population density in forests, compared to other continents such as Africa or Latin America (Chomitz & Buys 2007). About 450 million people depend on forests for their livelihood in Asia (Liu et al. 2005), but resource degradation, land-use conflicts, large-scale conversion of forest land to industrial cropland and inequality remain widespread in the region. Even though net forest area increased between 2000 and 2010 by around 14 million hectares in Asia (Yasmi et al. 2010), deforestation, which is mainly caused by illegal logging (Liu et al. 2005), is still a critical problem in many countries such as Indonesia and Myanmar. As seen in Table 18.1 and Figure 18.1, over the last two decades forest area has been reduced more seriously in forest-rich countries such as Indonesia and Myanmar than in some other countries such as Vietnam, where it increased as a result of forest rehabilitation and plantation programs (FSIV 2009). Considered for their value in storing carbon, many Asian forests are worth more as standing environmental assets than as cleared and burned fields. However, forest holders are currently unable to tap the financial value of these environmental benefits under existing global carbon finance systems (Chomitz & Buys 2007).

Rationale for regional forest governance

Many forest issues, such as forest fires, forest insects and disease, and forest products trade, cross national boundaries. Countries can pool resources to address problems that are too costly for one country to address effectively alone. Countries can create synergies by sharing information, experience and expertise, e.g. through research networks. Regional groups may carry more

Table 18.1 Land and forest area of major countries in East Asia for the last two decades

Country*	Item+/year	1991	1996	2001	2006	2011
CN	LA	932,742	932,748	932,749	932,749	932,749
	FA	159,127	169,057	180,209	195,807	209,624
ID	LA	181,157	181,157	181,157	181,157	181,157
	FA	116,631	107,063	99,099	97,172	93,747
MY	LA	32,855	32,855	32,855	32,855	32,855
	FA	22,298	21,905	21,451	20,803	20,369
MM	LA	65,354	65,354	65,354	65,347	65,329
	FA	38,783	36,608	34,559	33,011	31,463
PH	LA	29,817	29,817	29,817	29,817	29,817
	FA	6,625	6,898	7,172	7,446	7,720
TH	LA	51,089	51,089	51,089	51,089	51,089
	FA	19,495	19,222	18,983	18,913	18,987
VN	LA	32,549	32,549	31,109	31,007	31,007
	FA	9,599	10,780	11,995	13,221	13,941

Notes:

* CN indicates China, ID Indonesia, MY Malaysia, MM Myanmar, PH Philippines, TH Thailand and VN Vietnam.

+ LA = land area; FA = forest area. Values are in 1,000 hectares. FAOSTAT defines 'land area' as the total area of the country excluding area under inland water bodies. 'Forest area' is the land spanning more than 0.5 hectares with trees higher than 5 meters and a canopy cover of more than 10 percent, or trees able to reach these thresholds in situ. It does not include land that is predominantly under agricultural or urban land use.

Source: Statistics Division of the Food and Agriculture Organisation (FAOSTAT).

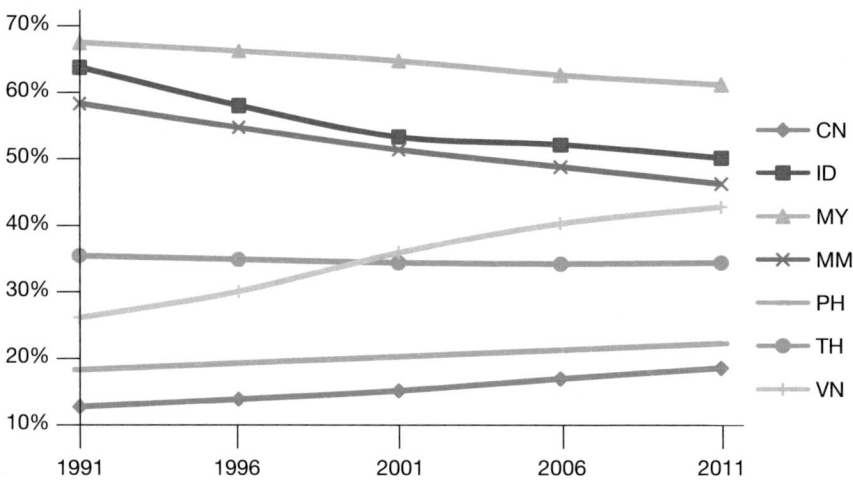

Figure 18.1 Percentage forest area of major countries in East Asia for the last two decades

Source: Calculated from Table 18.1. Percentage of forest area was obtained by dividing FA by LA.

political and economic weight than individual countries. However, regional initiatives are a complement, not a substitute, for work at the national level. For example, individual countries have laws and institutions enabling them to take actions, e.g., to levy taxes and enforce regulations, whereas regional groups do not usually have these powers (Martin 2004).

Empirically, regional-level forest-related processes have tended to be more flexible than global-level arrangements in their use of hard law and new governance instruments (Howlett & Rayner 2010). Global-scale processes have frequently become locked in debates over the desirability of legally binding commitments to slow forest conversion or promote sustainable forest management, which caused unease with mounting transaction costs of global regimes and stagnant global convention fatigue (e.g., Patt 2010). Meanwhile, various regional forums and non-governmental forest certification schemes have made significant progress in framing and implementing relatively comprehensive approaches.

Theoretically, in the multiple levels of environmental governance structures, the issue of regional governance has become salient in discussions of new regionalism, institutional interplay, authority migration, etc. (e.g., Andonova & Mitchell 2010; Batterbury & Fernando 2006; Biermann et al. 2009). For example, the conclusion of Kanie (2004) was that paying more attention to vertical linkages among global, regional, national, and local levels will reduce the loss of resources that is produced through friction between the levels. In this connection, a possibility for improved synergy between levels may exist in enhancing the regional level as a device that bridges the gap between national and global levels, as exemplified by the cases in the next section.

Genesis of regional forestry institutions

FAO regional forestry commissions

Regional forest governance dates back at least to World War II. In 1945, the Food and Agriculture Organization (FAO) was created with responsibility within the UN system for forests, which account for approximately 4% of the FAO budget (Humphreys 2006: 22). By resolution of the third session of the Conference of FAO in 1947, the first Regional Forestry Commission was established in Europe (Koné et al. 2004). By 1961, each major geographical region of the world had its own forestry commission. The regional commissions, comprising the respective FAO member countries of the region, meet every two years to address the most important forestry issues in the region at both a policy and a technical level. They serve as a link between global FAO fora, the United Nations Forum on Forests (UNFF), and national implementation (Rayner et al. 2010: 164).

By the time of the first global-level meeting of the FAO Committee on Forestry (COFO) in 1972, 50 regional forestry commission meetings had already been held. By the end of 2013, the regional commissions will have met 155 times, compared with 21 meetings of COFO and a combined 18 meetings of the Ad Hoc Intergovernmental Panel on Forests (IPF), the Intergovernmental Forum on Forests (IFF) and the UNFF. As shown in Table 18.2, the simple fact is that more international dialogue on forests historically has taken place at the regional level than at the global level (Koné et al. 2004).

Although several controversies undermined the credibility of the FAO in the early 1990s, the depth of the FAO's reach, with its regional forestry commissions and its field-level work, surpasses other institutions (Tarasofsky 1999: 8). Although the Asia-Pacific Forestry Commission (APFC) has done only limited work specific to illegal logging and illegal trade in timber and timber products (Schloenhardt 2008: 42), it is more instrumental in other areas. For example,

Table 18.2 Regional forestry commissions – facts and figures, December 2013

Commission	Year of first meeting	Number of sessions	Number of members (observers)
European Forestry Commission	1948	37	40
Latin American and Caribbean Forestry Commission	1949	28	35 (1)
Asia-Pacific Forestry Commission	1950	25	33 (1)
Near East Forestry Commission	1955	20	27 (5)
African Forestry and Wildlife Commission	1960	18	49 (8)
North American Forest Commission	1961	27	3
Total		155	187

Source: FAO, updated from Koné et al. (2004).

the Asia–Pacific Forest Invasive Species Network (APFISN) was launched as a cooperative alliance of the 33 member countries in the APFC at the 20th session of the APFC, held in Fiji in 2004, and the APFISN actively works to control forest invasive species in Asia (Rayner et al. 2010: 164).

However, these processes are limited in their ability to address drivers outside the forest sector (McDermott et al. 2010). Some issues – for example, climate change or international trade – cannot be addressed sufficiently at the regional level because the problems themselves are global. Concerning the destructive force to forests that exists outside of them, Brown (2001) asks: '[M]any of the root causes of unsustainable forest management have their origin in other sectors and how could a forest agreement control these?'

Forest law enforcement and governance

Although exact figures are difficult to state, illegal logging is rarely omitted as one of the most obvious forces underlying the Asian regional forest policy making processes (e.g., Liu et al. 2005; Potter 2008). Illegal logging, although not a new phenomenon, has increased greatly over the past decade or so and has many manifestations in Asia. In Indonesia, for example, processing capacity and industrial demand for wood products far outweigh the supplies of timber from legal sources, such as concessions, which means that much internal trade is technically illegal. This is especially true in provinces such as Riau, where large pulp and paper companies continue to obtain fiber mainly from natural forests (Potter & Badcock 2001).

Humphreys (2006: 149) records how the role an individual plays can be critical in creating international institutional arrangements by the example of Jan McAlpine, who first raised the issue of illegal logging at the IPF. McAlpine, then a US State Department official, did not consider it worthwhile pursuing illegal logging in the UN negotiating institutions, which were perceived as time consuming, poorly equipped for dealing with problems on the ground and tend to have a culture of defensiveness. She decided to focus on regions, and first approached the World Bank, which agreed to work with the US on the issue and then worked with the UK's Department for International Development in proposing to the World Bank a ministerial conference on illegal logging that the Bank would co-host. They settled on East Asia as the region in which to launch the first regional process. The Indonesian government agreed to host a ministerial conference on illegal logging in Bali.

The outcome of the East Asia Forest Law Enforcement and Governance (FLEG) Ministerial Conference in Bali in 2001, the Bali Declaration, is the most comprehensive regional agreement to date addressing the specific characteristics of the illicit trade in timber (Schloenhardt 2008: 43), and the first intergovernmental statement to elaborate political measures to address illegal logging (Humphreys 2006: 150). In the Declaration, the ministers agreed to 'take immediate action to intensify national efforts, and to strengthen bilateral, regional and multilateral collaboration to address violations of forest law and forest crime, in particular illegal logging, associated illegal trade and corruption, and their negative effects on the rule of law' (World Bank 2001).

The Bali meeting spawned other regional FLEG processes such as Africa FLEG and the Europe and North Asia FLEG, which includes China and Russia to address the illegal trade between the two countries (Humphreys 2006: 161). Also, the issue of illegal logging has become much more pronounced in the international society in recent years (Yasmi et al. 2010: 41). For example, strengthening country capacities to address illegal logging was successfully negotiated into the list of objectives of the International Tropical Timber Agreement (ITTA) in 2006 (Hajjar & Innes 2009). Perhaps most importantly, by recognizing the need for mutually reinforcing actions from both producer and consumer nations, the Bali Declaration can be credited with inspiring action that eventually led to the amendments to the US Lacey Act (2008), the EU's new Timber Regulation (2010), and the Australian Illegal Logging Bill, all of which introduced penalties for those caught engaging in the trade in illegally sourced wood products (Canby 2011).

Yet, of course, there have also been limitations and challenges such as overlapping forest laws at national and subnational levels between civil and criminal laws, between forest and non-forest laws; access to transparent, reliable and timely information; market demand (i.e., not enough differential in price between legal and illegal logs) and so on (Pescott & Durst 2010). Most of these problems still persist.

European forest law enforcement, governance, and trade

The EU's Forest Law Enforcement, Governance and Trade (FLEGT) process arose from dissatisfaction with the lack of progress in tackling the problem of forest degradation through multilateral institutions. In fact, none of the World Bank FLEG processes generated binding commitments among the participating countries, or the creation of systematic mechanisms for monitoring progress toward their agreed aims (Overdevest & Zeitlin 2012). The EU FLEGT initiatives use access to the lucrative EU markets as an incentive to promote responsible forest governance in exporting countries. The EU's strategy in these efforts is in the form of negotiations with individual exporting countries in Asia and Africa to create voluntary partnership agreements (VPAs) that amount to de facto binding law (Bernstein et al. 2010). By linking the improvement of FLEG to regulation of trade (T) and by obtaining the consent of exporting countries, the centerpiece of the FLEGT is the bilateral VPAs to establish licensing systems for the export of legally harvested wood to the European market (Overdevest & Zeitlin 2012).[2]

Although it had started as a bilateral scheme, the EU FLEGT soon established a regional structure with Asia, as the European Commission (EC) launched the 'Regional Support Program for the EU FLEGT Action Plan in Asia' in 2008 (Global Witness 2009).[3] The motivation for regional focus of the program is to complement the bilateral actions that are already taken through VPAs between the EU and timber-producing countries. That is, because of their bilateral nature, VPAs are not able directly to regulate the significant intraregional trade or trade from partner countries to third countries. FLEGT Asia intends to complement VPAs on a regional level by addressing fundamental sector challenges (EU 2010).[4]

For instance, following the logging ban in China in 1998, as the source of forest products for the Chinese market has shifted from China mainly to Southeast Asian countries, the impact on Southeast Asia has rapidly increased unsustainable and often illegal production and export for the market. Exporting of furniture and plywood from China is also growing to the EU, UK and elsewhere using imported and often illegally harvested timber from Southeast Asia. It is argued that it is difficult to interrupt the continuing deforestation in Southeast Asia because (1) the profits from exporting forest products from Southeast Asia to the Chinese market and the profits for Chinese firms that use the forest products to produce plywood and furniture for export to developed countries are substantial and (2) there is a lack of political will at all levels to interrupt these chains of trade and flows of profit (Lang & Chan 2006).

In this context, a separate EU–China Bilateral Coordination Mechanism (BCM) on FLEG was signed in 2009, providing a framework for cooperation on issues linked to FLEG (EC & ECO 2010). But even this bilateral mechanism between China and the EU is pursuing a regional collaboration, as delegates called on 'the EU and China to actively contribute to *regional* FLEG processes' in their conclusions reached at the EU–China FLEG conference, held in Beijing in 2007 (Global Witness 2009: 104).

The executing agency, the European Forest Institute (EFI), hosts the EU FLEGT Facility under which FLEGT Asia operates. The overarching goals include building synergies and collaborating with existing regional forest governance partners in Asia such as the Asia Forest Partnership, and the Asia-Pacific Network for Sustainable Forest Management and Rehabilitation (APFNet) (Pescott & Durst 2010); these are discussed later.

Recent developments in East Asia

The development of the regional forest governance mechanisms described so far provides an illustration of how actors from outside of Asia can make a decisive impact on the policies of Asian institutions and countries. For example, the World Bank FLEG East Asia was, in fact, possible because there had been a commitment made by the US to the G8 summit. The need for action to address illegal logging was first recognized in the 1997 'proposals for action' of the IPF, and then included in the G8 Action Program on Forests (1998–2002) (McDermott et al. 2010). The Action Program served as a notice of intent that some G8 countries were serious about addressing illegal logging, and foremost among them was the US. At the 2000 G8 summit in Okinawa the US announced that it was planning a more ambitious initiative in Asia, which turned out to be the FLEG ministerial conference in Bali the following year (Humphreys 2006: 148). In short, there is a cause-and-effect link between G8 (or the US) and the World Bank FLEG (e.g., Davidson 2007), which implies that the FLEG East Asia originated from outside Asia.

Likewise, the EU FLEGT Action Plan, adopted in 2003, is an expression of policy commitments made by the EU and producer partner countries within the framework of the G8 Action Program on Forests. Historically, the basis for the EU FLEGT Action Plan is the council resolution of December 15, 1998, on a 'Forestry Strategy for the European Union,' which is a non-binding instrument defining the policy basis for a new forest strategy in the EU (Glück et al. 2010). In contrast, Asia-based institutions such as the Association of Southeast Asian Nations (ASEAN) and Asia Pacific Economic Cooperation (APEC), and East Asian countries such as China, Japan and Korea have begun to organize themselves in the recent development of regional governance of Asian forests.

Asia Forest Partnership

A few months after the Bali conference in 2001, the second session of the UNFF produced a Ministerial Declaration and decided to transmit it to the World Summit on Sustainable

Development (WSSD). The Plan of Implementation of the 2002 WSSD called for 'political commitment to achieve sustainable forest management by endorsing it as a priority on the international political agenda, taking full account of the linkages between the forest sector and other sectors through integrated approaches' (WSSD 2002), which was quoted from the UNFF Ministerial Declaration. At the 2002 WSSD, two major new regional initiatives on forests were announced – Congo Basin Forest Partnership and Asia Forest Partnership (AFP) (Martin 2004).

Holding its first meeting three months later in Japan, the AFP aimed to promote sustainable forest management in Asia through addressing five urgent issues: control of illegal logging, control of forest fires and rehabilitation and reforestation of degraded lands, as well as crosscutting issues of good governance and forest law enforcement and developing capacity for effective forest management (Davidson 2007). The public–private membership of AFP includes governments, inter-governmental organizations, civil society groups, the private sector and intersectoral partnerships (Visseren-Hamakers & Glasbergen 2007).

At present the AFP is a soft process, while FLEG has hard political support in the form of the 2001 Bali Declaration and commitment from developed nations. FLEG's greatest strength is its concentrated focus on legality and law enforcement, while the AFP has taken on additional issues, as stated earlier. Through the involvement of Malaysia, AFP has stronger support from the regional timber industry (Davidson 2007). The fact that the Malaysian Timber Certification Council (MTCC), an intersectoral partnership, is a member of the AFP, another intersectoral partnership, is cited as a good example of the increasing complexity of the forest biodiversity governance system (Visseren-Hamakers & Glasbergen 2007).

The AFP was initiated by Japan's Ministry of Foreign Affairs in close association with Indonesia's Ministry of Forestry. By the time it was launched a core group of four 'lead partners' had emerged. The other two were a non-governmental environmental group, the Nature Conservancy and the Center for International Forestry Research (CIFOR), both participating through programs in Indonesia. Funding and staffing for AFP activities have come from the partners themselves, with the governments of Japan and Indonesia providing most of the support. The four lead partners act in consensus to organize the process, while actions can be proposed by any of the partners. If the partners agree, work plans are developed and implemented, with one partner taking responsibility to facilitate the process. Progress does not depend on actions by all of the partners and can proceed relatively independently once the partners have agreed to adopt the work plan (Sizer 2004). The four leading partners have made a stronger commitment to the advancement of the partnership, but do not have any more authority or rights than other partners (Lee 2004).

The AFP does have strong ownership within the region, with Japan and Indonesia particularly active, and with regular input from Malaysia and others. It is also attracting steadily increasing funding from various sources, including the partners themselves (Sizer 2004). Given the recent development of the Asian FLEG process, however, the AFP can be seen as an attempt by Japan to reclaim control over Asian forest dialogue at the expense of FLEG, which may be perceived as Anglo-American driven (Humphreys 2006: 152).

Asia-Pacific Network for Sustainable Forest Management and Rehabilitation

At the 15th APEC Economic Leaders' Meeting in 2007 in Australia, the president of China, Hu Jintao, made the following remarks during his speech:

> From 1980 to 2005, China made tremendous efforts to carry out afforestation and
> strengthen the rehabilitation and management of forests and it has developed expertise in

this respect. We are ready to share such expertise with other APEC members. I propose the Asia-Pacific Network on Forest Rehabilitation and Sustainable Management be set up. This will provide a platform for APEC members to share best practices, conduct policy dialogue, carry out personnel training on forest rehabilitation and management, and work together to promote rehabilitation and growth of forests, increase carbon sink and mitigate climate change in the Asia Pacific region. (Hu 2007)

The outcome of the meeting, the Sydney APEC Leaders' Declaration on Climate Change, Energy Security and Clean Development, set 'an APEC-wide aspirational goal of increasing forest cover in the region by at least 20 million hectares of all types of forests by 2020 – a goal which if achieved would store approximately 1.4 billion tonnes of carbon, equivalent to around 11% of annual global emissions (in 2004)' (APEC 2007). Along the same line, the establishment of the Asia-Pacific Network for Sustainable Forest Management and Rehabilitation (APFNet) was agreed at the meeting, and was incorporated in the Sydney Declaration 'to enhance capacity building and strengthen information sharing in the forestry sector' (APEC 2007). The agreed APEC-wide goal on forests is significant, both for the political benchmarks it establishes and for the fact that, for the first time, the goal is now shared by both key developed and developing economies in the region (Douglas 2008).

The shared mission of the APFNet is to promote sustainable forest management in the Asia-Pacific region through capacity building, information sharing, regional policy dialogues and pilot projects. Its three main objectives are to promote forest rehabilitation, reforestation and afforestation in the region to contribute to the increasing forest cover in the APEC region, to strengthen sustainable forest management and improve forest quality including climate change mitigation and adaptation, and to improve the productive capacity and socioeconomic benefits of forest ecosystems. APFNet was formally launched with an International Symposium on Sustainable Forest Management, held in Beijing in 2008.

It is also noteworthy that the APFNet, still in its infancy, set the stage for a new specialized organ called 'Experts Group on Illegal Logging and Associated Trade' to the body of APEC. To seek opportunities to meet the forest-related goals in the Sydney Declaration, China again proposed at the 18th APEC Economic Leaders' Meeting to host the first APEC meeting of ministers responsible for forestry in September 2011 in Beijing in collaboration with the APFNet. At the first APEC Meeting of Ministers of Forestry, the ministers endorsed the terms of reference for APEC Experts Group on Illegal Logging and Associated Trade, which held its first meeting in February 2012 in Moscow.

One of the implications from such a series of forest-related institutionalizations in APEC might be that it illustrates how the paradigm of the international organization on forests has changed since its early days. APEC, the incubator of the APFNet, is a regional economic grouping dedicated to promoting free trade and investment among its member economies such as the EU and the North American Free Trade Agreement (NAFTA) (e.g., Park & Lee 2009). Given the primary goal of trade and investment, forestry was one of the 15 sectors for early voluntary sectoral liberalization (EVSL) proposal to APEC, which was endorsed at the 5th APEC Economic Leaders' Meeting in Vancouver in 1997. Although the APEC members finally failed to reach an agreement for further progress of EVSL due to sharp divisions within the organization,[5] APEC was originally perceiving forest products as a sector that should eliminate tariffs just like manufactured goods under the then General Agreement on Tariffs and Trade (GATT) (Park & Lee 2009; Ravenhill 2007). In APEC, forests originally were perceived as products for sale, but by 2007 were seen obviously as common properties to be conserved.

Asian Forest Cooperation Organization

Asian Forest Cooperation Organization (AFoCO) is a regional organization with a goal of facilitating forest cooperation among Asian countries to prevent deforestation and forest degradation and promoting sustainable forest management on ecological, environmental, and economical aspects. There are four key areas of cooperation in the AFoCO – (1) forest rehabilitation and prevention of forest disasters, (2) strengthening forest contributions to addressing climate change, (3) sustainable forest management and (4) forestry technology transfer (KFS 2010).

Just as the AFP was initiated by Japan and the APFNet by China, South Korea has been taking the lead in establishing the AFoCO. At the ASEAN-Korea commemorative summit in 2009, Korea proposed the establishment of an action-oriented regional organization that is focused on forest greening and rehabilitation. It was well received by the heads of state from ASEAN member countries, as recorded in their joint summit statement as follows:

> We agreed to endeavour to strengthen our cooperation in the context of the United Nations Framework Convention on Climate Change (UNFCCC), especially on the Reducing Emissions from Deforestation and Forest Degradation in Developing Countries (REDD) initiative, enhancement of sustainable forest management, wasteland restoration, and promotion of industrial forestation. In this regard, we appreciated the ROK's proposal to establish an 'Asian Forest Cooperation Organisation.' (ASEAN 2009)

Although it is very young at the time of this writing, the AFoCO is already undertaking several projects such as improving capacity on forest restoration in Cambodia. Most of them are bilateral projects between one of ASEAN member countries and Korea, but some are on a regional basis such as reclamation, rehabilitation, and restoration of degraded forest ecosystems through pilot testing, exchange of expertise and capacity development. This seemingly early start of AFoCO's functional operation is attributable to the fact that Korea had already been collaborating with ASEAN countries for environmental issues for more than a decade (Lee et al. 2012).

AFoCO came into being as a result of ASEAN-Korea Environmental Cooperation Project (AKECOP), which has been active since its launch in 2000. The AKECOP consists of two main parts – research, and education and training. For example, there were two onsite field research programs conducted by Korean scientists at experimental forest of the University of the Philippines at Los Baños in the Philippines, and eight regional research projects independently conducted by the eight member countries of ASEAN (Lee 2004).[6] Also, more than 20 ASEAN students have graduated from Korean universities under its master's and doctoral programs (Lee et al. 2012). There is obviously continuity of this focus from AKCOP to AFoCO.

But in terms of institutional capabilities, perhaps the most prominent difference of AFoCO from AFP or APFNet is that AFoCO will have a legally binding force with the participating countries. The AFoCO is established on the basis of intergovernmental multilateral arrangement involving ASEAN member states, Korea and other Asian countries.[7] Through a negotiation process that took more than two years since 2009, the agreement was concluded at the 14th ASEAN-Korea summit in Bali in 2011 (ASEAN 2011) and entered into force in 2012, which led to the launching of the secretariat in Seoul in September 2012 (Lee 2012). The AFoCO distinguishes itself from AFP or APFNet by having a legal standing of the organization to provide accountability or stability and sound regional representation in dealing with global forestry agenda, carrying out actions (KFS 2010). According to the Korea Forest Service, AFoCO is the first intergovernmental organization dedicated to forestry cooperation in Asia.

Overall trends

From the historical development in regional governance of Asian forests, what kind of pattern can be deduced? Table 18.3 shows a chronology of the regional structures of Asian forest governance with the key drivers. At least three features are identifiable from the evolution of the Asian regional forest governance as follows.

Asianization

As seen in the cases of recent development, regional governance of Asian forests has relatively recently been initiated and implemented by Asian countries, while the earlier models such as World Bank FLEG East Asia and the EU FLEGT Asia were all operated by the institutions outside of Asia. Financial support for the AFP is provided by Japan for the most part. The APFNet secretariat is hosted by China and maintains its website. Korea is hosting the AFoCO secretariat and responsible for 90% of the funding. Even though initiated by the World Bank, FLEG has been considerably transferred to Asia. ASEAN Regional Knowledge Network on FLEG was established in 2008 to encourage the use of regional knowledge networks to better inform ASEAN decision makers (Dam & Savenije 2011). This could actually be linked to the concept of self-governance such as what Ostrom (1999) describes in that 'forest users in many locations have organized themselves to vigorously protect and, in some cases, enhance local forests' (Ostrom 1999: 3). How could this transition from outside to inside Asia be explained?

Above all, Asia-based institutions such as ASEAN and APEC are now full-fledged to be used as a platform for starting or even experimenting with new policy tools. When ASEAN was established in 1967, for example, environmental management was not expressly recognized as a concern (Koh 1996). Against a global background of major international conferences such as the Stockholm Conference on the Human Environment in 1972, environmental issues were first inscribed on the ASEAN agenda in 1977, after the then-five member states adopted the Treaty of Amity and Cooperation (Elliott 2012). ASEAN integrated the environment into its complex system of regional consultations on economic, social, technical and scientific development (Koh & Robinson 2002), and adopted its first regional policies on the environment in 1978 in the form of a subregional environment program (ASEP) (Elliott 2012).

In fact, the 1985 ASEAN Agreement on the Conservation of Nature and Natural Resources is possibly the very first treaty in which 'sustainable development' is mentioned *eo nomine* (Schrijver 2008: 103). It is remarkable that the article 6 in particular addresses the conservation of the forest cover on lands, reforestation and afforestation planning, forest species diversity and forestry management plans on the basis of ecological principles (ASEAN 1985). A number of

Table 18.3 Chronological summary of the regional structures of Asian forest governance

Starting year	Governance institution	Key driver	Legally binding?
1950	Asia-Pacific Forestry Commission	FAO	No
2001	FLEG East Asia & Pacific	World Bank	No
2002	Asia Forest Partnership	Japan	No
2008 (September)	Asia-Pacific Network for Sustainable Forest Management and Rehabilitation	China, APEC	No
2008 (November)	FLEGT Asia	EU	Yes
2012	Asian Forest Cooperation Organisation	Korea, ASEAN	Yes

Source: Authors.

regional treaties contain general provisions on rational or sustainable use of tropical forests; of these only the 1985 ASEAN Agreement requires a serious commitment to forest protection in a broader environmental context, and it is not in force (Birnie et al. 2009: 694–695).

Also for the APEC, from the beginning, sustainable development was, at least rhetorically, part of the group's agenda. APEC's dedication to this objective was made explicit in the APEC Economic Leaders' Economic Vision Statement in 1993, and once again reaffirmed in 1995 in the APEC Economic Leaders' Declaration for Action (Carrapatoso 2008). Although these 'statements' or 'declarations' have not yet blossomed in terms of implementation, let alone measurable improvements in addressing environmental problems in the region, the seeds of environmental cooperation at APEC were still germinating. This is why it has been argued for long that regional economic integration must be complemented by the creation of regional frameworks for sustainable resource and ecosystem use (e.g., Ivanova & Angeles 2006).

After World War II, when the Bretton Woods system including the World Bank and the UN agencies such as the FAO were formalized, and while a European regional institution, the Council of Europe was created in 1949, driven by the continental political climate in favor of unity among European countries, there was no such counterpart in Asia. But now regional organizations are available for Asian countries and forest-related policymakers.

However, this trend of Asianization in regional governance of Asian forests does not have to be interpreted solely as the domain of liberal institutionalists who focus on how international cooperation can be sustained in anarchy or of constructivists who are oriented to 'imagining' an Asian community based on the region's rhetoric. The reality of Asian institutionalization is that it is still at a stage where it can be well understood as an extension and intersection of national interest and power rather than as an objective force in itself (Gill & Green 2009), as the individual countries such as China, Japan and Korea are behind each of the APFNet, AFP and AFoCO as drivers.

Legally binding nature

The cases studied here consistently show that they have evolved from non-legally binding processes to legally enforceable mechanisms. It is crucial that development of regional and international standards help give substance to the array of laws on paper (Crawford 1992: 45), particularly when they need greater enforceability to systematically address the causes and consequences of, for example, the illegal trade in timber and timber products in Asia and beyond.

As explained already, part of the reason for the emergence of EU FLEGT was lassitude from non-binding commitments among the participating countries in the World Bank FLEG processes. The VPA of the EU FLEGT is a legally binding agreement between the EU and an individual timber exporting country, which once agreed, will include commitments from both parties to halt trade in illegal timber with a license scheme to verify the legality of timber exported to the EU. Whereas the World Bank FLEG processes are primarily supply-side approaches to reduce illegal logging at source in timber producing countries, the EU FLEGT promotes both supply-side measures by providing assistance to developing countries, and demand-side measures to curtail the trade of illegally logged timber to the EU (Humphreys 2006: 156).

Similarly, among the Asia-based institutions, the AFoCO, the most recent model, is equipped with a legally binding nature. The AFP meetings have yet to produce any binding measures or even declarations (Schloenhardt 2008: 44). Launched at the WSSD in 2002 as a Type II partnership for sustainable development, which is voluntary, networked, multistakeholder and multisectoral as opposed to a more traditional international treaty style of Type I partnership (e.g., Bäckstrand 2006), the AFP is far from a legally binding mechanism. The APFNet is also focused on information sharing or regional policy dialogues without a legally binding nature just like APEC. In contrast, the AFoCO chose a legally binding format for independent and sustainable operation of the secretariat. With the legally binding agreement, the AFoCO advertises itself as an action–oriented organization that implements result-based projects based on proven technologies and sound management systems (KFS 2010).

However, given the so-called 'ASEAN Way,' which emphasizes consensus and respects the principles of noninterference, the hard law basis for both the EU FLEGT Asia and the AFoCO is noticeable. The principles of national sovereignty and noninterference are usually listed as obstacles to Asian regionalism or regional integration, sometimes even working against environmentalists or human rights activists (e.g., Frost 2008: 219). If such decades-old principles of ASEAN are still demonstrable, what made the ASEAN countries exert some flexibility in working with the EU FLEGT and the AFoCO?

In the case of the EU FLEGT, a closer look at the way in which the VPAs work discloses that they not only challenge but also reinforce the sovereignty of producer countries. Although the EU must risk being viewed as international interference in the process of VPAs for FLEGT, at least two sets of incentives are made available to the producer countries. First, resources, ideas, knowledge and technology are shared with existing groups, and the creation of new groups or coalitions is facilitated. Second, the coalitions of companies, activists, governments and aid agencies coalesce around market incentives to promote baseline 'legality verification' as a means for reinforcing domestic sovereignty. In this way, a producer country in Southeast Asia such as Indonesia comes to agree with the EU that they are aimed at ensuring that products produced in any particular country conform to that country's domestic requirements (Bernstein et al. 2010).

It should also be noted that there are more signals of ASEAN countries' movement away from the traditional interpretations and practice of noninterference and toward arrangements in which enforceable standards are set (Cole & Jensen 2009: 252), particularly when it comes to environmental issues. For example, the ASEAN Agreement on Transboundary Haze Pollution, signed in 2002, demonstrates that, in a crisis situation, ASEAN members can rally together to reach consensus on a hard law instrument (Koh & Robinson 2002). It requires parties of the Agreement to cooperate in developing and implementing measures to monitor transboundary haze pollution, control sources of fires, exchange information and technology, and provide mutual assistance. ASEAN countries are coming to the view that giving up a degree of sovereignty to some extent is itself an exercise of sovereignty (Acharya 2009: 184).

Justification

As many observers of Asian regionalism or multilateralism find that the architecture of government-driven institutionalization has been substantially overlapping and redundant (e.g., Frost 2008: 147; Gill & Green 2009: 12), it appears to be difficult to see how each of the forest governance mechanisms from APFC to AFoCO justifies its own birth and existence. Although these organizations are all primarily focused on forest governance, they are at the same time observably differentiated, to some extent at least, in terms of their missions and geographic coverage.

As illustrated in Figure 18.2, although the six forest governance institutions discussed in this chapter cover 49 Asian governments, only five states – Indonesia, Malaysia, Philippines, Thailand and Vietnam – are covered by all six institutions. As much as there is overlapping among the institutions in geographic boundaries, the covered countries are also extensively

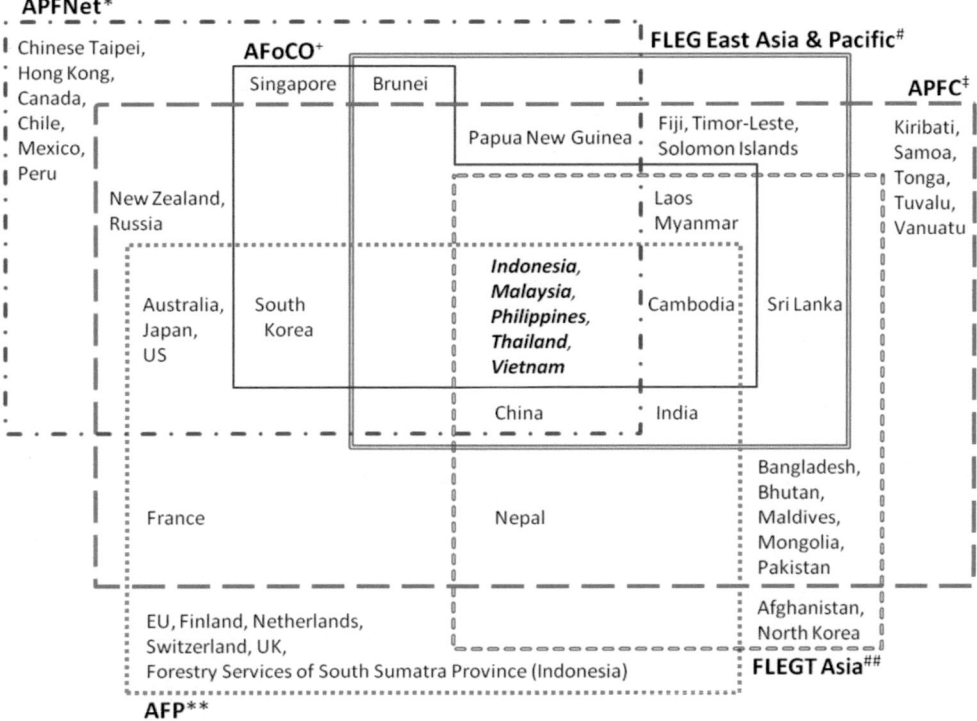

Figure 18.2 Geographic coverage of regional governance of Asian forests

Source: Authors, adapted from Gill & Green (2009).

distributed, which allows the structures of forest governance to have some degree of distinctiveness. The distinctiveness becomes more visible when a similar comparison is made for the major Asian regional institutions such as APEC, ASEAN and East Asian Summit, which show much more redundancy than discrepancy. However, since some of them such as AFoCO are willing to invite more countries as their new members, it is too early to say that the differentiated boundaries will remain the same.

The six institutions also show some deviation in their emphasis on the missions. While the FLEG, FLEGT and AFP are all focused on law enforcement against illegal logging, the AFP is broader in its objectives, which include capacity building and financing for sustainable forest management. Similarly, both the APFNet and the AFoCO share the same objectives of promoting SFM and enhancing forest carbon stocks, but the overall goal of APFNet is mainly expanding forest cover in the region while the AFoCO is closer to the Korean concept of green growth, which aims for continued economic growth and environmental progress at the same time. Another difference is that AFoCO is aiming at forest technology transfer and forest disaster prevention, while APFNet mentions biodiversity conservation.

Conclusion

East Asia is an ideal focus for students who attempt to explain the development of regional environmental governance. Sometimes it was a test bed as the World Bank FLEG initially unfolded there prior to the FLEG processes in other regions. At other times the region served as a common extractive resource pool for local institutions such as ASEAN or APEC. Forest governance at a regional level in East Asia has been evolving from global institutions' experimental projects to Asian institutions' collaborative partnerships. Governments, wood products firms and retailers in East Asia have felt pressure from markets and policymakers in Europe and the US, as well as from transnational organizations, to address illegal logging and associated trade issues. This pressure has had a useful role in moving the issues up the political agenda in the region, and as the argument has been made for a decade that Asian regional ownership of the process is important if it is to be sustained (Sizer 2004), institutionalization of forest governance is now led and financed by Asian countries.

In East Asia, the institutional authority of regional forest governance has progressed from soft law to hard law, as shown in Table 18.3. This trend seems to be in agreement with the emerging and increasingly accepted principles of international forest law such as the principle of common concern, although a tension still exists between the firmly rooted claims of sovereignty over forests and the emerging global norms (Brunee & Nollkaemper 1996). The trend is particularly salient, given the ASEAN tradition of noninterference based on the respect for sovereignty and APEC's non-binding approach. With overlapping boundaries and redundant missions, the regional forest governance institutions examined here may be seen as trying to differentiate themselves from one another by having various priorities on their goals and membership composition.

Have these institutions been effective in protecting or managing forests in the region? Given their short history, it seems premature to answer the question definitively. Poor forest governance and corruption are extensive problems of long standing, and their impact on institutions, government revenues lost, poverty and unsustainable forestry is huge. That is why, for better forest governance, a multipronged approach is needed involving biophysical, social and cultural instruments (Kishor & Damania 2007). In fact, such institutions are but one of many dimensions to ideal solutions.

This chapter infers some general trends from several major cases of regional governance of Asian forests. Other extant models such as ASEAN Social Forestry Network are not covered here. A careful look at these and other cases will be needed for further research.

Bibliography

Acharya, A. (2009) 'The Strong in the World of the Weak: Southeast Asia in Asia's Regional Architecture', in B. Gill & M. Green (Eds.), *Asia's New Multilateralism: Competition, Cooperation and the Search for Community* (Columbia University Press).

Agrawal, A., Chhatre, A. & Hardin, R. (2008) 'Changing Governance of the World's Forests', *Science* 320(5882), 1460–1462.

Andonova, L.B. & Mitchell, R.B. (2010) 'The Rescaling of Global Environmental Politics', *Annual Review of Environment and Resources* 35(1), 255–282.

Asia-Pacific Economic Cooperation (APEC) (2007) *Sydney APEC Leaders' Declaration on Climate Change, Energy Security and Clean Development*, 9 September.

Association of Southeast Asian Nations (ASEAN) (1985) Agreement on the Conservation of Nature and Natural Resources, Kuala Lumpur. Available at http://www.aseansec.org/1490.htm

ASEAN (2009) Joint Statement of the ASEAN-Republic of Korea Commemorative Summit, Jeju Island, Korea, 2 June.

ASEAN (2011) 'Advancing Forestry Cooperation in International Year of Forest 2011', ASEAN press release, 18 November. Available at http://www.aseansec.org/26733.htm

Bäckstrand, K. (2006) 'Multi-stakeholder Partnerships for Sustainable Development: Rethinking Legitimacy, Accountability and Effectiveness', *European Environment* 16(5), 290–306.

Batterbury, S.P.J. & Fernando, J.L. (2006) 'Rescaling Governance and the Impacts of Political and Environmental Decentralization: An Introduction', *World Development* 34(11), 1851–1863.

Bernstein, S., Cashore, B., Eba'a Atyi, R., Maryudi, A. & McGinley, K. (2010) 'Examination of the Influences of Global Forest Governance Arrangements at the Domestic Level', in J. Rayner, A. Buck & P. Katila (Eds.), *Embracing Complexity: Meeting the Challenges of International Forest Governance* (IUFRO).

Biermann, F., Pattberg, P., van Asselt, H. & Zelli, F. (2009) 'The Fragmentation of Global Governance Architectures: A Framework for Analysis', *Global Environmental Politics* 9(4), 14–40.

Birnie, P., Boyle, A. & Redgwell, C. (2009) *International Law & the Environment*, 3rd edn (Oxford University Press).

Brown, K. (2001) 'Cut and Run? Evolving Institutions for Global Forest Governance', *Journal of International Development* 13, 893–905.

Brunee, J. & Nollkaemper, A. (1996) 'Between the Forests and the Trees – An Emerging International Forest Law', *Environmental Conservation* 23(4), 307–314.

Canby, K. (2011) 'Forest Governance in Asia', in *Forest Matters: Make it Work* (EU FLEGT Facility).

Carrapatoso, A.F. (2008) 'Environmental Aspects in Free Trade Agreements in the Asia-Pacific Region', *Asia Europe Journal* 6(2), 229–243.

Chomitz, K.M. & Buys, P. (2007) *At Loggerheads? Agricultural Expansion, Poverty Reduction, and Environment in the Tropical Forests* (World Bank).

Cole, W. & Jensen, E. (2009) 'Norms and Regional Architecture: The Impact of Regional Multilateralism on Governance and Democracy', in G. Bates & M. Green (Eds.), *Asia's New Multilateralism: Competition, Cooperation and the Search for Community* (Columbia University Press).

Colfer, C.J.P. & Pfund, J. (Eds.) (2011) *Collaborative Governance of Tropical Landscapes* (Earthscan, CIFOR).

Colfer, C.J.P., Dahal, R.D. & Capistrano, D. (Eds) (2008) *Lessons from Forest Decentralization: Money, Justice and the Quest for Good Governance in Asia-Pacific* (Earthscan).

Crawford, J. (1992) 'The Role of Transnational Environmental Law in Protecting the Environment of Asia and the Pacific', *Asia-Pacific Law Review* 1, 32–50.

Dam, J. van & Savenije, H. (2011) *Enhancing the Trade of Legally Produced Timber. A Guide to Initiatives* (Tropenbos International).

Davidson, J. (2007) 'Forest Law Enforcement and Governance (FLEG)', in L. Elliott (Ed.), *Transnational Environmental Crime in the Asia-Pacific* (Australian National University Press).

Dietz, T., Ostrom, E. & Stern, P. (2003) 'The Struggle to Govern the Commons', *Science* 302, 1907–1912.

Douglas, A. (2008) APEC in 2007–2008. UNISCI Discussion Paper, No. 16.

Elliott, L. (2012) 'ASEAN and Environmental Governance: Strategies of Regionalism in Southeast Asia', *Global Environmental Politics* 12(3), 38–57.

European Commission & EuropeAid Cooperation Office (EC & ECO) (2010) FLEGT Asia Regional Programme: Supporting Responsible Trade for Asia's Forests, European Union.

European Union (EU) (2010) 'EU Launches New Programme for Forest Governance in Asia', press release, 25 January.

Forest Science Institute of Vietnam (FSIV) (2009) Vietnam Forestry Outlook Study. Asia-Pacific Forestry Sector Outlook Study II Working Paper No. APFSOS II/WP/2009/09, FAO Regional Office for Asia and the Pacific.

Frost, E. (2008) *Asia's New Regionalism* (Lynne Rienner Publishers).

Gill, B. & Green, M. (2009) 'Unbundling Asia's New Multilateralism', in B. Gill & M. Green (Eds.), *Asia's New Multilateralism: Competition, Cooperation and the Search for Community* (Columbia University Press).

Global Witness (2009) A Disharmonious Trade: China and the Continued Destruction of Burma's Northern Frontier Forests, Global Witness London.

Glück, P., Rayner, J., Cashare, B., et al. (2005) 'Changes in the Governance of Forest Resources', in G. Mery, et al. (Eds.), *Forests in the Global Balance: Changing Paradigms*, World Series Vol. 17 (IUFRO).

Glück, P., Angelsen, A., Appelstrand, M., Assembe-Mvondo, S., Auld, G., Hogl, K., et al. (2010) 'Core Components of the International Forest Regime', in J. Rayner, A. Buck & P. Katila. (Eds.), *Embracing Complexity: Meeting the International Forest Governance*, World Series Vol. 28 (IUFRO).

Hajjar, R. & Innes, J.L. (2009) 'The Evolution of the World Bank's Policy Towards Forestry: Push or Pull?', *International Forestry Review* 11, 27–37.

Howlett, M. & Rayner, J. (2010) 'Overcoming the Challenges to Integration: Embracing Complexity in Forest Policy Design through Multi-level Governance', in J. Rayner, A. Buck & P. Katila (Eds.), *Embracing Complexity: Meeting the International Forest Governance*, World Series Vol. 28 (IUFRO).

Hu, J. (2007) Speech at the 15th APEC Economic Leaders' Meeting, Sydney, Australia. Available at http://www.apfnet.cn.

Humphreys, D. (2006) *Logjam: Deforestation and the Crisis of Global Governance* (Earthscan).

Ivanova, A. & Angeles, M. (2006) 'Trade and Environment Issues in APEC', *Social Science Journal* 43(4), 629–642.

Kanie, N. (2004) 'Global Environmental Governance in Terms of Vertical Linkages', in N. Kanie & P.M. Haas (Eds.), *Emerging Forces in Environmental Governance* (United Nations University Press).

Kishor, N. & Damania, R. (2007) 'Crime and Justice in the Garden of Eden: Improving Governance and Reducing Corruption in the Forestry Sector', in *The Many Faces of Corruption* (World Bank).

Koh, K.L. (1996) Selected ASEAN Documents on the Environment, July (1), Asia-Pacific Centre for Environmental Law.

Koh, K.L. and Robinson, N.A. (2002) 'Regional Environmental Governance: Examining the Association of Southeast Asian Nations (ASEAN) Model', in D.C. Esty & M.H. Ivanova (Eds.), *Global Environmental Governance: Options and Opportunities* (Yale School of Forestry and Environmental Studies).

Koné, P.D., Durst, P., Prins, C., Marx Carneiro, C., Abdel Nour, H.O. & Kneeland, D. (2004) 'In the Beginning, There Were Six Regional Forestry Commissions', *Unasylva* 218(55), 10–17.

Korea Forest Service (KFS) (2010) Asian Forest Cooperation Organization, Working Paper for AFoCO.

Lang, G. & Chan, C.H.W. (2006) 'China's Impact on Forests in Southeast Asia', *Journal of Contemporary Asia* 36(2), 167–194.

Lee, D.K. (2004) 'Information Networking for New Partnerships in Forestry', *Journal of Forest Research* 9(4), 299–305.

Lee, D.K. (2012) 'The Forest Sector's Contribution to a "Low Carbon, Green Growth" Vision in the Republic of Korea', *Unasylva* 63(239).

Lee, Y., Lee, Y. & Lin, H. (2012) 'AFoCO: a Model of Regional Forest Cooperation in Asia', 산림과학 공동학술대회.

Liu, C., Lobovikov, M., Murdiyarso, D., Oka, H. & Youn, Y.C. (2005) 'Paradigm Shifts in Asian Forestry', in G. Mery, R. Alfaro, M. Kaninnen & M. Lobovikov (Eds.), *Forests in the Global Balance* (IUFRO).

McDermott, C.L. Humphreys, D., Wildburger, C. & Wood, P. (2010) 'Mapping the Core Actors and Issues Defining International Forest Governance', in J. Rayner, A. Buck & P. Katila (Eds.), *Embracing Complexity: Meeting the International Forest Governance*, World Series Vol. 28 (IUFRO).

Martin, R.M. (2004) 'Regional Approaches: Bridging National and Global Efforts', *Unasylva* 55(218), 3.

Ostrom, E. (1999) Self-Governance and Forest Resources, CIFOR Occasional Paper No. 20, Bogor, Indonesia, CIFOR.

Overdevest, C. & Zeitlin, J. (2012) 'Assembling an Experimentalist Regime: Transnational Governance Interactions in the Forest Sector', *Regulation & Governance*.

Park, S. & Lee J. (2009) 'APEC at a Crossroads: Challenges and Opportunities', *Asian Perspective* 33(2), 97–124.

Patt, A.G. (2010) 'Effective Regional Energy Governance – Not Global Environmental Governance – is What We Need Right Now for Climate Change', *Global Environmental Change* 20(1), 33–35.

Pescott, M.J. & Durst, P.B. (2010) 'Reviewing FLEG Progress in Asia and the Pacific', in M.J. Pescott, P.B. Durst, et al. (Eds.), *Forest Law Enforcement and Governance: Progress in Asia and the Pacific* (Asia-Pacific Forestry Commission (APFC)).

Potter, L. (2008) 'Governance, Tenure and Equity in Asia-Pacific Forests', in C. Colfer, G.R. Dahal & D. Capistrano (Eds.), *Lessons from Forest Decentralisation – Money, Justice and the Quest for Good Governance in Asia Pacific* (Earthscan).

Potter, L. & Badcock, S. (2001) The Effects of Indonesia's Decentralization on Forests and Estate Crops in Riau Province: Case Studies of the Original Districts of Kampar and Indragiri Hulu, Case Study Nos. 6 and 7, Bogor, Indonesia, CIFOR.

Rametsteiner, E. (2009) 'Governance Concepts and their Application in Forest Policy Initiatives from Global to Local Levels', *Small-Scale Forestry* 8(2), 143–158.

Ravenhill, J. (2007) From Poster Child to Orphan: The Rise and Demise of APEC, UNISCI Discussion Papers No. 13.

Rayner, J., Buck A. & Katila P. (Eds.) (2010) *Embracing Complexity: Meeting the International Forest Governance*, World Series Vol. 28 (IUFRO).

Schloenhardt, A. (2008) *The Illegal Trade in Timber and Timber Products in the Asia–Pacific Region* (Australian Institute of Criminology, Australian Government).

Schrijver, N. (2008) *The Evolution of Sustainable Development in International Law: Inception, Meaning and Status* (Martinus Nijhoff).

Sizer, N. (2004) 'Regional Approaches to Tackle Illegal Logging and Associated Trade in Asia', *Unasylva* 55(218), 40–44.

Tarasofsky, R. (1999) 'Assessing the International Forest Regime: Gaps, Overlaps, Uncertainties and Opportunities', in R. Tarasofsky (Ed.), *Assessing The International Forest Regime* (IUCN).

van Bodegom, A.J., Klaver, D., van Schoubroeck, F. & van der Valk, O. (2008) 'FLEGT Beyond T: Exploring the Meaning of Governance Concepts for the FLEGT Process', Wageningen University & Research Centre.

Visseren-Hamakers, I.J. & Glasbergen, P. (2007) 'Partnerships in Forest Governance', *Global Environmental Change* 17, 408–419.

Westoby, J.C. (1989) *Introduction to World Forestry* (Blackwell).

World Bank (2001) Bali Declaration. Available at http://go.worldbank.org/Y5E5KHNZT0.

World Summit on Sustainable Development (WSSD) (2002) 'Plan of Implementation'. Available at http://www.un.org/jsummit/html/documents/summit_docs.html.

Yasmi, Y., Broadhead, J., Enters, T. & Genge, C. (2010) Forestry Policies, Legislation and Institutions in Asia and the Pacific: Trends and Emerging Needs for 2020, Asia-Pacific Forestry Sector Outlook Study II, Working Paper No. APFSOS II/WP/2010/34, Bangkok, Thailand, FAO.

Acronyms

AFoCO	Asian Forest Cooperation Organization
AKECOP	ASEAN-Korea Environmental Cooperation Project
APEC	Asia Pacific Economic Cooperation
APFC	Asia-Pacific Forestry Commission
APFISN	Asia-Pacific Forest Invasive Species Network
APFNet	Asia-Pacific Network for Sustainable Forest Management and Rehabilitation
ASEAN	Association of Southeast Asian Nations
ASEM	Asia-Europe Meeting
BCM	Bilateral Coordination Mechanism (between European Union and China on Forest Law Enforcement and Governance)
COFO	Committee on Forestry (of the United Nations Food and Agriculture Organization)
EC	European Commission
EFI	European Forest Institute
EVSL	early voluntary sectoral liberalization
FAO	Food and Agriculture Organization
FLEG	Forest Law Enforcement and Governance
FLEGT	Forest Law Enforcement, Governance and Trade
GATT	General Agreement on Tariffs and Trade
IFF	Intergovernmental Forum on Forests (1997–2000)
IPF	Intergovernmental Panel on Forests (1995–1997)
ITTA	International Tropical Timber Agreement
KFS	Korea Forest Service
NAFTA	North American Free Trade Agreement
NLBI	non-legally binding instrument on all types of forest
REDD	reducing emissions from deforestation and forest degradation in developing countries

UNCED	United Nations Conference on Environment and Development
UNFCCC	United Nations Framework Convention on Climate Change
UNFF	United Nations Forum on Forests
VPA	voluntary partnership agreement
WSSD	World Summit on Sustainable Development

Notes

1   For the concepts of forest governance, see Agrawal et al. (2008), Colfer & Pfund (2011), Glück et al. (2005), Rametsteiner (2009), van Bodegom et al. (2008), among others.
2   In Asia, a VPA was signed with Indonesia in 2011 and negotiations are currently underway with Malaysia, Vietnam and Papua New Guinea (EC & ECO 2010).
3   The FLEGT Asia regional office is in Kuala Lumpur, Malaysia. It is focused on three objectives:

  • to facilitate the collection, analysis and dissemination of new and existing research/information relevant to the implementation of the EU FLEGT Action Plan in Asia
  • to strengthen FLEGT-relevant institutions and initiatives in Asia
  • to develop mechanisms, tools and increased capacity for cooperation between FLEGT-related enforcement agencies in the region (EU 2010).

4   The FLEGT Asia also affiliated with the Asia Europe Meeting (ASEM), which had a conference on 'Forests, Forest Governance and Forest Products Trade: Scenarios and Challenges for Europe and Asia', held in May 2010 in Phnom Penh, Cambodia, to address trends in EU-Asia forest area, production and trade, scenarios for forest ownership, and governance as well as non-forest land use and their impacts on forests. Technical inputs were provided by the EU-EFI FLEGT Asia Program (http://www.aseminfo board.org/).
5   The division was that many of APEC's Asian members such as Japan had problems with the EVSL program when the US was consistently pushing for trade liberalization. In part, these derived from domestic political constraints arising from the proposed liberalization of heavily protected sectors. The outcome of frequently acrimonious talks was a decision by APEC leaders at their next meeting in Kuala Lumpur in 1998 to refer the proposal to the WTO, essentially abandoning the EVSL process in APEC. The failure eventually damaged the credibility of APEC (Park & Lee 2009; Ravenhill 2007).
6   The eight countries are Cambodia, Indonesia, Laos, Malaysia, Myanmar, the Philippines, Thailand, and Vietnam (Lee 2004).
7   The official title is 'Agreement between the Governments of the Member States of ASEAN and the Republic of Korea on Forest Cooperation.'

19

Wildlife consumption

Cultural and environmental values in China and Southeast Asia

Kyle Swan and Kirsten Conrad

1 Introduction

A common way of looking at environmental problems suggests that they are almost exclusively the result of bad things people do. Solutions to these problems begin with philosophers, conservation activists and policy advocates articulating a moral vision that explains what is wrong with people polluting an area or (over-)using a resource, and, especially in the case of non-human animals, what is wrong in even viewing them as a resource to be used for human ends. Environmentalists work to communicate a set of values, which provides the basis for rules and policies that will protect environmental goods by discouraging or preventing people from doing the bad things. In other words, they *apply ethics*. They attempt to make things *right*.

Contrast this approach to environmental problems with one that says they arise out of conflict, disagreement or, more generally, a failure of interested parties to coordinate with each other. David Schmidtz has identified three such sources of environmental conflict (Schmidtz 2002a: 417–418). A conflict in *use* occurs when my consumption of an environmental good interferes with yours, producing an externality or a commons tragedy. A perhaps more fundamental source of environmental conflict is conflict in *values*. Again, many deny that environmental goods are appropriately seen as merely resources to be used, even at sustainable levels. For example, many environmentalists advance a moral vision requiring a non-instrumental, bio-centric or eco-centric approach to environmental goods – a preservationist ethic – and attempt to undermine instrumental, anthropocentric approaches – a conservationist ethic (Varner 1998). Anthropocentric theorists locate the value of environmental goods in the value they have to human beings. According to this view, there is no value in nature without conscious, human valuers. Eco-centric theorists argue that environmental goods have value apart from their value or usefulness to human beings. On preservationist views, we are subject to a standing obligation to justify our use of environmental goods. Preservationists may disagree about how stringent the justificatory burden is, but they will typically affirm that the fact that some use merely satisfies a desire or is relatively commonplace in some culture isn't sufficiently weighty to render that use legitimate. Accordingly, at least sometimes, and probably more often than people think, human ends should be sacrificed in order to preserve environmental goods. Therefore, preservationism conflicts with a tradition common in Chinese-influenced culture areas that sees consumption of wild fauna as basic and necessary to a 3,000 year-old set of systematized medicine practices and related cosmological beliefs (Coggins 2003). In this sort of environmental conflict, everyone sees themselves as taking a principled stand against the injustice of the other side. On one hand, environmentalists are working to protect vulnerable species from a global network of poachers and traffickers. On the other hand, practitioners of traditional medicines in Asian countries reject the authority of Western environmentalists to impose alien values that interfere with their conception of a balanced and healthy life. Finally, Schmidtz shows that a conflict in *priorities* can also lead to environmental problems. If rules and policies designed to regulate wildlife consumption are imposed without regard for the values and commitments of people at the local level who are to be bound by them, then the moralism of environmental advocates may be at cross-purposes with the legitimate aim of protecting environmental goods.

We apply this understanding of environmental conflicts to wildlife protection policy debates in China and other Chinese-influenced culture areas in Asia, such as Vietnam, Laos and Cambodia, where attitudes towards wildlife consumption are more permissive and several key species are utilized in a variety of cultural practices. Ben Davies quotes TRAFFIC's James Compton as saying, 'China is a like a vacuum cleaner. It is the single greatest threat to wildlife in the whole of Asia' (Davies 2005: 34). According to Susan Shen, 'The single most important fact hampering wildlife conservation in China is the traditional use of wild animals for medicinal purposes, meat and skins' (Shen et al. 1982: 340). The tiger policy debate in China is especially instructive. China joined the Convention on International Trade in Endangered Species (CITES) in 1981 and in 1993 banned domestic trade in tiger derivatives to bring itself into compliance. In the same year the Chinese government also removed tiger from the official ingredients list of medicines, some of which are used in traditional Chinese medicine (TCM). Poaching, however, continues to deplete the wild tiger population worldwide under the ban. Its supporters continue to press positive arguments that the consequences for tigers would be disastrous if the ban is lifted, but environmental philosophers in the West provide much of the normative justification for it and similar trade restrictions. Yet the values they invoke are not universally accepted. Many TCM practitioners support the use of substitute products for tiger derivatives based on their endorsement of a vaguely Taoist and/or Buddhist value of living in harmony with nature, but, so far, this has failed to translate into a broad-based acknowledgement of the moral status of tigers among Asian consumers of TCM and there is ample evidence of instrumentalist approaches to the value of tigers, and other environmental goods, in Chinese culture. A great deal of attention is directed towards the tiger in debates about wildlife consumption because of its status as a charismatic mega-fauna species. Yet the rhinoceros, the Asiatic black bear and the snow leopard are also included in the CITES Appendix I list of species in which commercial trade is prohibited. With the Siberian musk deer, an Appendix II species, all are utilized in TCM and other culturally based practices, luxuries, novelties and charms in various parts of Asia.

We argue that CITES trade restrictions wrongly disregard local cultural values in the conflict between TCM practices and Western environmentalism over values and priorities. The argument will proceed in three stages. In section 2, we present an overview of distinctive ways that environmental values conflict with many cultural values and practices in Asia. In section 3, we provide an overview of the background and impact of the current approach to the problem of wildlife consumption in Asia through the implementation of CITES trade bans in the region. In section 4, we attempt to outline the sort of considerations we think are relevant to an environmental policy regime governing threatened and at-risk species in the region. We conclude our discussion in section 5. A preservationist 'no-use' approach to policy might save wildlife if the relevant values are shared among a large enough percentage of the population, but such an approach goes predictably awry otherwise (Schmidtz 2002b: 321). Policy solutions to environmental problems, to be genuine solutions, have to be compatible with the cultural attitudes and practices of the time and place under consideration. Until and unless cultural attitudes towards wildlife consumption in Asia change, policymakers should rethink CITES-style trade bans.

2 Environmental and Asian cultural values in conflict

Identifying the sort of conflict we find ourselves in can help to mitigate or resolve it. A conflict in use might call for a relatively straightforward empirical or technical debate concerning the most efficient way to manage a resource. Usually, a set of rules, typically property rules, is sufficient to address this sort of conflict and induce a sufficient level of care and sustainable use. Matters are often more difficult if the conflict concerns whether or not nature or wildlife is even the sort of thing that is appropriately regarded as a resource or commodity. Some of these debates in value theory have become intractable and all are subject to irremediable disagreement. This fact subtly shifts the terms of the debate and what it will take to resolve the conflict and protect environmental goods. We will turn to this issue in section 4; this section makes the relatively uncontroversial point that the immorality of wildlife consumption is not so clear to everyone concerned. Preservationist environmental values are in conflict with Asian cultural values.

Environmental ethics is an 'applied' branch of moral philosophy. Like other areas of applied ethics, environmental ethics involves applying philosophical analysis to practical moral controversies. But environmental ethicists are often involved in a different project in that they are not simply applying well-worn ethical theories and principles to some practical domain in the way, say, business and medical ethicists do. Many environmental ethicists instead find themselves wondering whether extant theories of moral value are adequate to speak to environmental problems. What sort of theory could ground the idea that the natural world is a direct object of moral consideration? Schmidtz suggests that environmental ethics is 'more like highly theoretical meta-ethics, except with real world examples. So someone trots out his theory of value, and in environmental ethics, you get to bring the conversation down to earth (as it were) by saying things like, does that apply to trees?' (Leiter 2005). According to Ip Po-Keung, 'the major task of environmental ethics is the construction of a system of normative guidelines governing man's attitudes, behavior, and action toward his natural environment. The central question to be asked is: how ought man, either as an individual or as a group, to behave, to act, toward nature?' (Ip 1983: 335). Investigating questions like these has led many environmental ethicists to engage a set of distinctions familiar in discussions of environmental philosophy between instrumental and non-instrumental value, anthropocentric and bio-centric value, conservationism and preservationism and individualism and holism (see, e.g., Katz 1991; O'Neill 1992). Moreover, many philosophers take their conclusions about these theoretical matters to have straightforward practical implications. For example, Robert Elliot has argued that 'wild nature has intrinsic value,' because it is *natural*, 'which gives rise to obligations both to preserve it and to restore it' (Elliot 1992: 138).

Such arguments invoke the ideals of Western environmentalists. Are there reasons to think that they are normative for people in Asian countries such as China, Japan, South Korea, Malaysia and Vietnam where, according to Felix Cheung, TCM is deeply rooted and widespread (Cheung 2011: S82)? There is predictable Western skepticism and dismissiveness of TCM, which smacks of mysticism and pseudoscience to many medical professionals trained according to Western approaches to disease, illness and prevention. This even somewhat sugarcoats their attitudes, which more typically involve thinly veiled contempt, especially for remedies that contain animal products. If preservationists are right that the use of animal products must be justified with good reasons in order to be legitimate, then TCM practitioners would minimally have to show that the ban of animal products would deprive them of significant health benefits. Scientifically backward beliefs are not appropriate objects of toleration on this view. Richard Harris reports that after the US certified, in accordance with the Pelly Amendment, that China violated CITES protections in the 1990s, he had the sense that some of the indignation expressed was less about China failing to enforce trade restrictions and more about 'issues of values: [. . .] what were those old-fashioned Chinese doing consuming tiger and rhinoceros products in the first place?' (Harris 1996: 324). More recently, the online edition of *Nature Outlook* published an article on the conflict between wildlife preservation advocates and TCM, which elicited the following reader comment and illustrates the prevailing attitude: 'Jean Smiling Coyote said: We have to change the cultures which include the idea that there is some medicinal value to various parts of rhinos, tigers, and other now-endagered [sic] animals. Where did these ideas come from? [. . .] Just because some practice is part of an ancient culture not our own, doesn't mean it's acceptable' (Graham-Rowe 2011).[1]

According to Ramachandra Guha, much of Western environmentalism amounts to a kind of cultural imperialism that discounts the values and commitments of local communities (Guha 1989). To the extent that wildlife advocates acknowledge the importance of understanding Asian cultural attitudes and values, it is to demonstrate the error in their ways (Harris 1996: 324). This proves to be an incredibly tough sell. The growing illicit market in animal products is a demand-driven phenomenon. The products are sought largely for medicinal products, but also for iconic cultural symbols of status and, especially in China and Asian Tiger economies, recently acquired wealth. Both of these markers of Asian culture are enabled by relatively accommodating values governing the permissible use of wildlife. Where Western preservationists endorse the intrinsic moral status and non-anthropocentric value of wildlife, Chinese-influenced culture areas manifest a distinctively instrumentalist and human-centered approach to such value.

Paul Harris cites the Chinese government's approach to sustainable development as evidence of this, which mandates that environmental considerations take a backseat to economic growth (Harris 2004: 147, 156). This was true under Mao as well as subsequent reformers (Kobayashi 2005). The way that Chinese and other governments in Asia have prioritized economic growth has contributed to an ecological nightmare (Harris, P.G. 2008). Richard Harris provides other evidence of instrumentalism in Chinese attitudes towards wildlife conservation. He describes a primary school reader that doubles as an environmental consciousness–raising tool for children. In the reader, Mr. Lin, an environmental engineer, addresses a group of children's concerns about wildlife:

> These animals can be dangerous, but they are also beneficial! Take the tiger, for example. People call it 'King of the Mountain', but one could also say it's quite a treasure. The children protest, 'But tigers threaten people!' Wanting to appear reasonable, Mr. Lin responds, 'Yes, that's true, but the benefits to people from tigers are also great. [. . .] The entire body of a tiger is a treasure! Why, one could say that the tiger is a drug store capable of curing 100 ills!' (Harris 1996: 308)

This attitude is found even among Chinese who study wildlife. Harris relates his meeting with a mammalian taxonomist who 'allowed that he would himself use tiger bone for medicinal purposes, given the chance' (Harris 1996: 326, n. 30).

This evidence of anthropocentrism and instrumentalism in Asian thought stands in stark contrast to recent attempts by environmentalists to uncover in Eastern philosophy and religion normative principles more accommodating of preservationist goals. For example, philosophers have cited Taoist, Buddhist and Confucian traditions in support of prescriptions to live in balanced harmony with nature and to respect – revere – the interconnectedness of all living things (see Callicott 1987; Tucker 1991). Perhaps the rapacious depletion of flora and fauna in Asian countries can be explained as a recent development antithetical to their cultural traditions. However, Guha claims this story is based on a selective reading of Eastern traditions and 'does considerable violence to the historical record' (Guha 1989: 77). According to Heiner Roetz, anthropocentrism and instrumentalism in contemporary Chinese attitudes towards nature are nothing new and owe little to the influence of either Western Marxism or Western economic liberalism; both of these systems of thought are compatible with established cultural norms in Asia permitting the subjugation of nature for human purposes, which predate both (Roetz 2010). Roetz writes:

> It was the typical occupation in pre-dynastic times of the early rulers and cultural heroes, who represent the self-understanding of Chinese civilization. Huang Di 'deforested the mountains and dried out the swamps' (*Guanzi* 84, p. 414). Shun 'burned out the swamps and slew the wild animals (literally, the "numerous plagues" *qun hai*)' (ibid.), and Yi 'burned down the mountains and the swamps, causing the animals to flee and hide themselves' (*Mengzi* 3A4). (Roetz 2010: 201–202)

None of this should be surprising. As Roetz notes, 'For more than three millennia China has been one of the most intensively cultivated regions of the world. It has gone the way of all highly advanced civilizations, a way that is marked by the constant expansion of agricultural and otherwise utilizable areas at the expense of the original flora and fauna' (Roetz 2010: 201). Being comfortable with the aggressive utilization of natural resources to advance social goals has been a winning cultural strategy (whether it will continue to be so is another matter). It would be more remarkable if there were an ancient society with an aversion to it that survived.

However, even if the preservationist credentials of Eastern thought were vindicated, there may be little reason to expect much to come of it. As Holmes Rolston writes, 'a test of the power of Eastern thought will be to see how environmental problems are resolved in industrialized Eastern nations' (Rolston 1987: 189). As we have observed, they tend to be resolved in the direction of enhanced economic growth or other human goals. Moreover, it is unclear what advice Eastern thought would give to deal with a specific problem if, say, Taoism were operationalized:

> It may be right to say repeatedly 'More *yin*; more *yin*' in making environmental decisions, but this is a little like saying 'More love; more love' in making social decisions. The advice is sound enough, but unless one has a more sophisticated model to explain what adding *yin* or love means in the making of nitty-gritty decisions, and unless one can work the new attitude into either policy regulations or the moral calculus, nothing comes of it. (Rolston 1987: 180–181)

In other words, a moral vision is not a decision-making procedure. An institutional approach that focuses on feasibility and conflict resolution is necessary, too. This is an especially important lesson in contexts where there is no univocal moral vision. This does not mean that Asian cultural attitudes towards wildlife consumption will prove recalcitrant in the long run, and there have been some changes. For example, many practitioners of TCM now encourage the development and use of alternative synthetic or non-endangered ingredients from the traditional lists of remedies (Animals Asia 2012). However, these attitudes have yet to fully trickle down into the preferences of consumers of TCM. So long as this is the case, even if Western preservationist convictions about the non-instrumental value of wildlife are correct, it may be a mistake to impose certain restrictions on wildlife use. Implementing a policy in the sense of passing legislation is not the same thing as implementing a policy in the sense of bringing it about that people act in the prescribed ways. Policymakers cannot always predict how people will react to their rules. As Schmidtz says, 'People decide for themselves. We have to ask what their values are, what their priorities are, and what could lead people with such values and priorities to act in environmentally benign ways' (Schmidtz 2002a: 420).

3 Wildlife consumption in Asia

The conflict between preservationists and conservationists is more than theoretical. In this section, we cite evidence of its impact on the practice of wildlife protection. The gap between a policy's intent and design, on the one hand, and its results, on the other, highlights the importance of a policy's fit with local cultural values. We provide some background and examples of this gap generated by the prevailing wildlife policy approach, which largely ignores cultural values. We focus on the principal international wildlife treaty, CITES, and its level of effectiveness in protecting five species: tiger, rhinoceros, Siberian musk deer, Asiatic black bear and snow leopard. We then examine in more detail some of the traditional cultural uses of wildlife and their importance to people in the region.

CITES is an international trade agreement among signatory countries (called Parties). The purpose is to protect wild species of plants and animals from unsustainable international trade. The treaty was drafted in 1963 by the International Union for Conservation of Nature (IUCN), the world's largest global conservation network, and 80 countries adopted it in 1973. CITES went into force in 1975 and to date has been adopted by 180 Parties. A fundamental assumption of CITES is that excessive, unregulated commercial trade is harmful to wild species.

CITES operates by placing a species on one of three Appendices, or lists, depending on the degree of protection two-thirds of voting Parties deem necessary. Species that are considered to be at risk of extinction in the wild are listed on Appendix I. This listing amounts to a trade ban in wild-caught specimens and these species can only be traded for educational or scientific purposes (Swanson 2000: 136). In these cases, the treaty requires export and import permits certifying that the species will not be used for commercial purposes. Species listed on Appendix II are not necessarily threatened, but Parties judge that trade in specimens must be controlled in order to ensure that use does not threaten the survival of the species in the wild. A Party may ask for assistance from CITES in regulating trade in any species under its jurisdiction that it considers at risk by requesting that the species be listed on Appendix III. Parties will also sometimes protest a listing by entering a reservation, where they refuse to comply with the restrictions attached to that listing. CITES relies on the Parties to fund, manage, monitor and issue reports regarding implementation of the Convention. Each Party is obligated to designate independent CITES management and scientific authorities within its government. Any import, export, or re-export of species listed on an Appendix requires a license.

The teeth in CITES come through trade sanctions and wildlife trade bans. For example, the United States threatened to impose trade sanctions on South Korea and China for failing to police trade in rhino horn and tiger parts and on Japan for registering a reservation to a CITES listing. It imposed trade sanctions on Taiwan in 1994, which were lifted in 1996 (Ellis 2005: 221). CITES has suspended wildlife trade for countries that fail to enact requisite legislation or engage in significant levels of trade in banned species. Currently, in East Asia suspensions are recommended for Laos and Vietnam (CITES 2012).

The underlying assumption behind CITES restrictions appears to be that when trade is prohibited, demand for that particular species will abate, shift to a substitute, or be eliminated through enforcement. But international trade regulation policies only work when they take account of characteristics of both producer and consumer countries. The populations must embrace implementation on ideological, moral, or economic grounds, or disincentives to trade must be effective. Some claim that CITES is somewhat paternalistic and was implemented without considering how developing countries could maintain their wildlife or how people who depend on wildlife for their livelihoods would fare (Swanson 2000: 136). At the time countries adopted CITES, unregulated commercial trade was seen as a problem; few saw trade as a potential tool for conserving wild species. The experience of time, however, shows that within Asia many populations of species, despite the protection afforded by CITES, are in general decline. The IUCN has singled out poaching as either the predominant or a contributing factor.

No consensus exists about whether trade bans have helped or hindered protection of wildlife. Defenders of CITES claim that the situation today would be worse had the Convention not been implemented. Much of the current debate is about whether trade bans are likely to produce positive conservation results. Since CITES was adopted, several countries in Asia have undergone significant economic development. In China and Vietnam, the new middle and upper income classes have become driving forces behind trade in endangered species. They constitute a new market seeking derivatives of elephants, tigers, bears and rhinos as collectables to signal status and prestige (Mongabay.com 2012). The common wisdom is that, to combat this market force and protect species threatened with extinction, Parties need to enhance the effectiveness of trade bans through better enforcement and greater political will. Others argue that trade bans have made matters worse by restricting supply, rendering trade more profitable. In these circumstances, species are essentially converted into a non-renewable resource, thereby removing economic incentives to develop them (Conrad 2012: 245). Additionally, from a TCM perspective, less access to animal products contributes to diminished public health. Meanwhile, arguments continue concerning whether commerce can be a potentially positive tool for wildlife protection, rather than the chief threat.

The tiger, *Panthera tigris*, illustrates the complexity of the dilemma. Most subspecies were listed on Appendix I in 1977, so international trade has been illegal since the inception of CITES ('t Sas-Rolfes 2000).[2] Tigers occur only in Asia and the species is classified as endangered by the IUCN. Three of the nine subspecies have become extinct in the last 100 years and a fourth, the South China tiger, *P. t. amoyensis*, is possibly extinct in the wild. Tiger census figures are very controversial and have only recently become rigorously undertaken, but IUCN reports about 2,200 breeding adults, down from an estimated total population of 100,000 in 1900. During that same period of time, tiger habitat has also declined by 50 percent. The overall population has declined by over 50 percent in the last three tiger generations (or, 21 to 27 years; the IUCN uses species generation to measure trends). Although habitat loss poses a serious long-term threat to the continued existence of wild tigers, illegal trade in high-value tiger body parts poses the more immediate threat. Both wild-caught and captive-bred tigers are traded throughout East Asia. Among other range states, tigers are smuggled into East Asia from India and several Project Tiger reserves have been 'poached out'. While some argue that without the trade ban wild tigers would be even worse off, clearly the CITES trade ban has been far from universally successful in eliminating demand and protecting the tiger.

Rhinoceros present a similar situation. All five species were listed on Appendix I by 1977, but limited trade is allowed as some African sub-populations were subsequently moved to

Appendix II ('t Sas-Rolfes 2000). Three of the five species occur in Asia, and the other two in Africa. According to IUCN estimates, the Sumatran rhino, *Dicerorhinis sumatrensis*, numbers fewer than 250 individuals, and the Javan population, *Rhinocerous sondaicus*, is at fewer than 50. Both species are classified as critically endangered as populations have either shrunk by more than 80 percent over the last three generations (20 years) or survive only in isolated areas. The primary threat to these two rhino species is excessive poaching for its horn. The Indian one-horned rhinoceros, *Rhinoceros unicornis*, is vulnerable. Habitat loss and poaching threaten its survival. In Africa, the white rhino, *Ceratotherium simum*, is near threatened and the black rhino, *Diceros bicornis*, is critically endangered. Both inhabit private and public land, and both are suffering heavy and increasing poaching losses. In 2011, 448 rhinos were illegally killed in South Africa, and 668 more in 2012 (Mongabay.com 2013). Vietnam is thought to be the major destination for African rhino horn (Milliken & Shaw 2012). The demand for rhino horn is such that over the last 10 years 65 horns have been stolen from museums, zoos and game ranches, sometimes by armed robbery (Milliken & Shaw 2012: 65). Here, too, the protection afforded by CITES is not working.

Despite being listed on Appendix I since 1979, populations of Asiatic Black Bear, *Ursus thibetanus*, are declining throughout Asia, with the possible exception of Japan. Actual population data are not available, but the IUCN estimates that the world population has declined by 30 to 49 percent over the last 30 years and that this rate will continue. It is listed as vulnerable. Habitat loss for the species is most severe in the southern portions of its range. According to the IUCN, however, the major threat to bears in China and Southeast Asia is trade. Bears are valued for their bile, which has medicinal properties. Captive breeding of bears occurs in China, South Korea, Laos, and Vietnam so that bile can be removed on a regular basis. The IUCN has reported that bears and bear cubs are taken from the wild to establish or supplement captive populations. Here, too, CITES has not stopped illegal poaching and trade.

Listed on Appendix I in 1975, the snow leopard, *Panthera unica*, is classified by the IUCN as endangered because populations have declined by at least 20 percent over the last 16 years, or two generations. Due to their extensive ranges, thin density and hard-to-reach habitat, snow leopards are extremely difficult to census; their total population is currently estimated to be between 3,500 and 7,000 (International Snow Leopard Trust 2012). The reasons for their decline are habitat loss, loss of prey species, and poaching. Snow leopard pelts, stoles and hats are sold in markets in China and Mongolia, including even government-owned department stores and official historical sites. Its meat, nails, and bone are traded; the last are valued for their medicinal properties that some consumers substitute for tiger bone. Despite legal protection, poaching is considered a major threat to the snow leopard.

Musk is among the most important ingredients in traditional medicines and is found only in the genus *Moschus spp.*, or musk deer. There are at least seven recognized species, five of which are distributed in China and listed on Appendix II. Other animals produce a musky odor, but the musk deer alone produces true musk. Synthetic alternatives are available and used in the perfume industry, but there is no official substitute on traditional medicine ingredients lists. There are 398 patented Chinese medicines using musk and it is also an ingredient in 165 prescription medicines (Parry-Jones & Wu 2001: 4). The Siberian musk deer, *Moschus moschiferus*, is classified as vulnerable because wild population densities appear to have decreased throughout its range. According to the IUCN, a population in Mongolia declined by 84 percent between 1990 and 2000. In China, estimates put the decline from 3 million in the 1950s to between 200,000 and 300,000 in the 1990s. The principal threat to musk deer is illegal hunting, which continues despite the protection afforded by CITES and China's Wild Animal Protection Law.

Existing CITES regulations to protect wildlife in Asia have not reversed these worrying population trends. We have offered the explanation that such policies ignore the values of those who are expected to implement and follow the rules. Yet conservation practices compatible with Asian cultural values exist. Indeed, some practices developed alongside those values. Asian peoples implemented measures to address issues of scarcity and protection of natural resources long before CITES came into effect. These include rules restricting hunting and designating protected areas, as well as institutional capacity building, and captive breeding. In Mongolia, hunting seasons were designated by the time of Marco Polo, and in 1778 the first protected area was established. China's first nature reserve was established in 1956 (Coggins 2003: 14). Japan started setting up protected areas in the 1930s (Japan Ministry of Environment 2012).

The most controversial conservation practice is captive breeding. In 1984, North Korea established bear farms, which were subsequently adopted by Vietnam, to guarantee a stable supply of bile. China, which has a long history of domesticating wildlife, has established farms or breeding bases for tiger, rhino, musk deer and bears. Chinese musk deer farms were started more than 50 years ago, but the animals are difficult to raise in captivity and musk farms have thus far not been profitable (Parry-Jones & Wu 2001: 19). Tigers are also bred commercially in Laos, Vietnam and Thailand. Captive breeding is controversial because farmed animals are often kept in poor living conditions. Additionally, some reject farming as a solution because they believe it puts further pressure on wild populations by increasing demand (Cameron 2009). Others have raised doubts about this concern ('t Sas-Rolfes & Conrad 2010).

The long history of consumptive use of wildlife in Asia is directly at odds with strict trade bans. Consumptive use is connected with longstanding cultural beliefs and attitudes of people in the region. They value wildlife, not because it is wild or natural, but because it is useful. Asia has an immense richness of flora and fauna; within this environment humans and wildlife have co-existed for thousands of years. It should not be surprising that this long cultural association has resulted in deep-rooted dependency on, and ongoing demand for, animal products (Coggins 2003: 9). Wildlife consumption and use in the region have taken many forms: medicinal and therapeutic uses and symbolic uses to suggest attributes like strength, wealth or status. Animals have been used to detect poison, for entertainment, talismans and even as a unit of value or currency. As one Chinese biologist put it:

> Since the beginning of human history, man has exploited mammals for his own purposes, i.e. providing him with food, clothing, and medicine, as well as sport and recreation and even an outlet for human artistic expression. Today, mammals are exploited for the same purposes as in the past and will continue to be vitally important to man far into the future.
> (Sheng et al. 1999: 1)

Chinese culture has had a strong influence over Asia and the instrumentalist approach towards wildlife has been embraced throughout Indochina, Malaysia, Indonesia and the Philippines – wherever the Chinese diaspora has occurred, as well as Japan, Korea and Mongolia. And this, of course, is the typical approach – there is nothing particularly Chinese or Asian about it. Preservationist approaches are novel, even in the West, despite attempts to link these values to older traditions.

The most significant manifestation of the instrumentalist approach in Asia is TCM. TCM refers to a set of traditional medicines researched and practiced in Asia for at least 3,000 years (Ellis 2005: 34). Felix Cheung reports that 60 percent of the population of Hong Kong and mainland China has visited a TCM practitioner and 'anywhere from 60% to 75% of the popu-lations of Taiwan, Japan, South Korea and Singapore use traditional medicine at least once a year' (Cheung 2011: S83). Wildlife has been an essential ingredient. Chris Coggins writes that in China wildlife is widely believed 'to be endowed with healing properties greater than the biochemical sum of its parts' (Coggins 2003: 13). Traditional medicinal practitioners use musk for diseases of the liver, loins and heart, as well as in the treatment of delirium, stroke, skin infection and sore throats (Parry-Jones & Wu 2001). Rhino horn is used to reduce fever, calm convulsions, stop nosebleeds and prevent strokes. The tiger is considered a 'treasure trove' of medicinal substances. Every part of the tiger is considered to have some medicinal, therapeutic, protective, spiritual, or status-conferring benefit. Bear bile is used to treat diabetes and eye problems (Davies 2005: 74). The active ingredient in bear bile, ursodeoxycholic acid, has been shown to be effective in the treatment of liver disease (Angulo 2002).

After joining CITES, each Party developed and promulgated the requisite domestic legislation and designated management and scientific authorities. Officially, laws and regulations are in place to protect most species, as required by CITES, although 'major prosecutions for wildlife crime are still rare' (Nowell 2012: 10). Much of this amounts to mere 'paper embracement' where Parties enact the requirements but do not prioritize enforcement, even when intelligence and documented evidence of ongoing trade is readily available. In many parts of Asia anti-trade laws are mostly 'aspirational' or symbolic (Harris, R.B. 2008: 94). Parties also have other ways to avoid interference with trade. For example, they lodge reservations – Japan did so regarding whales in the 1980s. Sometimes Parties have applied to CITES for permission to participate in one-off auctions of stock-piled animal products. In 2007, some African countries were authorized to sell ivory to China and Japan. In 2006, the Chinese government undertook an internal investigation into the legalization of domestic trade in farmed tiger derivatives. And, the September 2012 meeting of the IUCN World Conservation Congress in South Korea included a Chinese TCM contingent lobbying against a motion that would phase out farming of bears for their bile.

We conclude from these and countless other examples of resistance to CITES implementation, as well as its results, that 'no-use' trade bans suffer from a lack of cultural fit with Asian beliefs and values regarding wildlife. This has been detrimental to species covered by CITES restrictions. In the next section, we argue that considering and incorporating these beliefs and attitudes is essential for the survival of vulnerable and endangered species, as well as morally required.

4 Environmental policy, as if other people matter

Richard Harris outlines three wildlife protection strategies: legal, educational and economic. He concludes that, while aspects of all three are ultimately necessary for successful conservation, Westerners have tended to overemphasize the first, have unwarranted faith in the second, and have underutilized the potential of the third (Harris 1996: 306). Advocates of continued trade-ban enforcement have often invoked the 'precautionary principle' to motivate ruling out the third strategy: absent a scientific consensus and given high levels of uncertainty about the effects of a policy, in order to avoid significant and irreversible harm, advocates of the policy bear the burden of justification (Cooney & Dickinson 2005: 287). However, in light of the evidence adduced in section 3, it is unclear that the precautionary principle favors CITES-style trade bans ('t Sas-Rolfes & Conrad 2010: 22–23). Top-down legal rules that conflict with the values and cultural commitments of those who are expected to implement and follow them are likely to meet with resistance. In the context of wildlife protection, this means that poaching and illegal trade will continue and may increase.

Legislation of one-size-fits-all constraints, which follow from a preservationist theory of environmental value, plausibly violates the precautionary principle precisely because one size does not fit all. As we have shown, Chinese-influenced culture areas in Asia generally reject the theory of environmental value from which these legal constraints are thought to follow. The laws will therefore require significant levels of interference, force and coercion in order to motivate conforming behavior. Penalties for violating them must be such that violators' evaluations of the expected costs are greater than expected benefits. This evaluation includes their estimation of the likelihood of being caught, so the state must have significant police powers, and demonstrate the internal discipline to avoid corruption in using them, in order for poachers and users of illicit animal products to discern a credible commitment on the part of the state to rooting out violations. Therefore, enforcement is costly and potentially wasteful. Regulators usually have very little to go on to determine the optimal level of penalties and investment in protection measures. Whether a given level of enforcement costs is efficient ultimately depends on the value of the environmental good, and there is no market for the effective enforcement of trade bans. Determining the social costs and benefits of enforcing them with more or less stringency and severity is, therefore, a difficult matter and it is unsurprising that CITES regulations have a low rate of compliance.

The potential of what Harris called the economic approach is based on the observation that permitting wildlife to be used as a commodity introduces economic incentives that help align the interests of wildlife with the interests and values of the people whose decisions affect it most. This is done most effectively by integrating the valued aspects of wildlife into the day-to-day lives of these people. There are many examples of this approach outside the Asian context. In Kenya, a system of revenue sharing with local tribes by state wildlife parks has had positive conservationist consequences (Schmidtz 2002b: 323ff.) Zimbabwe's CAMPFIRE program has had similar success incentivizing conservation by granting local communities the autonomy to decide what it will do with surplus numbers of wildlife and empowering them to monitor poaching activity. In the Philippines, Amanda Vincent's efforts to make seahorse fishing near Handumon sustainable relied on devising fishing methods and incentive mechanisms that encourage fishermen to manage seahorses as a renewable resource (Hasnas 2009: 123ff.) These and other examples of community-based resource management represent alternatives to top-down statutory enforcement. They tap into the economic interests of local people and, in addition, are compatible with their culturally based beliefs and values.

Ronald Coase's most important contribution to this general set of issues was to emphasize common law solutions to resolve disputes over resources (Coase 1960). Common law solutions are tied to cultural norms and so avoid many of the problems with enforcing statutory regulations devised by outsiders. The common law characteristically invokes customary rules. Customary rules are established and time honored, having evolved 'bottom up' within a social and historical process, rather than imposed top down. They have evolved through a long process of trial-and-error in addressing actual, historical disputes. They reflect the negotiated practices of a people as they found ways to cooperate and live among each other in relative peace. Legal scholar John Hasnas summarizes the broad historical outlines of this process of endogenous social learning:

> Conflicts frequently arose that would result in violence or otherwise disrupt communal life and undermine cooperative activities. This created strong social incentives to find nonviolent, nondisruptive methods of resolving such conflicts. The members of the community responded to disruptions by pressuring disputants to voluntarily negotiate settlements and by facilitating such negotiations by acting as mediators. As certain types of negotiated settlements (through trial and error) proved successful and were repeated, the members of the community came to expect that similar disputes would be resolved similarly, and they began to base their behavior on these expectations.
>
> (Hasnas 2009: 111–112)

Rules governing the use of a resource, like wildlife, would have currency because they became acknowledged by all the relevant parties to be authoritative. Members of the group recognized the individual benefits of converging on a common set of rules that guide people's decisions and interactions with others – a social morality. It is a morality that is justified in the relevant sense to all who have a stake in the social enterprise. They will have internalized the rules and, therefore, are more likely to be guided into conformity with them, and even to help enforce them. Such a morality is the imprint of a people's history.

According to Elinor Ostrom, this is largely how communities have resolved conflicts in resource use that threatened to significantly undermine many of the benefits that people derive from living in society with others. The puzzling thing about social problems deriving from conflicts in use is not that they occur; it is that they do not occur more often. Narratives explaining how norms governing resource use developed and became internalized address part of this puzzle. The process of devising legitimate rules is devolved to the local level. Even enforcement of these rules has historically relied on community monitoring and peer controls (Ostrom 1990). The examples of wildlife protection measures we cited above in section 3 that developed in the Asian cultural context to manage their natural resources fit the general description of customary practices that protected the interests of wildlife by integrating their valued aspects into the day-to-day lives of people at the local level.

But legislation imposed from above that clashes with these de facto norms introduces disruptions and impedes the ongoing evolution of the social morality to deal with novel problems. Statutory regulations are usually fairly inflexible and difficult to change or revoke should they prove unsuccessful because they usually come with bureaucratic agencies with entrenched interests in the regulatory regime. Individuals within these agencies rationally aim to maximize their budgets and expand their regulatory reach (Niskanen 1971). Statutory regulations are also more likely to fail to take advantage of the history of a people and glean from their collective wisdom ways the group has settled on to address past disputes. Rather, the statutes are more likely to be devised by individuals further removed from the local situation and who do not know enough about matters peculiar to the local culture. Therefore, they are more likely to include provisions the community simply does not see the point of or which conflict with highly valued forms of life.

A further implication of this analysis is that top-down constraints will, as Harris says, take the form of *fa*, rather than *li*, in the motivational psychology of the affected populations (Harris 1996: 312). That is, when they reject the values the statutory regulations are based on, their motivation to adhere to them will be tied to fear of detection and punishment rather than a genuine internalization of the relevant norms. CITES requirements, in effect, hold many people in Asia accountable to values that they do not acknowledge and to which they have no connection. This is plausibly morally illegitimate (Gaus 2011: Ch. I). It is presumptuous because it substitutes the judgments of Western environmentalists for their own considered views. And this seems oppressive and authoritarian because it disrespects their equal moral standing to settle on ways of living that answer to their own conception of what is good and right. The idea that all individuals are free and equal in moral status is central to modern moral and political theory. The claim is that (1) people are free in the sense that there are basic reasons to avoid coercing them and it requires special justification in order to be legitimate; and (2) people are equally capable moral reasoners in the sense that they have sufficient agency to reflect and reasonably determine questions of value for themselves. To coerce compliance with beliefs and values that people reasonably reject is disrespectful. The coercion has not been justified.

The environmentalist response at this point might be, 'But the values of TCM practitioners who use animal products don't merit respect.' We suspect many environmentalists harbor this judgment, but it implies that the use of animal products in TCM is beyond the limits of reasonable pluralism in values. This seems immoderate. One may have, given her own values and commitments, adequate reason to think that such use of animals is wrong, but it is another matter entirely to issue demands to others from that perspective about what they must not do. One's own values and commitments are not sufficient to justify interfering with those of others. Given the importance all accord to living their lives in ways consistent with their deepest values, preservationists should be wary of resorting to moralistic legislation that interferes with the commitments of TCM practitioners (after all, TCM practitioners may feel that the views of preservationists do not merit respect). And, to suggest a connection with the arguments just presented, given the track record of top-down legislation in protecting wildlife, it seems counter-productive to try.

The point here is that, as David Schmidtz puts it, 'morality is more than one thing' (Schmidtz 2002b: 328). Preservationism may play an important role in *the morality of personal aspiration*. As a first-personal set of highly valued projects and ideals, it orders and structures the choices, projects, decisions, and reasons for acting of individual adherents. But preservationism makes a poor basis for *the morality of interpersonal constraint*. This 'morality' is the set of social rules that establishes limits to each individual's project pursuit such that all are legitimately held accountable to them. Schmidtz concludes, 'Even given that preservationism is acceptable as a personal ideal, it remains a bad idea to create institutions that depend on people who do not share that ideal to take responsibility for realizing it' (Schmidtz 2002b: 328).

5 Conclusion

Western environmentalists continue to press familiar 'no-use' approaches to the problem of vulnerable species based largely on a commitment to preservationist ideals and the consequent notion that instrumentalist approaches are strictly impermissible. They generally trivialize or denounce the concerns of TCM consumers who, from the point of view of their own ideals, have decisive reason to reject the coercive policies required by CITES. No-use policies are legitimate, according to preservationists, because those who use and consume animal products must justify their use in order for it to be permissible, and the use of threatened, vulnerable and endangered species in traditional remedies for which there is little evidential support of their efficacy within modern medicine does not begin to meet the justificatory burden. Therefore, TCM consumers of remedies that contain such animal products are appropriately marginalized. Their views are false and they do bad things. But even if much of this is right – even if wildlife has non-instrumental value and TCM is largely bogus – the beliefs and values of TCM consumers should not be dismissed. Even those who share the preservationist morality of personal aspiration should worry about implementing it as a basis for interpersonal constraint in a global context where not everyone shares it. In light of this context, we have argued that the policies most environmentalists recommend based on their beliefs and values are authoritarian and counterproductive. We have, in effect, argued that the justificatory burden should be reversed. In order for coercive policies to be legitimate – that is, unless it is a matter of simply bossing other people around – they must be justified in terms that all have sufficient reason to endorse. TCM consumers, however, have reason to reject the proposed restrictions on the use of animal products. To impose these restrictions anyway is authoritarian as it fails to treat them as moral equals. Moreover, given their reasons to reject no-use restrictions, it is little surprise that this approach has failed to stop or reverse declining numbers of target species. Without an appropriate cultural fit, preservationist approaches may be incompatible with realizing preservationist goals. Instrumental approaches that build on local and mutually acceptable rules are the basis for a more successful wildlife protection regime.

Bibliography

Angulo, P. (2002) 'Use of Ursodeoxycholic Acid in Patients with Liver Disease', *Current Gastroenterology Report* 4(1), 37–44.

Animals Asia (2012) 'TCM Symposium Highlights Endangered Species'. Available at http://www.animalsasia.org/index.php?UID=NMEDLQFFAU5.

Callicott, J.B. (1987) 'Conceptual Resources for Environmental Ethics in Asian Traditions of Thought: A Propaedeutic', *Philosophy East and West* 37(2), 115–30.

Cameron, A. (2009) 'Saving the Wild Tiger: Enforcement, Trade and Free Market Folly', Environmental Investigation Agency discussion document. Available at http://www.eia-international.org.php5-20.dfw1-1.websitetestlink.com/wp-content/uploads/reports182–1.pdf.

Cheung, F. (2011) 'Made in China', *Nature Outlook* 480, S82–S83.

CITES (2012) 'Reference Lists'. Available at http://www.cites.org/eng/resources/ref/suspend.php.

Coase, R. (1960) 'The Problem of Social Cost', *Journal of Law and Economics* 3, 1–44.

Coggins, C. (2003) *The Tiger and the Pangolin* (University of Hawai'i Press).

Conrad, K. (2012) 'Trade Bans: The Perfect Storm for Some Species', *Tropical Conservation Science* 4(3), 245–254.

Cooney, R. & Dickinson, B. (2005) 'Precautionary Principle, Precautionary Practice: Lessons and Insights', in R. Cooney & B. Dickinson (Eds.), *Biodiversity and the Precautionary Principle* (Earthscan).

Davies, B. (2005) *Black Market: Inside the Endangered Species Trade in Asia* (Earth Aware Editions).

Elliot, R. (1992) 'Intrinsic Value, Environmental Obligation and Naturalness', *Monist* 75(2), 138–160.

Ellis, R. (2005) *Tiger Bone & Rhino Horn* (Island Press).

Gaus, G. (2011) *The Order of Public Reason* (Cambridge University Press).

Graham-Rowe, D. (2011) 'Biodiversity: Endangered and in Demand', *Nature Outlook* 480, S101–S103. Available at http://www.nature.com/nature/journal/v480/n7378_supp/full/480S101a.html.

Guha, R. (1989) 'Radical American Environmentalism and Wilderness Preservation: A Third World Critique', *Environmental Ethics* 11, 71–83.

Harris, P.G. (2004) '"Getting Rich is Glorious": Environmental Values in the People's Republic of China', *Environmental Values* 13, 145–165.

Harris, P.G. (2008) 'Environmental Perspectives and Behavior in China', *Environment and Behavior* 38(5), 5–21.

Harris, R.B. (1996) 'Approaches to Conserving Vulnerable Wildlife in China: Does the Colour of the Cat Matter – if it Catches Mice?', *Environmental Values* 5, 303–334.

Harris, R.B. (2008) *Wildlife Conservation in China* (M. E. Sharpe).

Hasnas, J. (2009) 'Two Theories of Environmental Regulation', *Social Philosophy and Policy* 26(2), 95–129.

International Snow Leopard Trust (2012) 'Snow Leopard Fact Sheet'. Available at http://www.snowleopard.org/downloads/snow_leopard_fact_sheet_english.pdf.

Ip, P.K. (1983) 'Taoism and the Foundations of Environmental Ethics', *Environmental Ethics* 5, 335–343.

Japan Ministry of Environment (2012) 'Parks Index'. Available at http://www.env.go.jp/en/nature/nps/park/parks/index.html.

Katz, E. (1991) 'Ethics and Philosophy of the Environment: A Brief Review of the Major Literature', *Environmental History Review* 15(2), 79–86.

Kobayashi, Y. (2005) 'The "Troubled Modernizer": Three Decades of Chinese Environmental Policy and Diplomacy', in P.G. Harris (Ed.), *Confronting Environmental Change in East and Southeast Asia: Eco-Politics, Foreign Policy, and Sustainable Development* (United Nations University Press).

Leiter, B. (2005) 'New "Cooperative" Philosophy PhD Program at North Texas/UT Arlington'. Available at http://leiterreports.typepad.com/blog/2005/08/new_cooperative.html.

Milliken, T. & Shaw, J. (2012) *The South Africa Vietnam Rhino Horn Trade Nexus* (TRAFFIC).

Mongabay.com (2012) 'Yuppies Are Killing Rhinos, Tigers, Elephants'. Available at http://news.mongabay.com/2012/0907-yuppies-consuming-wildlife.html.

Mongabay.com (2013) 'Rhino Poaching Hits New Record in 2012'. Available at http://news.mongabay.com/2013/0111-rhino-poaching-record.html.

Niskanen, W. (1971) *Bureaucracy and Representative Government* (Atherton).

Nowell, K. (2012) *Wildlife Crime Scorecard* (World Wildlife Fund).

O'Neill, J. (1992) 'The Varieties of Intrinsic Value', *Monist* 75(2), 119–137.

Ostrom, E. (1990) *Governing the Commons* (Cambridge University Press).

Parry-Jones, R. & Wu, J. (2001) *Musk Deer Farming as a Conservation Tool in China* (TRAFFIC East Asia).

Roetz, H. (2010) 'On Nature and Culture in Zhou China', in H.U. Vogel, G. Dux & M. Elvin (Eds.), *Concepts of Nature: A Chinese–European Cross-Cultural Perspective* (Brill).

Rolston, H. (1987) 'Can the East Help the West to Value Nature?', *Philosophy East and West* 37(2), 172–190.

Schmidtz, D. (2002a) 'Natural Enemies: An Anatomy of Environmental Conflict', in D. Schmidtz & P. Willott (Eds.), *Environmental Ethics: What Really Matters, What Really Works* (Oxford University Press).

Schmidtz, D. (2002b) 'When Preservationism Doesn't Preserve', in D. Schmidtz & P. Willott (Eds.), *Environmental Ethics: What Really Matters, What Really Works* (Oxford University Press).

Shen, S., Ables, E.D. & Xiao, Q.Z. (1982) 'The Chinese View of Wildlife', *Oryx* 16, 340–347.

Sheng, H., Ohtaishi, N. & Lu, H. (1999) *The Mammals of China* (China Forestry Publishing House).

Swanson, T. (2000) 'Developing CITES: Making the Convention Work for All of the Parties', in J. Hutton & B. Dickson (Eds.), *Endangered Species, Threatened Convention* (Earthscan).

't Sas-Rolfes, M. (2000) 'Assessing CITES: Four Case Studies', in J. Hutton & B. Dickson (Eds.), *Endangered Species, Threatened Convention* (Earthscan).

't Sas-Rolfes, M. & Conrad, K. (2010) 'Making Sense of the Tiger Farming Debate'. Available at http://www.asiacat.org.

Tucker, M. (1991) 'The Relevance of Chinese Neo-Confucianism for the Reverence of Nature', *Environmental History Review* 15(2), 55–69.

Varner, G. (1998) *In Nature's Interests?* (Oxford University Press).

Acronyms

CITES	Convention on International Trade in Endangered Species of Wild Fauna and Flora
IUCN	International Union for Conservation of Nature
TCM	traditional Chinese medicine
TRAFFIC	Trade Records Analysis of Flora and Fauna in Commerce

Notes

1   The comment was submitted on 22 December 2011 and is available at http://www.nature.com/nature/journal/v480/n7378_supp/full/480S101a.html.
2   *P. t. altaica*, the Amur tiger, moved to Appendix I in 1987.

Part VI
Energy and climate change

20

Drivers of climate change in East Asia

The energy dilemma

Bo Miao

Introduction

Although scientific debate over climate change still lingers, the Intergovernmental Panel on Climate Change (IPCC) has found that the evidence of human impact on climate change is 'unequivocal' (IPCC 2007). There are six greenhouse gases (GHGs) that contribute to the phenomenon of climate change, namely, CO_2, CH_4, N_2O, HFCs, PFCs and SF_6. While CO_2 accounts for more than two-thirds of all the GHGs, the other five non-CO_2 gases have much greater global warming potential. For instance, the capacity to heat the atmosphere of one ton of methane is equivalent to 21 tons of CO_2; for N_2O, 271 times equivalent; and the other three minor GHGs are even stronger (IPCC 2001). The international community has used CO_2e (equivalent to a given volume of CO_2) to calculate GHG emissions for the sake of convenience.

Where do these GHGs come from? Generally, burning fossil fuels for electricity and transportation, deforestation and methane from agriculture and livestock all contribute to climate change (IPCC 2001). The biggest driver in East Asia, however, is the huge consumption of fossil fuels to sustain economic development, especially that of China (Zhang 2010). China is now the world's top energy consumer, and coal, the most carbon-intensive fossil fuel, meets more than two-thirds of China's energy demand (British Petroleum 2012). Burning coal is the biggest contributor to GHG emissions in China. The rapidly increasing fleet of vehicles also adds more pressure to curtail the soaring GHG emissions (NDRC 2007a). Emissions from other sources such as agriculture and waste are relatively small compared to these two leading contributors.

Various studies indicate that China has surpassed the US and become the largest GHG emitter in terms of annual contribution as early as 2007, although Beijing has never officially confirmed it (Netherlands Environmental Assessment Agency 2007). Whereas China's historical contribution to climate change is not as significant as that of the US or EU, its per capita GHG emissions are rapidly approaching the world average (International Energy Agency 2011). Taking into account the population in China, the need to contain the soaring GHG emissions is great. However, the heavy reliance on fossil fuels, in particular coal, to power the economy makes tackling climate change a daunting, albeit laudable, task for the Chinese government.

Policymakers in Beijing are not blind to the heavy constraints imposed by coal and other fossil fuels. Neither have they underestimated the urgency and difficulty in transitioning into a low-carbon society. Can China effectively alter the main driver of climate change and diversify the emitting sources? What policies has it adopted to reduce the reliance on carbon-intensive energies and moderate the rapidly growing GHG emissions? How have those programs performed? This chapter aims to address these questions.

In the mean time, GHG emissions from other East Asian countries are also important. Japan and South Korea, two of the most affluent economies in East Asia, both contribute significantly to climate change. However, the sources of their GHG emissions are notably different from those in China due to a variety of reasons. The energy and climate polices they formulate deserve no less attention. It is therefore interesting to explore the similarities and differences of the energy and GHG profile of the three major East Asian countries.

China

China's continuing dependence on coal has a long history. As a country rich in coal reserves, the most carbon-intensive fossil fuel has served as the cheapest energy to boost China's economy for decades. Since the 1990s, China's coal use has more than tripled, making it the largest consumer of coal in the world. In 2011, China alone represented almost half of the world's coal consumption (British Petroleum 2012). In the same year, coal satisfied 69 percent of national energy demand, while oil contributed 18 percent, natural gas more than 4 percent, and renewable energy such as hydro, wind, solar and nuclear together less than 8 percent (National Bureau of Statistics of China 2012).

Most of China's CO_2 comes from energy use and burning coal is the single leading contributor. Living in a coal-dominant and coal-constraining world has led us to wonder whether China's coal-dominant energy structure is likely to change fundamentally in the foreseeable future. Or put another way, will coal continue to be the largest driver for climate change in China? Answers can be explored by addressing the following questions:

1   Will the development of more efficient and clean coal technology such as carbon capture and storage reduce significantly CO_2 emissions from coal power plants?
2   Can less carbon-intensive fossil fuels like oil and natural gas play a bigger role to sustain China's economy?
3   Is it possible for alternative energy such as wind, solar, hydro and nuclear power to replace coal as China's main energy supplier?

The next three sections will examine these questions respectively.

Development of coal industry

China is still building more coal-fired power plants at the pace of two per week. Some study shows that coal-fired plants in China have outnumbered those in the US, UK, and India combined (Asia Society & Pew Centre on Global Climate Change 2009: 20). On the one hand, the government has been consistently making efforts to retire small, outdated coal power plants and achieved great success; on the other hand, the road to develop and adopt advance clean coal technology has been exciting but bumpy.

Cases of installing more efficient coal power plants in China are constantly reported (Miao 2012). It is estimated that the average efficiency of the coal-fired fleet will be increased from

32 percent in 2005 to roughly 40 percent by 2030 should more supercritical units be built as planned (Asia Society & Pew Centre on Global Climate Change 2009: 28). Improving efficiency means more electricity can be generated from the same amount of input and therefore less CO_2 will be emitted.

However, it should be noted that advanced technologies such as integrated gas combined cycle (IGCC) are only adopted by a small number of power plants due to mainly the concern of installing and operation costs (Bradsher 2009a). Overcoming the heavy financial constraints is no easy task. So is the mustering of sufficient political will to deploy more clean coal technology. It is equally important to note that the small, less efficient coal power plants will remain in operation for a long period of time in order to meet China's energy thirst. It is a tough job for Beijing to reduce significantly the GHG emissions from coal consumption.

Then what about the development of carbon capture and storage (CCS)? Can the deployment of this zero-carbon technology help sustain the expansion of coal-fired power plants? CCS is a technology that captures CO_2 either before combustion or after combustion, compresses the captured CO_2 and transports it through pipelines for storage in deep, underground geological formations such as depleted oil fields (Asia Society & Pew Centre on Global Climate Change 2009: 29). The CCS is promising in controlling CO_2 emissions because should it be adopted, a coal-fired power plant will virtually emit zero GHG. The CCS technologies have been widely researched by a number of nations, but many of them still remain at a preliminary stage and there are only a few small-scale demonstration projects.

China is also devoting resources to the CCS research with other international partners. In collaboration with the Australian Commonwealth Scientific and Industrial Research Organization (CSIRO), Huaneng Power Group (one of China's largest electricity generators) started running a 3,000-ton post-combustion carbon capture pilot project near Beijing in 2008 (EA 2009: 107). The CO_2 captured is not stored but used for beverage production. Other proposed demonstration projects including GreenGen (a 400-megawatt IGCC plant with CCS to be added by 2020) and near zero emission coal are also under planning (Asia Society & Pew Centre on Global Climate Change 2009: 29).

The major hurdle for wide application of CCS is cost. It is estimated that the electricity produced by a coal power plant that uses CCS would be 75 percent to 100 percent more expensive than the electricity produced by a conventional coal power plant (IEA 2009: 106). Neither can the concern of 'energy penalty' of running the capture equipment be ignored. With current CCS technology, the energy that is required to capture the CO_2 is not trivial and may reduce a plant's combustion efficiency by roughly one-third. While transportation from power plant to storage site is a further costly complication for carbon sequestration, the storage of the captured CO_2 is another problem.

While some initial assessment of China's storage capacity has been conducted, it is too early to conclude that China would be able to store the CO_2 generated by its enormous number of coal-fired power plants. Indeed, the volume of CO_2 currently generated by burning coal, when compressed for storage, would be much larger than any current storage sites. The issue gets further complicated as the storage sites in many cases are usually very far from existing or planned power plants. The magnitude of the CO_2 emitted by all China's coal power plants would undoubtedly make the transportation and storage quite a challenge.

What effects burying CO_2 would have on aquifers also remains an open question. The risk involved in the potential leakage of stored CO_2 cannot be underestimated, either. The leaked CO_2 could be lethal. Insurance companies would not act unless scientific research can convincingly ensure that leakage would not be a problem. Without the financial assurance from the insurance companies, it is unlikely that large-scale CCS projects that normally demand huge amounts of investment will kick off. In addition, some researchers have argued that burying CO_2 from coal-power plants could increase the emissions of other pollutants such as NO_x and SO_2, casting a shadow over this highly acclaimed technology (Barry 2008).

Indeed, whereas the international community has admitted that technologies such as CCS are key to the continual use of coal, it is widely agreed that large-scale promotion of CCS would not be commercially viable until at least 2030 (IEA 2009). China is actively participating in the research and development of this clean coal technology, but it seems unrealistic to expect the development of this technology to be soon enough to hold back the soaring CO_2 from China's coal power units.

Securing more oil and natural gas

Compared to coal, oil and natural gas is less carbon intensive. If China can use more oil and natural gas to support the economic development rather than coal, the emissions of GHG will be correspondingly reduced in a relative sense. The question is: can China consistently secure more oil and natural gas supplies, domestically or overseas?

The answer is not hard to detect. China's domestic oil production has increased very slightly in the past decade while its oil consumption has risen sharply. The growing gap between supply and demand has made China the world's third largest oil importer. In 2011, imported oil met more than half of China's domestic need (British Petroleum 2012).

Can China boost domestic oil production in the future? No new major oil field was discovered in China in the past two decades and Xinjiang and Tibet are the key areas for future oil development (IEA 2009: 77). Taking into account the political and geographical complexities in these regions, it seems unlikely that China will significantly increase domestic oil supply by using existing technologies. There are also few technological innovations that profoundly improve the efficiency of oil extracting. One of China's own research organisations predicts that China's domestic crude oil production would probably peak around 2015 and then start to decline (ERI 2004). Nonetheless, the technology of coal-to-liquid (CTL) deserves special attention as it has the potential to bring more oil to the domestic market.

Coal could be used to produce oil substitutes, as has occurred in the past in Germany during World War II, and in 'apartheid' South Africa during the period of economic sanctions. However, coal gasification uses large amounts of energy, thus producing much less net energy than oil, and hence, while it may offer short-term energy substitutions to replace oil, it would also contribute heavily to GHG emissions and other forms of pollution. Thus, massive resort to coal as oil supply dwindles would not be a politically feasible or responsible solution over the longer term.

China has been prudent in developing CTL. The central government issued a temporary moratorium on all new CTL projects except for the two demonstration projects by the state-owned Shenhua Group in late 2008 (Nature 2008). It was actually the third time that the National Reform and Development Commission (NDRC) temporarily held up CTL projects since 2006. The reasons are both economic and environmental.

The technological and economic risks associated with CTL are so high that they have induced Beijing to remain cautious when approving large-scale CTL projects. CTL projects normally require huge inputs of energy, coal and water resources. It is reported that four to five tons of coal are normally required to produce one ton of oil. Even under the Shenhua project, which has the best conservation efficiency technology, three tons of high-quality coal are still needed for converting into one ton of oil. As for the environmental costs, the need for enormous amounts of water for manufacturing and the emitting of GHG during the transformation and later consumption of the coal-based products has to be considered, in particular when China is facing increasing international pressure to moderate its GHG emissions. It is estimated that eight to nine tons of water are required for the production of one ton of oil. The water consumption in the indirect coal gasification technology offered by the South African CTL giant Sasol is even higher – 1.5 times that in the direct gasification technology.

For these reasons, it is safe to say that the prospects of CTL in China are still uncertain. Meanwhile, it is reported that China's largest coal corporation, Shenhua, the only company that has been granted the approval to continue with its CTL projects, has made breakthroughs in core CTL technologies. The Inner Mongolia project is declared to be a successful case of applying the method of directly gasifying coal underground. Shenhua is also working closely with Sasol on the technology of indirectly gasifying coal in the Ningxia demonstration projection (Caijing Net 2009). However, even if these technologies can be developed as planned, they will still face huge obstacles for its deployment for the above economic and environmental reasons.

If domestic oil supply remains relatively stable, can China's effort to obtain more oil from overseas resources change the game? Most of China's oil imports come from the Middle East and Russia. However, China is currently unable to guarantee the safe passage through the Strait of Hormuz and the Strait of Malacca for oil ships from the Middle East to China. The relationship between China and Russia is so complicated that no Chinese politician would dare to rely solely on oil supply from Russia to sustain China's economy. Importing oil from regions with sensitive politics such as Sudan and Angola is also troublesome. The silence over the Darfur genocide that China maintained a few years ago invited heavy criticism from the international community. The later splitting of the Sudan government then cast shadows over the stable flow of oil from this region to China.

China is also exploring other mechanisms of securing more overseas oil. For instance, China signed long-term oil-for-loan deals with Venezuela by which the oil exporter could repay its loan in the form of oil rather than cash (Ellsworth & Parraga 2012). While the innovative trading forms break new ground for China to acquire more oil from Latin America, they are exposed to the risk of political instability and regime change.

Indeed, China's effort to hunt for overseas oil supply has encountered various obstacles, even though it does channel more oil to China (Miao & Lang 2010). Then what about the prospects for obtaining more natural gas for China?

Natural gas is the cleanest energy among the fossil fuels and also the least carbon intensive. However, China is not rich in natural gas reserves, ranking only 13th in the world by 2011 (British Petroleum 2012). Although discoveries of new gas resources are reported from time to time, it is difficult for China's indigenous gas production to meet the fast growing demand. For instance, while the Chinese government is developing its natural gas infrastructure and plans to reform the energy price system to promote the use of natural gas, gas-fired power plants only account for a very small proportion of China's power plant fleet due to the cost disadvantage when compared to coal. The massive West–East Gas Pipeline was completed in 2005, but only relieved partially the thirst for natural gas by rich coastal regions in China. Even with optimistic predictions, the total natural gas-fired power generation would represent only a small portion of China's primary energy consumption by 2020 (IEA 2007).

Can acquiring gas from overseas sources enrich China's natural gas profile? China has started to import liquefied natural gas (LNG) from Australia to meet the rising demand for natural gas for some southern provinces since 2006 (IEA 2009: 80). It also tries to secure stable gas supplies from countries like Indonesia, Russia and Kazakhstan. However, natural gas supplies from these countries are influenced by global supply fluctuation as well as constrained by transport stability (McKinsey & Company 2012). The conflicts over the exploration of oil and natural gas in the

South China Sea among China and some Southeast Asian countries also illustrate the difficulty in expanding significantly the supply of these fossil fuels.

The latest technological innovations in exploiting shale gas in the US might increase the prospects for natural gas exploration (Chazan 2012). It is possible for China to benefit from the technological breakthrough, but the difficulty in developing directional drilling and hydraulic fracturing technologies on its own or acquiring them from technology holders in the US is not trivial. The GHGs that are emitted when extracting shale gas may also dim the prospect of tackling climate change (Harvey 2012). Oil and natural gas are cleaner than coal and less carbon intensive. Nonetheless, they cannot replace coal to power China's economy due to the limited domestic reserves and great uncertainty in ensuring a stable and constant flow from overseas sources.

Development of alternative energies

The track record of China's alternative energy development is impressive. The announcement of a target of 15 percent of primary energy from renewable energy by 2020 has stimulated the rapid development of the renewable industry in China (NDRC 2006). As renewable energy emits virtually no GHG, its increasing share in China's overall energy supply will certainly help China reduce its contribution to climate change and moderate the dominating role of coal in the generation mix. Let us take a close look at the development of wind energy, solar energy, hydropower and nuclear power in China, respectively (These are the main types of alternative energy in China but this list is far from exhaustive.)

The development of wind industry in China is particularly interesting. In mid-2012, China surpassed the US and built the world's largest wind power capacity. Its current installation is 62 gigawatts (GW), equivalent to the electricity needed to light up all Australia. Beijing has set a further target of expanding the generation capacity to 100 GW by 2015 and 200 GW by 2020 (NDRC & ERI 2012). Despite the encouraging numbers, wind energy has experienced difficulties in trying to play a bigger role in cutting China's heavy reliance on coal. It is reported that almost one-third of wind turbines sit idle.

There are good reasons for the rapid expansion of wind industry in China: first, government provides significant subsidies and preferential financial policies for the production and installation of wind turbines. Second, Beijing has put in place laws and policies, in particular the Renewable Energy Law, to promote renewable energy including wind energy. Third, investing in wind farms could gain carbon credits through the Clean Development Mechanism under the Kyoto Protocol.

However, the reasons that undermine the promotion of wind energy are no less powerful. First, the national grid operator lacks adequate commercial incentive to purchase electricity generated by wind farms as it is much more expensive than that from coal-fired power plants. Second, the construction of a transmission network such as ultra high-voltage lines that transfers the wind power to remote end users has severely lagged behind. Third, the current economic downturn also makes grid operators more reluctant to connect and dispatch the costly and unpredictable wind power.

The opportunities and obstacles are equally important for China's future development of wind energy. The story of promoting solar energy in China is no less interesting. Most of China's land area has abundant solar energy. However, although the installed capacity of solar photovoltaic (PV) power has steadily grown in the past decade, solar energy currently only plays a very insignificant role in powering China (NDRC & ERI 2012). Compared to the rapid expansion of wind power, the development of solar energy is relatively slower mainly because the cost of producing electricity from solar power is much higher than from other sources.

The Chinese government is not insensitive to the great potential of harnessing solar energy. It has put in place a series of policies to prioritize the promotion of solar industry. For example, the Ministry of Finance promulgated 'Application guidelines for demonstration projects of solar photovoltaic building' in 2009, offering significant subsidies for solar PV projects (Ministry of Finance 2009). Nevertheless, it is not easy to overcome the cost constraints and commercialize demonstration projects in China.

It should also be noted that China is the world's largest producer of PV cells but more than 90 percent of the products are sold to Europe and the US (Miao 2012). The financial difficulties that China's solar industry has experienced since 2008 demonstrate the vulnerability of the industry and point out the great need to nurture the domestic market. To make things more complicated, the US and EU are accusing China of dumping solar panels, and China has also initiated similar allegations against the EU to the WTO (Beattie & Chaffin 2012). A similar legal dispute between China and US occurred over US allegations that China provided illegal subsidies to wind turbine manufacturers (*Wall Street Journal* 2012). The development of wind and solar energy, albeit impressive, faces many such obstacles.

What about hydropower? With Three Gorges Dam coming into operation in 2007, hydroelectricity provided 16 percent of the nation's power generation in 2010 (NBS 2012). As more dams are proposed to be constructed, it is likely that China will boost its hydropower capacity in the future (NDRC 2007b). However, there are three concerns that may plague the development of hydro industry.

First, hydropower construction has to pay up to 30 different taxes and fees that impose heavy financial constraints on hydropower projects (IEA 2009: 83). Second, how properly to address issues of population displacement and ecological conservation associated with large-scale hydropower projects will continue to afflict the government, developers and the public. The construction of Three Gorges Dam has illustrated how hard it is to strike a proper balance among these competing interests. Third, the severity of transnational disputes over dam construction on cross-border rivers such as Lancang (Mekong) River and Nu (Salween) River cannot be underestimated.

Bearing these concerns in mind, it is safe to say that while hydropower will continue to diversify China's energy portfolio, it is unable to change significantly the coal-dominant generation mix.

Another important alternative energy in China is nuclear power. China has a long history in developing this clean energy, and in 2010 nuclear power met roughly 2 percent of national electricity need (NBS 2012). The government has set the world's most ambitious target to build more than 20 new nuclear reactors and provide up to 5 percent of national power generation by 2020. Development in the past two decades also helped China to break away from constraints that have long inhibited the construction of more nuclear reactors. For instance, China has secured a stable flow of nuclear material by conducting commercial uranium trade with Australia and jointly exploring uranium sources with the government of Niger (IEA 2009: 81). It works closely with international partners such as France to develop and deploy advanced nuclear technology. In addition, Beijing abolished the monopoly in uranium development by China National Nuclear Corporation and introduced more players into the game.

The central government's ambition of developing nuclear power is echoed by local governments. Cities are competing to host the construction of nuclear reactors. However, the Fukushima Daiichi nuclear accident in 2011 disrupted China's nuclear expansion plan, which was not resumed until late 2012 but with much more moderate goals and more stringent conditions. A booming nuclear industry in China is still taking time to mature.

Taken together, the development of alternative energy in China is impressive, but there is far to go before alternative energy can make a significant dent in China's rising overall power demand (Miao & Lang 2010). It is reported that production of coal grew even faster than that of non-fossil fuels in 2011 (Twenty-First Century Net 2012). With domestic production remaining relatively stable, China's effort to secure more oil and natural gas supplies from overseas sources is unlikely to alter fundamentally the coal-dominant energy mix. Burning coal will continue to be the largest contributor to China's GHG emissions in the short to medium term. Although progress has been made in developing clean and efficient coal technologies, renovating China's coal-fired power plant fleet is no easy task.

Improving energy efficiency can also help reduce China's contribution to climate change. China's energy efficiency has been notoriously lower than that of many other economies, meaning that China consumes more energy and emits more GHGs for producing the same amount of economic output. Beijing set a 20 percent energy-intensity reduction target in the Eleventh Five-Year Plan (2005–10) and claimed that it has successfully achieved the goal. Studies show that the 20 percent energy-intensity improvement could have produced an annual reduction of over 1.5 billion tons of CO_2 by 2010 (Lin et al. 2008). A further reduction target has been put forward by the central government to keep improving China's energy performance. Indeed, energy conservation has particular importance for China as it not only moderates its soaring GHG emissions under a business–as–usual scenario but also strengthens China's energy security.

GHG emissions from transportation

Aside from burning coal, transportation is an increasingly important contributor to China's GHG emissions. The rapidly growing fleet of vehicles across Chinese cities has drastically driven up the demand for oil and other energy. It is reported that emissions from the transportation sector have become the second largest GHG source in China (State Council 2011a). Can emissions of CO_2 from transportation be contained?

China adopts a very stringent fuel economy standard for light-duty passenger vehicles, only next to the EU and Japan (Oliver et al. 2009). The stringency is, in part, due to the concern of reducing reliance on imported oil. However, while a stricter fuel standard makes the combustion of oil more efficient, the saved energy could be easily offset by more driving. It is difficult for efficient engines alone to slow down the rapid pace of CO_2 emissions from China's transportation. Can development of cleaner vehicles, for example the electric vehicle (EV), help?

Unlike the hybrid car that still consumes oil and emits GHG, the electric car produces no emissions at the tailpipe. In most models, it does not even have a tailpipe, making it virtually zero carbon. Many have expressed interest in developing EVs including China. Beijing put forward the goal of becoming a world leader in EVs by 2012 with the aim of creating jobs, reducing urban pollution and decreasing oil dependence (Bradsher 2009b). In order to achieve the goal, the government has allocated large sums of funds to EV research. It also ran a pilot program in 13 cities that offers a subsidy of up to US$8,800 to each EV that joins the taxi fleet or is purchased by the local government agencies. Complementary infrastructure such as charging stations for EVs is also under construction in big cities such as Beijing, Tianjin and Shanghai. China also set an ambitious aim of boosting its production of hybrid and electric vehicles from 2,100 in 2008 to 500,000 by 2011. Unfortunately, only 6,000 EVs were produced by the end of 2011 (McKinsey & Company 2012).

Auto companies in China have been active in promoting EVs. The Shenzhen-based BYD Auto, China's Tianjin Qingyuan Electric Vehicle Company and Hafei Auto Group developed several EV models and passed strict safety tests. However, how to commercialize EVs in the market remains an open question – even with government subsidies, the retail price of these EVs is still much higher than that of their gasoline-engine counterparts. There are also other obstacles. For instance, the demand for electricity in China is so huge that it may leave little room to provide sufficient electricity to recharge the battery for the large-scale promotion of EVs.

Electric vehicles also have an eco-label – the well-to-wheel CO_2 emissions of EVs are always lower than those of conventional cars. The question is, as EVs replace the burning of gasoline and diesel by the burning of coal to produce electricity, will it create great 'emissions saving' for China?

The answer is closely related to the emission intensity of China's existing electricity infrastructure. As previously discussed, 80 percent of China's electricity is produced by coal. It could be expected that most of the electricity that will be used to recharge an EV's battery would be provided by coal-fired power plants. Bearing in mind that many of China's coal power plants are still inefficient, using electricity coming from carbon-intensive fossil fuels would negate to a great extent the climatic benefits brought by the efficiency advantages of EV. A McKinsey & Company report states that 'given China's reliance on coal-fired plants for electricity, electric vehicles today only have a 19 percent carbon abatement potential over current internal combustion engine technologies' (Gao et al. 2008).

However, EVs could produce better environmental benefits if more renewable energy is introduced to the grid. By diversifying the energy source to fuel cars, China would be able to achieve as much as 49 percent carbon abatement potential. However, it is a long way for China to realize the high carbon abatement potential since its coal-dominant energy mix will remain for a long period of time.

To sum up, coal will continue to be the largest driver of climate change in China, while emissions from transportation are increasing drastically. In China's Twelfth Five-Year Plan (FYP), Beijing sets a 40–45 percent carbon intensity reduction target by 2020. The central government also published 'The Working Plan on the GHG Emissions Reduction during the Twelfth FYP' and put forward a 17 percent reduction goal by 2012. Beijing then allocated the 17 percent reduction target among province-level governments and links officials' careers with the achievement of the reduction target (State Council 2011b).

Even so, China has to deal with the dilemma of sustaining a developing economy which depends crucially on GHG-emitting processes. Climate change has brought to the fore a new global discourse on issues relating to how to strike the proper balance between environment and economy in a coal-dominant and coal-constraining world, and how to address other challenges that are inextricably linked with climate change such as energy security at the same time.

Japan and South Korea

Japan and South Korea are both important contributors to climate change, ranking respectively the fourth and 10th globally in 2008 in terms of annual GHG emissions (World Resources Institute 2012). While emissions in Japan rose slightly from 1990 to 2007, they have started to decline in the past few years (UNFCCC 2012); South Korea has more than doubled its GHG emissions since 1990 with no sign of stopping (Ministry of Environment of the Government of South Korea 2012). It is predicted that Japan's emissions might flatten or even reverse, but South Korea is likely to have a difficult time containing its GHG-emitting activity (Meltzer 2011). CO_2 is the largest source of GHGs in both countries, representing more than 80 percent of all six GHGs.

Where do these GHGs come from? In the case of Japan, energy use is the biggest contributor, accounting for roughly 90 percent of all GHG emissions. To be specific, emissions from industry have conventionally been the largest single source for GHGs but sharply declined, in particular after 2008 when the economic downturn hit Japan. Transportation is the second largest source but emissions from it only increased slightly over the years. These two leading sources together contribute more than half of Japan's GHG emissions in 2009. Emissions from commercial buildings and homes rose rapidly and accounted for another 30 percent of the overall profile. GHGs from non-energy sources such as industrial process, waste, and agriculture were relatively small and remained so since 1990 (Ministry of Environment of the Government of Japan 2012).

In the case of South Korea, more than 80 percent of GHGs come from energy use in 2008. Emissions from electricity and heat are the single largest contributor, followed by emissions from manufacturing and construction. Transportation is an increasingly bigger contributor, while emissions from other non-energy sectors remain relatively small (Ministry of Environment of the Government of South Korea 2012).

What are the similarities and differences in drivers for climate change in Japan and South Korea? Similarities are actually more significant than differences in the GHG profile of the two countries. Although Japan's overall annual emission is almost twice as big as that of South Korea, energy use, in particular electricity and heat, is the single largest contributor in both countries. Within the energy sector, emissions from transportation are second only to industry in both economies. Other sources such as agriculture and waste, although they do contribute to climate change, remain on the periphery. Indeed, both Japan and South Korea are affluent OECD countries bearing similar geographic heritages. After decades of economic development, their current GHG profiles are also very similar.

As energy use is the leading GHG contributor in both Japan and South Korea, it is necessary to examine their energy portfolio. What type of energy has sustained their economies? Can they deploy more low-carbon or zero-carbon energy in the future?

The energy scenario in Japan and South Korea is astoundingly similar. Both nations have very limited fossil fuel reserve and rely heavily on imported coal, oil and natural gas. Conventional fuels together represent more than 80 percent of total primary energy supply by 2009, although coal is never the biggest supplier. Nuclear power was a major energy source for both countries until the Fukushima Daiichi accident changed the energy scenario in 2011. Renewable energy including hydro, wind, solar and geothermal only meets a small fraction of energy demand, as of 2009, in both Japan (3.6 percent) and South Korea (1.5 percent) (IEA 2012).

It can be seen that Japan and South Korea bear remarkable similarities in energy structure and GHG profile. So are they adopting similar energy and climate policies?

Both countries put forward ambitious GHG reduction targets. Japan, as a nation undertaking binding obligations under the Kyoto Protocol, aims to cut its GHG emissions by 25 percent below 1990 levels by 2020, and 80 percent by 2050. South Korea, as one of the few OECD countries that bears no mandatory reduction obligation under Kyoto Protocol, also pledges to cut 30 percent of its GHG emissions from projected levels by 2020.

In order to deliver the climate commitment, Japan promoted the Basic Energy Plan in 2010, which aims to boost the use of nuclear power to 50 percent of the generation mix and renewable energy to 20 percent by 2030. Unfortunately, the Fukushima Daiichi accident made it a political nightmare to rely more on nuclear power. Almost all of the nuclear reactors stopped operation and leave the vacuum to be filled by other alternative energy. Japan then turned to the development of renewable energy by releasing a Renewable Energy Feed-In Tariff Law in late 2011. It is also experimenting with innovative mechanisms to induce various stakeholders to control their GHG emissions. In 2010, Japan introduced the world's first mandatory emissions trading program that targets energy end users such as commercial buildings and factories in Tokyo (Tokyo Metropolitan Government 2010).

However, it should be noted that Japan's energy and climate polices are more complicated than they appear. The prospects for meeting the country's reduction target under the Kyoto Protocol has remained gloomy until the 2008 economic crisis. Since then, emissions from industry dropped sharply, putting Japan back on track to fulfill its climate commitment. Neither was Japan a big supporter for the continuation of the Kyoto Protocol, the only legally binding international climate agreement that was named after one of its cities.

Introducing more low-carbon energy to sustain Japan's economy becomes more difficult if nuclear power is taken out of the energy mix. How alternative energy can fill the proposed 50 percent supply gap left by nuclear power is certainly a big challenge for the Japanese government. Large-scale development of wind, solar and geothermal energy faces various constraints in Japan (Meltzer 2011). Fossil fuels then might play a more important role in powering the nation than they would otherwise. Taking into account all of these factors, although Japan is one of the most energy-efficient countries and has built a fleet of relatively new and efficient coal-fired generators, reducing GHG emissions remains a tough task for the Japanese government (EIA 2012).

South Korea is no less active than Japan in formulating ambitious energy and climate policies to curtail its rapidly growing GHG emissions. As a country that witnessed the fastest growth in carbon emissions among all OECD countries since 1990, South Korea is trying to deploy more low-carbon energy into the power generation mix (Climate Connect 2011). It put forward a Green Growth Strategy in 2010 which sets targets for carbon emission reduction and promotion of renewables (11 percent of total primary energy supply from renewable energy by 2020). A series of initiatives including a new renewable energy portfolio standard have been implemented to pursue these goals (Yale Center for Environmental Law & Policy 2012).

Apart from projects that directly improve energy efficiency or develop zero-carbon energy, one climate initiative is particularly interesting. In mid-2012, South Korea enacted an economy-wide emissions trading program that sets a mandatory cap on the country's carbon emissions, making it the first country in Asia to pass a binding climate law (Andreassen 2012). The climate legislation covers over 60 percent of South Korea's GHG emissions and regulates some 470 firms through a cap-and-trade system. Although the climate law will not come into full operation until 2015, it is expected to help put the country on track to fulfill its green growth commitment.

Why did South Korea take bold climate action, but not Japan or China? For one thing, South Korea is under great pressure from the international community to reverse the increasing trend in carbon emissions as it is a member of the OECD countries. Domestically, the government also faces constant petitions from sub-national governments, NGOs and the general public for stronger climate actions. Besides, there are no influential 'climate skeptics' in South Korea as there are in Japan (Gray 2012).

Indeed, both Japan and South Korea are proposing ambitious polices to diversify energy supply sources and give more weight to the development of renewable energy. The need is more urgent for Japan since nuclear power has become a politically unpalatable option for the government to deploy more low-carbon energy. They are also both experimenting with innovative approaches such as emissions trading to curtail carbon emissions.

Comparison and summary

Although most GHG emissions come from energy use, drivers of climate change are notably different in China, South Korea and Japan. While burning coal is the single largest contributor to China's GHG emissions, emissions from coal use represent a much smaller portion in the carbon portfolio of the two other major East Asian economies. Less carbon-intensive fossil fuels like oil and natural gas meet more than half of the energy demand in Japan and South Korea. China's thirst for oil and natural gas is growing, but constrained by limited domestic supply and unstable overseas flow. As for nuclear power, it played a much more important role in powering Japan and South Korea until the Fukushima Daiichi accident. For China, whereas it receives less than 2 percent of its electricity from nuclear reactors, its determination to expand the use of this low-carbon energy seems unwavering.

China's ambition to promote renewable energy such as wind, solar and biomass is also gaining solid ground. It has installed the world's largest wind power capacity and produces more PV cells than any other nation, although the efforts to introduce more wind and solar power to China's generation mix are plagued by a variety of problems. South Korea's development plan for renewable energy is relatively less ambitious than that of China, but supported by a strong climate law and other energy initiatives. Japan originally planned to increase the share of renewable energy to 20 percent by 2030, and probably needs a bigger contribution from renewables to fill the gap left by nuclear power. However, obstacles such as high financial costs and constraints may inhibit the large-scale expansion of low-carbon energies, in particular solar power (Meltzer 2011).

It is important to note that coal will continue to sustain China's economy for a long period of time, despite the government's effort to diversify the energy mix, and we can predict a steady rise in carbon emissions from the combustion of coal. Transportation is becoming an increasingly important GHG emitter in China and adds more pressure to China's climate effort. While it has not yet contributed as much to GHG emissions and climate change as transportation in Japan and South Korea, the aggregate amount of carbon emissions from China's growing fleet of vehicles is not small. Emissions from commercial buildings and homes in China are also significantly lower than those in Japan and South Korea.

Indeed, China, Japan and South Korea live in different stages of development and possess varying capacity to deal with climate change. While Japan promised to cut its carbon emissions by 6 percent below their 1990 level by 2012 under the Kyoto Protocol, China and South Korea have no binding climate obligations. Nonetheless, all three nations have committed to voluntary long-term emission reduction targets. Living in a coal-dominant and coal-constraining world, governments in East Asian countries have to face the challenge of climate change, and figure out how to balance it with other equally critical challenges such as energy security.

Bibliography

Andreassen, J. (2012) 'South Korea's New Climate Law Signals Growing Global Momentum to Curb Climate Change', Environmental Defense Fund. Available at http://blogs.edf.org/climatetalks/2012/05/02/south-koreas-new-climate-law-signals-growing-global-momentum-to-curb-climate-change/.

Asia Society & Pew Centre on Global Climate Change (2009) 'Common Challenge, Collaborative Responses: A Roadmap for US–China Cooperation on Energy and Climate Chang'. Available at http://www.pewclimate.org/US-China.

Barry, P. (2008) 'Carbon Sequestration Frustration', *Science News*, 13 August. Available at http://www.sciencenews.org/view/generic/id/35181/title/Carbon_sequestration_frustration.

Beattie, A. & Chaffin, J. (2012) 'China Takes Solar Power Dispute to WTO', *Financial Times*, 5 November Available at http://www.ft.com/cms/s/0/b5b8a1cc-2768–11e2–8c4f-00144feabdc0.html#axzz2CRkap6wr.

Bradsher, K. (2009a) 'China Outpaces US in Cleaner Coal-Fired Plants', *New York Times*, 10 May.

Bradsher, K. (2009b) 'China Vies to be World's Leader in Electric Cars', *New York Times,* 1 April.

British Petroleum (2012) 'BP Statistical Review of World Energy', June.

Caijing Net (2009) 'NDRC Pauses CTL New Projects'. Available at http://www.caijing.com.cn/2008–09–04/110010415.html.

Chazan, G. (2012) 'Shale Gas Boom Helps Slash US Emissions', *Financial Times*, 23 May. Available at http://www.ft.com/intl/cms/s/0/3aa19200-a4eb-11e1-b421–00144feabdc0.html#axzz2CRkap6wr.

Climate Connect (2011) 'Fact Sheet: South Korea Climate Change Policies'. Available at http://www.climate-connect.co.uk/Home/sites/default/files/Fact%20Sheet%20South%20Korea%20Climate%20Policy.pdf.

Ellsworth, B. & Parraga, M. (2012) 'Venezuela Expands China Oil-for-loan Deal to $8 Billion', *Reuters*, 22 May. Available at http://www.reuters.com/article/2012/05/23/us-venezuela-china-idUSBRE84M01M20120523.

Energy Information Administration (EIA) (2012) International Energy Statistics 2008: Energy Efficiency Measured as Total Primary Energy Consumption Per Dollar of GDP, Energy Information Administration.

Energy Research Institute (ERI) (2004) *China's Energy Development Strategy and Policies: Energy Structure and Optimisation* (ERI).

Gao, P., Wang, A. & Wu, A. (2008) 'China Charges Up: The Electric Vehicle Opportunity', McKinsey & Company. Available at http://www.mckinseychina.com/wp-content/uploads/2008/10/the_electric_vehicle_opportunity.pdf.

Gray, L. (2012) 'Most Climate Change Sceptics in US, UK and Japan', *Telegraph*, 4 October. Available at http://www.telegraph.co.uk/earth/environment/climatechange/9588019/Most-climate-change-sceptics-in-US-UK-and-Japan.html.

Harvey, F. (2012) 'Shale Offers Freedom and Security – But it Could be a Trap', *Guardian*, 15 November Available at http://www.guardian.co.uk/environment/2012/nov/15/shale-gas-freedom-security-tra.

Intergovernmental Panel on Climate Change (IPCC) (2001) *Climate Change 2001: The Scientific Basis. Contribution of Working Group I to the Third Assessment Report of the Intergovernmental Panel on Climate Change* (Cambridge University Press).

International Energy Agency (IEA) (2007) *World Energy Outlook 2007–China and India Insights* (OECD/IEA).

IEA (2009) *Cleaner Coal in China* (IEA).

IEA (2012) *IEA Energy Statistics: South Korea, Japan and China* (IEA).

IPCC Working Group II (2007) *Climate Change 2007: Impacts, Adaptation and Vulnerability* (Cambridge University Press).

Lin, J., Zhou, N., Levine, M. & Fridley, D. (2008) 'Taking Out 1 billion tons of CO_2: The Magic of China's 11th Five-year Plan?', *Energy Policy* 36(3), 954–970.

McKinsey & Company (2012) 'Recharging China's Electric Vehicle Aspirations: A Perspective on Revitalizing China's Electric Vehicle Industry'. Available at http://www.mckinseychina.com/wp-content/uploads/2012/04/McKinsey-Recharging-Chinas-Electric-Vehicle-Aspirations.pdf.

Melter, J. (2011) 'After Fukushima: What's Next for Japan's Energy and Climate Change Policy?', Brookings Institution. Available at http://www.brookings.edu/~/media/research/files/papers/2011/9/07%20after%20fukushima%20meltzer/110907_japaneseenergypolicy_final.pdf.

Miao, B. (2012) 'China's Climate Policy and Foreign Diplomacy', in E. Kavalski (Ed.), *Ashgate Research Companion to Chinese Foreign Policy* (Ashgate).

Miao, B. & Lang, G. (2010) 'China's Emissions: Dangers and Responses', in C.L. Tracy (Ed.), *Handbook of Climate Change and Society* (Routledge).

Ministry of Environment of the Government of Japan (2012) 'Environmental Statistics 2012'. Available at http://www.env.go.jp/en/statistics/contents/index_e.html.

Ministry of Environment of the Government of South Korea (2012) 'Environmental Statistics Yearbook 2011'. Available at http://eng.me.go.kr/board.do?method= view&docSeq=10608&bbsCode=law_law_statistics&currentPage=1&searchType=&searchText=.

Ministry of Finance (2009) 'Application Guidelines for Demonstration Projects of Solar Photovoltaic Buildings'. Available at http://www.gov.cn/zwgk/2009–04/20/content_1290550.htm.

National Bureau of Statistics of China (NBS) (2012) *China Statistical Yearbook 2011* (China Statistics Press).

National Development and Reform Commission (NDRC) (2006) The Medium to Long Term Development of Renewable Energies, issued by NDRC under the Authority of Renewable Energy Law of the People's Republic of China.

Nature (2008) 'News in Brief', *Nature* 455, 11 September.

NDRC (2007a) China's National Climate Change Programme, Beijing, NDRC.

NDRC (2007b) Medium-and-long-term Development Plan for Renewable Energy in China, Beijing, NDRC.

NDRC & ERI (2012) *Report on China's Renewable Energy Industry Development* (Chemical Industry Press).

Netherlands Environmental Assessment Agency (2007) 'China Now No. 1 in CO_2 Emissions; USA in Second Position'. Available at http://www.pbl.nl/en/news/pressreleases/2007/20070619Chinanowno1 inCO2emissionsUSAinsecondposition.

Oliver, H.H., Gallagher, K.S., Tian, D.L. & Zhang, J.H. (2009) 'China's Fuel Economy Standards for Passenger Vehicles: Rationale, Policy Process, and Impact', *Energy Policy* 39(11), 4720–4729.

State Council (2011a) 'White Paper on China's Climate Change Policy and Action 2011', 22 November. Available at http://www.gov.cn/jrzg/2011–11/22/content_2000047.htm.

State Council (2011b) 'The Working Plan on the GHGs Emissions Reduction during Twelfth Five-Year Plan', 1 December. Available at http://www.gov.cn/zwgk/2012–01/13/content_2043645.htm.

Tokyo Metropolitan Government (2010) 'Tokyo Cap-and-trade Program: Japan's First Mandatory Emissions Trading Scheme'. Available at http://www.kankyo.metro.tokyo.jp/en/attachement/Tokyo-cap_and_trade_program-march_2010_TMG.pdf.

Twenty-First Century Net (2012) 'Energy Structure Adjustment Failed in 2011 and Non-fossil Fuel Provided even Less Energy', 3 February. Available at http://www.21cbh.com/HTML/2012-2-3/3NMDY5 XzM5ODg3Ng.html.

United Nations Framework Convention on Climate Change (UNFCCC) (2012) National Greenhouse Gas Inventory Data for the Period 1990–2010.

Wall Street Journal (2012) 'China Group Slams U.S. Wind-turbine Tariffs', 20 July. Available at http://online.wsj.com/article/SB10000872396390444226904577558254078212684.html.

World Resources Institute (2011) Climate Analysis Indicators Tool (CAIT) version 9.0., Washington, DC, WRI.

Yale Center for Environmental Law & Policy (2012) 'Climate Policy & Emissions Data Sheet: South Korea'. Available at http://envirocenter.yale.edu/uploads/pdf/South_Korea_Climate_Policy_Data_Sheet.pdf.

Zhang, Z.X. (2010) 'China in the Transition to a Low-carbon Economy', *Energy Policy* 38(11), 6638–6653.

21

Vulnerabilities to climate change

Adaptation in the Asia-Pacific region

Benjamin K. Sovacool

Introduction

Climate change is a substantial security concern not only because direct flooding and natural disasters can damage infrastructure, disrupt the delivery of imported commodities, and destroy crops, but also because it can have severe impacts on food security and nutrition, health, and environmental refugees, which can all lower the income base of Asian countries and add to government debt, further complicating attempts at sound policymaking. Under a business as usual scenario, the IPCC forecasts that emissions of all greenhouse gases will rise from 52.5 billion tons today to 140 billion tons by 2100 – almost a threefold increase, and also a trend discussed in other chapters in this book. The atmospheric concentration of carbon dioxide reached 396 parts per million (ppm) in 2012 and 407 ppm when all greenhouse gases are accounted for, and will rise to 1,410 ppm for all gases by 2100 under business as usual. Depending on assumptions and particular models, such emissions and concentrations could result in a temperature change of 4.81° Celsius by 2100.

Although climate change is certainly a global phenomenon, in many ways it is becoming an Asian problem. Figure 21.1 shows annual tons of carbon dioxide emissions from fuel combustion divided by total national population for selected Asian countries, and it indicates emissions have more than doubled from 1990 to 2010 for countries such as Cambodia, Indonesia, Malaysia, Myanmar, Thailand, Vietnam, and China. Figure 21.2 also indicates that when changes in land use are included, five of the top 10 emitters of greenhouse gases – China, Indonesia, Russia, India, and Japan – already reside in Asia.

Unfortunately from a climate standpoint, the greenhouse gases *already* emitted will threaten Asian countries with a staggering list of negative consequences. But how do these impacts differ in scale and scope? Which countries are more vulnerable than others? And, lastly, how ought they best respond and adapt to the impacts of climate change?

Asia's vulnerability to climate change

Because the Asia-Pacific region is home to a diverse mix of countries with different geographic, economic, political, and cultural traits, and because vulnerability is conditioned by both the

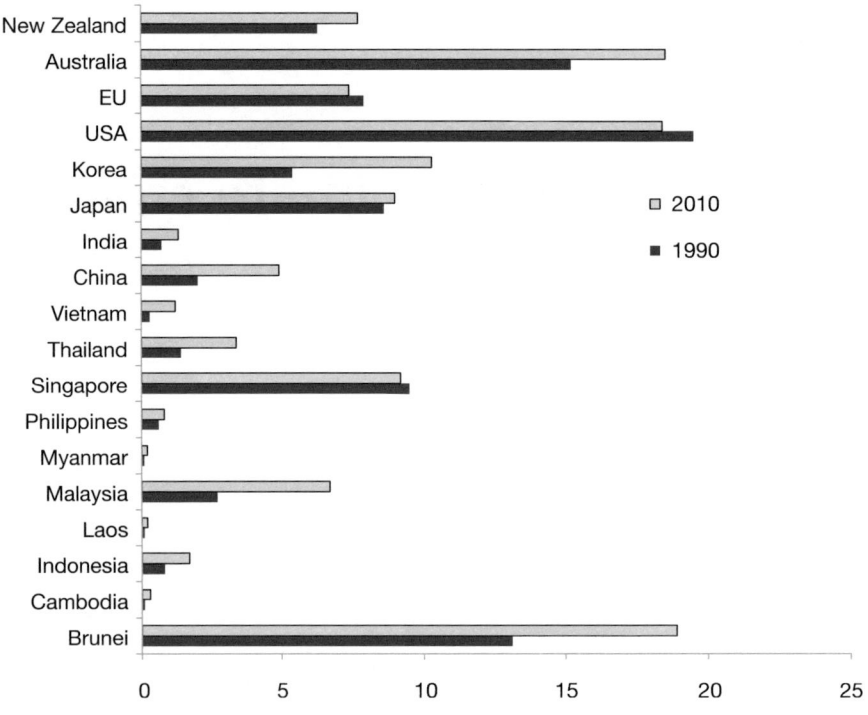

Figure 21.1 Per capita energy-related carbon dioxide emissions (metric tons), 1990 and 2010
Source: Sovacool et al. (2011).

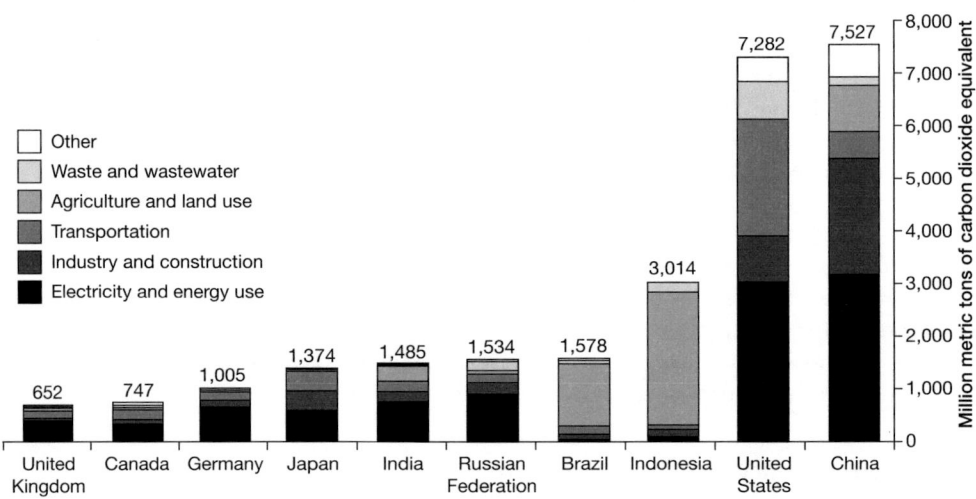

Figure 21.2 Share of greenhouse gas emissions by top 10 countries, 2010
Source: Brown & Sovacool (2011).

nature of a threat and the resilience of those experiencing it, each Asian nation will experience climate change differently. For instance, because of their unique climatology, low per capita incomes, and changing patterns of urbanization, Indonesia, the Philippines, Thailand, and Vietnam are expected to lose 6.7 percent of combined gross domestic product (GDP) by 2100 if temperatures change as the IPCC predicts, more than twice the rate of global average losses (Asian Development Bank 2009). Even uniform changes in climate will not affect Asian countries equally, with Cambodia, Laos, the Philippines, Mekong River Delta, central Thailand, and Sumatra and Java in Indonesia more at risk than wealthier countries such as Brunei or Singapore (Singapore Ministry of the Environment and Water Resources 2008; Yusuf & Francisco 2009).

China and India would see a potentially larger range of impacts. They could lose between 1 and 12 percent of their annual GDP coping with climate refugees, changing disease vectors, and failing crops (CNA 2009; Economics of Climate Adaptation Working Group 2009; RUSI 2009). One study forecasts a 37 percent reduction in national crop yields by 2050 in China if current climate trends continue (McMichael 2007). Some states, such as Maharashtra, India, are projected to suffer greater drought that will likely wipe out 30 percent of food production, inducing $7 billion in damages among 15 million small and marginal farmers (Economics of Climate Adaptation Working Group 2009). In India as a whole, farmers and fishers will have to migrate from coastal areas as sea levels rise and they confront heat waves lowering crop output and manage declining water tables from saltwater intrusion (CNA 2009).

Bangladesh is prone to a multitude of floods, droughts, tropical cyclones and storm surges. Fifteen percent of its 162 million people live within one-meter elevation from high tide. In 1991, a particularly devastating cyclone with winds stronger than 200 kilometers an hour and a tidal surge of 6 meters claimed 140,000 lives and induced $240 million in damages (Asian Development Bank 2005a). Such climatic vulnerabilities are only compounded by a high incidence of poverty and heavy reliance on agriculture and rural forestry. Studies have found that rising sea levels could place more than 40 million Bangladeshis at direct risk of saltwater intrusion of water supplies for drinking and irrigation, and the ever present occurrence of floods from drainage congestion and severe storms (Government of Bangladesh 2009; Government of Bangladesh and United Nations Development Program 2008). During the monsoon season in 2004 in Bangladesh, for example, flooding placed 60 percent of the country under a solid pool of water mixed with industrial and household waste. More than 20 million people suffered shortages of water, skin infections, and communicable illnesses (Rawlani & Sovacool 2011).

In Bhutan, the acceleration of glacial melting has compounded the risk of glacial lake outburst floods (GLOFs). Glacial lakes there hold tens of millions of cubic meters of water and can release high volumes in minutes, devastating valleys and communities downstream. Major sectors of the economy involve agriculture, livestock, and forestry, but these have become situated in close proximity to flood paths. One inventory identified 25 glacial lakes at 'high risk' of a GLOF with 12 located in the Pho Chhu and Chamkhar Chu sub-basins, home to more than 40 Bhutanese villages and towns with tens of thousands of residents (Dorji et al. 2001; National Environment Commission, Royal Government of Bhutan 2008). Melting glaciers could flood river valleys in Kashmir and Nepal to the point where more than 180 million people could be at risk from the flooding and from subsequent epidemics and starvation (Brown & Sovacool 2011).

In Cambodia, droughts and floods have already caused substantial human and crop losses and are widely viewed as a prelude to more extreme weather. Rice, Cambodia's largest crop by volume and value, is forecasted to suffer yield losses of 5 percent over current levels in 2020 under IPCC scenarios. Annual rainfall is projected to increase in some areas, but when coupled with increased variability and ambient temperatures, yield losses will worsen through 2080 and potentially turn Cambodia into a net rice importer (Cambodian Ministry of Environment 2006; Ponlok 2009).

Geographic and geophysical traits, such as its small size, low elevation, narrow width, and dispersed nature of coral islands and reefs make the Maldives especially vulnerable to rainfall flooding and ocean-induced flooding. About half the country's human settlements are within 100 meters of the shoreline, along with almost three-quarters of its critical infrastructure, including airports, power plants, landfills, and hospitals. The Maldives is the 'flattest country on earth' and 'extremely vulnerable' to climate change, so much so that 85 percent of its geographic area could be underwater by the year 2100 if sea levels rise under more extreme projections (Khan et al. 2002). Severe weather events from 2000 to 2006 already flooded 90 inhabited islands at least once and 37 islands repeatedly. Sea swells in 2007 inundated 68 islands in 16 atolls, destroyed 500 homes, and necessitated the evacuation of 1,600 people (Sovacool 2011, 2012a, 2012b).

Complicating this picture is the issue of water availability and climate change, which is slowly but steadily altering precipitation and water patterns. For instance, if global warming induces the rise in sea level that many climatologists and scientists expect, the intrusion of saltwater could contaminate freshwater aquifers – possibly reducing worldwide potable water supplies by 45 percent (Smith & Ibakari 2007). Warmer temperatures resulting from global climate change will also increase energy demands in urban areas and require more intensive air-conditioning loads, in turn raising the water needs for power plants. Hotter weather also increases the evaporation rates for lakes, rivers, and streams, and thus accelerates the depletion of reservoirs, and causes more intense and longer lasting droughts as well as more wildfires – which in turn need vast quantities of water to fight and control them (Sovacool & Sovacool 2009).

Another extremely negative trend is the acidification of the ocean, which will threaten Asian fish stocks and coral reefs. Scientists expect the greatest degree of ocean acidification, caused by greenhouse gas emissions, to occur in the Atlantic, Pacific, and Arctic seas, each crucial summer feeding grounds for billions of organisms, and each home to fish that migrate across the Asia-Pacific (Carrell 2009). Recent scientific studies warn that climate change will likely lead to numerous local extinctions and drastic species turnovers (removal and/or extinction from an area) affecting more than 60 percent of all marine biodiversity, as well as declines in the vitality of coral reefs due to bleaching, diseases, and tropical storms – with roughly one-third of all coral reefs at risk of becoming extinct (Gosling et al. 2011). If allowed to run its course, such acidification could turn the shining seas surrounding Asia (and, indeed, other continents) into a 'carbon cesspool' (Lamirande 2011).

Although these vulnerabilities are great, perhaps the most severe climate change impacts will befall small island developing states. Small island countries in the Pacific sit at the ever present mercy of natural disasters, especially cyclones and storm-induced floods that can damage energy infrastructure and lower national incomes. Figure 21.3 shows that since the 1950s, the quantity and magnitude of natural disasters throughout the Pacific have increased significantly. Table 21.1 also illustrates that a selection of Pacific island countries have had no fewer than 257 disasters from 1950 to 2008, which have caused $6.8 billion in damages (Asian Development Bank 2005b; World Bank 2009).

In the Solomon Islands, the Ministry of Environment, Conservation and Meteorology has warned that 'energy production, utilization, conversion and transportation' have and will continue to be negatively affected by 'droughts, floods, fires, storm surges, and cyclones' (Ministry of Environment, Conservation and Meteorology 2008). In Samoa, the earthquake and tsunami in September 2009 greatly damaged the Electric Power Corporation's (EPC) generation and distribution assets in the southern and eastern coastal areas of Upolu, Manono, and Savii. Damages included toppled power poles and fittings, cracked transformers, and destroyed hydroelectric dams (Electric Power Corporation 2010). With assets of only $163 million (hardly enough to

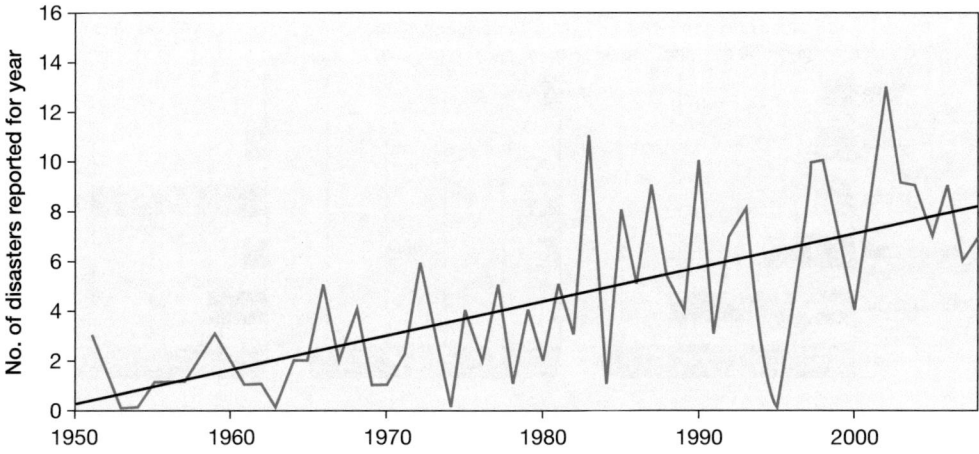

Figure 21.3 Number of natural disasters reported in the Pacific Islands Region, 1950–2008
Source: World Bank (2009).

Table 21.1 Estimated economic and social impact of disasters in selected Pacific Island countries, 1950–2008

Country	No. of disasters	Losses (US$ 2008)	Average population affected (%)		Average impact on GDP (%)	
			Disaster years	All years	Disaster years	All years
American Samoa	6	237,214,770	5.81	0.61	7.76	0.82
Cook Islands	9	47,169,811	5.13	0.63	3.48	0.43
Fiji	43	1,276,747,934	5.39	2.74	3.48	0.78
French Polynesia	6	78,723,404	0.53	0.04	0.31	0.02
FSM	8	11,915,993	6.20	0.65	0.82	0.09
Guam	10	3,294,869,936	1.97	0.28	10.13	1.42
Kiribati	4	0	29.19	1.54	0.00	0.00
Marshall Islands	3	0	6.40	0.22	0.00	0.00
New Caledonia	15	69,623,803	0.14	0.03	0.09	0.02
Niue	6	56,461,688	73.15	7.70	80.88	8.51
Papua New Guinea	58	271,050,690	0.69	0.36	0.14	0.07
Samoa	11	930,837,187	21.15	3.71	16.97	2.98
Solomon Islands	21	39,215,686	2.93	0.98	0.52	0.17
Tokelau	4	4,877,822	39.70	2.79	N/A	N/A
Tonga	12	129,344,561	21.32	3.37	5.76	0.91
Tuvalu	5	0	3.19	0.28	0.00	0.00
Vanuatu	36	406,402,255	5.33	2.06	3.78	1.46

Notes: The term 'natural disasters' includes droughts, earthquakes, epidemics, floods, landslides, severe storms, volcanic eruptions, wave surges and wildfires.

Source: World Bank (2009).

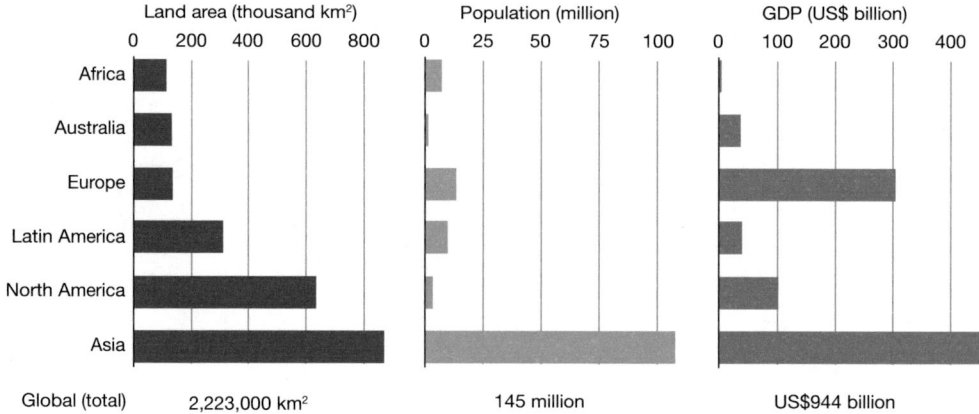

Figure 21.4 Population, area and economy directly affected by a one-metre rise in sea level
Source: USAID (2010).

build a single large power plant) and a net operating profit of $2.1 million per year, the EPC has little revenue to draw from to address these types of damage. In Fiji, unexpected shortfalls in water have forced the country's hydroelectric dams to operate below full capacity, increasing reliance on diesel imports and precipitating increases in electricity tariffs (Fiji Electricity Authority 2012).

One wide-ranging survey of climate impacts in Asia-Pacific from the United States Agency for International Development (USAID 2010) puts the 'Asian' extent of these risks into perspective. That study predicted, among other things:

- Accelerated river bank erosion, saltwater intrusion, crop losses, and floods in Bangladesh that will displace at least 8 million people and destroy up to 5 million hectares of crops.
- More frequent and intense droughts in Sri Lanka, crippling tea yields and reducing national foreign exchange and lowering incomes for low-wage workers.
- Higher seal levels inundating *half* of the agricultural lands on the Mekong Delta, causing food insecurity throughout Cambodia, Laos, and Vietnam.
- Increased ocean flooding and storm surges inundating 130,000 hectares of farmland in the Philippines affecting the livelihoods of 2 million people.
- Intensified floods in Thailand placing more than 5 million people at risk and causing $39 billion to $1.1 trillion in economic damages by 2050.

Indeed, that study concluded that Asia-Pacific will be subject to more land degraded, people displaced, prosperity threatened, and economies disrupted from sea level rises than any other part of the planet, as Figure 21.4 illustrates. Already, Asia-Pacific was home to 85 percent of deaths and 38 percent of global economic losses due to natural disasters from 1980 to 2009 (UNESCAP 2012).

Lessening vulnerability through adaptation

Community-based adaptation measures offer perhaps the best tool for enabling Asian countries to minimize the impending impacts of climate change. The term 'adaptation' describes adjustments in natural or human systems in response to the impacts of climate change (Ayers & Forsyth 2009). The Intergovernmental Panel on Climate Change (2007) defines 'adaptive capacity' as the ability of a system to adjust to climate change (including climate variability and extremes) to moderate potential damages, to take advantage of opportunities, or to cope with the consequences. Adaptation is 'community-based' when implemented by local stakeholders, and it is sometimes called 'anticipatory adaptation' when it tries to preempt particular risks (Boyle & Dowlatabadi 2011). Most adaptation efforts are targeted towards enhancing 'resilience,' the amount of disturbance a local system, climatic or social, can absorb and still remain within the same state (Folke 2006). Table 21.2 illustrates three of the most salient types of resilience.

Adaptation efforts are necessary if communities are to respond to drastic changes in climate once tipping points, such as acidification of the ocean, alteration of the Gulf Stream, or thawing permafrost are crossed, and adaptation can also have a high relevance regarding slow or gradual changes in climate (Jerneck & Olsson 2008). Furthermore, adaptation efforts tend to be 'win–win situations,' for they not only improve resilience to climate change but often spill over into ancillary benefits such as economic stability, improved environmental quality, community investment, and local employment (Economics of Climate Adaptation Working Group 2009).

But what types of adaptation project best reduce vulnerability? To provide a partial answer, consider ongoing and diverse projects in five extremely vulnerable Asian countries: Bangladesh, Bhutan, Cambodia, the Maldives, and Vanuatu. These five countries are arguably the 'most at risk' to climate change due not only to their geographic and geophysical traits (such as lying in a floodplain, such as Bangladesh, or being incredibly flat, such as the Maldives), but also due to their economic and political traits. The United Nations (2005) classifies all five of these countries as having 'least developed country' status, meaning they have high rates of poverty, poor social indicators related to education and health, and economic vulnerability. In 2010, for instance, each of these countries had average per capita incomes, when adjusted for purchasing power parity, of *less* than $1,000 per year.[1]

In Bangladesh, the Ministry of Environment and Forests is aiming to reduce the vulnerability of coastal communities to the impacts of climate change by sponsoring afforestation in four *upazilas* (translated as sub-districts) in the coastal districts of Barguna and Patuakhali (Western region), Chittagong (Eastern region), Bhola (Central region) and Noakhali (Central region). Project managers selected sites on the basis of their expected vulnerability and also through public participation. The project has four primary components. The first is implementing interventions that generate income and couple afforestation with community livelihood. The second is enhancing national, sub-national, and local capacities of government authorities and sectoral planners so that they better comprehend climate risk dynamics in coastal areas and implement

Table 21.2 Dimensions of resilience and adaptive capacity

Type of resilience	Explanation
Infrastructural	Refers to the assets, infrastructure, technologies, or 'hardware' in place to ensure the delivery of services that could be disrupted by climate change (such as electricity or water)
Institutional	Refers to the endurance of an institution or set of institutions, usually government ministries or departments, in charge of planning and community and infrastructural assets
Community	Refers to the cohesion of communities and the livelihoods of the people who compose them

appropriate risk reduction measures. The third is reviewing and revising coastal management practices and policies. The fourth is developing a functional system for the collection, distribution and internalization of climate change related data (Sovacool 2013).

In Bhutan, the government launched a project to tackle disaster risks that arise from glacial lake outburst floods (GLOFs). It has three primary components. A component focused on lowering of lake water levels is being undertaken by the Department of Geology and Mines (DGM) to reduce the risk of GLOFs at two glacial sites in the Himalayas. So far mitigation work by DGM has focused only on one lake, Thorthormi, where it is aiming to reduce the lower lake's water level by five meters, enough to eliminate hydrostatic pressure on its unstable moraine dam. An early warning component is being led by the Department of Energy, and a third community awareness component is attempting to increase the knowledge of climate change among community leaders and rural policymakers (Sovacool 2013).

Planners in Cambodia, by contrast, are focusing on building adaptive capacity for water management and agriculture. They are enhancing the ability of local government and communities to integrate long-term climate risks into policy and decision making related to subsistence farming and rice paddy production. Many of the communities living in the targeted districts in Preah Vihear and Kratie practice subsistence farming and are reliant on agriculture for their livelihoods. Adaptation efforts are focusing on educating these farmers and local leaders about climate change, and also strengthening infrastructure such as irrigation channels and ponds (Sovacool 2013).

The Maldivian government is integrating climate change risk management into formal planning processes. It is funding demonstration projects on four islands that promote a suite of different infrastructural improvements including beach nourishment, coral reef propagation, land reclamation, and community relocation. The project is also creating 'composite risk reduction plans' to be integrated with coastal protection and adaptation measures. Disaster risk profiles are being created for the four demonstration islands, revised and updated as scientific knowledge about climate change and sea level rise accumulates. These are to be synthesized into a national level 'multi-hazard early warning system' (Sovacool 2013).

The Republic of Vanuatu, an archipelago of 83 islands in the South Pacific, offers a particularly vivid final example of locally driven adaptation projects. Efforts include developing and implementing land use guidelines that consider natural hazards and climate risks, establishing early warning systems to cope with flooding, rainwater harvesting, and saltwater desalinization. Three smaller pilot project sites have been initiated on Pele Island, South Efate, and South Santo prioritizing the introduction of climate-resistant strains of crops, adapting livestock to more extreme weather, and promoting community land-use plans so farmers can share resources and arable farming sites. Projects have also emphasized the sharing of information among subsistence farmers about low tillage and soil stabilization methods, and how to reduce coastal erosion (Richmond & Sovacool 2012).

Taken as a whole, this suite of five adaptation projects is catalyzing three sets of distinct benefits: the improvement of national infrastructure, the enhancement of institutional capacity, and the betterment of community assets.

Each of the five projects enhances physical and infrastructural resilience in some way. Bangladesh's coastal forest today is almost a monoculture of mangroves. These monoculture forests have a limited ability to mitigate the impacts of climate change as they have been prone to pest outbreaks, deforestation, and logging. Historically, Bangladesh had a 500-meter buffer of mangroves to reduce the shocks of incoming storms and monsoons that has now been reduced to 12 to 50 meters in most locations. Stem borer attacks, a pest, have felled thousands of hectares and illegal deforestation and logging have worsened the situation. The project in Bangladesh addresses this problem and sponsors 6,000 hectares of community-based mangrove plantations, 500 hectares of non-mangrove mount plantations, about 220 hectares of dykes, and more than 1,000 kilometers of embankments. The Bangladesh project also develops early warning information and disaster preparedness systems in vulnerable areas to protect at least twenty villages and towns.

In Bhutan, planners are improving early warning systems and draining glacial lakes. Previously, the Bhutanese Department of Energy managed only a single station in Thanza, which housed two people with a wireless radio set, a single satellite phone that monitored glacial lake water levels, and (in all likelihood) copious amounts of coffee. The problem is that the two people did not always report for work, have fallen asleep, and could have been killed by the GLOF itself. Under the project, the government will replace the manual system with an automatic one composed of gauges monitoring glacial lake bathymetry (depth) as well as sensors along rivers connected to automated sirens. The project will also eventually expand the automated warning system to cover more glacial lakes.

In Cambodia, infrastructural resilience will be improved by the construction and rehabilitation of retention ponds, canals, dykes, and reservoirs that due to years of neglect are currently in disrepair. Instead of repairing these irrigation systems using design parameters derived from historical hydrological patterns, the project aims to integrate climate forecasts into their rehabilitation so that the infrastructure can withstand future climatic events such as droughts or floods.

In the Maldives, planners have moved away from exclusively building capital-intensive sea walls and tetrapods – a four legged concrete structure that breaks the impact of waves – to bolster infrastructural adaptation by replenishing natural sea ridges, planting mangroves and vegetation on shorelines, and raising the height of water storage tanks so they are no longer susceptible to sea swells and saltwater intrusion. The government has started propagating new coral reefs around Thulusdhoo and Kudhahuvadhoo and adopting beach nourishment activities to mitigate flooding.

In Vanuatu, on Epi Island investments are being made to repair roads, bridges, and wharves at risk to sea level rise, and also to erect coastal walls to reduce the severity of storm surges.

Adaptation efforts prioritize not only infrastructure but also improving institutions and propagating standards of good governance. In Bangladesh, the government provides free training sessions for local level administrators in disaster management and also facilitates input from civil society and community members in the formulation of state and national policies and regulations. In Bhutan, training of government planners is intended to build institutional capacity. The project has sponsored the training of geologists and employment for civil engineering work, and funded the creation of community-based disaster management committees, whose job it is to highlight hazards and form district disaster management teams at village levels. In Cambodia, their project is encouraging community development plans based on long-term climate forecasts and scenarios, budgeting for water resources investments that are appropriate given anticipated risks. In addition to devolving ministerial functions to local levels where possible, the project has shifted responsibility for planning onto community groups. In the Maldives, institutional capacity is being strengthened through the training of government officials in risk analysis, hazard mitigation, and land use planning. By 2014 the goal is to train at least 12 senior decision makers and planners from national ministries in Malé as well as all senior decision makers in four provinces and atolls. Part of this component involves participating with local island leaders to share knowledge and learn about local efforts. In Vanuatu, the government is sponsoring consultations with stakeholders to learn about options to bolster sector-level resilience measures which attempt to improve adaptive capacity in the agriculture, tourism, marine resources, water resources and forestry sectors.

Finally, each adaptation project enhances community and social resilience. In many parts of the coastal forests of Bangladesh, the average annual per capita income is less than $130, one-third the national average, rendering people completely dependent on wetlands and coastal forests to meet their subsistence needs (Matthew 2007). To counter this incentive to damage forests for their survival, the LDCF project is disbursing revenues to vulnerable coastal communities so that they can diversify income sources and occupational training. One especially innovative dimension of this component is its focus on the 'Triple F' model of 'forest, fish, and food.' The coastal communities most vulnerable to rising sea levels – the places where mangroves need to be planted and forests replenished – are also those where farming and forestry are the primary sources of income. The 'FFF' model attempts to maintain community livelihood and adapt to climate change at the same time by integrating aquaculture and food production within reforested and afforested plantations.

In Bhutan, a community awareness subcomponent is being implemented in Punakha, Wangdi and Bumthang. Officials are creating a zoning map to mark several safe evacuation areas and extremely unsafe areas, and setting up emergency operation centers at district administration offices to enable them to better handle crises. Communities are being trained in their response to calamities and emergency situations using mobile phones and radio broadcasts in addition to traditional sounding gongs and bells from monasteries. These efforts will give communities a better understanding of the risks and hazards surrounding GLOF occurrences. This information also enables communities to better plan for where to locate infrastructure, homes, and farmland.

In Cambodia, in addition to devolving ministerial functions related to adaptation efforts to local levels where possible, millions of dollars of funds have been transferred to fund the agricultural adaptation projects selected by village planning committees. Training and educational sessions have also been offered to communes expressing an interest in learning about climate change.

In the Maldives, planners are attempting to increase awareness of climate change in the outer atolls. One Maldivian official told the author that the project will 'help decentralized adaptation investment planning so that each island decides what to spend its own budget on, therefore creating incentive for islands to "pick best value for the money" so that they have resources left to improve community welfare in other ways.' The program will also send 'training teams' to remote islands to 'create awareness among the community so that they can take stock of existing vulnerabilities and soft adaptation measures.'

In Vanuatu, a 'standard climate kit' with key messages about climate change has been designed and distributed to communities around Port Vila and is being scaled up for distribution in rural areas, where it will be given to tribal chiefs.

Conclusions and implications

What can we take away from both Asia's vulnerability to climate change and the adaptation efforts that some of its countries are pursuing? This section of the chapter presents five key takeaway conclusions.

First, due to the developing nature of most of its economies, population density along its coastlines, scope and extent of biodiversity hotspots, and other factors, Asia-Pacific may very well be the region of the world most at risk to the future impacts of climate change. Economists expect Asian countries to lose a larger proportion of their GDP addressing climate change. Small island developing states like the Maldives and Vanuatu may not even exist by the end of this century if sea levels rise as expected, and species extinctions could occur throughout the tropical rainforests of Indonesia and Malaysia (including Borneo) and throughout the region's network of coral reefs. Even large countries such as China and India are likely to suffer from climate refugees, accelerated disease epidemics, and declining food security. These issues remind us that no country throughout the region will be immune to the consequences of a changing climate, even those that have contributed few greenhouse gas emissions to the global atmosphere.

Second, adaptation efforts in the least developed countries of Asia highlight the necessity of viewing resilience to climate change as multidimensional. As Table 21.3 summarizes, Bangladesh is not only sponsoring dykes and mangrove plantations, it is incentivizing agriculture and aquaculture to improve community income and training local officials. Bhutan is not only altering the physical shape of glacial lakes and rivers, building shelters, and creating an early warning system, but educating public and private leaders about emergency preparedness and climate risks. Cambodia is not only experimenting with crops and rehabilitating canals and ponds, but educating provincial officials and empowering local villagers to decide on infrastructure investments. Maldivian planners are not only thickening coastal vegetation and nourishing coral reefs, but decentralizing planning and disbursing funds directly to local communities so that they can decide what is best for them. Vanuatu planners are hardening coastal infrastructure, soliciting feedback from stakeholders and civil society, and enhancing the informational awareness of indigenous peoples in rural areas. Their efforts remind us that adaptation may work best not by improving technology alone, but by seamlessly strengthening three types of adaptation – infrastructural, organizational, and social – to bolster ecosystems, communities, and human organizations.

Table 21.3 Efforts in least developed Asian countries and their contributions to adaptation

Country	Infrastructural adaptation	Organizational adaptation	Social adaptation
Bangladesh	Mangrove plantations, mound plantations, dykes, and embankments; early warning system	Capacity building through training courses for local government officials in forestry and organizational change through setting up new functional departments	Coupling of forestry programs to income generation through forest products, fish and food
Bhutan	Lowering glacial lake levels; deepening river channels; early warning system; climate shelters	Workshops for government officials at the nodal level	Community training in search and rescue, evacuations and first aid
Cambodia	Climate proofing of canals and communal ponds; experimentation with crop variation and diversity	Education sessions for provincial and local officials	Local empowerment over prioritization of climate-proofing schemes
Maldives	Sea walls; replenishment of sea ridges; mangrove afforestation; beach nourishment; coral reef propagation; repositioning of water tanks	Decentralization of adaptation planning and management to local political units	Community control over adaptation investments
Vanuatu	Roads; bridges; port infrastructure; sea walls	Consultation of adaptation options with community stakeholder	Dissemination of information kits to tribal leaders

Such a finding has been confirmed by a few recent studies. A research team from the World Resources Institute (McGray et al. 2009) investigated 135 case studies of adaptation efforts in developing countries, and noted that a combination of three types of adaptive effort were most useful:

- Building responsive capacity, such as improving communication between institutions, or enhancing the mapping or weather monitoring capability of a government institution.
- Managing climate risks, such as disaster planning, researching drought resistant crops, or climate proofing infrastructure.
- Confronting climate change, such as relocating communities or repositioning infrastructure in response to flooding or glacial melting.

Similarly, the World Resources Institute, in collaboration with United Nations Development Programme, United Nations Environment Programme, and World Bank (2008), argued that three dimensions to resilience exist and must be promoted synergistically. Ecological resilience refers to the disturbance an ecosystem can absorb without changing into a different structure or state. Disturbances can be natural, like a storm, or human induced, such as deforestation. Social resilience refers to the ability of a society to face internal or external crises and still cohere as a community and possess a sense of identity and common purpose. Economic resilience refers to the ability for an economy to recover from shocks, and often entails having a diversified economic base composed of members with a variety of different skills.

However, multidimensional resilience also entails risks; degradation or destruction along one dimension of resilience can affect the others – the influence can be both positive and negative. Depleting a forest, for instance, could reduce ecological resilience that in turn creates fewer jobs (affecting economic resilience) and erodes the community's social resilience (by causing a high proportion of migration or dissention within the community). Conversely, improved ecological resilience can improve rents and revenue from logging (economic resilience) and also improve business skills and connection with markets (social resilience). Resilience can be reactive, making the present system resistant to change, or proactive, creating one that is capable of adapting to change (Klein et al. 2003). The key challenge for future adaptation efforts will be promoting different types of resilience – infrastructural, institutional, community – that do not tradeoff with each other, where improving one type is not to the detriment of the others

Third, this chapter demonstrates the value of a functions-based approach to resilience and adaptive capacity rather than an asset-based one. Community or social assets – things like higher wages or better technology – are useless if communities do not have the skills or capacity to use them. Knowledge and assets must be coupled with capacity and improved governance. This creates a more fluid and messy picture of adaptation, but also one that is more realistic. Assets remain only potential until communities leverage them, and adaptation programs must find ways to improve living standards.

Fourth, some types of adaptation project and knowledge are community specific whereas others transcend community types. For example, in building rainwater harvesting facilities, coastal walls, or water desalination plants, it can be difficult to transfer technology between communities within the separated islands of the Maldives and Vanuatu, or the dense rural areas of Cambodia, or the rugged mountainous areas of Bhutan, because these are fixed assets and some, like coastal walls, apply only to coastal communities. Knowledge about climate change, crop varieties, livestock, soil, and community land use, however, can travel by word of mouth and are applicable to virtually every community in each country.

Fifth, although a less certain conclusion than the first four (it needs more research to confirm it), some of the experiences in Bhutan, Cambodia, and Vanuatu imply that smaller scale projects that involve communities in risk assessment, evaluation of adaptation options, project implementation and project appraisal have a greater likelihood of being accepted, and effective. This way, individuals are better suited to understand their vulnerabilities to climate change and see adaptation projects as legitimate. They are more democratic than large-scale projects chosen somewhat arbitrarily by consultants and ministers. And given stronger local support for new facilities and equipment, they are more likely to be maintained over the long term as adaptation measures are seen as a community asset.

Acknowledgments

Research presented in this chapter contributes to the Nordic Centre of Excellence for Strategic Adaptation Research (NORD-STAR), which is funded by the Norden Top-level Research Initiative sub-program 'Effect Studies and Adaptation to Climate Change'.

Bibliography

Asian Development Bank (2005a) 'Coastal Greenbelt Project (Loan 1353-BAN[SF]) in the People's Republic of Bangladesh', ADB, October.

Asian Development Bank (2005b) *Climate Proofing: A Risk-Based Approach to Adaptation* (ADB).

Asian Development Bank (2009) *The Economics of Climate Change in Southeast Asia: A Regional Review* (ADB).

Ayers, J. & Forsyth, T. (2009) 'Community-based Adaptation to Climate Change: Strengthening Resilience through Development', *Environment* 51(4), 22–31.

Boyle, M. & Dowlatabadi, H. (2011) 'Anticipatory Adaptation in Marginalized Communities Within Developed Countries', in J.D. Ford & L. Berrang-Ford (Eds.), *Climate Change Adaptation in Developed Nations: From Theory to Practice, Advances in Global Change Research* 42, 461–473.

Brown, M.A. & Sovacool, B.K. (2011) *Climate Change and Global Energy Security: Technology and Policy Options* (MIT Press).

Cambodian Ministry of Environment (2006) National Adaptation Programme of Action to Climate Change (NAPA), Ministry of Environment, Phnom Penh.

Carrell, S. (2009) 'Ocean Acidification Rates Accelerating', *The Hindu*, 11 December.

CNA (2009) 'Climate Change, State Resilience, and Global Security', CNA Conference Center, Alexandria, Virginia, 4 November.

Dorji, W., Bajracharya, S.R., Kunzang, K., Gurung, D.R., Joshi, S.R. & Mool, P.K. (2001) 'Inventory of Glaciers and Glacial Lakes and Glacial Lake Outburst Floods Monitoring and Early Warning Systems in the Hindu Kush-Himalayan Region Bhutan', ICIMOD, UNEP.

Economics of Climate Adaptation Working Group (2009) *Shaping Climate-Resilient Development: A Framework for Decision-Making* (Climate Works Foundation).

Electric Power Corporation (2010) 'Annual Report 2009–2010', EPC.

Fiji Electricity Authority (2012) 'Annual Report 2011', Fiji Electric Company.

Folke, C. (2006) 'Resilience: The Emergence of a Perspective for Social-ecological Systems Analyses', *Global Environmental Change* 16(3), 253–267.

Gosling, S.N., Warren, R., Arnell, N.W., Good, P., Caesar, J., Bernie, D., et al. (2011) 'A Review of Recent Developments in Climate Change Science. Part II: The Global-scale Impacts of Climate Change', *Progress in Physical Geography* 35(4), 443–464.

Government of Bangladesh (2009) Bangladesh Climate Change Strategy and Action Plan, Ministry of Environment and Forest (MOEF), Dhaka.

Government of Bangladesh & United Nations Development Program (2008) Project Document: Community Based Adaptatin through Coastal Afforestation in Bangladesh, UNDP, 10 December, Dhaka Office, Project ID, PIMS 3873.

IPCC (2007) *Climate Change 2007: Impacts, Adaptation, and Vulnerability. Contribution of Working Group II to the Fourth Assessment Report of the Intergovernmental Panel on Climate Change* (Cambridge University Press).

Jerneck, A. & Olsson, L. (2008) 'Adaptation and the Poor: Development, Resilience, and Transition', *Climate Policy* 8, 170–182.

Khan, T., Quadir, D., Murty, T.S., Kabir, A., Aktar, F. & Sarker, M. (2002) 'Relative Sea Level Changes in Maldives and Vulnerability of Land Due to Abnormal Coastal Inundation', *Marine Geodesy* 25, 133–143.

Klein, R.J.T., Nicholls, R.J. & Thomalla, F. (2003) 'Resilience to Natural Hazards: How Useful is This Concept?', *Environmental Hazards* 5, 35–45.

Lamirande, H.R. (2011) 'From Sea to Carbon Cesspool: Preventing the World's Marine Ecosystems from Falling Victim to Ocean Acidification', *Suffolk Transnational Law Review* 34, 183–217.

McGray, H., Hammill, A., Bradley, R., Schipper, E.L. & Parry, J.E. (2009) *Weathering the Storm: Options for Framing Adaptation and Development* (World Resources Institute).

McMichael, A.J. (2007) 'Seeing Clearly: Tackling Air Pollution in China', *Lancet* 370, 927–928.

Matthew, R.A. (2007) 'Climate Change and Human Security', in J.F.C. DiMento & Doughman, P. (Eds.), *Climate Change: What It Means for Us, Our Children, and Our Grandchildren* (MIT Press).

Ministry of Environment, Conservation and Meteorology (2008) Solomon Islands National Adaptation Program of Action, Honiara, November.

National Environment Commission, Royal Government of Bhutan (2008) Bhutan National Adaptation Program of Action.

Ponlok, T. (2009) 'Climate Change in Cambodia: What Does it Mean to Cambodia & National-Based Adaptation to Climate Change', Presented at: Climate Change Repercussions in Rural Cambodia, 19 November, Phnom Penh.

Rawlani, A. & Sovacool, B.K. (2011) 'Building Responsiveness to Climate Change through Community Based Adaptation in Bangladesh', *Mitigation and Adaptation Strategies for Global Change* 16(8), 845–863.

Richmond, N. & Sovacool, B.K. (2012) 'Bolstering Resilience in the Coconut Kingdom: Improving Adaptive Capacity to Climate Change in Vanuatu', *Energy Policy* 50, 843–848.

Royal United Services Institute (RUSI) (2009) *Socioeconomic and Security Implications of Climate Change in China* (CNA).

Singapore Ministry of the Environment and Water Resources (2008) Singapore's National Climate Change Strategy, Singapore, March.

Smith, J. & Ibakari, M. (2007) 'Data Mining and Spatiotemporal Analysis of Extreme Precipitation and Northern Great Plains Drought', Presentation at the 2007 Geological Society of America Meeting, Denver, Colorado, 28–31 October.

Sovacool, B.K. (2011) 'Conceptualizing Hard and Soft Paths for Climate Change Adaptation', *Climate Policy* 11(4), 1177–1183.

Sovacool, B.K. (2012a) 'Perceptions of Climate Change Risks and Resilient Island Planning in the Maldives', *Mitigation and Adaptation of Strategies for Global Change* 17(7), 731–752.

Sovacool, B.K. (2012b) 'Expert Views of Climate Change Adaptation in the Maldives', *Climatic Change* 114(2), 295–300.

Sovacool, B.K. (2013) *Energy and Ethics: Justice and the Global Energy Challenge* (Palgrave).

Sovacool, B.K. & Sovacool, K.E. (2009) 'Preventing National Electricity–Water Crisis Areas in the U.S.', *Columbia Journal of Environmental Law* 34(2), 333–393.

Sovacool, B.K., Mukherjee, I., Drupady, I.M. & D'Agostino, A.L. (2011) 'Evaluating Energy Security Performance from 1990 to 2010 for Eighteen Countries', *Energy* 36(10), 5846–5853.

UNESCAP (2012) *Low Carbon Green Growth Roadmap for Asia and the Pacific* (UNESCAP).

United Nations (2005) 'The Criteria for the Identification of the LDCs'. Available at http://www.un.org/special-rep/ohrlls/ldc/ldc%20criteria.htm.

USAID (2010) *Asia Pacific Regional Climate Change Adaptation Assessment: Final Report Findings and Recommendations* (USAID).

World Bank (2009) *Preparedness, Planning, and Prevention: Assessment of National and Regional Efforts to Reduce Natural Disaster and Climate Change Risks in the Pacific* (World Bank).

World Resources Institute (WRI) in collaboration with United Nations Development Programme, United Nations Environment Programme, and World Bank (2008) *World Resources 2008: Roots of Resilience – Growing the Wealth of the Poor* (WRI).

Yusuf, A.A. & Francisco, H.A. (2009) *Climate Change Vulnerability Mapping for Southeast Asia* (IDRC).

Note

1 Although the Maldives 'graduated' from its least developed country status in 2011.

22

Impacts of climate change

Challenges of flooding in coastal East Asia

Faith K.S. Chan, Daniel A. Friess, James P. Terry and Gordon Mitchell

Introduction

Currently, more than half of the Asian population live in coastal areas, especially in vulnerable deltas and coastal cities (Fuchs et al. 2011; Woodroffe 2010). More than 325 million inhabitants are living in coastal low-lying flood prone areas in East Asia alone (McGranahan et al. 2007). Many of these areas are predicted to be vulnerable under near future climate change (e.g., sea level rise), with millions of people and their economic assets exposed to floods and storms (Ward et al. 2011). Seto (2011) projected that in the next few decades, most population increase will take place in the exposed deltas, estuaries, coastal zones and coastal cities of Asia (and Africa) due to better employment and education opportunities. Rapid socioeconomic trends amplify the possible consequences of future floods, with increasing urban populations and greater financial capital invested in the flood-prone coastal zone. Hanson et al. (2011) found that more than 20 cities in East Asia will be highly exposed to coastal flood risks in the 2070s. Wilby and Keenan (2012) expect the frequency, intensity and duration of extreme precipitation events to increase as a result of climate change. Many Asian coastal areas are suffering from an increasing frequency of typhoons, rainstorms and storm surges from the West Pacific (Webster et al. 2005). At the same time, various deltas and coastal cities also experience (often anthropogenically induced) land subsidence (Syvitski 2008).

In recent years, numerous coastal areas throughout Asia, especially deltas, have been impacted by severe floods. For example, Cyclone Nargis in 2008 inundated to 75 km inland in the Irrawaddy Delta in southern Myanmar (Terry et al. 2012), causing 146,000 casualties and economic losses over US$17 billion (Syvitski et al. 2009). Natural disasters also have huge global economic impacts: the central Thailand floods in 2011 caused serious economic damage of > 100 billion baht (equivalent to US$4 billion) (BBC 2011), and together with the Tohoku earthquake and ensuing tsunami in Northern Japan, had a cascading effect on global supply chains (Shibahara 2011). This chapter aims to understand the vulnerability to flood risk and to assess how current climate change adaptation and flood risk management might be improved in the East Asian region (China, Hong Kong) specifically. East Asia is of particular relevance to discussions of flood risk vulnerability due to high (and rapidly growing) coastal populations, historical evidence for coastal inundation events, and an expected increased vulnerability in the future associated with global climate change. The discussion focuses on the causes of coastal vulnerability and evidence of climate change impacts on coastal populations, and then presents an in-depth case study for the Pearl River Delta. Ultimately, we ask the question, can international flood risk management experiences offer valuable insights to reduce coastal vulnerability in East Asia?

Causes of coastal vulnerability and evidence of climate change impacts on coastal populations

Geography and demographic change in East Asia

The deltas and coastlines of East Asia are home to large populations. Yeung (2009) reported how most of Asia's megacities (those with a population near or in excess of 8 million) are located on coastal or deltaic areas. In East Asia, the Yangtze and Pearl River Deltas support populations of more than 75 million and 40 million respectively (Gu et al. 2011). These areas contribute an important proportion of national economic and industrial development, with manufacturing of electronics, automobiles and textiles. Rapid coastal urbanization and industrialization has caused huge internal migration from rural to continental areas (Long et al. 2009) and resulted in a large and significant 'floating' migrant population (Bailey 2010).

National economic policy has driven the rapid urbanization of East Asia's coastal regions, for example China's 'open door [economic] policy' in 1979, transformed its previously agrarian coastal regions into industrial economies, catalysed by the establishment of special economic zones (SEZs) (Yeung 2010). SEZs are mainly located in three major deltaic areas in China, such as the Bohai economic zones in the Yellow River Delta and periphery, the Yangtze River Delta and the Pearl River Delta. Pudong in Shanghai and the Suzhou development zones in the Yangtze River Delta are both success stories of the last few decades. The Pudong New Area (PNA) had a population of 1.5 million with a GDP of about RMB6 billion during the 1990s (Yeung & Sung 1996). Now the PNA is the symbol of China's economic reform, with GDP exceeding RMB92 billion in the 2000s. At an annual growth rate of > 18%, it helped Shanghai become the most important economic hub in East China (Zhang et al. 2012). Similarly, Suzhou also exhibits accelerated development as a high-tech industrial zone (Wei et al. 2009). The city has attracted more than 7,500 foreign enterprises, with investments totalling some RMB210 billion in the 2000s. Recent figures suggest the GDP of Suzhou has reached over RMB670 billion (Gu et al. 2011), and is now one of the most rapidly growing cities in China. These examples illustrate how economic reform and urbanization influence rapid demographic change in the region.

Several countries in SE Asia are following similar development policies and economic reforms. The Vietnam and Myanmar governments have adopted SEZs in the Mekong, Red River and Ayeyarwady deltas, with resulting rapid urbanization in their coastal cities (Seto 2011). Yangon on the Ayeyarwady delta recorded a population growth of at least 22% every decade since the 1960s (United Nations 2010). While Tokyo was the only megacity in Asia during the 1950s, eight megacities have emerged in East Asia during the 2000s, seven of which are located in coastal areas (Osaka-Kobe, Shanghai, Jakarta, Manila, Seoul, Guangzhou, Shenzhen and Hong Kong). It is projected that other coastal megacities will emerge in the region by 2015 (Yeung 2009).

Consequences of climate change: flood risk in East Asian coastal areas

Climate change and flood risk

East Asian coastal regions already experience a high incidence of extreme events such as typhoons and storm surges (Mendelsohn et al. 2012), the negative effects of which are generally more prevalent in coastal areas than inland (Ericson et al. 2006). Climate change will worsen flood risks from both landward and seaward directions. Landward influences on flood risk relate to higher rainfall from large storms. The frequency and intensity of storms and other extreme events in the West Pacific region have increased from the 1970s to 2000s, a trend that may continue further owing to climate change (Hanson et al. 2011; Prudhomme et al. 2010; Wilby & Keenan 2012). There is strong and clear evidence that elevated greenhouse gas concentrations will contribute to greater volumes and intensification of precipitation events (Min et al. 2011). This is expected to significantly increase annual mean river discharge and annual maximum monthly discharge, equating to a higher annual probability of the one-in-a-100-year flood event for many large drainage basins (Milly et al. 2002). Studies elsewhere have already quantified the links between increased human greenhouse gas contributions and heightened flood risk; for example, greater precipitation and runoff directly attributable to climate change increased flood occurrence in England and Wales by up to 90% in the year 2000 (Pall et al. 2011).

From the seaward side, a rise in global mean sea level is escalating the flood risk for low-lying coastal areas. While predictions vary widely, recent estimates suggest a possible global sea level rise of 150 cm to 190 cm by the end of the century (Vermeer & Rahmstorf 2009). Rising seas cause a range of effects that are factors in flood risk, such as coastal submergence, erosion and ecosystem loss (Nicholls et al. 2011). A rising sea level also raises the baseline for storm surges driven ashore by extreme meteorological events (Ericson et al. 2006), which may result in more frequent inundation or overtopping of sea defences that are designed for lower level scenarios. Moreover, for the Western North Pacific Ocean, annual and decadal cycles in the genesis points, migratory paths and maximum intensities of typhoons should be anticipated (Feng & Terry 2012; Terry & Feng 2010), which, in turn, influence the likelihood of coastal inundation by systems that eventually make landfall along East Asian coastlines.

Of course it is important to remember that the physical impacts of climate change are just one component of overall 'flood risk'. Risk may also increase owing to anthropogenic land uses and socio-economic factors relating to rapid urbanization and population densities in coastal zones (see below), as well as by the fluvial and coastal consequences of climate change already mentioned.

Climate change, human influences and outlook

Anthropogenic land use exacerbates the magnitude of sea level rise, and thus overall coastal vulnerability. Today, many coastal and deltaic regions in East Asia are confronting land subsidence in response to declines in fluvial sediment loads reaching the coastal zone (caused by upstream river damming), land compaction and resource extraction (e.g., petroleum, gas and groundwater). On the east coast of China, annual rates of land subsidence in Tianjin exceeded 11 cm during the 1980s owing to groundwater extraction, with cumulative subsidence rates exceeding 1 m in the last decade alone. Subsidence affected an area of 60,000 km^2, with a maximum subsidence of 3.9 metres recorded at Tianjin since the 1950s (Xu et al. 2008). Tang et al. (2008) reported that much of central Shanghai is now 2 m below mean sea level, with the CBD area now reliant on structural flood protection measures. Recent research has revealed that the city is the most vulnerable to coastal floods compared with eight other global coastal cities (including Rotterdam, Osaka, Manila and Dhaka) (Balica et al. 2012). Bangkok faces similar issues. Over-extraction of groundwater in the Chao Phraya Delta has caused subsidence of > 2 m since the 1970s, with an annual subsidence rate of 10 cm/yr. Partly in consequence, the shoreline of Bangkok has receded by several kilometres (Chen & Saito 2011).

Rapid urbanization also causes negative geomorphic impacts on coastal and deltaic areas. Sediment input has declined due to the diversion of river channels and the construction of levees, artificial riverbanks and upstream dams (Syvitski & Kettner 2011). These practices limit the sediment supply needed to maintain deltas and associated wetland habitats (Yang et al. 2006) hence aggradation, the natural increase in surface due to sediment deposition, is reduced. Syvitski and Saito (2007) estimated that sediment loads have reduced by > 70% in the deltas of the Yellow and Yangtze Rivers and > 90% in the Pearl River Delta. Most of East Asian deltas are therefore sinking in response to these human and natural influences (Syvitski et al. 2009).

All of these factors exacerbate flood risks from storm surge and sea level rise for East Asian coastal cities, 16 of which are ranked in the top 20 from 136 global port cities at risk (Hanson et al. 2011). Moreover, coastal megacities will continue to experience rapid socioeconomic growth (see Table 22.1). For example, Qingdao on the East coast of China will continue developing as a technological hub (Yeo et al. 2011) and economic assets there may increase to US$600 billion. However, Gu et al. (2011) cautioned that the Yangtze River Delta has insufficient flood protection infrastructure, and populated East Asian coastlines all need better preparation for the future (Nicholls 2011). There is an important and urgent need for appropriate flood risk management strategies to be developed for East Asia.

The case of East Asian mega-deltas

Current challenges of mega-deltas in East Asia – the PRD case

The PRD is located in Southern Guangdong province of China, covering 11 cities including the Hong Kong and Macau Special Administrative Regions. The agricultural delta before the 1980s has now been transformed into China's major industrial and economic hub. For example, the fishing town of Shenzhen has grown into a megacity with over 15 million people (Yeung 2009). Other PRD cities have similarly recorded a five- to tenfold growth in population within the last three decades (Vogel et al. 2010). Remarkably, the PRD covers < 1% of China's landmass (41,698 km²) (Yang 2006), but contributes 20% of its national GDP, being called the 'world's factory' by economic commentators (Yeung 2010). Reforms and huge investments have attracted international trade and a large labour force through migration (Bailey 2010). The PRD now has a population of some 60 million (Marsden 2011) and is the world's most densely populated delta with > 7,500 people per km² (Syvitski & Saito 2007). UN-Habitat (2008) projected that the population may increase to over 120 million by 2050. This extraordinary growth has led to rapid urbanization and concomitant vulnerability to both inland and coastal flooding.

Climate change and flood vulnerability in the PRD

Inland floods

The PRD catchment is mostly characterized by steep hills and floodplains, comprising > 20% and 70% of the land area respectively (Cheng 2005). Inland floods in the PRD occur in summer from May to September (Dou & Zhao 2011; Zhang et al. 2010), when 80% of the annual rainfall (2,200 mm) arrives, often as intense precipitation associated with Pacific typhoons. The

Table 22.1 Selected East Asian coastal megacities ranked top 40 globally in 136 port cities in terms of population and economic assets exposed to coastal flood risk at present and (projected) in the 2070s

Coastal megacities	Country	Located on the deltas (yes/no)	Current population exposed to flood risk	Future population exposed to flood risk	Rank in exposed population (2070s)	Current exposed assets (US$ billions)	Future exposed assets (US$ billions)	Rank in exposed assets (2070s)
Guangzhou	China	Yes, Pearl River Delta	2,718,000	10,333,000	4	84.17	3,357.72	2
Ho Chi Minh City	Vietnam	Yes, Mekong Delta	1,931,000	9,216,000	5	26.86	652.82	16
Shanghai	China	Yes, Yangtze River Delta	2,353,000	5,451,000	6	72.86	1,771.70	5
Bangkok	Thailand	Yes, Chao Phraya Delta	907,000	5,138,000	7	38.72	1,117.54	10
Yangon	Myanmar	Yes, Irrawaddy Delta	510,000	4,965,000	8	3.62	172.02	39
Hai Phong	Vietnam	Yes, Red River Delta	794,000	4,711,000	10	11.04	333.70	26
Tianjin	China	No	956,000	3,790,000	12	29.62	1,231.48	7
Khulna	Bangladesh	Yes, Ganges – Brahmaputra	441,000	3,641,000	13	4.41	177.86	38
Ningbo	China	Yes, Yangzte River Delta	299,000	3,305,000	14	9.26	1,073.93	11
Shenzhen	China	Yes, Pearl River Delta	701,000	749,000	18	21.7	243.29	31
Tokyo	Japan	No	1,110,000	2,521,000	19	174.29	1,207.07	8
Jakarta	Indonesia	No	513,000	2,248,000	20	10.11	321.24	27
Osaka-Kobe	Japan	No	1,373,000	2,023,000	21	215.62	968.96	13
Qingdao	China	Yes, Yellow River Delta	88,000	1,851,000	23	2.72	601.59	18
Nagoya	Japan	No	696,000	1,302,000	27	109.22	623.42	17
Hong Kong	China	Yes, Pearl River Delta	223,000	687,000	39	35.94	1,163.89	9

Source: Adapted from Hanson et al. (2011).

Hong Kong Observatory (HKO) noted that both the peak intensity and frequency of rainstorms in the PRD region have increased during the last century, a trend that is likely to continue in the near future (Lee et al. 2010). Particularly, the frequency of heavy rainfalls (> 100 mm in 24 hours) may increase the risk of flash flooding, particularly in the flood-prone Guangzhou, Shenzhen and Hong Kong megacities. The PRD has also experienced flooding of the main Pearl River during the wet season. Guangzhou has experienced significant inundation 24 times since the Ming dynasty (1368 AD), with 10 large flood events 1911 and 1983 (Weng 2007). The 1915 flood of the North and West Pearl Rivers was the most severe, a one-in-200 year event that displaced some 6 million people, and caused 100,000 deaths and injuries in the western PRD (Zhang & Wang 2007). More recently, over the period of 8 to 17 June 1994, flooding after > 600 mm precipitation from Typhoon Russ caused 102 deaths, 2,000 injuries, the inundation of more than 9,000 villages, 230,000 houses and 100,000 ha of farmland. The total economic loss stood at RMB 3.2 billion (Wong & Zhao 2001). Many embankments and dykes along the North Pearl River were breached and collapsed; prompting questions on whether existing flood protection structures will meet future needs (Huang et al. 2004).

A projected increase in extreme events is exacerbated by rapid urban development in the PRD, which increases surface runoff. The urban land cover of 30% in the delta in 1982 is now > 80%, mostly on the floodplains (Yeung 2010). Consequently, maximum urban flood discharge in Shenzhen has increased by nearly 13% in 10 years (Shi et al. 2007). Similarly, Typhoon Chanthu in July 2012 highlighted vividly how urban drainage systems in Hong Kong, designed for the one-in-50 year flood, may not be able to cope with future peak discharges (Chui et al. 2006). Fuchs et al. (2011) therefore questioned the appropriateness of further urban expansion in exposed Asian deltaic and coastal areas in the face of unpredictable climatic regimes.

Coastal floods

While preparing for, and managing, inland flood events, East Asian megacities are also vulnerable to coastal flooding. Vulnerability is caused, in part, by anthropogenic changes to the coastal zone. In the PRD, more than 3,720 km² of coastal land has suffered subsidence, especially in Macau, Zhuhai, Zhongshan, Shenzhen and Guangzhou cities. Much subsidence is triggered by construction on mollisol soils, which are often unstable with an organic rich profile, a calcareous base and become saturated easily (Xu et al. 2009). At the same time, rapid urbanization causes land scarcity and municipal governments have favoured land reclamation to meet demands, such as the large Shenzhen Bay reclamation project along the Shekou Peninsula (Li & Damen 2010). In some areas, reclamation has extended coastlines over 1 km seawards during the last decade (Hay & Mimura 2006). Unfortunately, most reclaimed lands were converted from coastal wetlands such as mangroves and saltmarshes, which can provide seawater storage or hydrodynamic attenuation during regular tidal cycles (e.g., Lacambra et al. 2013; Möller 2006). Reclamation activities also modify estuarine morphology, by introducing dry land where previously only wetland existed, which may likewise affect tidal dynamics in the PRD (Zhang et al. 2009).

Recent research showed that mean sea level rise in the PRD has risen with the rate at 26 mm per decade from 1954–2009, and noted a significant upraise during the 1990s (Figure 22.1) (Zhang et al. 2011b). Woo and Wong (2010) projected sea level would rise a further 200 mm in the PRD by 2050, exposing more than 2,000 km² of coastal low-lying areas to tidal inundation (Huang et al. 2004).

Since the PRD is located within a subtropical monsoon climatic zone, typhoons and storm surges are common during the wet season, as previously mentioned. For example, storm surges driven by typhoons Hagupit and Koppu in 2008 and 2009 inundated the low-lying coastal areas of Tai O town, flooding 100 properties (Figure 22.2). From 1991 to 2005, 41 storm surges of

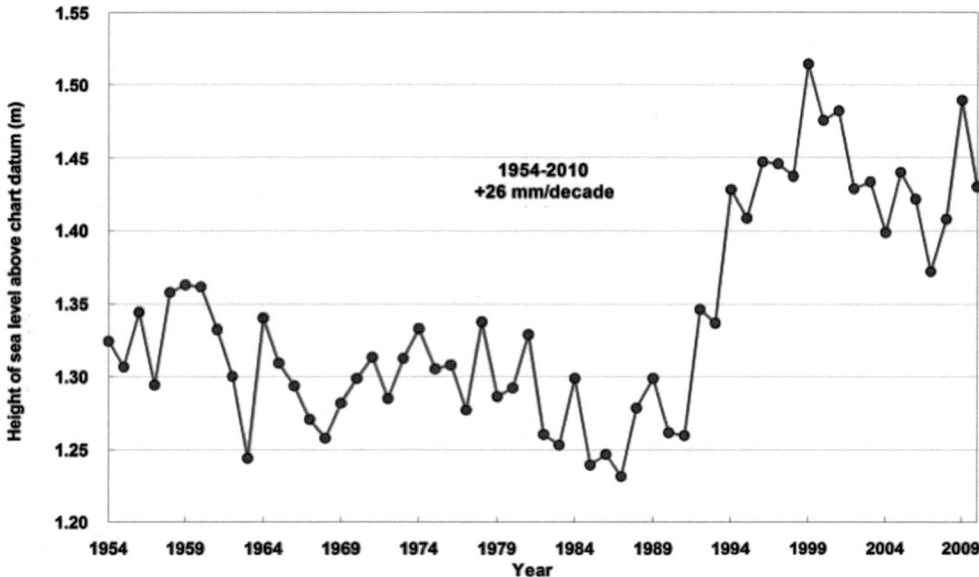

Figure 22.1 Annual mean sea level at North Point/Quarry Bay, 1954–2010

Source: Adapted from Zhoung et al. (2011).

two to three metres occurred in the PRD as a whole (Zhang et al. 2011a), while the HKO recorded over 10 surges higher than 1.5 m from 1954 to 2009 in Hong Kong alone (Lee et al. 2010). Typhoon Wanda in 1962 was a particularly severe event that generated a surge reaching four metres average mean sea level (Yim 1996). Historical events like this, which could recur, underscore the vulnerability of Hong Kong and other PRD cities to future typhoon-generated surges and associated coastal floods.

Current governance in climate change adaptations and flood risk management

In spite of the contextualization of flood risks facing the PRD as mentioned earlier, flood risk management (FRM) and climate change adaptation (CCA) are receiving little attention. More than 86% of the PRD coastal area relies on flood protection infrastructures (dykes and embankments); although only a limited proportion could withstand a one-in-100 year event (Cai et al. 2011). Moreover, if a projected sea level rise of 30 cm occurs by 2030, then a one-in-100 year storm surge would inundate 80% of the delta, with an estimated 1 million homes flooded, and economic losses exceeding RMB232 billion (Zhang 2009). These estimates notwithstanding, however, improving the current flood protection standards in diverse deltaic and estuarine areas would be costly (Woodroffe 2010).

Alarmingly, a recent governmental report on PRD strategic regional planning (Guangdong Province Housing & Urban–Rural Department 2011) addresses neither existing flood risks nor the possible effects of climatic change. Ng (2012) criticized the fact that regional CCA remains at the public consultation stage (EPD 2010), with limited consideration of implementing FRM. Past events have also shown that no institutions are specifically responsible for coastal flood mitigation. In Hong Kong, for instance, the Drainage Service Department (DSD) mainly deals

Figure 22.2 Tai O town flood after Typhoon Hagupit, 2008
Source: TVB.

with urban flood problems and their Stormwater Manual illustrates ad hoc approaches that are not based on strategic long-term plans that take into account climate change projections (DSD 2000). Similarly, Zhou and Cai (2010) noted how numerous land reclamation and development plans with the Shenzhen Bay area do not include coastal flood vulnerability, and it is more than apparent that the PRD and other Asian coastal cities need to address projected climate change extremes within CCA (Fuchs et al. 2011). Lack of central coordination of flood management may be an issue for many regions globally, although this has particularly dire consequences for a heavily urbanized and vulnerable area such as the PRD.

Regarding inland and river flood management, the PRD region has a long history of protection measures. Dykes and river channel diversion have been used for centuries since the Ming Dynasty (Weng 2007). In modern times, local governments continue to depend on hard engineering approaches. Hong Kong and Shenzhen authorities, for instance, rely mainly on river regulation through construction of artificial channels and embankments for flood protection against one-in-50 year events. Protection is aimed at economic assets such as railway terminals, luxury properties and government buildings (Chui et al. 2006). However, the channelized river silts up without frequent dredging, so reducing flood protection by 50% (Chan & Lee 2010). This demonstrates that engineering defences are insufficient, and that integrated FRM approaches incorporating 'soft' protection measures such as flood warning and risk mapping are necessary for urbanized cities (Ma et al. 2010). Hong Kong and Shenzhen authorities have applied flood risk modelling in the Drainage Master Plan (Chui et al. 2006), although certain aspects of the planning process cannot take advantage of this important information until it is released into the public domain. Overall, the PRD and most Asian coastal regions currently face tough challenges, with a lack of holistic FRM policy existing against a canvas of rapid socioeconomic growth and emerging climate change threats.

European experiences in flood risk management offer important lessons for East Asia

Growing threats of climate change in East Asia, particularly risks of more frequent floods affecting populous deltas and coastal cities, suggest that it may be wise to learn lessons from wider international experiences.

In the Netherlands, the Dutch people have lived with floods for centuries, as most of the country, including the large cities of Rotterdam and Amsterdam, lie near to or below sea level (Van Koningsveld et al. 2008). Consequently, Dutch authorities have learnt to use dykes to protect against high tides, windmills to pump water from floodplains, and have successfully reclaimed agricultural land from the sea since the thirteenth century (Wesselink et al. 2007). Nonetheless, unprecedented floods events still occur. The 1953 flood during a storm surge in the North Sea breached 900 dykes, inundated more than 1 million properties and caused 1,835 deaths (Gerritsen 2005; Vis et al. 2003). Afterwards, the Dutch government recognized that coastal measures to cope with the one-in-10,000 year return period event were needed (van Stokkom et al. 2005). The 1953 storm surge led to an unprecedented national-scale engineering works involving a complex system of dykes and surge barriers, requiring investment of €13 billion to date (Kabat et al. 2005). Indeed, the 1953 storm surge led to wholesale changes in coastal flood risk management throughout Northwest Europe. More recently, floods in the Rhine and Meuse rivers during 1993 and 1995 winter rainstorms (Tol et al. 2003) indicated how inland river floods are still a concern, and the government realized by the mid-1990s that flood protection standards would not be able to cope with expected changes in flood events due to climate change. Consequently, the '*Ruimte voor de River*' ('Room for the River') policy has: 1. encouraged provision of more space for water storage, 2. restricted further developments on floodplains, and 3. begun to manage flood risk strategically by planning within sustainable frameworks (van Herk et al. 2011). Some have voiced criticism over continued infrastructure building on risky areas and a reliance on engineering-based flood management approaches, though more recent policies such as '*Living with Water*' and '*Living in a Dynamic Delta*' are promoting resilience against flood risk through the better conservation of floodplains and wetlands (Wesselink et al. 2007). Hall and Penning-Rowsell (2010) support such practices as they also help to sustain nature and biodiversity, and promote good quality of watercourses, thus delivering multiple benefits consistent with the European Water Framework Directive (2000). This policy commits EU member states to achieve a good status, in both qualitative and quantitative terms, for all water bodies by 2015. Van Stokkom and Witter (2008) highlight that the latest Dutch FRM policy is still in a transitional state, and aims to address social justice by ensuring that all citizens benefit equally from flood protection measures. Importantly, the Dutch public have a right to understand their flood risk. This means that relevant risk and hazard information can be publicly accessed (e.g., data on flood return periods, flood histories and locations), which raises awareness and improve preparedness (Gersonius et al. 2011). As such, these practices fulfil requirements of the European Floods Directive (European Union 2007) that all EU member states openly provide relevant flood risk information to their citizens by 2015.

The 1953 North Sea floods that impacted the Netherlands also caused severe damage to Southeast England, with some 300 deaths and widespread inundation, including central London, and most notably Canvey Island in the Thames Estuary (Penning-Rowsell et al. 2006). London is another megacity, with a population around 10 million, and Lonsdale et al. (2008) estimated that an area of > 345 km^2, 480,000 properties, 1 million people and 2,000 km of transport links are exposed to flood risk. As a measure against storm surge threats to London, the current Thames Barrier began operation in 1983 (31 years after planning commenced), designed to protect up to a one-in-1,000 year event, a standard that at the time was greater than most flood defence measures in the UK. However, although the barrier may provide sufficient protection to meet 2035 climate change projections, flood vulnerability remains high owing to continuous population increase and economic growth along the Thames Estuary: this area has been slated for major development and redevelopment in the coming decades. Worryingly, the design life may well be reached sooner, as the original design in the 1950s–60s did not adequately account for accelerated sea level rise (Lonsdale et al. 2008).

In response, the Environment Agency (EA), the UK government agency responsible for CCA and FRM, has devised a strategic regional flood risk and development plan in the Thames Estuary, namely the 'TE2100' project (Environment Agency 2009) (Figure 22.3). This project is designed to address flood risk holistically, by applying integrated river basin management that includes 1. land use planning (e.g., restriction of new development in high-risk areas), 2. surface water management and soft FRM measures (e.g., improved emergency response, flood warning systems, and enhanced public participation), and 3. conservation of natural marshland areas (e.g., at the Thames river mouth) (Dawson et al. 2011). The plan deliberately takes a proactive and long-term approach, looking forward 80 years to the end of the century, and importantly considering both 'hard' and 'soft' flood protection measures. As the EA recognized, in order to deal effectively with emerging climate change impacts over coming decades, it is no longer sufficient, or cost effective, to rely on engineered flood defences and ad hoc solutions (Environment Agency 2009). Importantly, in the context of climate change, TE2100 was one of the first large-scale projects to recognize and appreciate uncertainties in climate projections, and incorporate them into the decision-making process (Kiker et al. 2011).

A third international example, also from the UK, may also have important implications for coastal flood risk management in East Asia. This example is primarily concerned with issues of coastal erosion and CCA, although it also provides important lessons for how a holistic FRM framework should operate. With an increasing appreciation that coastal processes cross administrative boundaries and can operate over large scales, the UK government embarked on a series of large-scale shoreline management plans (SMPs), designed as a 'large-scale assessment of the risks associated with coastal processes [that] helps to reduce risks to people and the developed, cultural and natural environment' (DEFRA 2006; Winn et al. 2003). SMPs are novel in that they were some of the first management plans where management boundaries were based on a 'behavioural systems approach', incorporating geomorphological characteristics (such as sediment movement and currents), as opposed to purely administrative boundaries. Therefore, they encompass processes and impacts of management decisions that may cross borders and affect neighbouring areas. SMPs advocate a combination of four management decisions along the coast, namely:

1   Advance the current defence line by building new (hard) defences further seaward.
2   Hold the current defence line by maintaining current (hard) defence standards.
3   No active intervention, where current defences are not actively maintained.
4   Managed realignment, where the coastline is (actively) allowed to move landwards.

A key aspect of SMPs is their use in attempting to predict a) future land uses, and b) the consequences of these various management interventions. Similar to the EU Water Framework Directive described previously, the results of the SMP process are publicly available, and public participation in this process is encouraged.

To summarize the Dutch and UK experiences, integrated, large-scale and holistic approaches have been developed that are transforming flood risk management in coastal areas experiencing growing threats from climate change and internal growth. However, to date, such approaches do not appear to have been coherently adopted within East Asian cities experiencing similar pressures. A number of important lessons should be learnt from this international experience, and lead to a set of key considerations for improved FRM in the PRD:

1   *Planning.* FRM must be an integral part of urban and economic planning. For example, DEFRA (2006) describe the close interplay between SMPs and the land use planning process. SMPs may recommend limited development in areas at risk from erosion or flooding, areas where managed realignment is likely to be implemented, or restrict development that may interfere with coastal processes. Like the TE2100 and SMP process, a mechanism for frequent review and update must also be included. In Europe, Directive 2007/60/EC (European Union 2007) (known as the 'Floods Directive') is driving such a review. The directive requires member states to produce a preliminary flood risk assessment, considering impacts on health, economic activity, cultural heritage and the environment. This preliminary assessment then guides more detailed modelling of areas at significant risk, considering the extent and depth of flooding under low, medium and high probability event scenarios. Flood risk management plans must then be established, by 2016, which communicate the flood risk to policymakers, developers, and the public, with a view to developing prevention, protection and preparedness measures.

2   *Participation.* This is an essential requirement if people and organizations are to make informed decisions about flood risk. Participation is evident in the UK and Dutch flood risk management cases described earlier (e.g., the Thames Gateway Partnership, TE2100 public consultation, publicly available information in the Netherlands and publicly available and searchable online flood risk mapping tools in the UK (Environment Agency 2012). Similarly the EU Floods Directive recognizes the importance of participation in the development of flood risk management plans, which needs community support if they are to be effective. It does not, however, prescribe how this should be achieved, and it is worth noting that participation comes in many forms, ranging from tokenistic information provision, through to true partnership with full interactive dialogue between parties, perhaps involving citizen juries or community champions (Arnstein 1969).

3   *Spatial scale.* The Delta Plan in the Netherlands required a national-scale approach to FRM; a municipal or *county*-level planning approach would be substantially less effective as floodplain and coastal zones are not closed units, but management changes in one locality may have knock-on impacts further along. The Delta Plan and UK SMPs are effective precisely because they take a large-scale approach. The EU Floods Directive promotes a coherent approach to spatial scale, by requiring a preliminary flood risk appraisal for each member nation, which informs more detailed local level assessment for identified at significant risk areas.

4   *Temporal scale.* For example, SMPs identify a combination of management interventions that may be required over the next 100 years (DEFRA 2006). Such a long-term viewpoint allows a proper treatment of issues relating to cost and uncertainty.

5   *Integration.* FRM must not solely focus on hard engineering works, but integrate other aspects such as soft defence methods, including vegetation (riparian, coastal). Large-scale shoreline management planning (such as in the UK) provides a planning framework to highlight areas where different management approaches (hard defences, soft defences) may be most appropriate. A holistic approach to FRM is required.

6   *Uncertainty.* Uncertainties inherent in climate change scenarios must be incorporated into the decision-making process. This is explicit in the Floods Directive, where low- to

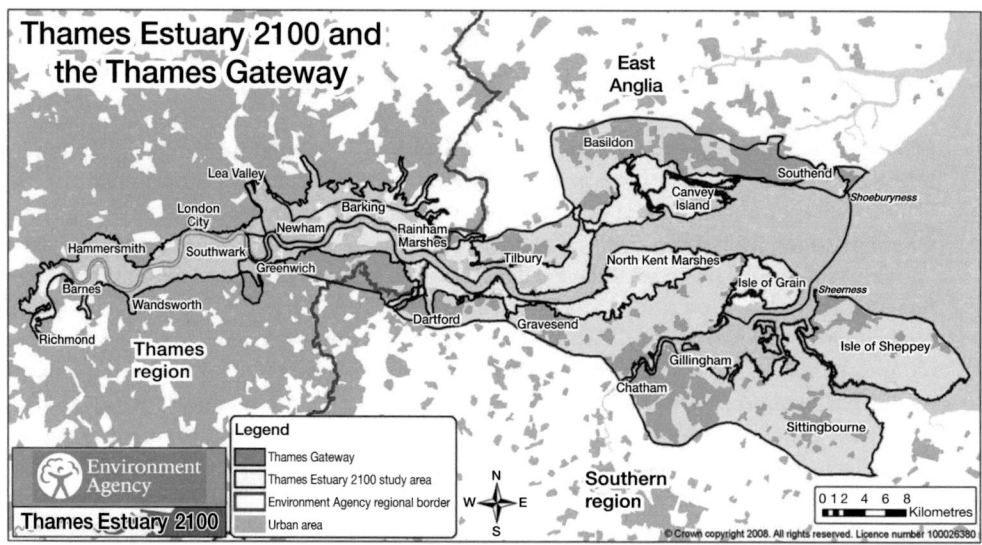

Figure 22.3 Thames Estuary 2100 project study area
Source: Environment Agency (2009).

high-probability events must be modelled. The financial (and political) cost of building defences for a scenario that may not occur is high; this approach requires *long-term* management plans that can absorb these 'sunken costs' (Kiker et al. 2011).

Many of the recommendations listed here require an overarching FRM institutional framework, which provides opportunities to consider multiple management and policy instruments over varying temporal and spatial scales. Previous experiences in Europe, prompted by natural disasters such as the 1953 North Sea storm surge, provide important lessons for the shape such a framework should take.

Conclusions

Across East Asia, coastal regions (deltas and cities) with large populations have become global hubs of socioeconomic activities. The PRD is a primary exemplar. Unfortunately, recent histories demonstrate how the region faces significant flood risks, while rapid economic and population growth continue at astonishing rates. At the same time, current climate change and sea level rise enhance the likelihood of intensive rainstorms and storm surges delivered by Western Pacific typhoons. Consequently, low-lying coasts and East Asian megacities in particular face magnified exposure to severe flood risk.

In spite of this, the East Asian coastal region suffers from limited implementation of holistic strategies that address CCA and FRM. Evidence suggests a tenacious devotion to traditional 'hard' (engineered) flood protection measures, practised in a piecemeal manner. Such approaches are costly and not economically sustainable; more strategic 'soft' management options have yet to be adopted. In this regard, lessons may be learned from experiences in the Netherlands and the United Kingdom where FRM has embraced multiple aspects of CCA, land use planning, awareness building and public engagement.

These EU practices may encourage the testing of similar approaches to improve flood resilience in East Asia. Although the EU is unlike East Asia in many ways, Asian nations should nonetheless strengthen regional collaboration in order to promote CCA within FRM policies. This seems a reasonable recommendation since most East Asian nations are facing similar challenges of balancing coastal development with adequate protection against inundation. Thus, heightened exposure to flood risk on East Asian coasts and deltas, against the backdrop of unabated rates of population and economic growth, emphasizes how integrating climate change adaptation and flood risk reduction is an imperative best tackled sooner rather than later.

Bibliography

Arnstein, S. (1969) 'A Ladder of Citizen Participation', *Journal of the American Planning Association* 35, 216–224.

Bailey, A. (2010) 'Population Geographies and Climate Change', *Progress in Human Geography* 35, 685–695.

Balica, S.F., Wright, N.G. & Meulen, F. (2012) 'A Flood Vulnerability Index for Coastal Cities and its Use in Assessing Climate Change Impacts', *Natural Hazards* 64, 73–105.

BBC (2011) 'Thailand Floods: Bangkok Braced as Drainage Begins'. Available at http://www.bbc.co.uk/news/world-asia-pacific-15399421.

Cai, Y.P., Huang, G.H., Tan, Q. & Chen, B. (2011) 'Identification of Optimal Strategies for Improving Eco-resilience to Floods in Ecologically Vulnerable Regions of a Wetland', *Ecological Modelling* 222, 360–369.

Chan, S.N. & Lee, J.H.W. (2010) 'Impact of River Training on the Hydraulics of Shenzhen River', *Journal of Hydro-environment Research* 4, 211–223.

Chen, Z. & Saito, Y. (2011) 'The Megadeltas of Asia: Interlinkage of Land and Sea, and Human Development', *Earth Surface Processes and Landforms* 36, 1703–1704.

Cheng, X. (2005) 'Changes of Flood Control Situations and Adjustments of Flood Management Strategies in China', *Water International* 30, 108–113.

Chui, S.K., Leung, J.K.Y. & Chu, C.K. (2006) 'The Development of a Comprehensive Flood Prevention Strategy for Hong Kong', *International Journal of River Basin Management* 4, 5–15.

Dawson, R.J., Ball, T., Werritty, J., Werritty, A., Hall, J.W. & Roche, N. (2011) 'Assessing the Effectiveness of Non-structural Flood Management Measures in the Thames Estuary under Conditions of Socio-economic and Environmental Change', *Global Environmental Change* 21, 628–646.

DEFRA (2006) *Shoreline Management Plan Guidance. Volume 1: Aims and Requirements* (Department for Environment, Food and Rural Affairs). Available at http://www.defra.gov.uk/publications/files/pb11726-smpg-vol1-060308.

Dou, H. & Zhao, X. (2011) 'Climate Change and its Human Dimensions based on GIS and Meteorological Statistics in Pearl River Delta, Southern China', *Meteorological Applications* 18, 111–122.

DSD (2000) *Stormwater Drainage Manual* (Drainage Services Department, Hong Kong).

Environment Agency (2009) *TE2100 Plan Consultation Document, Thames Estuary 2100* (Environment Agency).

Environment Agency (2012) 'Am I at risk of flooding?' Available at http://www.environment-agency.gov.uk/homeandleisure/floods/31650.aspx.

EPD (2010) Agreement No. CE 45/2007 (EP): A Study of Climate Change in Hong Kong – Feasibility Study, Hong Kong: Environmental Protection Department, Hong Kong SAR Government.

Ericson, J.P., Vörösmarty, C.J., Dingman, S.L., Ward, L.G. & Meybeck, M. (2006) 'Effective Sea-level Rise and Deltas: Causes of Change and Human Dimension Implications', *Global and Planetary Change* 50, 63–82.

European Union (2000) Directive of the European Parliament and of the Council Concerning Establishing a Framework for Community Action in the Field of Water Policy (2000/60/EC), Brussels, European Union.

European Union (2007) Directive 2007/60/EC on the Assessment and Management of Flood Risks, Brussels, European Union.

Feng, C.C. & Terry, J.P. (2012) 'Exploratory Spatial Analysis of Typhoon Characteristics in the North Pacific Basin', in J.P. Terry & J.R. Goff (Eds.), *Natural Hazards in the Asia–Pacific Region: Recent Advances and Emerging Concepts* (Geological Society of London).

Fuchs, R., Conran, M. & Louis, E. (2011) 'Climate Change and Asia's Coastal Urban Cities', *Environment and Urbanization Asia* 2, 13–28.

Gerritsen, H. (2005) 'What Happened in 1953? The Big Flood in The Netherlands in Retrospect', *Philosophical Transactions of the Royal Society A* 363, 1271–1293.

Gersonius, B., Veerbeek, W., Subhan, A., Stone, K. & Zevenbergen, C. (2011) 'Toward a More Flood Resilient Urban Environment: The Dutch Multi-level Safety Approach to Flood Risk Management', in K. Otto-Zimmermann (Ed.), *Resilient Cities* (Springer).

Gu, C., Hu, L., Zhang, X., Wang, X. & Guo, J. (2011) 'Climate Change and Urbanization in the Yangtze River Delta', *Habitat International* 35, 544–552.

Guangdong Province Housing and Urban–Rural Department (2011) Regional Cooperation Plan on Building a Quality Living Area Consultation Document, Guangzhou, China, Guangdong Provincial Government.

Hall, J.W. & Penning-Rowsell, E.C. (2010) *Setting the Scene for Flood Risk Management* (Wiley-Blackwell).

Hanson, S., Nicholls, R., Ranger, N., Hallegatte, S., Corfee-Morlot, J., Herweijer, C., et al. (2011) 'A Global Ranking of Port Cities with High Exposure to Climate Extremes', *Climatic Change* 104, 89–111.

Hay, J. & Mimura, N. (2006) 'Supporting Climate Change Vulnerability and Adaptation Assessments in the Asia-Pacific Region: An Example of Sustainability Science', *Sustainability Science* 1, 23–35.

Huang, Z., Zong, Y. & Zhang, W. (2004) 'Coastal Inundation Due to Sea Level Rise in the Pearl River Delta, China', *Natural Hazards* 33, 247–264.

Kabat, P., Van Vierssen, W., Veraat, J., Vellinga, P. & Aerts, J. (2005) 'Climate Proofing The Netherlands', *Nature* 438, 283–284.

Keokhumcheng Y., Tingsanchali, T. & Clemente, R.S. (2012) 'Flood Risk Assessment in the Region Surrounding the Bangkok Suvarnabhumi Airport', *Water International* 37, 201–217.

Kiker, G.A., Munoz-Capena, R., Ranger, N., Kiker, M. & Linkov, I. (2011) 'Adaptation in Coastal Systems: Vulnerability and Uncertainty within Complex Socioecological Systems', in I. Linkov & T.S. Bridges (Eds.), *Climate. NATO Science for Peace and Security Series C: Environmental Security* Part 4, 374–500.

Klijn, F., Niehuis, P.H. & Pedroli, G.B.M. (2004) 'Preface: Towards Sustainable Flood Risk Management in the Rhine and Meuse River Basins', *River Research and Applications* 20, 227.

Lacambra, C., Friess, D.A., Spencer, T. & Möller, I. (2013) 'Bioshields: Using Mangroves to Reduce Disaster Vulnerability and Improve Livelihoods', in F. Renaud, K., Sudmeier-Rieux & M. Estrella (Eds.), *The Role of Ecosystems in Disaster Reduction* (United Nations University Press).

Lee, B.Y., Wong, W.T. & Woo, W.C. (2010) Sea-level Rise and Storm Surge – Impacts of Climate Change on Hong Kong, Hong Kong, Hong Kong Observatory, Hong Kong SAR Government.

Li, X. & Damen, M.C.J. (2010) 'Coastline Change Detection with Satellite Remote Sensing for Environmental Management of the Pearl River Estuary, China', *Journal of Marine Systems* 82, S54–S61.

Long, H., Zou, J. & Liu, Y. (2009) 'Differentiation of Rural Development Driven by Industrialization and Urbanization in Eastern Coastal China', *Habitat International* 33, 454–62.

Lonsdale, K., Downing, T., Nicholls, R., Parker, D., Vafeidis, A., Dawson, R., et al. (2008) 'Plausible Responses to the Threat of Rapid Sea-level Rise in the Thames Estuary', *Climatic Change* 91, 145–169.

Ma, J., Tan, X. & Zhang, N. (2010) 'Flood Management and Flood Warning System in China', *Irrigation and Drainage* 59, 17–22.

McGranahan, G., Balk, D. & Anderson, B. (2007) 'The Rising Tide: Assessing the Risks of Climate Change and Human Settlements in Low Elevation Coastal Zones', *Environment and Urbanization* 19, 17–37.

Marsden, S. (2011) 'Assessment of Transboundary Environmental Effects in the Pearl River Delta Region: Is There a Role for Strategic Environmental Assessment?', *Environmental Impact Assessment Review* 31, 593–601.

Mendelsohn, R., Emanuel, K., Chonbayashi, S. & Bakkensen, L. (2012) 'The Impact of Climate Change on Global Tropical Cyclone Damage', *Nature Climate Change* 2, 205–209.

Middelkoop, H., Van Asselt, M.B.A., Van T Klooster, S.A., Van Deursen, W.P.A., Kwadijk, J.C.J. & Buiteveld, H. (2004) 'Perspectives on Flood Management in the Rhine and Meuse Rivers', *River Research and Applications* 20, 327–342.

Milly, P.C., Wetherald, R.T., Dunne K.A. & Delworth T.L. (2002) 'Increasing Risk of Great Floods in a Changing Climate', *Nature* 415, 515–517.

Min, K.-S., Zhang, X., Zwiers F.W. & Hegerl, G.C. (2011) 'Human Contribution to More-intense Precipitation Extremes', *Nature* 470, 378–381.

Möller, I. (2006) 'Quantifying Saltmarsh Vegetation and its Effect on Wave Height Dissipation: Results from a UK East Coast Saltmarsh', *Estuarine, Coastal and Shelf Science* 69, 337–351.

Ng, M.K. (2012) 'A Critical Review of Hong Kong's Proposed Climate Change Strategy and Action Agenda', *Cities* 29, 88–98.

Nicholls, R.J. (2011) 'Planning for the Impacts of Sea Level Rise', *Oceanography* 24, 144–157.

Nicholls, R.J., Marinova, N., Lowe, J.A., Brown, S., Vellinga, P., de Gusmão, D., et al. (2011) 'Sea-level Rise and its Possible Impacts given a "beyond 4°C world" in the Twenty-first Century', *Philosphical Transactions of the Royal Society A* 369, 161–181.

Pall, P., Aina, T., Stone, D.A., Stott, P.A., Nozawa, T., Hilberts, A.G., et al. (2011) 'Anthropogenic Greenhouse Gas Contribution to Flood Risk in England and Wales in Autumn 2000', *Nature* 470, 382–385.

Penning-Rowsell, E., Johnson, C. & Tunstall, S. (2006) 'Signals from Pre-crisis Discourse: Lessons from UK Flooding for Global Environmental Policy Change?', *Global Environmental Change* 16, 323–339.

Prudhomme, C., Wilby, R.L., Crooks, S., Kay, A.L. & Reynard, N.S. (2010) 'Scenario-neutral Approach to Climate Change Impact Studies: Application to Flood Risk', *Journal of Hydrology* 390, 198–209.

Rohling, E.J., Grant, K., Hemleben, C., Siddall, M., Hoogakker, B.A.A., Bolshaw, M., et al. (2008) 'High Rates of Sea-level Rise During the Last Interglacial Period', *Nature Geosciences* 1, 38–42.

Seto, K.C. (2011) 'Exploring the Dynamics of Migration to Mega-delta Cities in Asia and Africa: Contemporary Drivers and Future Scenarios', *Global Environmental Change* 21, S94–S107.

Shi, P.-J., Yuan, Y., Zheng, J., Wang, J.-A., Ge, Y. & Qiu, G.-Y. (2007) 'The Effect of Land Use/Cover Change on Surface Runoff in Shenzhen Region, China', *CATENA* 69, 31–35.

Shibahra, S. (2011) 'The 2011 Tohoku Earthquake and Devastating Tsunami', *Tohoku Journal of Experimental Medicine* 223, 305–307.

Syvitski, J.P.M. (2008) 'Deltas at Risk', *Sustainability Science* 3, 23–32.

Syvitski, J.P.M. & Saito, Y. (2007) 'Morphodynamics of Deltas under the Influence of Humans', *Global and Planetary Change* 57, 261–282.

Syvitski, J.P.M. & Kettner, A. (2011) 'Sediment Flux and the Anthropocene', *Philosophical Transactions of the Royal Society A: Mathematical, Physical and Engineering Sciences* 369, 957–975.

Syvitski, J.P.M., Kettner, A., Overeem, I., Hutton, E.W.H., Hannon, M.T., Brakenridge, G.R., et al. (2009) 'Sinking Deltas Due to Human Activities', *Nature Geosciences* 2, 681–686.

Tang, Y.-Q., Cui, Z.-D., Wang, J.-X., Yan, L.-P. & Yan X.-X. (2008) 'Application of Grey Theory-based Model to Prediction of Land Subsidence due to Engineering Environment in Shanghai', *Environmental Geology* 55, 583–593.

Terry, J.P. & Feng, C.C. (2010) 'On Quantifying the Sinuosity of Typhoon Tracks in the Western North Pacific Basin', *Applied Geography* 30, 678–686.

Terry, J.P., Winspear, N. & Cuong, T.Q. (2012) 'The "Terrific Tongking Typhoon" of October 1881 – Implications for the Red River Delta (Northern Vietnam) in Modern Times', *Weather* 67, 72–75.

Tol, R.S.J., Van Der Grijp, N., Olsthoorn, A.A. & Van Der Werff, P.E. (2003) 'Adapting to Climate: A Case Study on Riverine Flood Risks in the Netherlands', *Risk Analysis* 23, 575–583.

UN-Habitat (2008) State of the World's Cities Report, Nairobi, United Nations.

United Nations (2010) *World Urbanization Prospects: The 2009 Revision* (United Nations).

Van Herk, S., Zevenbergen, C., Ashley, R. & Ruke, J. (2011) 'Learning and Action Alliances for the Integration of Flood Risk Management into Urban Planning: A New Framework from Empirical Evidence from The Netherlands', *Environmental Science & Policy* 14, 543–554.

Van Koningsveld, M., Mulder, J.P.M., Stive, M.J.F., Van Der Valk, L. & Van Der Weck, A.W. (2008) 'Living with Sea-Level Rise and Climate Change: A Case Study of the Netherlands', *Journal of Coastal Research* 24, 367–379.

Van Stokkom, H.T.C., Smits, A.J.M. & Leuven, R.S.E.W. (2005) *Flood Defense in The Netherlands – A New Era, a New Approach* (Routledge).

Van Stokkom, H.T.C. & Witter, J.V. (2008) 'Implementing Integrated Flood Risk and Land-use Management Strategies in Developed Deltaic Regions, Exemplified by The Netherlands', *International Journal of River Basin Management* 6, 331–338.

Vermeer, M. & Rahmstorf, S. (2009) 'Global Sea Level Linked to Global Temperature', *Proceedings of the National Academy of Sciences* 106, 21527–21532.

Vis, M., Klijn, F., De Bruijn, K.M. & Van Buuren, M. (2003) 'Resilience Strategies for Flood Risk Management in The Netherlands', *International Journal of River Basin Management* 1, 33–40.

Vogel, R.K., Savitch, H.V., Xu, J., Yeh, A.G.O., Wu, W., Sancton, A., et al. (2010) 'Governing Global City Regions in China and the West', *Progress in Planning* 73, 1–75.

Ward, P., Marfai, M., Yulianto, F., Hizbaron, D. & Aerts, J. (2011) 'Coastal Inundation and Damage Exposure Estimation: A Case Study for Jakarta', *Natural Hazards* 56, 899–916.

Webster, P.J., Holland, G.J., Curry, J.A. & Chang, H.-R. (2005) 'Changes in Tropical Cyclone Number, Duration, and Intensity in a Warming Environment', *Science* 309, 1844–1846.

Wei, Y.H.D., Lu, Y. & Chen, W. (2009) 'Globalizing Regional Development in Sunan, China: Does Suzhou Industrial Park Fit a Neo-Marshallian District Model?', *Regional Studies* 43, 409–427.

Weng, Q. (2007) 'A Historical Perspective of River Basin Management in the Pearl River Delta of China', *Journal of Environmental Management* 85, 1048–1062.

Wesselink, A.J., Bijker, W.E., De Vriend, H.J. & Krol, M.S. (2007) 'Dutch Dealings with the Delta', *Nature and Culture* 2, 188–209.

Wilby, R.L. & Keenan, R. (2012) 'Adapting to Flood Risk under Climate Change', *Progress in Physical Geography* 36, 348–378.

Winn, P.J.S., Young, R.M. & Edwards, A.M.C. (2003) 'Planning for the Rising Tides: The Humber Estuary Shoreline Management Plan', *Science of the Total Environment* 13(30), 314–16.

Wong, K.-K. & Zhao, X. (2001) 'Living with Floods: Victims' Perceptions in Beijiang, Guangdong, China', *Area* 33, 190–201.

Woo, W.C., & Wong, W.T. (2010) Sea-level Change – Observations, Causes and Impacts, Hong Kong, Hong Kong Observatory, Hong Kong SAR Government.

Woodroffe, C.D. (2010) 'Assessing the Vulnerability of Asian Megadeltas to Climate Change Using GIS', in D.R. Green (Ed.), *Coastal and Marine Geospatial Technologies* (Springer).

Xu, Y.-S., Shen, S.-L., Cai, Z.-Y. & Zhou, G.-Y. (2008) 'The State of Land Subsidence and Prediction Approaches Due to Groundwater Withdrawal in China', *Natural Hazards* 45, 123–135.

Xu, Y.-S., Zhang, D.-X., Shen, S.-L. & Chen, L.-Z. (2009) 'Geo-hazards with Characteristics and Prevention Measures along the Coastal Regions of China', *Natural Hazards* 49, 479–500.

Yang, C. (2006) 'The Geopolitics of Cross-boundary Governance in the Greater Pearl River Delta, China: A Case Study of the Proposed Hong Kong-Zhuhai-Macao Bridge', *Political Geography* 25, 817–835.

Yang, S., Li, M., Dai, B., Liu, Z., Zhang, J. & Ding, P. (2006) 'Drastic Decrease in Sediment Supply from the Yangtze River and its Challenge to Coastal Wetland Management', *Geophysical Research Letters* 33, L06408.

Yeo, G.-T., Roe, M. & Dinwoodie, J. (2011) 'Measuring the Competitiveness of Container Ports: Logisticians' Perspectives', *European Journal of Marketing* 45, 455–470.

Yeung, Y.-M. (2009) 'Mega-Cities', in C. Xa, K., Rob & T. Nigel (Eds.), *International Encyclopedia of Human Geography* (Elsevier).

Yeung, Y.-M. (2010) 'The Further Integration of the Pearl River Delta', *Environment and Urbanization Asia* 1, 13–26.

Yeung, Y.-M. & Sung, Y.W. (1996) *Shanghai: Transformation and Modernization under China's Open Policy* (Chinese University of Hong Kong Press).

Yim, W.W.S. (1996) 'Vulnerability and Adaptation of Hong Kong to Hazards under Climatic Change Conditions', *Water, Air, & Soil Pollution* 92, 181–190.

Zhang, H. & Wang, X.-R. (2007) 'Land-use Dynamics and Flood Risk in the Hinterland of the Pearl River Delta: The Case of Foshan City', *International Journal of Sustainable Development & World Ecology* 14, 485–492.

Zhang, J. (2009) 'A Vulnerability Assessment of Storm Surge in Guangdong Province, China', *Human and Ecological Risk Assessment* 15, 671–688.

Zhang, J., Wang, L. & Wang, S. (2012) 'Financial Development and Economic Growth: Recent Evidence from China', *Journal of Comparative Economics* 40, 393–412.

Zhang, Q., Xu, C.-Y. & David Chen, Y. (2010) 'Variability of Water Levels and Impacts of Streamflow Changes and Human Activity within the Pearl River Delta, China', *Hydrological Sciences Journal* 55, 512–525.

Zhang, Q., Zhang, W., Chen, Y. & Jiang, T. (2011a) 'Flood, Drought and Typhoon Disasters During the Last Half-century in the Guangdong Province, China', *Natural Hazards* 57, 267–278.

Zhang, W., Yan, Y., Zheng, J., Li, L., Dong, X. & Cai, H. (2009) 'Temporal and Spatial Variability of Annual Extreme Water Level in the Pearl River Delta Region, China', *Global and Planetary Change* 69, 35–47.

Zhang, Y., Xie, J. & Liu, L. (2011b) 'Investigating Sea-level Change and its Impact on Hong Kong's Coastal Environment', *Annals of GIS* 17, 105–112.

Zhou, X. & Cai, L. (2010) 'Coastal and Marine Environmental Issues in the Pearl River Delta Region, China', *International Journal of Environmental Studies* 67, 137–145.

23

East Asia's renewable energy strategies

Low carbon developmentalism in the making?

Christopher M. Dent[1]

Introduction

The twin challenges of climate change and energy security are compelling East Asian states to develop stronger renewable energy and other green energy sectors as part of their low carbon development strategies. Energy has always played a fundamental role in economic development, and the cleaner and more sustainable nature of renewable energy (RE) systems makes them crucial to securing low carbon futures. The RE sector has expanded significantly worldwide over the last few years, and in East Asia more quickly than any other region. Renewable energy has furthermore become one of the defining features of East Asia's new industrial policy and 'new developmentalism', founded more generally on new configured forms of state capacity shaped in response to various challenges confronting the region's nations. Studying the recent progress of East Asia's RE sector provides useful insights into these key developments in East Asia's political economy, and the region's prospects for transition towards low carbon development: a particularly difficult challenge given the many high carbon-intensive aspects of East Asia's economy.

This chapter first discusses the evolving nature of state capacity and industrial policy in East Asia to set the broad context of debate on where renewable energy policies in the region are currently situated. Key trends in RE sector development and the renewable energy policies of East Asian states are then analysed. This is followed by a series of case studies on the role of renewables in the recently launched macro-development plans (e.g., China's 12th Five-Year Plan, Japan's New Growth Strategy) underpinning East Asia's new developmentalism. The chapter then considers how and why different approaches to RE policy have emerged in East Asia, to what extent the promotion and expansion of East Asia's RE sector is part of a new industrial policy paradigm and new developmentalism, and what the study of East Asian policies on promoting renewable energy can tell us about the region's broader approach to low carbon development.

It is argued that East Asian states will strongly support the development of their RE sectors over forthcoming years, albeit in quite different ways. The March/April 2011 Fukushima incident in Japan and the subsequent re-evaluation of nuclear power in East Asia have conferred even greater attention to renewables in delivering safe and sustainable energy from a very long-term strategic perspective (Len 2011), albeit with certain scale, environmental and other constraints taken into account when considering significant RE sector expansion. While the promotion of renewable energy is an integral element of East Asia's recent macro-development plans and new developmentalism generally, these same plans and certain associated policies suggest that East Asian states will simultaneously continue to significantly promote high carbon and ecologically damaging industrial activities. This undermines the low carbon credentials of East Asia's new developmentalism. Nevertheless, it is argued that like all other nations East Asian states have only just embarked on what is likely to prove a very long transition to low carbon development, and that the notable recent and future expected expansion of the region's RE sector constitutes a helpful initial contribution to the process.

The emergence of East Asia's 'new developmentalism'

To understand the expansion of East Asia's renewable energy sector and the institutional contexts of its RE policies, it is essential to consider the role played by the state in the region's dynamic economic development generally. With roots in 'late industrialisation' theory and the German Historic School (Gerschenkron 1962; Weber 1947), the developmental state concept was first applied to Japan (Johnson 1982) and thereafter to other economies in the region, especially South Korea, Taiwan and Singapore (Amsden 1989; Gold 1986; Low 2001; Rodan 1989; Woo-Cumings 1999). Developmental states are fundamentally capitalist economies with three general attributes: first, *state institutions* with the capacity to strategise and plan a transformative economic development path, with institutional and planning co-ordination led by a 'pilot agency'; second, *strategic policies* to operationalise these plans and realise their transformative economic goals that change over time in accordance with developmental progress; third, *developmental partnerships* forged by the state primarily with the business sector but also with society, whereby non-state actors (mostly firms) are co-opted into transformative economic projects through various institutionalised consultative mechanisms.

Meanwhile, the socialist market concept has been applied to the communist states of China and Vietnam, where the state too has strategic economic planning institutions and policies but where the state plays a substantial interventionist role in the economy through more direct control of markets and the means of production, yet still harnessing the power of private enterprise to help realise strategic economic objectives albeit through less consultative means than in developmental states. The state's capacity to bring about prosperity-generating transformative change is a core issue that links both concepts, and the notion of 'state capacity' has been deployed more commonly to examine the positive role played by governments in economic management (Weiss 1998).

Industrial policy principally concerns the promotion of industries deemed essential to the nation's long-term economic welfare and prosperity, and has been an essential component of the macro-development plans of East Asia's developmental states and socialist market economies that, in turn, comprise broader strategic blueprints for future economic development. This is at least the intent and role of most industrial policy, although it has often been exploited to serve the political machinations of 'predatory state' decision makers and leaders in East Asia, e.g., Marcos in the Philippines. Japan was the first East Asian country to practise industrial policy and specifically refer to it by name, *sangyo seisaku* (Chang 2010; Johnson 1982). Today, Japan's Ministry of Economy, Trade and Industry defines the main purpose of industrial policy as 'the act of policy intervention to cope with market failures in resource distribution under the pricing system', and to 'use various measures to influence resource distributions among industries and to control, restrain and promote certain economic activities of private sector companies' (METI 2010a: 2). According to Chang (2010), industrial policy measures have typically included: various types of tax incentive; state co-ordination of complementary and competing investments; promotion of infant industries to attaining competitive scale economies; regulation of technology imports and foreign direct investment; provision of venture capital finance to help incubate new technologies; and export promotion measures.

Theoretically, the rationale for both industrial policy and state support of the RE sector is mainly premised on public good and externality arguments. While the market mechanism can account for the 'private' and purely price-determined costs and benefits arising from economic transactions, it can fail to capture their wider 'social' costs and benefits (therein negative and positive externalities respectively) that affect society. Public actions undertaken by the state are thus required in such circumstances to both minimise negative externalities and maximise positive externalities. For example, state policies and investments in the expansion of renewable energy installations and the development of new RE technologies will reduce carbon emissions, lead to cleaner air, and mitigate society's supply risk dependency on exhaustible fossil fuel resources. These state actions constitute the provision of public goods where the market (i.e., private enterprise) alone is unable to sufficiently deliver these welfare-enhancing outcomes, and more broadly placing the economy and society on track to a low carbon future.

Left purely to the market, RE sectors would not become adequately commercialised to compete on price with fossil fuel generated energy. Strategic industry theory has provided further intellectual basis for industrial policy, whereby the state contributes to covering proportionately high initial capital costs (e.g., infrastructure, new technology research) to help commercialise infant industries with significant futures based on the promise of substantial long-term returns for the economy and public welfare (Lall 2003; Rodrik 2004; Schmitz 2007).

However, globalisation, ascendant neo-liberalism and socio-technological change created pressures and new conditions on both developmental state and industrial policy practice in East Asia from the 1980s and 1990s onwards. The combination of globalising forces, technological developments and strengthening neo-liberal orthodoxy made national economic spaces more permeable through the growth of transnational systemic linkages (e.g., production networks and supply chains), and more liberalised and integrated national markets. This has presented certain challenges for governments when defining and controlling national development projects at both the macro-economy and industry levels. Moreover, stronger adopted neo-liberal oriented policies in East Asia led to less proactive state economic management and a corresponding greater faith placed in market mechanisms (Radice 2008). The concurrent rise in the influence of business and civil society over economic policy formation also compelled these governments to adapt and re-balance their developmental partnerships accordingly (Holden & Derneritt 2008; Lee & Park 2009; Pirie 2008; Weiss 2010).

As the 2000s progressed, however, the increasingly evident risks associated with globalisation and neo-liberalism (e.g., financial crises principally caused by deregulation, commodity price volatility), combined with emergent global challenges of climate change and acute energy security predicaments reinvigorated state capacity in East Asia and elsewhere, or at least refocused state capacity in response to these risks (Gills 2010; Weiss 2010). This does not imply negating market forces to deliver optimal welfare outcomes, rather a stronger institutionalisation of market order where states seek to make markets work better rather than supplant them (Khan & Christiansen 2010; Mok & Yep 2008). This forms an important underlying basis of East Asia's new developmentalism, and a new evolutionary phase of state capacity and industrial policy in East Asia (Dittmer 2007; Hayashi 2010; Holden & Derneritt 2008; Lim 2010; Stubbs 2009).

Given that socialist market states such as China and Vietnam have continued to devise their development plans more or less as usual through this period, this change especially applies to the region's developmental states, where new state institutions have been formed and new strategic economic plans devised to pursue new transformative goals in the context of the aforementioned risks and challenges, and hence often strongly linked to low carbon development. East Asia's socialist market economies are also incorporating similar goals into their most recent strategic plans as is later discussed, the renewable energy sector playing a notable role. Some state capacity analysts have made a case for the resilience of East Asian developmental statism based on the embeddedness of its formal institutions, and socio-cultural practices and norms, and thus also contend that developmental states have simply evolved institutionally, technically and relationally in response to changing social, economic and political conditions (Lee & Park 2009; Stubbs 2009). In sum, East Asia's new developmentalism is based on adapted technical and relational methods, and refocused (and in many cases re-invigorated) state capacity set on achieving low carbon development goals and other new transformative objectives.

Largely due to globalisation, industrial policy itself is far less focused on protecting domestic firms from foreign competition and more on developing locally embedded, high value-added activity with particular emphasis on industry or technology cluster formation, where 'clustering' may be spatially concentrated in certain geographic zones. Multinational enterprises tend to anchor their core activities in such zones, which may be considered locational centres of competitive advantage due to infrastructural, human capital, supply chain logistical and other kinds of asset situated here (Dicken 2011). Clustering may also concern closely related and mutually reinforcing industry and technology based activities and competences – this often arising too in particular geographic zones – and it is common for states to facilitate both aspects of cluster development.

The RE sector is playing a significant role in East Asia's new industrial policy and new developmentalism for three main reasons. First, it forms an integral element of low carbon development by providing cleaner, more sustainable energy systems. Second, it is in the green energy sector that all East Asian countries possess a notable level of industrial capability and realistic potential, which cannot be said for advanced energy efficiency technologies, fuel cell technology or alternative energy vehicles where mainly speaking only the most developed parts of the East Asian region enjoy noteworthy strengths. Third, and with reference to Kondratiev long-wave development cycle theory,[2] renewable energies are positioned on various new high-technology frontiers as discussed in the following section, e.g., wind energy's technology cluster links with aerospace and nanotechnology. Green technology may itself be considered a contender on which the next long-wave cycle of advanced development is based. Many RE sub-sectors may still be considered infant industries or at least having 'stunted childhoods' due to past lack of investment. They are certainly young in a commercial sense (except large hydro) compared to the centenarian oil and gas industries, and thus public good and strategic industry theory arguments are relevant in terms of state support for raising RE industry scale capacities to commercially viable levels.

Thinking of renewable energy sectors as 'industries' may appear somewhat oxymoronic due to the strong connotations that 'industry' has with energy-intensive and pollutive forms of material-based productive activity. Certainly, most mainstream industries (e.g., textiles, steel and automobiles) remain firmly associated with high carbon energy industrialisation and economic 'modernisation' per se, whereby nature's power and resources are subjugated to the desire for material outputs that 'satisfy' human demands. Thus, the extent to which East Asia's 'new developmentalism' represents a new post-modernist alignment of economic development and ecological balance is a key subject for debate.

Developments in renewable energy globally and in East Asia

Renewable energies are derived from replenishable natural processes, sources or phenomena, such as wind, solar, geothermal, hydropower, ocean, biomass and biofuels. Their ability to produce clean energy – whether for electricity generation, thermal heat, or motive power – makes renewables an essential element of low carbon development and climate change strategies. For nations depending heavily on imported energy fuels (e.g., Japan, South Korea), renewables have the added advantage of being inherently indigenous energy sources, thus helping mitigate foreign supply risks in addition to environmental risks.

Over 2005 to 2011, worldwide renewable energy power capacity increased from 931 gigawatts (GW) to 1,360GW and from 15.6 percent of total world electricity generation to 20.3 percent (Table 23.1). In terms of global primary energy consumption (electricity, motive power, thermal heat), renewables accounted for 16.7 percent of the total in 2010, with traditional biomass being the most prominent RE sector (10.0 percent of the global total) while nuclear power accounted for 2.8 percent and fossil fuels 81.2 percent. From 2008 to 2011, renewables also accounted for half of all investment in new energy generation capacity worldwide (REN21 2013).

Table 23.1 shows that wind and solar photovoltaic (PV) have been the fastest growing and now most significant new RE sectors, although the long established hydroelectric sector remains by far the most dominant, still taking over 80 percent of the renewables total worldwide. Nevertheless, wind and solar energy are now scaled-up and fully commercialised industries with the combined potential to provide more off-grid and grid-connected electricity than any other RE sector in the decades to come. The pace of techno-innovatory advances made in wind and solar energy has been staggering (Lee et al. 2009), and in many countries these two sectors are competing on near parity terms with fossil fuels in power generation (REN21 2013). They also have cluster linkages to various other high-tech sectors. In wind energy's case, this includes nanotechnology (composite materials), aerospace (aerodynamics and advanced materials), energy storage electronics (advanced batteries) and meteorology software. Solar energy meanwhile has close links with advanced materials and chemicals, satellite guidance technology, electronics, lasers, and high-grade concentrators used in space technology applications. Bioenergy (comprising biomass and biofuels) is another important sector both for electricity generation as well as transportation and thermal energy. Many renewables remain, though, relatively small scale. For instance, global installed capacity for geothermal was only 11.2GW in 2011, around half that of China's Three Gorges Dam plant. Ocean energy (wave and tidal) is still a largely experimental micro-sector with a global power capacity of just 0.5GW in 2011, as is concentrating solar power (CSP) with only 0.7GW total installed capacity.

There are, however, a number of fundamental constraints and contentious issues associated with renewable energies. First, the intermittent power problem means for example that on-shore wind turbines normally operate at 20 to 40 percent of their technical capacity, solar power only produces energy during daytime hours, and ocean power is dependent on wind speeds, currents and tidal movements. A second related problem is the non-dispatchable nature of RE sectors like wind, solar photovoltaic and ocean power where energy output cannot be stored but normally only used at immediate time of generation. Hydroelectric, geothermal and biomass are able to circumvent this problem through tapping into constant energy streams, thus having more preditable load factors.

Due to intermittency and non-dispatchable issues, wind, solar and ocean energy are likely to remain as supplements to more consistent and predictable forms of energy generation. A third problem concerns various environmental and sustainability predicaments facing all

Table 23.1 Renewable energy sector development: world and East Asia overview

Sector	Electricity generation (GW installed capacity, world)			Main East Asian producers, 2011 (installed capacity level, world ranking)
	2005	2011	Added	
Wind	59.0	237.7	178.7	China 62.4GW (1) Japan 2.5GW (13) Taiwan 0.6GW (24)
Solar photovoltaic	5.4	70.0	64.6	Japan 4.9GW (3) China 3.1GW (6) South Korea 0.8GW (12)
Concentrating solar power	0.4	1.8	1.4	—
Hydroelectric	816.0	970.0	154.0	China 213.0GW (1) Japan 27.2GW (9)
Geothermal	9.3	11.2	1.9	Philippines 1.9GW (2) Indonesia 1.2GW (3) Japan 0.5GW (8)
Biomass	44.0	72.0	28.0	China 4.0GW (3) Thailand 1.3GW (9)
Wave	0.0	0.0	0.0	Sector still in experimental stage Japan one of the world's technological leaders
Tidal	0.3	0.5	0.2	South Korea 0.25GW (1)
Renewables sector total	931 (15.6% total)	1,320 (18.0% total)	389	China 282GW (1) Japan 39GW (7)
Nuclear sector	835	885	50	
Fossil fuel sector	4,201	4,599	398	
All sectors	5,967	6,804		
Bioethanol	33bn litres	86bn litres		China 2.1bn (3); Thailand 0.4bn (8)
Biodiesel	4bn litres	19bn litres		Indonesia 0.7bn (8); Thailand 0.6bn (9)

Source: Energy Information Administration (2010); EPIA (2012); GWEC (2012); IEA (2014); REEEP (2014); REN21 (2006, 2010, 2013).

renewables, such as acute ecological problems caused by large hydroelectric dams, emissions from biomass and biofuel combustion, visibility and noise nuisances related to wind farms, spatial scale constraints (especially for wind and solar farms), and the dependency of certain sectors on finite natural materials, e.g., lithium for solar panels, rare earth minerals for wind turbine magnets. Although such constraints and issues are becoming increasingly apparent as renewables are scaled-up, rapid technological advances in many RE sectors are simultaneously providing solutions to them. For instance, solar PV panel prices are currently a fraction of what they were just a few years ago and their power generation costs have also fallen drastically.

Most renewable energy sectors have grown faster in East Asia than anywhere else worldwide, and China deserves special attention in this regard. By 2011 it had developed the world's largest capacity for renewable electric power at 282GW, the United States coming a distant second at 147GW and Japan seventh with 39GW (REN21 2013). Over recent years China has been responsible for around 40 percent of new additions to global renewable energy capacity. In 2010, China overtook the US as having the world's largest installed capacity for wind energy. Meanwhile Japan is positioned third worldwide on installed solar PV capacity and is a top 10 producer for biomass, hydroelectric and geothermal. The Philippines and Indonesia hold second and third global positions respectively behind the US on geothermal, while South Korea has emerged as a world leader in tidal energy and an emerging key player in solar PV. Furthermore, there has been a marked shift in the manufacture of RE equipment to East Asia, and particularly to China, which by the early 2010s was producing 40 percent of the world's solar PV products, 35 percent of all wind turbines and around 80 percent of solar hot water collectors. Japanese companies remained technological leaders in wind, solar PV and bioenergy sectors.

Renewable energy policy in East Asia

There are now well over a hundred nations that pursue renewable energy policies and those from East Asia have been among the most active (REN21 2013). While RE policy in East Asia is mostly focused on business and industry – and hence its primary classification as industrial policy – it can extend to society generally by involving households and communities as both grid-connected and off-grid renewable energy producers.

Japan was the first country from the region to initiate a discernible RE policy from the early 1970s onwards, mainly in response to the 1973/74 oil crisis. Other nations from the region followed later that decade and into the early 1980s: the Philippines in 1978, Singapore (1979), Malaysia (1981), and South Korea (1985). China, Indonesia and Thailand first developed their RE policies in the 1990s, and Vietnam from 2001 (IEA 2014; REEEP 2014). RE policy instruments can be categorised as follows:

- *Regulatory mandates* establish legally binding requirements on firms to undertake particular actions. For example, renewable portfolio standards (RPS, which have been gradually introduced into East Asia) and ad hoc mandatory obligations oblige energy producers to source certain amounts of their electricity generation from renewable energies, and mandatory obligations may also work on a similar principle in the construction and biofuels industries, or some other stipulation.
- *Direct financial support instruments* comprise state subsidies, grants, loans, and capital investment in RE sector plants and infrastructure. Subsidies and grants have remained a particularly important feature of RE policy for most East Asian nations.
- *Market-based instruments* adapt or use the market mechanism to provide a variety of different financial incentive measures. These include competitive bidding, tradable permits or certificates and various kinds of tax incentives. Feed-in-tariff (FIT) systems have proved increasingly popular, whereby the government pays premium tariff rates to normally small-scale suppliers of renewable energy under long-term contracts.

These policy instruments also form an essential part of increasingly substantive and coherent *strategic planning* on RE sector development in East Asia. In the industrial policy context, governments have over time promoted particular renewables over others or adopted a broad multi-sector approach, where renewables in general have been selected for promotion. Development targets are normally set in terms of actual installed power capacity, percentage share of total energy production, or market size or share. While East Asian nations do not possess the world's highest level targets on renewables' share of the total energy mix, their level of strategic ambition is very high when compared to similar income-level countries worldwide (REN21 2013).

South Korea and Taiwan have set the broadest range of RE sector-specific targets, followed by China, Indonesia, the Philippines and Thailand (see Tables 23.2 and 23.3). All nations (except Singapore) have set such targets for at least one or two specific sectors as well as an overall RE sector target in terms of power capacity or share of the energy mix. Japan and South Korea failed to meet more or less all its past targets during the 2000s. While China well exceeded its wind and solar energy targets over the same time, it failed to realise its general 11th FYP non-fossil energy target of 10 percent of primary energy consumption by 2010, achieving only 8.4 percent, and this including nuclear.

Up to the mid-2000s, RE policy in East Asia had developed in a somewhat piecemeal and ad hoc manner but greater coherence came after governments in the region introduced their strategic master plans specifically for renewables, namely:

- China's Medium and Long-Term Development Plan for Renewable Energy to 2020
- Japan's Cool Earth Energy Innovative Technology Plan to 2050
- Malaysia's Renewable Energy Development – 9th Malaysia Plan 2006–2010, and thereafter the National Renewable Energy Policy and Action Plan – 10th Malaysia Plan 2011–2015
- Philippines' Renewable Energy Policy Framework 2003–2013
- South Korea's New Renewable Energy Medium-Term Plan 2010–2015
- Taiwan's New Energy Policy to 2025
- Thailand's Alternative Energy Development Plan 2008–2022
- Vietnam's Renewable Energy Master Plan 2011–2015 – 9th Five-Year Plan 2011–2015.

This corresponded to the more programmatic integration of strategic planning on renewables into broader macro-development plans and consequently becoming a key feature of East Asia's new developmentalism. Renewables have too become increasingly salient in East Asia's national energy strategy plans and national climate change plans. This marks a growing conflation regarding strategic planning on energy security, climate change, industrial policies and macro-development, with renewable energy being a pivotal link between all these policy domains.

Concerning the *main technical emphasis* of RE policy generally, developed nations (Japan, South Korea, Singapore) have fostered a more *technology-oriented* approach, in which the main emphasis has been on research, development and dissemination (RD&D) support to improve firms' technological and innovatory competitiveness, and the technical efficacies of RE installations. This is consistent with the technology cluster approach – consistent with past developmental state practice – where the state has promoted the development of core 'new energy' technologies to be diffused through multiple industrial sectors, such as Japan and South Korea's investments in fuel cells for the automotive and ICT industries. The development of energy efficiency and saving technologies has also been strongly prioritised by the countries. In sum, this may be considered as much an innovation policy as an industrial policy. The region's developing nations (China and most of Southeast Asia) have, though, adopted a more *installed capacity-oriented* RE policy, in which realising production or power generation based goals are the priority (see Table 23.2). These countries generally lack new technology development capacity, although China, Malaysia and Thailand are increasingly looking to strengthen their RE sector innovative capabilities.

Table 23.2 Renewable energy policy in East Asia: overview

| | General policy instruments | | | | | | | | Strategic planning | | | | | | | | | | | | | | |
| | Regulatory mandates | | Direct financial support | | Market based | | | | Target setting | | | | | | | | | Strategic development plans | | | Main technical emphasis | |
	Renewable portfolio standards (RPS)	Mandatory obligations (e.g. product level)	Subsidies, grants, loans	Capital investment (infrastructure, etc.)	Competitive bidding	Tradeable permits or certificates	Feed-in tariffs (FiT)	Tax incentives	Solar	Wind	Hydro (large or general)	Hydro (small)	Geothermal	Biomass	Biofuels	Ocean (wave, tidal)	Waste-to-energy	General all sectors target	Renewable energy master plan (all sectors)	Sector-specific renewable energy plans	Technology oriented – strong emphasis on R&D	Installed capacity oriented – strong emphasis on expanding installed RE generation capacity
China	●	●	●	●	●		●	●	●	●	●	●		●	●	●	●	●	●	●	●	
Indonesia	●	●	●	●		●	●	●	●	●			●	●	●			●	●	●	●	
Japan	●		●	●		●	·	·	●									●	●	●	●	
Malaysia	●		●	●			●	●	●		·							●		·		
Philippines	●	●	●		●	●	●	●	·	●	●	●	●	●	●	·		●	●	●	●	
Singapore		●	●				●	●										●		·		
South Korea	●	●	●	●	●		●	●	●	●								●	●	·	●	
Thailand	●		●				●	●	●	●	●			·				●		●	●	
Vietnam				●			●	●		·		●		●	●			●	●	●	●	

Note: Small dot indicates relatively minor significance of the factor in question, such as an FIT system that only applies to one RE sector (e.g. solar PV in Japan) but not others.

Source: IEA (2014); Martinot & Li (2010); Olz & Beerepoot (2010); REEEP (2014); REN21 (2013); author research.

Table 23.3 East Asia's renewable energy and carbon emission targets

Government	General renewables	Hydro	Wind	Solar	Biomass	Biofuels	Geothermal	Ocean	Fuel cells	Carbon emission reduction targets
Northeast Asia										
China	Non-fossil fuel, primary energy (inc. nuclear): 11.4% (2015), 15% (2020); non-fossil fuel (exc. hydro) electricity: 1% (2010), 3% (2020)	284GW (2015); 300GW (2020)	100GW on-grid, inc. 5GW offshore (2015); 150GW, inc. 30GW offshore (2020)	15GW, inc. 1GW CSP (2015); 20GW (2020 – unofficial)	30GW (2020)	5 million tonnes bioethanol used (2011–2015)				Energy consumption per GDP unit: –16% (2011–2015); CO$_2$ emission per GDP unit: –17% (2011–2015), –40 to –45% (2011–2020)
Japan	Primary energy: 10% (2020); non-hydro electricity: 1.63% (2014); inc. hydro electricity: 10% (2014)			4.8GW (2010), 14GW, 5.3m homes (2020), 53GW (2030)						15% reduction by 2020 from 2005 emission levels
Mongolia	Electricity: 20–25% (2020)									
South Korea	General energy: 13,016GWh, 2.9% (2015); 21,977GWh, 4.7% (2020); 39,517GWh, 7.7% (2030)	Large: 3,860GWh (2030); small: 1,926GWh (2030)	16,619GWh (2030), 2GW offshore (2020)		Forest: 2,628GWh (2030)	Biogas: 161GWh (2030)	2,803GWh (2030)	6,159GWh (2030)		30% reduction compared to business-as-usual scenario by 2020

continued . . .

Table 23.3 Continued

Government	General renewables	Hydro	Wind	Solar	Biomass	Biofuels	Geothermal	Ocean	Fuel cells	Carbon emission reduction targets
Taiwan	Electricity: 9,952MW, 14.8% total (2025), new: 6,600MW (2025); 12,502MW, 16.1% total (2030)	2,502MW (2030)	Onshore 1.2GW (2030) offshore 3GW (2030)	3,100MW (2030)	1,400MW (2030)		200MW (2030)	600MW (2030)	500MW (2030)	Return to 2005 emission level (2020), return to 2000 level (2025)
Southeast Asia										
Indonesia	Primary energy: 25% (2025); Electricity: 15% (2025)	1.3GW new (2015), 2GW (2025)	300MW (2014)	2GW (2014)	400MW new (2015), 810MW (2025)	Transportation: 5% (2025); biodiesel: 20% diesel (2025); bioethanol: 15% gasoline (2025)	12.6GW (2025)			26% from business-as-usual level by 2020 based on domestic funding, or 41% by 2020 based on international funding (US$18.2 billion)

Country						
Malaysia	Primary energy: 11% (2020), 14% (2030), 36% (2050) Electricity: 15,236.3MW (2030, triple 2010 capacity level)	8,729	1,250MW (2020)	1,065MW (2020)	1MW (2030)	16% from projected 2020 business-as-usual level
Philippines		2,378MW (2030)	285MW (2030)	306.7MW (2030)	3,467MW (2030)	
Singapore		1MW (2030)			70.5MW (2030)	
Thailand	Primary energy: 20.3% (2022), 14.1% electricity (2022)	1.2GW (2022)	PV: 2GW (2022); thermal: 100 ktoe (2022)	3,630MW (2022); thermal: 8,200 ktoe (2022)	Biogas: 600MW; bioethanol: 9m ltrs/day (2022); biodiesel: 6m ltrs/day (2022)	30% from 2009 level by 2020, 42 million tonnes CO_2 equivalent in 2020
Vietnam	Primary energy: 5% (2020), 8% (2025), 11% (2050); electricity: 5% (2020), 241MW p/year average new (2006–2015), 160MW p/year (2016–2025), 4,050MW new (2025)	1,608MW (2022)			Various for bioethanol and biodiesel (2015, 2020)	

Source: Energy Information Administration (2010); EPIA (2012); GWEC (2012); IEA (2014); Olz & Beerepoot (2010); REEP (2014); REN21 (2006, 2010, 2013).

Case studies on East Asia's new developmentalism

Introduction

In this section, I examine how and to what extent renewable energy industries are forming an integral part of the recent macro-development plans of East Asian states, these plans being the main programmatic basis of the region's new developmentalism. This will provide useful insights into the links and relationships between evolving forms of state capacity, new industrial policy and strategic thinking on low carbon development in the region, as well as further analysis on the different approaches taken by East Asian states on promoting the RE sector. Four case study nations have been chosen for this purpose – Japan, China, South Korea and Singapore – comprising a representative sample from the region in terms of development/income level, size, energy security predicaments, and political economy profile. Furthermore, these nations constitute almost 90 percent of the East Asian economy in gross domestic product terms, and hence also well represent the region from an economic significance perspective.

Japan

Japan was East Asia's pioneer developmental state (Johnson 1982). During the 1990s, Japan's industrial policies became less focused on specific sectors and more on macro-structural reform and developing generic technology clusters, this accordingly shaping the technology–oriented nature of Japan's RE policy (METI 2010a). Into the 2000s, Japan pursued in METI's own words a 'productivity-oriented economic policy that was based on excessive market fundamentalism and was overly tilted toward the supply side' (METI 2010b: 2), where industrial policy both lacked substance and had limited impact, and in which no coherent or lasting macro-development strategy was devised. However, most recently a new developmental approach was introduced in 2010 by the Japanese government's New Growth Strategy (NGS) that 'aims to achieve economic growth by turning the problems faced by the economy and society into opportunities for creating new demand and employment' (ibid.: 2).

A core element of the NGS strategy is to strengthen Japan's industrial-technological capabilities in seven strategic areas up to 2030, environment and energy ('green innovation') being the first listed, the other six being medical/health care, finance, science and technology nation, tourism/local revitalisation, human resources, and Asian regional economic integration. These are underpinned by 21 national strategic projects, one of which under 'green innovation' is the expansion of the renewables sector into a JPY10 trillion (US$110 billion) market by 2020. The NGS is augmented by the Industrial Structure Vision (ISV) that sets broad aspirational objectives for diversifying Japan's industrial base towards a multi-sector low carbon economy that includes a relatively prominent role for renewables (METI 2010c). It also incorporates the installed capacity target on solar PV (14GW by 2020, 53GW by 2030) from the aforementioned Cool Earth Energy Innovative Technology Plan.

The main goal of the NGS is the 'creation of a low-carbon society through a comprehensive policy package including new systems design, systems changes, new regulations, and regulatory reform, and to support the rapid spread and expansion of environmental technologies and products' (METI 2010b: 20). This transformation would include 'measures to support the spread and expansion of renewable energies (solar, wind, small-scale hydroelectric, biomass, geothermal, etc.) by expanding the electric power feed-in tariff system' (ibid.: 21–22). More generally, NGS employs other policy methods and measures such as public–private partnerships and consultation mechanisms, promoting corporate and industrial clusters in strategic technology areas, establishing

Japanese industrial and technology standards as international ones in 'energy management', and the construction of a new research system based on the industrial-academic-government research complexes. The NGS's overarching low carbon development goal was to reduce worldwide greenhouse gas emissions by at least 1.3 billion tons of CO_2 equivalent by 2030 using Japanese technology: the global scope target derives from the technological contributions made by Japanese companies worldwide, and reflects the aforementioned technological and innovatory leadership of Japanese firms in many renewable and other green energy sectors. It could be argued that renewable energy is overshadowed in the NGS to some degree by other green energy industries, especially alternative energy vehicles and fuel cells. Up until the 2011 Fukushima disaster, the nuclear power sector was additionally afforded high priority under the environment and energy strategic area, although this is now under re-evaluation.

China

Although less centrally planned and directed than before, China's Five-Year Plans (FYPs) are still devised largely through a complex intra-state process where private and civil sector agencies have limited influence. Managing China's increasingly acute energy security challenges has become a central theme of national strategic planning. Concurrently, fostering sustainable energy systems and economic development has also been afforded greater priority, as first substantively operationalised in the 11th FYP (2006–2010) and continued in the current 12th FYP (2011–2015), its other overarching aims being to achieve more equitable income growth, promote domestic consumption, improve social infrastructures and safety nets, and strengthen China's innovatory capabilities (NDRC 2011). Generally speaking, there is growing acknowledgement in China of the imperative to find low carbon solutions to the country's many development challenges (CCICED 2010; Wang & Watson 2009).

The 11th FYP incorporated China's Renewable Energy Development Targets set at the start of the plan period in 2006 of increasing RE's contribution of primary energy consumption to 15 percent by 2020 (up from 7.5 percent achieved by 2005), with various sector-specific targets set, for example creating 30 new large-scale wind farms by 2010. Meanwhile, the Medium and Long-Term Development Plan for Renewable Energy launched in 2007 committed US$263 billion of public investment in RE sectors up to 2020 while simultaneously making numerous references to harnessing the market in achieving its goals (NDRC 2007a). Later, the 12th FYP (2011–2015) included a RMB4 trillion (US$610 billion) funded programme to promote seven strategic emerging industries (SEIs) for 'clean' development and a new industry base, namely: new generation ICT, energy-saving and environment protection, new energy, biotechnology, high-end equipment, new materials and alternative energy cars. These are intended to form the new future backbone of the Chinese economy with the aim of increasing their GDP share from the current 5 percent to 8 percent by 2015 and 15 percent by 2020 (NDRC 2011).

Renewables are specified as a key component of the 'new energy' SEI sector that in addition includes nuclear power (40GW additional capacity by 2015) and large hydro (50 percent increase in installed capacity by 2015). Under the SEI programme, China plans to enhance its new technological and innovation capabilities in wind energy, solar PV, fuel cells and electric vehicles as part of a broader strategy of moving China from 'world factory' to an 'innovative hub' economy. The persistent fixation with expanding installed generation capacity regarding renewables notwithstanding, there is also a clear indication in the current FYP that China intends also to develop over time a more technology-oriented RE policy. The 12th FYP also set China's current sector-specific targets on renewables (see Table 23.3).

South Korea

South Korea was one of East Asia's most ardent developmental states until it dismantled its FYP framework in 1993 and thereafter shifted towards a more neo-liberal economic ideology and policy approach which persisted throughout most of the 2000s (Lim 2010). However, a return to developmentalism came with the introduction of the 'Green Growth Strategy' (GGS) macro-development programme in 2008 under the Lee Myun-bak administration. The GGS is a medium and long-term strategic plan to create a low carbon society and green growth economy (Jung & Ahn 2010; KEITI 2009). Its primary transformative goal is to make South Korea the world's 'seventh green power' by 2020, and the 'fifth green power' by 2050, to be achieved through new FYPs co-ordinated by the newly formed Presidential Committee on Green Growth, comprising governmental, business and civil societal representatives. The plans are based on three main strategic approaches and 10 strategic projects implemented across them. Renewables are included under the first strategic approach (*measures for climate change and securing energy independence*) and first strategic project (*reducing fossil fuel use*), where specific targets are set of increasing renewable and new energy supply use from 2.7 percent in 2009 to 3.8 percent by 2013, and then a target extending beyond the five-year period of 6.1 percent by 2020, and 11.5 percent by 2030 (Kwon 2010; Moon 2010).

GGS targets and other objectives on renewable energy are co-ordinated with South Korea's National Energy Plan 2008–2030. Renewable energy industries are also noted under the second strategic approach (*creation of new engines for economic growth*) and the associated strategic project of developing green technologies, specific reference made here to solar, biomass and fuel cell sub-sectors from the 27 core green technologies listed overall, as well as strengthening South Korea's new energy RD&D capacity over the plan period.

The core strategies and transformative goals of the GGS provide the basis for the Ministry of Knowledge Economy's (MKE) new industrial policy, where 17 strategic sectors are identified for development, renewable energy placed under the 'green technology' cluster category. The third and last strategic approach (*improving life quality and Korea's international status*) envisions South Korea to become an international leader on green growth and low carbon development. The country's environmental groups have, though, criticised the programme for lacking ambitious objectives on renewable energy development and including nuclear energy under 'green growth'.[3] Moon (2010) additionally is critical of its primary focus on economic growth and subsequent neglect of green society.

Singapore

The Singapore state has maintained high levels of developmental strategy making and has fostered close developmental partnerships with foreign MNEs in particular given the vital importance of transnational capital in the economy. In many ways, Singapore's government has been at the forefront of new developmentalist thinking in East Asia, introducing its knowledge-based economy strategy in 1999 devised to make the city-state a stronger innovation hub, the centrepiece being the new Biopolis biotechnology cluster. During the 2000s, Singapore's developmentalism gradually focused on low carbon economy objectives, although renewables did not enjoy the same high profile as in many other East Asian countries. A key reason for this is the inherent resource and spatial constraints facing small nations like Singapore in developing a broad RE sector.

The aggregation of numerous micro-RE projects has been a more viable approach for the city-state, especially the integration of RE technologies in building construction such as rooftop solar installations. Singapore's advancing innovatory capabilities have furthermore led renewables to be subsumed into wider green technology and industry clusters, and the technology-oriented approach of its RE policy has always been strong (IEA 2014; REEEP 2014). In 2003, the Singapore Green Plan 2012 was launched, a 10-year strategy that included measures on directly promoting renewables, albeit not that substantially. A few years before this plan ran its full course, the government introduced its new Sustainable Singapore Blueprint (SSB) macro-development plan, its primary goal of reducing the energy intensity of GDP by 35 percent from 2005 levels to 2030.

The SSB was formulated by an Inter-Ministerial Committee on Sustainable Development in broad consultation with public and private/civil society representatives but has only made publicly available the plan's general guiding principles and priority areas with scant detail on associated policy instruments. Moreover, its objectives are largely to be achieved through resource efficiency measures rather than expanding installed RE capacities. The SSB's relatively narrow portfolio on renewables is mostly based on programmatic integration with the Comprehensive Blueprint for Clean Energy (CBCE) launched two years earlier in 2007. The CBCE is a S$350 million (US$200 million approx.) multi-sector industrial and innovation strategy founded on the 'alternative energy' strategic area of Singapore's Economic Development Board's current industrial master plan, there being 20 such areas in total. The main aim of the CBCE is to make Singapore a global hub for developing and testing new clean energy technologies, this subsequently providing a new area of exports for the economy, contributing S$1.7 billion to Singapore's GDP (0.75 percent). Solar power is conferred special priority given Singapore's tropical sunbelt position but wind, biomass, fuel cells, energy efficiency, and carbon services are other clean energy technologies targeted for development. Noteworthy initiatives include the Solar Capability Scheme (SCS), which supports new innovative design and integration of solar panels in building construction, and the new Solar Energy Research Institute of Singapore (SERIS), which draws on the city-state's research strengths in nano-science, silicon thin-film technology and semiconductor processing.

From old to new developmentalism?

While it is evident that the goal or low or lower carbon development forms a core aim of East Asia's new developmentalism, it is also vies with other strategic imperatives such as strengthening national economic competitiveness. Furthermore, many macro-development plans contain programmes or promote industries that are inherently high carbon based and still rooted in the modernist 'old developmentalism' of the past. For example, the expansion of physical infrastructures, export-oriented manufacturing, and an emphasis on state direct financial support remain defining characteristics of China's 12th FYP, while its promotion of higher levels of domestic consumption somewhat contradict the plan's low carbon objectives. It should too be noted that China's energy policy is expanding on all fronts at relatively similar paces. For example, over the 12th FYP period (2011–2015) the Chinese government expects the number of coal-fired power stations to increase by 33 percent, a growth in capacity equivalent to the total current EU capacity in this sector. At the same time, the Plan aims to reduce the nation's electricity generation dependence on coal from the present 70 percent to 62 percent, meaning a comparative faster expansion of other sectors such as wind energy. Elsewhere in the region, conventional energy and chemicals are among the 20 strategic areas that comprise Singapore's latest industrial master plan, which includes the further expansion of the city-state's already huge petrochemical complex and the construction of a large new liquefied natural gas terminal. In South Korea, the GGS must simultaneously contend with MKE industrial policy for upgrading the traditional 'flagship industries', including the energy-intensive sectors of shipbuilding and steel.[4]

East Asia's national energy strategies and systems are still generally very dependent on securing fossil fuel resources. This is indicative of the broader problem of how many of East Asia's core industrial structures, policies and practices remain high carbon development geared. In Indonesia, for instance, the continued heavy subsidisation of fossil fuel prices makes it difficult for the nation's geothermal sector to seriously compete in the electricity generation market. Furthermore, many countries persist with old industrial policies involving mass-scale ecological damage (e.g. Indonesia and Malaysia's palm oil industries) whilst rolling out low carbon development strategies (Gunningham 2011). Despite the 2011 Fukushima disaster, many East Asian countries may still maintain their ambitious plans on nuclear power sector development, and moreover include nuclear under the policy rubric of green or clean energy. In sum, East Asia's 'modernist' industrialised development is likely to remain entrenched for some considerable time given the enormity of the structural change required to establish the broad foundations of low carbon development.

East Asia's developing countries especially face difficult challenges moving toward low carbon development given both the high proportionate GDP costs of replacing old or rudimentary equipment and infrastructure, and their relative lack of indigenous technological and innovatory capacity. Furthermore, there are domestic political pressures to meet more immediate socioeconomic needs (e.g., poverty alleviation, provision of basic welfare and utility services) than to prioritise environmental-related goals. This being said, socioeconomic and environmental problems are increasingly conflating in developing country regions. Many of East Asia's major cities (most notably China's) are subject to acute levels of pollution, causing chronic health problems on a mass scale and palpably deteriorating societal welfare. Even in relatively poor and authoritarian states there has been mounting civil unrest regarding pollution and a growing acknowledgement of problems caused by pursuing economic growth whatever the costs.

Conclusion

By most calculations, East Asia is having more of an impact on global energy security and climate change than any other region, and the magnitude of this impact is widely expected to grow in years and decades to come. The expansion of renewable energy practice in East Asia is vital to addressing energy security and climate change challenges faced by the region as well as at the global level. This chapter has examined the development of renewable energy sectors and policy in East Asian countries, and how this has become the basis of a formative 'new developmentalism', where the transformative objectives of low carbon development have been central to revitalised forms of state capacity. As has been shown, renewable energy and green energy more generally have become the focus of an emerging 'clean industry policy' paradigm that is embedded in recent strategic macro-development plans, such as Japan's New Growth Strategy, the Sustainable Singapore Blueprint, South Korea's Green Growth Strategy and the Strategic Emerging Industries component of China's 12th Five-Year Plan.

However, it has been noted that low carbon development is not the only main objective of East Asia's new developmentalism. Its associated macro-development plans have too included the simultaneous promotion of high carbon industrial activities, including conventional fossil fuel based sectors. Such apparent contradictions are likely to persist for some time given the structural dependencies of East Asian economies on high carbon-intensive industries for delivering material growth and prosperity, and that, like most others around the world, East Asian states have only just embarked on the very long transition to meaningful low carbon development, a process that most expect to take many decades if indeed it is to be achieved. In addition, East Asia's new developmentalism has tended generally to afford primacy to economistic objectives rather than environmentalism, to economic growth rather than sustainable development per se. This begs the question of just how far East Asia's new developmentalism marks a departure from the economic modernisation oriented policies of the past. It is perhaps better considered an intermediate concept through which *relatively lower* carbon development can be achieved over the medium term.

Moreover, achieving low carbon development requires deep social transformations, especially changing people's mindset towards creating a green society that cannot be realised solely through a top-down policy process. The 2011 Fukushima disaster may help provide greater societal and government support for renewables in East Asia in the shorter term but this cannot be assured over the long term. Nevertheless, by maintaining and improving their various forms of state capacity over time, East Asian states will be well positioned to sustain the significant recent advance of their renewable energy sectors and thereby further strengthen the low carbon development orientation of their new industrial policies, macro-development plans and strategic economic thinking. The deeper diffusion of renewable energy technologies in every element of society is also a critical socio-technical process to secure lower carbon outcomes, in East Asia and elsewhere.

Bibliography

Agency for Natural Resources and Energy/ANRE (2010) *The Strategic Energy Plan of Japan: 2010 Revision* (ARNE).

Amsden, A. (1989) *Asia's Next Giant: South Korea and Late Industrialisation* (Oxford University Press).

APCO Worldwide (2010) *China's 12th Five-Year Plan* (APCO Worldwide).

Chang, H.J. (2010) 'Industrial Policy: Can We Go Beyond an Unproductive Confrontation?', *Turkish Economic Association Discussion Paper* 2010/1.

China Council for International Co-operation on Environment and Development/CCICED (2010) *China's Pathway Towards a Low Carbon Economy* (CCICED).

Chu, Y.W. (2009) 'Eclipse or Reconfigured? South Korea's Developmental State and Challenges of the Global Knowledge Economy', *Economy and Society* 38(2), 278–303.

Dicken, P. (2011) *Global Shift* (Sage Publications).

Dittmer, L. (2007) 'The Asian Financial Crisis and the Asian Developmental State', *Asian Survey* 47(6), 829–833.

Energy Information Administration (2010) *International Energy Outlook 2010* (EIA).

European Photovoltaic Industry Association/EPIA (2012) *Global Market Outlook for Photovoltaics until 2016* (EPIA).

Gerschenkron, A. (1962) *Economic Backwardness in Historical Perspective* (Harvard University Press).

Gills, B. (Ed.) (2010) *Globalisation in Crisis* (Routledge).

Global Wind Energy Council/GWEC (2012) *Global Wind Energy Outlook 2012* (GWEC Secretariat).

Gold, T.B. (1986) *State and Society in the Taiwan Miracle* (M.E. Sharpe).

Gunningham, N. (2011) 'Energy Governance in Asia: Beyond the Market', *East Asia Forum Quarterly* 3(1), 29–30.

Hayashi, S. (2010) 'The Developmental State in the Era of Globalisation: Beyond the Northeast Asian Model of Political Economy', *Pacific Review* 23(1), 45–69.

Hayes, P. & von Hippel, D. (2006) 'Energy Security in Northeast Asia', *Global Asia* 1(1), 90–105.

Hirschl, B. (2009) 'International Renewable Energy Policy: Between Marginalisation and Initial Approaches', *Energy Policy* 37, 4407–4416.

Holden, K. & Derneritt, D. (2008) 'Democratising Science? The Politics of Promoting Biomedicine in Singapore's Developmental State', *Environment and Planning D* 26(1), 68–86.

International Energy Agency/IEA (2012) 'Global Renewable Energy: Policy and Measures Database', Paris, IEA. Available at http://www.iea.org/textbase/pm/?mode=re.

Johnson, C. (1982) *MITI and the Japanese Miracle* (Stanford University Press).

Jung, T.Y. & Ahn, J.E. (2010) 'Sowing the Seeds for Green Growth in Korea', *East Asia Forum*, 13 December.

Khan, S.R. & Christiansen, J. (Eds.) (2010) *Towards New Developmentalism: Market as Means rather than Master* (Routledge).

Korea Environmental Industry and Technology Institute/KEITI (2009) *Korea's Green Growth Vision and Eco-Innovation Policies* (KEITI).

Kwon, H.S. (2010) Research on the Way to Foster Leading New and Renewable Energy Industry on Greater Sphere Economic Area, KEEI Research Paper, 31 December, Seoul, Korea Energy Economic Institute.

Lall, S. (2003) 'Reinventing Industrial Strategy: The Role of Government Policy in Building Industrial Competitiveness', *QEH Working Paper Series* 111.

Lee. B., Iliev, I. & Preston, F. (2009) *Who Owns Our Low Carbon Future?: Intellectual Property and Energy Technologies* (Chatham House).

Lee, Y.H. & Park, T.Y. (2009) 'Civil Participation in the Making of a New Regulatory State in Korea: 1998–2008', *Korea Observer* 40(3), 461–493.

Len, C. (2011) 'Rethinking Nuclear Power After Fukushima', *East Asia Forum*, 25 March.

Lim, H. (2010) 'The Transformation of the Developmental State and Economic Reform in Korea', *Journal of Contemporary Asia* 40(2), 188–210.

Low, L. (2001) 'The Singapore Developmental State in the New Economy and Polity', *Pacific Review* 14(3), 411–441.

Martinot, E. & Li, J.F. (2010) 'Renewable Energy Policy Update for China', *Renewable Energy World*, 21 July. Available at http://www.renewableenergyworld.com/rea/news/article/2010/07/renewable-energy-policy-update-for-china.

Ministry of Economy, Trade and Industry/METI (2010a) *The History of Japan's Industrial Policies* (METI).

Ministry of Economy, Trade and Industry/METI (2010b) *The New Growth Strategy: Blueprint for Revitalising Japan* (METI).

Ministry of Economy, Trade and Industry/METI (2010c) *The Industrial Structure Vision 2010 (Outline)* (METI).

Mok, K.H. & Yep, R. (2008) 'Globalisation and State Capacity in Asia', *Pacific Review* 21(2), 109–120.

Moon, T.H. (2010) 'Green Growth Policy in the Republic of Korea: Its Promises and Pitfalls', *Korea Observer* 41(3), 379–414.

National Development and Reform Commission/NDRC (2007a) *Medium and Long-Term Development Plan for Renewable Energy in China* (NDRC).

National Development and Reform Commission/NDRC (2007b) *China's National Climate Change Programme* (NDRC).

National Development and Reform Commission/NDRC (2011) *China's 12th Five-Year Plan for Economic and Social Development* (NDRC).

Olz, S. & Beerepoot, M. (2010) *Deploying Renewables in Southeast Asia* (IEA).

Pirie, I. (2008) *The Korean Developmental State: From Dirigisme to Neo-Liberalism* (Routledge).

Radice, H. (2008) 'The Developmental State under Global Neoliberalism', *Third World Quarterly* 29(6), 1153–1174.

REN21 (2006) Renewables 2006 Global Status Report, Paris, REN21 Secretariat.

REN21 (2010) Renewables 2010 Global Status Report, Paris, REN21 Secretariat.

REN21 (2012) Renewables 2012 Global Status Report, Paris, REN21 Secretariat.

Renewable Energy and Energy Efficiency Partnership/REEEP (2012) *Policy and Regulation Review Database*, Vienna, REEEP. Available at http://www.reeep.org/9353/policy-database.htm.

Rodan, G. (1989) *The Political Economy of Singapore's Industrialisation: National State and International Capital* (Macmillan).

Rodrik, D. (2004) 'Industrial Policy for the Twenty-First Century', *KSG Working Paper Series* RWP04–047.

Schmitz, H. (2007) 'Reducing Complexity in the Industrial Policy Debate', *Development Policy Review* 25(4), 417–428.

Segal, A. (2010) 'China's Innovation Wall: Beijing's Push for Homegrown Technology', *Foreign Affairs* 28 September. Available at http://www.foreignaffairs.com/ articles/66753/adam-segal/chinas-innovation-wall?.

Sovacool, B.K. (2010) 'The Political Economy of oil and Gas in Southeast Asia: Heading Towards the Natural Resource Curse?', *Pacific Review* 23(2), 225–259.

Stubbs, R. (2009) 'What ever Happened to the East Asian Developmental State? The Unfolding Debate', *Pacific Review* 22(1), 1–22.

US–China Economic and Security Review Commission/USCESRC (2010) 2010 Report to Congress, Washington, DC, USCESRC.

Wang, T. & Watson, J. (2009) *China's Energy Transition: Pathways for Low Carbon Development* (Sussex Energy Group and Tyndall Centre for Climate Change Research).

Weber, M. (1947) *The Theory of Social and Economic Organisation* (Oxford University Press).

Weiss, L. (1998) *The Myth of the Powerless State: Governing the Economy in a Global Era* (Polity).

Weiss, L. (2010) 'The State in the Economy: Neoliberal or Neoactivist?', in J. Campbell, C. Crouch, P. Hull Kristensen, O.K. Pedersen & R. Whitley (Eds.), *Oxford Handbook of Comparative Institutional Analysis* (Oxford University Press).

Wesley, M. (Ed.) (2007) *Energy Security in Asia* (Routledge).

Wishnick, E. (2009) 'Competition and Co-operative Practices in Sino-Japanese Energy and Environmental Relations: Towards an Energy Security "Risk Community"?', *Pacific Review* 22(4), 401–428.

Woo-Cumings, M. (Ed.) (1999) *The Developmental State* (Cornell University Press).

World Wind Energy Association/WWEA (2010) World Wind Energy Report 2009, Bonn, WWEA Secretariat.

Notes

1   This chapter is a revised and updated version of an article that previously appeared in *Pacific Review* journal.

2   The economist Kondratiev theorised on how new phases of world economic development were essentially founded on new techno-industrial paradigms. For example, electricity generation in the late nineteenth century, Fordist mass production techniques in the early twentieth century, and computer and new information technologies in the late twentieth century.

3   *Korea Times*, 20 August 2008; *Korea Times*, 22 August 2008.

4   For details, see http://www.mke.go.kr/language/eng/policy/Ipolicies.jsp.

24

Explaining low carbon development in Asia

The case of China

Sarah Van Eynde, Bettina Bluemling and Hans Bruyninckx

Introduction

Asian emerging economies are among the world's most rapidly changing societies. As a result, those countries' energy demand has increased massively. In 2011, emerging economies accounted for all of the net growth of global energy consumption, with China alone accounting for 71 percent of global energy consumption growth (BP 2012). Considering their appetite for energy, Asian emerging economies, unsurprisingly, also account for the bulk of global greenhouse gas (GHG) emission growth. In 2011, global GHG emissions from fossil fuel combustion reached a record high of 31.6 gigatonnes (Gt), representing an increase of 3.2 percent over 2010 (IEA 2012). China made the largest contribution to the global increase, with its emissions rising by 720 million tonnes, or 9.3 percent, and per capita emissions approaching per capita emissions in the European Union (PBL 2012). India's emissions rose by 140 million tonnes, or 8.7 percent, thus becoming the fourth largest emitter behind China, the United States, and the European Union (IEA 2012).

In consideration of these trends, energy security and climate change concerns feature high on the political agenda of Asian emerging economies. A potential strategy to cope with both problems is so-called 'low carbon development' (LCD). It acknowledges that a balance must be achieved between environmental and development goals (Heggelund & Backer 2007). In order to do so, a number of Asian countries are taking measures to reduce the energy intensity of their economic growth and develop renewable energy sources. LCD (strategies) are mainly an instrument deployed at the national level and welcomed at a moment when international climate negotiations increasingly show signs of adopting a 'bottom–up' regime (Cai et al. 2012), thus increasing the importance of LCD to the countries' governments. However, caution is necessary. Since LCD strategies in Asian emerging economies often target the reduction in energy intensity (with continuing growth in consumption of fossil fuels), the question arises of whether 'low carbon' should be combined with 'development' when the emissions of carbon are still increasing albeit at a lower rate (Jackson 2009). Effectively reducing global emissions hence requires an overarching global framework and cooperation at a global scale since current national targets committed to or proposed in major economies tend to be not enough to achieve the

450 ppm CO_2 limit (recommended as the maximum allowable concentration of CO_2 in the atmosphere to avoid more than 2°C of global warming), regardless of good LCD intentions.

LCD appears to be a buzzword, popping up in numerous policy documents in both the developed world and developing world. Missing from the literature, however, are analyses that investigate how the concept of LCD is translated into action. In an attempt to fill that gap, this chapter describes and explores the nature of LCD strategies deployed by Asian emerging economies, exemplified by the case of China.

The remainder of this chapter is structured as follows. The first section discusses the emergence, scope and purpose of the concept of LCD and provides an overview of LCD initiatives in major Asian emerging economies. The second section elaborates on the case of China: in the first part, LCD will be explored against the backdrop of China's dependency on coal; the second part discusses nuclear energy development under the label of "clean energy"; and the third part explains the development of renewable energy sources in China.

Low carbon development in Asia

This section describes the emergence of the low carbon development (LCD) concept, elucidates its scope and purpose, and presents a number of examples of how Asian emerging economies implement the new development paradigm of LDC.

The concept of low carbon development (LCD): emergence, scope, and purpose

For the developing world, LCD is mainly interpreted as development that maintains growth, while simultaneously, reducing GHG emissions (Liu et al. 2011: 835), or more simply as 'using less carbon for growth' (Hu et al. 2011; Mulugetta & Urban 2010: 7546).[1] In light of the idea of the right to develop, the transition to a low carbon economy needs to be part of the economic development process, which includes major transformation and growth in the energy and transportation sectors and infrastructure development (Du & Zhou 2011). Those definitions clearly show that the concept of LCD has entered the lexicon of development, adding an important climate (mitigation) dimension (Bidwai 2012: 164). A number of developing and emerging economies hence have welcomed LCD as an alternative strategy to *voluntarily* reduce GHG emissions at the country, provincial or sector level, because it recognizes the right to develop as well as their growing responsibility for increasing GHG emissions (Zhang 2011).

A number of reports and academics emphasize the role of energy demand and supply in the decarbonization process of, for example, the power, industry, housing and transport sectors (Ahman et al. 2012; IIER 2011; IPCC 2011; WWF 2011; Xie et al. 2011). From a sector perspective, the energy production sector is playing a major role in the decarbonization process of (emerging) economies, as it represents the bulk of emissions (European Commission 2011; Walsh et al. 2011). Since energy consumption continues to increase rapidly, especially in emerging economies, replacing fossil fuels by low carbon technologies is of absolute importance. This replacement encompasses a technological, economic and political challenge, given that the technological capacities and systems necessary for decarbonisation need to be developed, demonstrated and commercially available for large-scale introduction in a relatively short period of time (Ahman et al. 2012). In that sense, target support to enable the R&D and market development of technologies plays an important role in guiding the design of decarbonization and has to complement current climate mitigation policies which determine the demand for such technologies (Fisher et al. 2011).

Acknowledging that targeting the reduction of GHG emissions in economic development by the development and usage of low carbon energy is only one aspect of a country's low carbon development process (Hu et al. 2011; Xie et al. 2011), this chapter will elaborate on LCD perceived as a strategy for energy saving and emission reduction in economic development, as well as on the related and required policy framework and instruments in China and other countries in the region (Cai et al. 2012; UNDP 2010; Zhuang 2007). The following questions will be addressed in the next section. What policies or initiatives do exist in Asian countries to accelerate or inhibit a shift towards LCD, i.e., to meet growth and development goals with proportionally less energy use (and associated CO_2 emissions)? What policies or initiatives do exist in Asian emerging economies to increase the share of energy from low carbon or zero carbon energy resources?

Low carbon development initiatives in Asian countries

The scale of economic growth and rising energy demand in a number of Asian emerging economies results in increasing country-based GHG emissions. Figure 24.1 presents total carbon dioxide (CO_2) emissions from the consumption of fossil fuels between 1985 and 2010 of selected Asian emerging economies. The choice to include China, India, Vietnam and Indonesia in the selection has two grounds. First, these countries' economies grew by more than 5 percent annually in terms of GDP during the last five years. Second, a comparison between the four countries sheds light on potential differences in LCD strategies.[2]

China's GDP grew by 9.4 percent on average between 2000 and 2010. In absolute numbers, China's country-based CO_2 emissions grew by 9 percent to 9.7 billion tones in 2011 and per capita emissions are approaching per capita emissions in Europe (PBL 2012; World Bank 2012). The rising per capita emissions in China are significant if we bear in mind that China accounts for around 20 percent of the global population (IEA 2012). In India, the GDP growth rate was

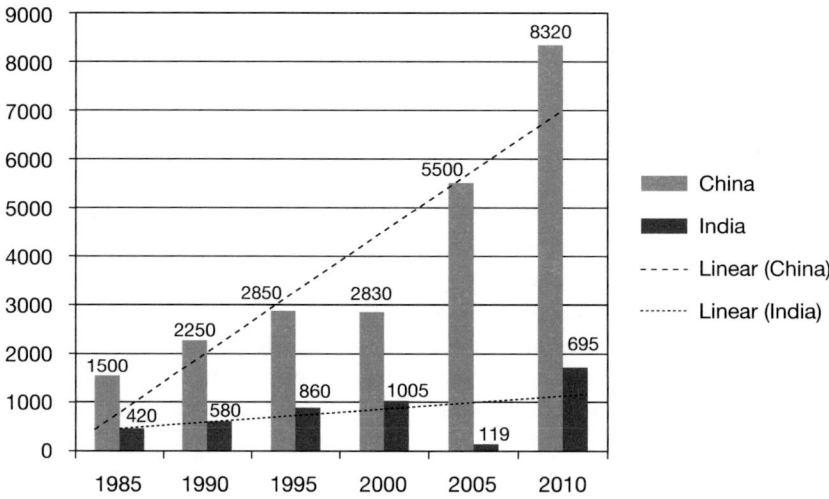

Figure 24.1 Total carbon dioxide emissions from the consumption of fossil fuel in China and India (million tonnes CO_2), 1985–2010

Source: Adapted from EIA (2012a).

about 7.4 percent on average during that same period. In absolute numbers, India's emissions grew by 6 percent to 2.0 billion tons in 2011 (PBL 2012; World Bank 2012).

While the absolute amount of GHG emissions in Indonesia and Vietnam is still minor compared to China's or India's amount (see Figure 24.1), (projected) GDP and GHG emissions growth rates in all the countries highlight the importance to decarbonize societies, above all because the necessary infrastructure for future development is under construction or will be built in the next decade(s) (Truong 2010). For example, taking into account the scale, speed and density of (future) Asian urbanization,[3] robust and sound city planning that includes decarbonization is a key challenge to policymakers (UN Habitat 2012). According to the Asian Development Bank (2012), Asia is home to more than half of the world's most polluted cities with an average per capita emissions growth rate of 97 percent versus only 18 percent for the world from 2000 to 2008.

Targeting energy in a decarbonization roadmap is of high importance in Asian emerging economies because of two reasons. First, energy demand is rapidly increasing in those countries. Second, at present, despite a number of differences in terms of energy type, the energy mix in Figure 24.2 shows that fossil fuels such as coal and oil, which are higher polluting and more carbon intensive than other energy alternatives, are currently responsible for the bulk of energy supply (ESMAP 2012). According to the 2009 WWF/Alliaz G8 Climate Change Scorecard, energy production is the major source of carbon emissions in China, accounting for 39 percent of all emissions (Ellis et al. 2009) (see Table 24.1).

Similar in all four countries, is the relatively small share of renewable energy in total energy consumption. In China and India, for example, renewable energy accounts for only 1 percent. In Vietnam and Indonesia the share of natural gas is significantly larger than in India and China. The share of hydro is the largest in Vietnam with 15 percent. If we compare the energy consumption by fuel type of those countries with the European situation (see Table 24.2), it is quite striking that oil and coal are also the main sources in the EU's total primary energy consumption. The share of nuclear and renewables, however, is significantly higher in the EU than in the Asian countries. Moreover, oil and coal also have the largest share in Japan's total primary energy consumption (see Table 24.2). Table 24.2 further shows that natural gas is the third largest energy source in Japan's energy mix, which is comparable to the Vietnamese and Indonesian situation. The share of renewables and nuclear is the same in the US and Japan. Their share of renewables is, again, relatively small.

A basic analysis of the energy mix of the Asian emerging economies clearly demonstrates that decarbonizing the power sector is an important step to be made within the LCD process. Both projected economic growth and growing GHG emission rates highlight the importance

Table 24.1 Total primary energy consumption by fuel type in China, India, Indonesia and Vietnam, 2011

Fuel type	China	India	Indonesia	Vietnam
Coal	70%	53%	30%	33%
Oil	18%	30%	43%	36%
Natural gas	4%	9%	23%	17%
Hydro	6%	5%	3%	15%
Renewables	1%	2%	1%	0%
Nuclear	1%	1%	0%	0%

Source: Adapted from BP (2012).

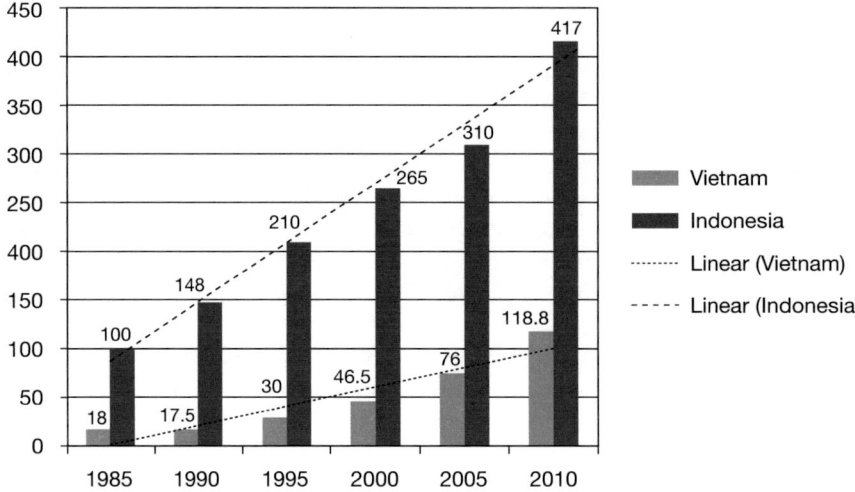

Figure 24.2 Total carbon dioxide emissions from the consumption of fossil fuel in Vietnam and Indonesia (million tonnes CO_2), 1985–2010

Source: Adapted from EIA (2012a).

Table 24.2 Total primary energy consumption by fuel type in the European Union (EU), Japan and the United States, 2011

Fuel type	EU	US	Japan
Coal	17%	22%	25%
Oil	38%	37%	42%
Natural gas	24%	28%	20%
Hydro	4%	3%	4%
Renewables	5%	2%	2%
Nuclear	12%	8%	8%

Source: Adapted from BP (2012).

of decoupling economic growth and growing emission rates in their development process. Decoupling economic growth and growing emission rates requires the implementation of climate change and technology related policies (Halsnaes & Garg 2011; Truong 2010). With regard to the power sector, low carbon policies in Asian emerging economies, that are compatible with continuing growth in energy consumption, address both the demand and supply side. Policies directed at the demand side aim to reduce energy intensity, while policies directed at the supply side aim to reduce the proportion of fossil fuels in the energy mix.

Table 24.3 presents an overview of low carbon policy targets in China, India, Vietnam and Indonesia, demonstrating that developing countries in Asia have begun to mainstream a low carbon development consideration in their planning process recently (ADB 2012).

All selected countries have energy efficiency and renewable energy targets, ranging from an increase of the share of renewable energy in the primary energy demand of 5.6 percent in Vietnam to 15 percent in China and India by 2020. Except for Vietnam, all selected countries have carbon intensity targets. Table 24.1 clearly demonstrates that the LCD strategy has been

Table 24.3 Low carbon targets for Asian emerging economies

Country	Renewable/clean energy (RE) targets	Eenergy efficiency (EE)	Carbon intensity (CI)
China	Increase the share of RE in the primary energy demand to 11.4% by 2015 and to 15% by 2020 (up from 8.3% in 2010)	16% EI reduction by 2015	17% CI reduction by 2015 from 2010 levels

40% to 45% CI reduction by 2020 from 2005 levels |
| India | Increase the share of RE to 15% by 2020 (up from 4% in 2010)

Increase PV to 20,000 MW by 2020 | 10,000 MW energy savings by 2020 | 20–25% carbon intensity reduction from 2005 levels by 2020 |
| Vietnam | Increase the share of RE to 3% by 2015, to about 5.6% by 2020, to 9.4% by 2030 (up from 3% in 2010) | Reduce the elasticity of electricity/GDP from 2 (2010) to 1.5 (2015), to 1 (2020). | — |
| Indonesia | Increase the share of RE (including nuclear) to 15% by 2025 | 1%/year energy intensity reduction until 2025 (from 2005 levels)

Reduce the elasticity of electricity/GDP to < 1% by 2025 | 36% to 41% emission reduction below BAU* levels |

Notes: * BAU refers to 'business-as-usual'. Carbon intensity refers to the volume of CO_2 emitted (tonnes) per unit of GDP. Energy intensity refers to the volume of energy consumed per unit of GDP. The absence of specific targets for particular issues does not imply absence of policy, rather the absence of a stated national target.

Source: Adapted from ADB (2012).

embraced by the selected countries and that a number of low carbon policies have been devised accordingly. The next section will elucidate the case of China focusing on the energy sector.

Low carbon development in China and its interdependencies

As a 'responsible developing country,' Chinese policymakers attach 'great importance to climate change issues' (State Council 2008: 2). Some policies have been devised in recent years to reduce domestic carbon emissions. They are laid down in the National Climate Change Program (NDRC 2007), the White Paper on 'China's Policies and Actions for Addressing Climate Change' (State Council 2008) and the 2011 White Paper on 'China's Climate Change Policies and Actions' (State Council 2011). In the 11th Five Year Plan, the government set the goal of reducing energy intensity by 20 percent over the period of 2006–2010, in the end achieving a reduction by 19.06 percent (Wu 2011). While China hence books some progress as a 'responsible developing country,' in its role as a developing country within global climate negotiations, the government strongly has been emphasizing the principle of 'common but differentiated responsibilities' (CBDR). China has been taking different positions and approaches to LDC (*ditan fazhan*).

According to Du and Zhou (2011), China's low carbon energy strategy consists of six core strategies: conserving energy, developing clean coal, securing natural gas in order to adjust the energy composition, developing hydropower and non-hydro renewables, developing nuclear energy, and developing a smart grid system and off-grid power using energy storage technologies. As a 'responsible developing country,' the Chinese government has devised domestic pathways for LCD. First, the government undertook measures to reduce the proportion of coal in China's energy mix. Second, under the prefix of 'clean energy,' it started to invest in nuclear power development. These two domestic pathways for LCD are introduced in the following two sections. Third, building on CBDR, China has used the clean development mechanism (CDM), one of the flexible mechanisms under the Kyoto Protocol, as a catalyst to develop its renewable energy sector. China now is a leading country for certain renewable energy technologies. The third section will show how China has become an international player in the competition for renewable energy technology markets.

Low carbon development against the background of China's coal dependency

In China, economic development is to a large extent based on energy from coal. With its extensive coal reserves, the country is by far the largest coal producer worldwide, producing more than three times more coal than the US, and even more than the other countries of the 'top 10 of coal producers' together (WCA 2011).[4] As of 2010, coal supplied nearly 67 percent of China's total primary energy supply; in 2009, 78.7 percent of China's electricity was produced from coal (Best & Levina 2012).

This dominance of coal within China's energy structure is not only a result of natural conditions, but also of China's energy security agenda. For a long time, energy security has played, and is still playing, a dominant role within the country's energy policy. In 2005, China's energy self-sufficiency rate was 96 percent, which can primarily be related to its reliance on domestic coal reserves (Liu & Sims Gallagher 2010). Coal is predicted to remain the dominant fuel source over the next two to three decades to come (Best & Levina 2012), estimated on a scale of 70 percent of total power generation (Zhang 2010). While China is able to supply the bulk of its coal needs, the increasing oil demand in China worries Beijing. China's energy mix reveals that oil is the second largest energy source (see Figure 24.2). China became a net importer of oil in the 1950s (Leung 2011). Imports are projected to account for 60 to 80 percent of China's oil consumption by 2020 (Downs 2006). Whereas the industrial sector is the biggest driver of China's coal demand, the transport sector became extremely reliant on oil because of the rapid growth of roads and transport, the dieselization of rail transport and the remarkable development of domestic air transport (Leung 2010: 940). China is the world's largest emitter of CO_2 (Liu & Sims Gallagher 2010; Zhang 2010). Considering the increasing share of coal and oil in China's energy mix, the country's economy hence is, and in future still will be, strongly coupled to high carbon emissions.

The government's approach to reducing emissions from coal production has been based on command and control instruments, e.g. the enforced closure of inefficient power plants, but has shifted towards a market-based approach as of recently. Coal prices were for a long time set by the government, yet since in 2006, they have been subject to market prices. As a consequence, coal prices rose by more than 50 percent from 2006 to 2008 (OECD/IEA 2012). However, the central government still tries to intervene in coal market prices through 'mediating contract negotiations, setting temporary price caps, and issuing warnings and directives on price-setting' (Yang et al. 2012: 2). From 2007 onwards, a coercive regulation was implemented, according to which coal power plants were to be closed based on their power capacity, date of construction, stage in their lifecycle and performance standards (Tian 2008; Zhang 2010). Further, small thermal power plants with a total capacity of 44.10 million kilowatts were closed down, in addition to iron, steel and cement factories in order to improve energy efficiency (Hou et al. 2011). Moreover, to achieve the 11th Five Year Plan's goal of a reduction in energy intensity by 20 percent, 'black-outs' of industries and cities were not uncommon (SEI-FORES 2012).

While these measures contributed to lowering energy intensity by 19.06 percent (Wu 2011), with a new compulsory target for further carbon intensity reductions by 17 percent until 2015 (relative to 2010 levels) in the 12th Five Year Plan, command and control instruments are not seen as appropriate and effective anymore. The government considers a higher reliance on market-based instruments such as carbon taxation and emission trading (Bachus & Cao 2013). Its goal is to establish a national carbon trading system by 2015. At the time of writing, pilots for a carbon trading scheme have been set up in seven provinces and cities, i.e., Beijing, Chongqing, Guangdong, Hunan, Shanghai, Shenzhen and Tianjin. Furthermore, sector-based carbon emission trading schemes are being developed. However, as of now, crucial conditions for the functioning of carbon markets are lacking: accurate emission measurement; a solid legal system that supports distributing emission rights and permits, trading rules, monitoring, collecting of emission data, verification, enforcement and punishment for non-compliance; and administrative capacity (SEI-FORES 2012).

Meanwhile, technology research and development for the use of cleaner coal and the development of efficient and clean power-generating technology are central measures in China's LCD (State Council 2008). The adoption of cleaner technologies accounted for about 69 percent of the 19 percent energy intensity reduction over the period 2006–2010 (CPI 2012). Technologies comprised higher efficiency coal-fired power plants, which were replacing the above mentioned small power plants. Furthermore, energy efficiency improvements in the industrial and building sectors contributed to energy intensity reductions, achieved foremost in the frame of the Top-1000 Energy Consuming Enterprises Program and the Ten Key Energy Conservation Projects (Hou et al. 2011; CPI 2011). Additionally, while there are still many stumbling blocks ahead before the technology of carbon capture and storage (CCS) could be fully commercialized, China heavily invests in research, development and demonstration of this technology (Best & Levina 2012). China here has the potential to become an important supplier of carbon capture technologies (ibid.). In coal-fired boilers supply (with the exception of high quality boilers), it already shows the highest export volume (Horbach et al. 2012).

Low carbon development based on 'clean energy'

Nuclear power has for a long time played a minor role in China. Only in 1991 was the first reactor connected to a grid, and, due to the government's inconclusiveness about the role of nuclear power in China, progress in developing nuclear power remained slow. In 2008, 11 nuclear power stations were in operation with a total installed capacity of 9.1 GWe (Zhang 2010), equaling a share of 1.9 percent electricity generation (Yan et al. 2011). Four further nuclear power reactors have been built up to the present, and 26 nuclear power reactors are currently under construction (He et al. 2012). The basis for this acceleration of nuclear power plant construction is the 2007 'Medium- and Long-term Nuclear Power Development Plan' of the National Development and Reform Commission (NDRC) which foresees an increase in nuclear power capacity to about 40 GWe by 2020. Even if, after the Fukushima nuclear crisis in March 2011, the government announced a temporary suspension of the approval process for new reactors pending a thorough safety review (EIA 2012b), political will is set for the development of nuclear power. The government frames nuclear power as a 'clean energy,' which can satisfy increasing energy needs and at the same time protect the environment (Government of China 2006). Different from 'sustainable energy,' which builds on renewable energies, 'clean energy' (*qingjie nengyuan*) is defined as energy without pollutant emissions, which includes nuclear power. Although the future of nuclear power on China's LCD roadmap may be uncertain, LCD hence to some extent brought nuclear power into the energy policy agenda in China under the prefix of 'clean energy.'

Renewable energy for low carbon development: China's new role

Within climate negotiations, Chinese policymakers and academia have been stressing the country's stage of development as a crucial condition for its pathway to LCD (see, e.g., State Council 2008; Wang & Zhuang 2011; Zhuang 2008). China would still be in the process of industrialization, which is why LCD is understood rather as a domestic *process* of emission reduction. This focus can also be seen in the 11th or 12th Five Year Plans' objective of reducing energy *intensity*, a goal that does not conflict as much with GDP growth as an absolute emission reduction target (SEI-FORES 2012). The graph in Figure 24.3 shows that, although China heavily invests in reducing energy intensity, total emissions are continuously increasing, demonstrating that absolute decoupling economic growth and GHG emissions growth did not occur yet.

In international negotiations, government representatives emphasized that an absolute cap on emissions and international binding targets are not an option in the foreseeable future (Wu 2011). LCD theory and practice is rather seen as a joint process of 'common maturation' (*gongtong chengzhang*) of developing and developed countries (Wang & Zhuang 2011: 137). One way of 'common maturation' has been the clean development mechanism (CDM), one of the flexible mechanisms of the Kyoto Protocol that enables industrialized countries to pay for projects that reduce, avoid or sequestrate GHG emission reductions in less developed countries, by buying certified emission reductions (CERs) that can be applied to meet their own GHG emission targets. As a developing country, China has made extensive use of the CDM to meet its own renewable energy development targets, which will be further shown later.

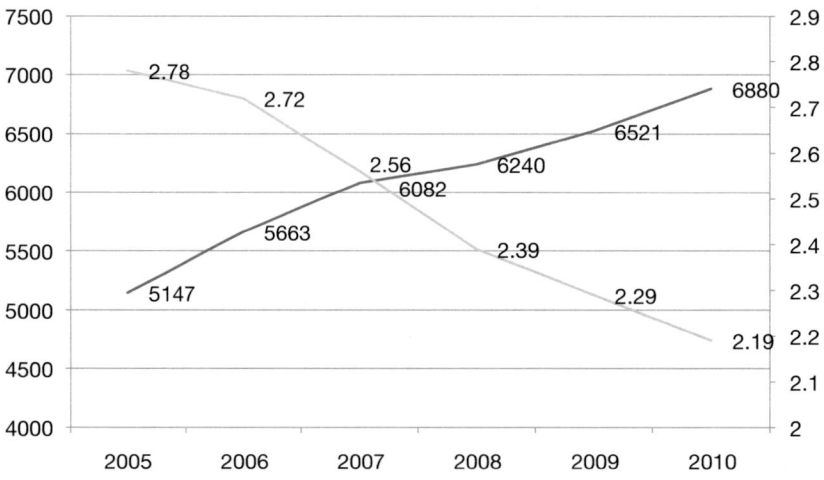

Figure 24.3 Relation between CO_2 emission intensity (right-hand scale, $tCO_2/10^4$yuan) and total CO_2 emissions (left-hand scale, $MtCO_2$) in China, 2005–2010

Source: Adapted from Qi (2012).

Renewable energy forms an important part of China's development program since 2002, when it was laid out in the 'Program of Action for Sustainable Development in China in the Early 21st Century' (a follow-up to China's 'Agenda 21 – White Paper on China's Population, Environment, and Development in the 21st Century'), with the objective to support the implementation of a sustainable development strategy. For some years, a number of programs and domestic projects were devised targeted at the development of renewable energies, until in 2006, the Renewable Energy Law (REL) came into effect. With the REL, renewable energy policy for the first time obtained an overarching framework. Subsequent to the law, the 2007 Development Plan for Renewable Energy set a target of 15 percent renewable energy supply in primary energy supply by 2020 to accelerate the process of decarbonization (see Table 24.3).

As of 2011, China has an estimated capacity of renewables of 70 GW (REN21 2011).[5] It therewith has the largest renewable power capacity in the world (with or without hydropower) (REN21 2011). Most of the 70 GW non-hydropower renewable energy capacity is coming from wind power, reaching an installed capacity of 62.4 GW. The remainder is rather equally distributed across biomass power and solar PV (REN21 2011). This rapid development of renewable energies, while anchored in the REL, has to a great extent to be understood in the context of China's participation in the CDM.

China is worldwide the country with the largest share of CDM projects (UNFCCC 2012). With 2,462 registered projects by 25 October 2012, it hosts far more projects than the country with the second most CDM projects, i.e., India with 922 projects. The large majority of CDM projects in China are renewable energy projects, which, based on the statistics of the National Development and Reform Committee (NDRC), account for 83.6 percent of Chinese projects (2,058 projects). The priorities for China with the CDM are wind power and hydropower projects.

Through the CDM, a considerable inflow of money supported the development of renewable energies in China. This inflow is orchestrated by the NDRC, which implements the CDM in China and at the same time looks after the admittance of renewable energy projects under the REL. After 2009, when fixed feed-in tariffs (FiT) for wind power were set (see later), these new tariffs were considered to provide only small profit margins, which is why wind projects had to apply for CDM registration (Zhang 2010). It is arguable that the NDRC deliberately kept tariffs low in order to maintain the possibility for wind project developers to apply for CDM registration (Bluemling & Mol 2012). As a consequence, the additionality of CDM projects has been questioned (Bluemling & Mol 2012).[6]

The CDM hence has been an important catalyst for China's renewable energy development, particularly hydropower and wind power. However, in terms of CO_2 emission reductions, its success has been questioned. For all Chinese CDM projects until 2010, Wu (2011) calculated that China offset about 431 mega tons of CO_2. However, for the year 2010 only, China's total carbon emissions were 8.95 gigatons.

While the CDM may not yield high emission reductions, its role as a catalyst for China becoming a leading country for renewable energy technologies seems, as we will show in the following, to be considerable. Renewable energies are supporting a development of China away from the 'workshop of the world' to an international player in the renewable energy technology market.

Wind power

The development of the on-grid wind power industry was established in 1988 in China with support from the Danish government (Liu & Kokko 2010). Chinese wind power policy, however, only started to develop in 1994 and progress in installed capacity was slow until the early 2000s (see Figure 24.4).

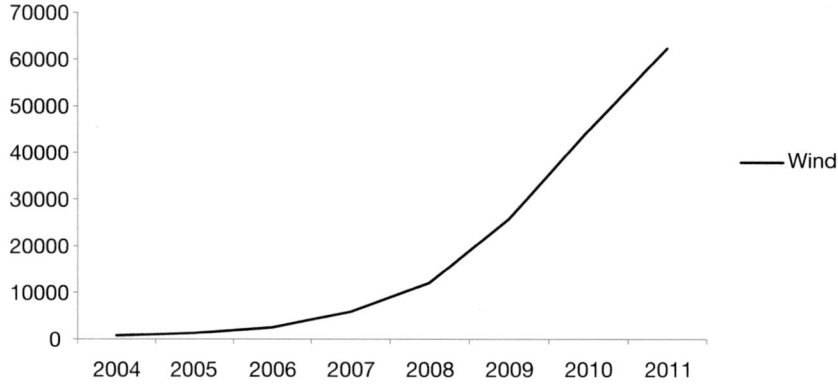

Figure 24.4 Wind power installed capacity development in China (MW)
Source: Adapted from GWEC (2012).

The Chinese government has not only been able to capitalize on the potential of technology transfer under the framework of the CDM, but did also launch a number of policies and policy instruments to promote the development of the Chinese wind power sector. Examples are subsidies, a competitive bidding system, a FiT, the 'buy Chinese' policy and Notice 124, known as the 'local content rule.' As shown in Figure 24.4, this has resulted in a tremendous increase of installed wind capacity and enabled Chinese wind companies to prepare for entering the global market. As of 2011, Chinese wind turbine manufacturers had a share of 26.7 percent in the global market, and the Chinese manufacturer Goldwind already ranks second in the list of global wind turbine manufacturers (REN21 2012). The down-side of these policies, addressed only at domestic Chinese wind companies, are trade frictions between China and other countries such as the US (Recharge News 2012). The rise of Chinese wind companies in the global market appears not to be happening without contention.

Hydropower

Hydropower is the most widely used renewable resource for electric power generation in China, and therewith plays a major role in China's effort of replacing fossil fuels and in its LCD (Cheng et al. 2012). Six percent of total primary energy consumption is covered by hydropower (see Figure 24.2). The total installed hydropower capacity was 213 GW by the end of 2010, ranking first in the world, even if only 22 percent of China's total generation capacity has been used (ibid.). There are more than 45,000 hydropower plants in China, composed of about 400 plants with more than 50 MW generation capacity, 85 plants with more than 300 MW and 32 plants with more than 1000 MW (ibid.).

Apart from being low carbon, hydropower is valued for bringing about further benefits like flood control, navigation, irrigation and drinking water storage, reforestation around dams, and support for the development of the rural economy (Räsänen et al. 2012; Tang et al. 2012; Zhao et al. 2012). However, the detrimental effects of hydropower are also multiple. The resettlement of dwellers has received a lot of criticism, resulting in landlessness, poverty, food insecurity and community disruption (Brown et al. 2008; Zhao et al. 2012). Apart from these overall social costs, ecological consequences are also manifold, with changes to the hydrological conditions through the construction of the dam and management of water flows, water eutrophication in the dam and downstream nutrient reduction, siltation, or changes in the aquatic ecosystem, leading to consequences for river wildlife habitats (Brown et al. 2008; Tang et al. 2012). Even if environmental impact assessment procedures have been established since 1989, some developers of small hydropower dams seem to have violated construction procedures, obscuring a coordinated exploitation of the resource (Zhao et al. 2012).

Downstream countries are furthermore affected by hydropower development in China. Nearly 59 percent of all economically exploitable hydro resources are located in Southwest China (Huang & Yan 2009). Within the Mekong River basin, China shares several rivers with Myanmar, Laos, Cambodia and Thailand, affecting all these countries with its dam constructions (Lebel et al. 2005; Xue et al. 2011).

Photovoltaic

For the Chinese government, the PV industry is of strategic importance, which is why it actively participates in the PV sector. A number of policy instruments targeted at the development of the PV industry were devised during the last decade (Huo & Zhang 2012; Van Eynde & Chang 2013). Examples are subsidy programs for grid connection and independent power production in remote areas, renewable energy projects commissioned by the Chinese government through a bidding program, a FiT, and the production of equipment for the industry (through state-owned enterprises or SOEs).

As a result of this state participation and propelled by the booming global market, Chinese solar companies were able to increase capacity, reduce costs, strengthen company management, expand their production scale and become credible players in the international market, which, in turn, led to an explosive growth of both installed PV capacity and the production of solar goods in China (Van Eynde & Chang 2013). Installed capacity is still small compared to cell production figures, but is now rising rapidly. As of late 2010, installed PV power capacity connected to the national grid was about 900 MW, which is about 0.1 percent of China's total power capacity (EIA 2011). The 2011 nationwide FiT for PV of 1 yuan per kWh as well as installation targets have provided long-term stability and incentives for the private sector, resulting in an explosion in the number of installations at the end of 2011 (Recharge News 2011). Although the Ministry of Finance has announced a 21 percent cut in the subsidies offered to demonstration solar projects, the industry is confident that the global outlook for PV in 2012 is positive (MOF 2012). This confidence is not ill founded as the central government has announced an increase in its domestic installation target from 15 GW to 21 GW by 2015 (*China Daily* 2012).

China is the world's largest producer of solar cells and responsible for over 60 percent of global production (REN21 2011). In just four years, the global solar manufacturing sector shifted from being geographically dispersed to one dominated by Chinese companies. In 2011, five Chinese companies were among the top ten module manufacturers of which three were in the top five (Lux Research 2012).

China's 'PV success story' has a number of implications for other countries in the region and for global market players in general. First, it can be argued that due to rising labor costs in China, manufacturing operations or services are to be outsourced to other countries such as for example Vietnam for the low(er) labor or material cost (Adams & Tran 2011; Chaponnière et al. 2008). Second, the global PV market has encountered a situation of overcapacity, due to increased production accompanied by decreased support schemes in Europe and the US, resulting in weaker demand. This situation has put enormous pressure on prices, with the result that at least half of the manufacturers are likely to consolidate (BNEF 2012). Third, companies outside China demand reciprocity in an open market. For foreign companies it is not easy to enter the Chinese domestic market due to strict 'house rules' that are often interpreted in

Beijing's favor (Chang et al. 2012; Garcia 2012; Van Eynde & Chang 2013). Fourth, and closely related, is trade dispute between China, on the one hand, and the US and EU, on the other hand, accusing China of selling its PV cells for well below market value (Bloomberg 2012; Euractiv 2012; Xinhua 2012). While some industry players highlight the existence of distorted factor costs in China such as capital, exemplified by the lavish credit lines from the government, others stress the importance of Chinese imports which have led to a lower price and helped the EU to meet its 2020 renewable energy targets.

Bioenergy

With a large proportion of the population still living in the countryside, biomass is a major resource for rural domestic energy use. As biomass (i.e., mainly firewood and straw stalks) is traditionally directly combusted, it is however a key source for environmental pollution (Ma et al. 2010). The Chinese government has for long been trying to promote a different use of biomass by encouraging the use of biogas digesters. In the 1970s already, biogas digesters were promoted extensively (Chen et al. 2010). The steepest increase in biogas digesters took place since 2003, with the implementation of the Rural Household Biogas State Project (He et al. 2011). As a result, by 2011, China was the world leader with 43 million household digesters (REN21 2011). Furthermore, by the end of 2009, nearly 2,000 large and medium-scale biogas digesters were installed at industrial enterprises; 22,570 digesters were installed at livestock and poultry farms; and 630 digesters were connected to municipal waste and sludge treatment facilities (REN21 2011: 35). China's total biogas power generating capacity was 800 MW by the end of 2010 (ibid.).

This steep development is a result of comprehensive domestic policy support, stretching over a FiT for electricity from biogas; loans without interest rates for medium- and large-scale biogas plants; or subsidy programs for building materials, equipment and for wages for technicians (see Jiang et al. 2011). However, in its support of biogas, the government has promoted and privileged biogas digester construction, and not so much follow-up maintenance services. Chen et al. (2010) found that only 60 percent of the biogas digesters have been working properly. Given the out-migration from the countryside, the question of follow-up maintenance will become even more important (Bluemling & Hu 2011). Furthermore, the contribution of domestic biogas systems to LCD has been contested. Gosens et al. (2012) found that households reduced the consumption of biomass fuels, but did not reduce the use of coal. Carbon emission reductions hence remain modest.

Different from wind and hydropower, the development of the PV sector and household biogas digesters has not made extensive use of the CDM. Even if biogas CDM projects exist, the transaction costs involved to coordinate a large number of households in their small-scale contribution to carbon emission reductions, makes the coordination of these projects a rather tedious task. Meanwhile, the Chinese government initiates projects to promote biogas technology for rural households in other countries (NDRC 2011). The province Guangxi has been playing a major role in the promotion of biogas digesters in Cambodia through the training of Cambodian officials and technical staff, as well as the construction of household biogas infrastructure (WXZG 2006).

Conclusion

Emerging economies in Asia have embraced LCD strategies to counter climate change and secure economic development. Although categorized as non-Annex 1 countries (developing countries), implying that Asian emerging powers do not have binding emission reduction targets, this chapter demonstrated that those countries have deployed targets and policy instruments to decrease energy intensity by unit of GDP and to increase the share of renewable energy.

China's role in global environmental governance has evolved significantly during the last five years, as it became the second largest economy of the world and the world's largest CO_2 emitter. As a 'responsible developing country,' building on 'common but differentiated responsibilities,' China has set ambitious low carbon targets such as to voluntarily increase the share of renewable energy, decrease carbon intensity and decrease energy intensity by unit of GDP. As shown in this chapter, the investigation of China's low carbon development roadmap cannot exclusively be understood as a process taking place inside China's national boundaries. LCD should be understood against the backdrop of China's domestic context and physical factors. In terms of the amount of emissions to be reduced, LCD in China strongly relates to a more efficient use of coal, which tends to remain the major energy source. However, in terms of qualitative changes in its coal-based development model, there are a number of policy documents targeted at the development of renewable energies, demonstrating their importance to the government. The development of China's renewable energy sector cannot be detached from the international climate regime. Categorized as a non-Annex 1 country, China made extensive use of the CDM to develop its renewable energy sector, in particular the wind and hydropower sector. The CDM is perceived as one of the major factors that enabled China to gain a leading position in renewable energy technologies.

Currently, Chinese wind turbine manufacturers had a share of 26.7 percent in the global market by 2011, and the Chinese manufacturer Goldwind already ranked second among global wind turbine manufacturers. By the same token, five Chinese companies were among the top ten module manufacturers for PV of which three were in the top five. In the PV sector, manufacturing operations have already been outsourced to other countries in the region where production costs are lower. However, renewable energy development does not only lead to outsourcing production, but also to intensified collaboration with countries in the region as in the case of biogas development, or in the case of hydropower, where Chinese dams impact downstream countries and Chinese companies at the same time built hydropower projects in Myanmar. Renewable energy technology development hence does not only help a shift away from the workshop of the world, but also aids the development of China's position in the region.

If one considers the multiple facets of LCD as demonstrated by this chapter, it can be argued that, on the one hand, international climate negotiations show signs of a 'bottom–up' regime because targets and policy instruments are voluntarily deployed at the national level. On the other hand, however, it appears that such a bottom–up regime can only be explained by incorporating both a regional and international dimension as exemplified by the case of China in this chapter. China perceives itself as a responsible developing country, building on common but differentiated responsibilities. As a result, China heavily invests in the development of its renewable energy industry with the help of swiftly evolving global value chains, international climate finance instruments and favorable physical conditions.

However, pressure from the international community is rising, demanding China to take up more responsibility and to commit to legally binding emission reduction targets. Domestic pressure is also intensifying as the country's growth miracle increasingly reveals its immense environmental cost. In the country's guidelines for the future, Chinese policy makers call China's government-directed and investment-driven economic model unbalanced and unsustainable. This rhetoric already calls for serious reflection on the operationalization of the concept of 'low carbon' in China. The future of China's low carbon strategies will depend, however, on how this rhetoric will be translated into action.

Sarah Van Eynde, Bettina Bluemling and Hans Bruyninckx

Bibliography

Adams, F.G. & Le Tran, A. (2010) 'Vietnam: From Transitional State to Asian Tiger? Issues of the Vietnamese Economic Transformation Experience', *World Economy*, 11(2).

ADB (Asian Development Bank) (2012) Green Urbanization in Asia: Key Indicators for Asia and the Pacific 2012, Report, Mandayulong City, ADB.

Ahman, M., Nikoleris, A. & Nilsson, L.J. (2012) Decarbonizing Industry in Sweden: An Assessment of Possibilities and Policy Needs, Report no. 77, September, Lund, Lund University, Department of Environmental and Energy Systems Studies.

Bachus, K. & Cao, J. (2013) 'Cap or Tax? Exploring the Potential for a Carbon Tax or Emission Trading in China', in H. Bruyninckx (Ed.), *The Governance of Climate Relations between Asia and Europe: Evidence from China and Vietnam as Key Emerging Economies* (Edward Elgar Publishing).

Best, D. & Levina, E. (2012) Facing China's Coal Future. Prospects and Challenges for Carbon Capture and Storage, Working Paper, OECD/IEA.

Bidwai, P. (Ed.) (2012) *The Politics of Climate Change and the Global Crisis* (Orient BlackSwan).

Bloomberg (2012) Solarworld-led Group Files China Anti-dumping Case in Europe, Press release, 23 July, London, Bloomberg.

Bluemling, B. & Hu, C.S. (2011) 'Vertical and System Integration instead of Integrated Water Management? Measures for Mitigating NPSP in Rural China', Conference paper, Conférence Gestion des ressources en eau souterraine, 14–16 March, Orléans,

Bluemling, B. & Mol, A.P.J. (2012) 'Clean Development Mechanism Implementation and Additionality in China: An Institutional Analysis', in L. da Costa Ferreira & J.A. guilhon Albuquerque (Eds.), *China and Brazil: Challenge and Opportunities* (Sao Paulo: Annablume).

BNEF (Bloomberg New Energy Financy) (2012) Bankruptcy or Acquisition Facing Half of the Manufacturers, Report in Renewable Energy Country Attractiveness Indices, May, Issue 33, London, BNEF in cooperation with Ernst & Young.

BP (2012) BP Statistical Review of World Energy, Report, June, London, BP.

Brown, P.H., Magee, D. & Xu, Y. (2008) 'Socioeconomic Vulnerability in China's Hydropower Development', *China Economic Review* 19, 614–627.

Cai, B., Wang, J., Yang, W., Liu, L. & Cao, D. (2012) 'Low Carbon Society in China: Research and Practice', *Advances in Climate Change Research* 3(2), 106–120.

Castro, P., Hayashi, D., Kristiansen, K.O., Michaelowa, A. & Stadelmann, M. (2011) Scoping Study – Linking RE Promotion Policies with International Carbon Trade (LINK), Scoping study, June, IEA-RETD.

Chang, P., Belis, D. & Bruyninckx, H. (2012) 'EU-China Climate Relations: The Clean Development Mechanism and Renewable Energy in China', in J. Wouters, T. de Wilde, P. Defraigne & J.-C. Defraigne (Eds.), *China, the European Union and Global Governance* (Edward Elgar Publishing).

Chaponnière, J.-R., Cling, J.-P. & Zhou, B. (2007) 'Vietnam Following in China's Footsteps: The Third Wave of Emerging Asian Economies', *WIDER Conference on Southern Engines of Global Growth*, Helsinki, 7–8 September.

Cheng, C.-T., Shen, J.-J., Wu, X.-Y. & Chau, K.-W. (2012) 'Operation Challenges for Fast-growing China's Hydropower Systems and Response to Energy Saving and Emission Reduction', *Renewable and Sustainable Energy Reviews* 16, 2386–2393.

China Daily (2012) China Raises 2015 Target for Installations, Press release, 4 July, Beijing, *China Daily*.

CPI (Climate Policy Initiative) (2011) Review of Low Carbon Development in China: 2010 Report – Executive Summary, Beijing, Tsinghua University, Climate Policy Initiative.

CPI (2012) Annual Review of Low-Carbon Development in China (2011–2012) – Executive Summary, Beijing, Tsinghua University, Climate Policy Initiative.

Downs, E. (2006) Brookings Foreign Policy Studies Energy Security Series: China, Report, December, Brookings Institution.

Du, X. & Zhou, D. (2011) 'China's Scientific, Green, and Low Carbon Energy Strategy' [中国的科学，绿色，低碳能源战略], *CNKI Journal* 1009, 1742.

EIA (US Energy Information Administration) (2011) *China Country Analysis* (EIA).

EIA (2012a) 'International Energy Statistics'. Available at http://www.eia.gov/cfapps/ipdbproject/IED Index3.cfm?tid=90&pid=44&aid=8.

EIA (2012b) 'Annual Energy Outlook 2012', Report, DOE/EIA-0383. Available at http://www.eia.gov/forecasts/aeo/pdf/0383%282012%29.pdf.

Ellis, K., Baker, B. & Lemma, A. (2009) Policies for Low Carbon Growth, Report, London, Overseas Development Institute.

EPIA (European Photovoltaic Industry Association) (2012) Global Market Outlook for Photovoltaics Until 2016, Report, May, Brussels, EPIA.

ESMAP (Energy Sector Management and Assistance Program) (2012) *Planning for a Low Carbon Future*, Low Carbon Growth Country Studies Program: Lessons Learned from Seven Country Study.

Euractiv (2012) EU Launches Solar Panel Probe into Chinese Dumping Claim, Press release, 6 September, Brussels, Euractiv.

European Commission (2011) *A Roadmap for Moving to a Competitive Low Carbon Economy in 2050* (European Commission).

Fisher, C., Torvanger, A., Shrivastava, M.K., Sterner, T. & Stigson, P. (2011) How Should Support for Climate-friendly Technologies be Designed?', Working Paper, CEPS Policy Brief No. 254, Brussels, Centre for European Policy Studies.

Garcia, C. (2012) 'Policies and Institutions for Grid-connected Renewable Energy: Best Practice vs. the Case of China', *Governance: An International Journal of Policy, Administration, and Institutions*, Early View, 21 August.

Gosens, J., Lu, Y., He, G., Bluemling, B. & Beckers, T.A.M. (2012) Sustainability effects of household-scale biogas in rural China', Energy Policy.

Government of China (Central People's Government of the People's Republic of China) (2006) '温家宝主持召开国务院常务会议审议并原则通过《核电中长期发展规划（2005–2020年）》', Wen Jiabao Zhuchi Zhaokai Guowuyuan Changwu Huiyi Shenyi Bing Yuanze Tongguo Hedian Zhong Changqi Fazhan Guihua [Wen Jiabao, chair of the State Council executive meeting, examined and approved the long-term nuclear power development plan 2005–2020], 1 July 2006, Beijing, Government of China.

GWEC (Global Wind Energy Council) (2012) Global Wind Report: Annual Market Update 2011, April, Brussels, GWEC.

Halsnaes K. & Garg, A. (2011) 'Assessing the Role of Energy in Development and Climate Policies – Conceptual Approach and Key Indicators', *World Development* 39(6), 987–1001.

He, G., Bluemling, B., Mol, A.P.J., Lu, Y., Tu, Q., Zhang, L., et al. (2011) 'The Pros and Cons of Centralized and Decentralized Biomass-to-energy Systems: Case of Shandong Province, China', Conference paper, Governance and Socioeconomics of Bioenergy in Rural Areas – Experiences in the EU and Asia, 22–23 May, Wageningen.

He, G., Mol, A.P.J., Zhang, L. & Lu, Y. (2012) 'Nuclear Power in China after Fukushima: Understanding Public Knowledge, Attitudes, and Trust', *Journal of Risk Research* 17(4), 1–17.

Heggelund, G. & Backer, E. (2007) 'China and UNP Environmental Policy: Institutional Growth, Learning and Implementation', *International Environmental Agreements: Politics, Law and Economics* 7(4), 415–438.

Horbach, J., Chen, Q., Rennings, K. & Vögele, S. (2012) Lead Markets for Clean Coal Technologies – A Case Study for China, Germany, Japan and the USA, Working Paper No. 12–063, Mannheim, Centre for European Economic Research (ZEW).

Hou, J., Zhang, P., Tian, Y., Yuan, X. & Yang, Y. (2011) 'Developing Low-carbon Economy: Actions, Challenges and Solutions for Energy Savings in China', *Renewable Energy* 36(11), 3037–3042.

Hu Y., Peng, Z. & Zhou, D. (2011) 'What is Low-carbon Development? A Conceptual Analysis', *Energy Procedia* 5, 1706–1712.

Huang, H. & Yan, Z. (2009) 'Present Situation and Future Prospect of Hydropower in China', *Renewable and Sustainable Energy Reviews* 13, 1652–1656.

Huo, M. & Zhang, D. (2012) 'Lessons from Photovoltaic Policies in China for Further Development', *Energy Policy* 51, 38–45.

IEA (International Energy Agency) (2012) Global Carbon Dioxide Emissions Increase by 1.0Gt in 2011 to Record High, Press release, 24 May, Paris, IEA.

IIER (Institute for Integrated Economic Research) (2011) *Low Carbon and Economic Growth: Key Challenges*, report, IIER: Meilen.

IPCC (Intergovernmental Panel on Climate Change) (2011) Special Report on Renewable Energy Sources and Climate Change Mitigation, Report prepared by Working Group III of the Intergovernmental Panel on Climate Change, Cambridge, Cambridge University Press.

Jackson, T. (2009) *Prosperity Without Growth* (Earthscan).

Jiang, X., Sommer, S. & Christensen, K.V. (2011) 'A Review of the Biogas Industry in China', *Energy Policy* 39(10), 6073–6081.

Lebel, L., Garden, P. & Imamura, M. (2005)'The Politics of Scale, Position, and Place in the Governance of Water Resources in the Mekong Region', *Ecology and Society* 10(2), 279.

419

Leung, G.C.K. (2010) 'China's Oil Use 1990–2008', *Energy Policy* 38(2), 932–944.

Leung, G.C.K. (2011) 'China's Energy Security: Perception and Reality', *Energy Policy* 39(3), 1330–1337.

Liu, C., Duan, M., Zhang, X., Zhou, J., Zhou, L. & Hu, G. (2011) 'Empirical Research on the Contributions of Industrial Restructuring to Low-carbon Development', *Energy Procedia* 5, 834–838.

Liu, H. & Sims Gallagher, K. (2010) 'Catalyzing Strategic Transformation to a Low-carbon Economy: A CCS Roadmap for China', *Energy Policy* 38(1), 59–74.

Liu, Y. & Kokko, A. (2010) 'Wind Power in China: Policy and Development Challenges', *Energy Policy* 38(10), 5520–5529.

Lux Research (2012) *Lux Research Reveals 2011 Top 10 Module Manufacturers* (Lux Research).

Ma, H., Oxley, L., Gibson, J. & Li, W. (2010) 'A Survey of China's Renewable Energy Economy', *Renewable and Sustainable Energy Reviews* 14(1), 438–445.

Ministry of Finance of the PRC(MOF) (2012), Notice on the 2012 Announced Golden Sun Programme, MOF.

Mulugetta, Y. & Urban, F. (2010) 'Deliberating on Low-carbon Development', *Energy Policy* 38(2), 7546–7549.

NDRC (National Development and Reform Commission) (2007) China's National Climate Change Programme, June, Beijing, NDRC.

NDRC (2011) Country Report on China's Participation in Greater Mekong Subregion Cooperation, 16 December, Beijing, NDRC in cooperation with Ministry of Foreign Affairs, Ministry of Finance, and the Ministry of Science and Technology.

OECD/IEA (2012) Energy Technology Perspectives – Tracking Clean Energy Progress, Report, Paris, International Energy Agency.

PBL (Netherlands Environmental Assessment Agency) (2012) Trends in Global CO_2 Emissions, Report, The Hague, PBL.

Qi, Ye (2012) 'China's Low-Carbon Transition Under no "Global Deal"', Presentation, The Way Forward After Durban: Chinese and European Responses to Climate Change, 16 March, Leuven.

Räsänen, T.A., Koponen, J., Lauri, H. & Kummu, M. (2012) 'Downstream Hydrological Impacts of Hydropower Development in the Upper Mekong Basin', *Water Resources Management* 26(12), 3495–3513.

Recharge News (2011) China Raises Solar Target to 15GW by 2015, Say Reports, Press release, 16 December, Oslo, Recharge News.

Recharge News (2012) Sinovel Ties Up With Mita-Teknik for Turbine Control Technology, Press release, 16 April, Oslo, Recharge News.

REN21 (2011) Renewables 2011 Global Status Report, Report, Paris, REN21 Secretariat.

SEI-FORES (2012) China's Carbon Emission Trading: An Overview of Current Development, Report, Stockholm, Stockholm Environment Institute and FORES.

State Council (2008) China's Policies and Actions for Addressing Climate Change, Beijing, State Council.

State Council (2011) White Paper on China's Climate Change Policies and Actions for Addressing Climate Change, 22 November, Beijing, State Council.

Tang, X., Li, Q., Wu, M., Tang, W., Jin, F., Haynes, J., et al. (2012)'Ecological Environment Protection in Chinese Rural Hydropower Development Practices: A Review', *Water Air Soil Pollution* 223, 3033–3048.

Tian, J. (2008) How the People's Republic of China is Pursuing Energy Efficiency Initiatives: A Case Study, Working Paper, ADB Economic Working Paper Series, Regional and Sustainable Development Department, Asian Development Bank.

Truong, P.T. (2010) A Comparative Study of Selected Asian Countries on Carbon Emissions with Respect to Different Trade and Climate Change Mitigation Policy Scenarios, Background Policy Paper for the East Asia Low Carbon Green Growth Roadmap Project under the East Asia Climate Partnership.

UNDP (United Nations Development Programme) (2010) China Human Development Report 2009–2010: China and a Sustainable Future: Towards a Low Carbon Economy & Society, Beijing, China Translation and Publishing Corporation, UNDP.

UNEP Risø Centre (2012) UNEP Risø CDM/JI Pipeline Analysis and Database, 1 October 2012, Roskilde, UNEP Risø Centre.

UNFCCC (2012) UNFCCC data. Available at http://cdm.unfccc.int/.

UN Habitat (United Nations Human Settlements Programme) *Sustainable Urbanization in Asia: A Sourcebook for Local Governments*, Nairobi: UN Habitat 2012.

Van Eynde, S. & Chang, P. (2013) 'Explaining the Development of Renewable Energy Policies in China: Comparing Wind and Solar', in H. Bruyninckx (Ed.), *The Governance of Climate Relations between Asia and Europe: Evidence from China and Vietnam as Key Emerging Economies* (Edward Elgar Publishing).

Waizi Xiangmu Zhongxin Guanli (WXZG) (2006) '柬埔寨沼气技术培训与推广项目又传捷报' Jianpuzhai Zhaoqi Jishu Peixun Yu Tuiguang Xiangmu You Chuan Jiebao' Waizi Xiangmu Guangli Zhongxin [Cambodia biogas technology training and promotion projects and successes], Nanning, Foreign Capital Project Management Center.

Walsh, S., Tian, H., Whalley, J. & Argarwal, M. (2011) 'China and India's Participation in Global Climate Negotiations', *International Environmental Agreements* 11, 261–273.

Wang, G. & Zhuang, G. (2011) '低碳经济的认识差异与低碳城市建设没模式' Ditan Jingji de Renshi Chayi Yu Ditan Chengshi Jianshe Moshi, *Xuexi Yu Tansuo* 193(2), 134–138.

WCA (World Coal Association) (2011) 'Coal Statistics'. Available at http://www.worldcoal.org/resources/coal-statistics/.

World Bank (2012) 'World Bank Statistics'. Available at http://data.worldbank.org/ indicator/NY.GDP.MKTP.KD.ZG/countries/1W-CN?display=graph.

Wu, Q. (2011) 'Policy and Politics of a Carbon Market in China', in J. Peetermans (Ed.), *Greenhouse Gas Market Report 2011: Asia and Beyond: The Roadmap to Global Carbon & Energy Markets* (International Emissions Trading Association).

WWF (2011) The Energy Report: 100 % Renewable Energy by 2050, WWF International in cooperation with Ecofys and OMA.

Xie Lijian, Suhong Zhou & Xiaopei Yan (2011) 'A Review and Outlook of the Study on Low-carbon Development in China and Overseas', *Human Geography* 21(1), 19–24.

Xinhua (2012) EU Launches Anti-dumping Investigation over Chinese PV Products, Press release, 6 September, Beijing, Xinhua.

Xue, Zuo, J. Paul Liu, Qian Ge (2011) 'Changes in Hydrology and Sediment Delivery of the Mekong River in the Last 50 Years: Connection to Damming, Monsoon, and ENSO', *Earth Surface Processes and Landforms* 36, 296–308.

Yan, Q., Wang, A., Wang, G., Yu, W. & Chen, Q. (2011) 'Nuclear Power Development in China and Uranium Demand Forecast: Based on Analysis of Global Current Situation', *Progress in Nuclear Energy* 53, 742–747.

Yang, C., Xuan, X. & Jackson, R.B. (2012) 'China's Coal Price Disturbances: Observations, Explanations, and Implications for Global Energy Economies', *Energy Policy* 51, 720–727.

Zhang, Z. (2010) 'China in the Transition to a Low-carbon Economy', *Energy Policy* 38 (11), 6638–6653.

Zhang, Z. (2011) 'Energy and Environmental Policy in China: Towards a Low-Carbon Economy', in W.E. Oates (Ed.), *New Horizons in Environmental Economics* (Edward Elgar Publishing).

Zhao, X., Liu, L., Liu, X., Wang, J. & Liu, P. (2012) 'A Critical Analysis on the Development of China Hydropower', *Renewable Energy* 44, 1–6.

Zhuang, G. (2007) *Low Carbon Economy: China's Development Road in the Background of Climate Change* (China Meteorological Press).

Zhuang, G. (2008) 'How Will China Move towards Becoming a Low Carbon Economy?', *China & World Economy* 16(3), 93–105.

Notes

1   While the dominant discourse in Asia is that of the right to develop and maintaining economic growth, another strand of discourse in the literature points out that 'economic growth' as a goal is unsustainable and should be replaced with a more sustainable goal and vision (see, for example, Jackson 2009). In their opinion, 'low carbon' is incompatible with continuing growth.

2   In particular, a comparison might identify differences between China and India perceived as the two major emerging economies in terms of population and GDP, and other 'smaller' emerging economies such as Vietnam and Indonesia.

3   While in 2010 the urban share of Asian population was still only 43 percent, by 2050, the urban share in Asia is projected to reach 63 percent, according to the Asian Development Bank (2012).

4   The other countries of the 'top 10 coal producers' are the USA, India, Australia, Indonesia, Russia, South Africa, Germany, Poland, and Kazakhstan (WCA 2011).

5   If we include hydro in the capacity of renewables the figure stands at installed capacity of 282 GW as of 2011.

6   'Additionality' is defined in the Kyoto Protocol as those reductions in GHG emissions that are additional to any that would occur in the absence of the certified CDM project activity (Kyoto Protocol, Article 12.5(c)). The wind power projects still manage to be validated as additional and registered by the EB because of certain stipulations within the international rules governing the CDM. With regard to national policies in host countries, the EB developed the E+/E– guidance. This guidance, 'Clarifications on the consideration of national and/or sectoral policies and circumstances in baseline scenarios' states that: 'national and/or sectoral policies or regulations that give comparative advantages to less emissions-intensive technologies over more emissions-intensive technologies (for example public subsidies to promote the diffusion of renewable energy or to finance energy efficiency programs) that have been implemented since the adoption by the COP of the CDM M&P (decision 17/CP.7, 11 November 2001) need not be taken into account in developing a baseline scenario' (UNFCCC 2005: Annex 3). This means that the feed-in tariffs do not have to be included in the baseline for the wind power sector in China. It is clear that the rationale behind this guideline is to prevent the CDM becoming an obstacle for initiatives undertaken by developing countries that contribute to climate mitigation. This guideline highlights a major deficiency in the CDM. The rule implies that projects can be registered and CDM money be generated for project activities that national policy already stipulates, but without it, the CDM would be a major obstacle for the development of climate mitigation policy in developing countries (Castro et al. 2011). One could wonder whether the E+/E– guidance does not lead to a situation in which the additionality analysis becomes counterproductive with respect to achieving real GHG emission reductions.

25

Nuclear energy in Asia

The end of the renaissance?

Rajesh Basrur, Youngho Chang and Swee Lean Collin Koh

The nuclear disaster at Japan's Fukushima Daiichi nuclear power plant (NPP) in the aftermath of the Tohoku earthquake and ensuing tsunami on 11 March 2011 has left a deep imprint, reviving the memory of the nuclear disasters at Three Mile Island (1979) and Chernobyl (1986). For many of those who have long been critical of the use of nuclear energy, this episode provides fresh ammunition for their advocacy against the nuclear option and justified their call for a shift to non-nuclear alternative energy sources. For those who view nuclear power as a key solution to dwindling supplies of hydrocarbon-based energy sources, the Fukushima disaster can be viewed as simply an aberration that has merely put a dent in the mainly sterling safety record of modern NPPs since Chernobyl.

What are the lessons of Fukushima? Only a preliminary and tentative understanding was possible then in the immediate aftermath of the disaster. First, the disaster arose from the simultaneous impact of a 9.0 magnitude earthquake and a 14-metre high tsunami. Both incidents were unanticipated despite scientific evidence that should have been taken into account while assessing potential threats (Ewing & Ritsema 2011). The International Atomic Energy Agency (IAEA) observed that the risk was 'underestimated' and called for regulatory efforts to aim for 'sufficient protection against infrequent and complex combinations of external events' (IAEA 2011). Second, there were a number of organizational failings. Plant operators avoided costly safety measures and were not corrected by an effective system of oversight (Onishi & Belson 2011; Onishi & Fackler 2011a). Third, deep distrust between the political leadership on the one hand and the bureaucracy and the NPP operators on the other led to delayed and erroneous responses (Onishi & Fackler 2011b). This preliminary understanding was only reinforced by an independent probe, launched by the Japanese Government, into the reasons behind the disaster.

According to the report published in 2012 by the National Diet of Japan Fukushima Nuclear Accident Independent Investigation Commission (NAIIC), the direct causes of the NPP accident were all foreseeable prior to the disaster. Instead, it found that the NPP was incapable of withstanding the earthquake and tsunami since the operator, Tokyo Electric Power Corporation, the regulatory bodies as well as the government body promoting the nuclear power industry all failed to correctly develop the most basic safety requirements, such as assessing the probability of damage, preparing for containing collateral damage from such a disaster, and developing evacuation plans for the public in the event of a serious radiation leakage (NAIIC 2012). This led the Commission to determine the Fukushima disaster to be man made, the unanticipated natural disaster of a 9.0 magnitude earthquake and the ensuing tsunami notwithstanding, and thus preventable.

These lessons are important because they will ultimately influence the decisions of policy makers as they grapple with the pulls and pushes that shape their efforts to achieve energy security. While perceptions of risk have already been altered since the disaster, nuclear energy may remain a significant option for many states (Tertrais 2011: 91–100). The outcome of the investigation, as well as recommendations furnished in the Commission's report, may well put renewed emphasis on nuclear safety regulatory measures in order to prevent a similar disaster in the future. Most significantly, however, it provided relief to nuclear energy proponents that the nuclear option is still viable after all. In this chapter, we argue that even though the global expansion of nuclear power-generating capacity has slowed since the Fukushima disaster, the quest for nuclear option remains very much alive in Asia. Countries of the Northeast Asia, South Asia and Southeast Asia sub-regions that have either nuclear energy expansion plans or are deemed to be major aspirants of nuclear energy are of interest in this chapter. We also furnish some general recommendations for policy makers of countries in the process or thinking of inducting nuclear power in their existing and future electricity generation mixes.

Asia's continued nuclear odyssey two years after Fukushima: a survey

Even though the international community was no doubt shocked by the Fukushima disaster, the initial reaction was measured amid calls among the anti-nuclear constituents to abolish nuclear power. Personalities representing international institutions were some of the first to speak up on the implications of Fukushima on the future of nuclear power. For instance, during a visit to Oslo just days after the fateful event, chief of the International Energy Agency, Nobuo Tanaka remarked that he was concerned about the effect the Fukushima disaster could have on support for nuclear energy given its important role in achieving both energy security and a low-carbon economy (*Agence France Presse* 2011a). The United Nations Secretary-General Ban Ki-moon commented that the Fukushima disaster justified time for a reassessment of the global nuclear safety regime (*Reuters News* 2011b). Mr Ban's remark then could be deemed as prudent, since it would be premature at that juncture – barely two weeks after the Fukushima disaster – to call for a phaseout of nuclear energy. Around a year later during an interview with Japanese press, Mr Ban remarked that nuclear energy can still remain a 'very important energy source' despite the Fukushima disaster, as long as nuclear safety standards are strengthened (*Jiji Press English News Service* 2012). In June 2011, as one of the indications that the nuclear option is not about to disappear from the world energy security agenda, all 33 countries attending a ministerial-level seminar held in Paris, including those possessing (Japan included) and seeking nuclear energy, pledged to boost cooperation to strengthen nuclear safety (*Kyodo News* 2011b).

Influential figures representing international governmental institutions, such as Mr Tanaka and Mr Ban, were certainly not alone in this regard. The nuclear industry was quick, and naturally so, to join in the defence of nuclear power. In September 2011, culminating from a three-year process aimed at finalising corporate standards and nuclear responsibility within the nuclear power industry, the world's NPP exporters inked the first ever six-principle code of conduct aimed at, among a list of objectives, raising NPP safety standards. The agreement, however, is not legally binding. Not all NPP exporters signed up to the code of conduct. For example, China

National Nuclear Corporation participated in the early talks but did not adopt the code. Iran was also not involved (*Agence France Presse* 2011d). The nuclear industry continued to step up efforts to revive enthusiasm for nuclear power even though post-Fukushima sales of NPPs took a dip. In late March 2012, World Nuclear Association director-general John Rich advocated nuclear power as being 'uniquely able to deliver on a global scale' both energy security and environmental protection, pushing for its central role in what he described as 'the global clean-energy revolution' (*Agence France Presse* 2012a). It would seem that from the discourse and actions undertaken to date by both international institutions and nuclear industry, there are keen efforts to keep the nuclear option alive in the world energy security agenda. That being said, however, attitudes towards nuclear energy do differ between governments across diverse regions.

Nuclear rethink and policy U-turns, influenced by the Fukushima disaster and ensuing public sentiments, appear to be the most pronounced in Western Europe. In July 2011 Germany confirmed its plans for complete nuclear energy phaseout by 2022 (*Reuters News* 2011c). The Swiss parliament upper house also followed suit three months later with an approval to phase out all NPPs over the next two decades (*Agence France Presse* 2011e). France, a major nuclear energy user in Western Europe, in the months after the Fukushima disaster determined not to abandon nuclear power, instead calling it a 'solution for the future' (*Agence France Presse* 2011c) even though later with a change of political leadership in September 2012 it decided to reduce reliance on nuclear power from the existing 75 percent of the total power generation mix to 50 percent by 2025 (*Reuters News* 2012a). Generally among the emerging economies of Eastern and Southern Europe in particular, pro-nuclear sentiments continue to be prevalent for the sake of sustaining economic development. Across the Atlantic Ocean, the United States continued to express faith in nuclear energy right from the immediate aftermath of the Fukushima disaster, with top leaders and officials of the incumbent administration defending the nuclear option (*Agence France Presse* 2011b; Wallsten & Yang 2011).

In Asia, the reaction towards the Fukushima nuclear disaster and the ensuing environmental fallout (which affected food supplies in the region) was first one of shock even if the tone appeared measured despite calls by anti-nuclear constituents, particularly those in Japan, to phase out nuclear energy. Among many existing nuclear energy users in Asia, one of the first moves undertaken in response to the disaster was a decision to place moratoriums on new NPP construction while existing NPPs were subjected to post-Fukushima inspections and strenuous review of nuclear regulatory mechanisms, particularly in the case of Japan, was conducted. The sentiments subsequently began to drift back to 'business as usual' even if governments in the region pledged to strengthen nuclear safety and security. One of such earliest indications was the meeting of the Forum for Nuclear Cooperation in Asia held in July 2011, when the 12 member countries' delegations said that their governments would delay the implementation of their nuclear power programmes due to decreased public support in the wake of the Fukushima disaster, but perhaps most notably, they also expressed their commitment to continue their nuclear power programmes while at the same time enhancing nuclear safety (*Kyodo News* 2011d). Compared to Western Europe, there are no plans instituted in Asia for a complete phaseout of nuclear energy even if some regional governments promised gradual scale-down of their reliance on nuclear as part of their national power generating mixes.

Still, even though the expansion of nuclear power generating capacity worldwide has slowed on the whole, Asia looks set to witness some of the most vibrant nuclear power expansion activities. The prevailing sentiment in Asia seems to be one which is predicated on national context-specific, broader socioeconomic development even if there is a widespread recognition of a need to strengthen nuclear safety and security measures. Figure 25.1 shows that whereas North America and Western Europe may see either slow growth or decline in nuclear energy use as part of the overall power generating mix at least until 2030, the Middle East and Asia-Pacific regions generally will see steady growth within the same timeframe. Of this group of countries, the Far East, which includes the major Northeast Asian economic powerhouses of China, Japan and South Korea, looks set to dominate nuclear energy expansion in Asia. The other Asian sub-region that may witness robust growth in nuclear power generating capacity is South Asia, where Bangladesh, India and Pakistan all have concrete plans to date since the Fukushima disaster.

The existing Asian NPP operators – mainly China, India, Japan and South Korea – are projected to continue expanding their nuclear power generating capacity within the next 15 years, leading the general expansion in the region. However, as Table 25.1 shows, by 2023 it might be conceivable for new regional nuclear energy aspirants to complete their plans, with Bangladesh, Indonesia and Vietnam joining the Asian NPP operators' 'club'. Before 2030, it might be conceivable for two more Southeast Asian countries – Malaysia and Thailand – to begin operating their own NPPs. In the more immediate future within the next five years, it appears that the following countries highlighted in Table 25.2 will lead the renewed growth of nuclear power capacity in Asia.

China

China looks set to experience the most substantial growth with up to 30 new NPPs expected to be operational within the 2012–2018 timeframe. In the immediate aftermath of the Fukushima debacle, China announced that even though lessons on nuclear safety ought to be learned from the disaster, it will not abandon nuclear power. Tian Jiashu, then nuclear safety director of China's Ministry of Environmental Protection, remarked that 'we're not going to stop eating for fear of choking', expressing the intent to continue with the NPP buildup, albeit a slowed one at that (*Associated Press Newswires* 2011a). Beijing's determination to stick to the nuclear option is well illustrated by its plan for a six-fold increase in the nuclear share of its national energy mix

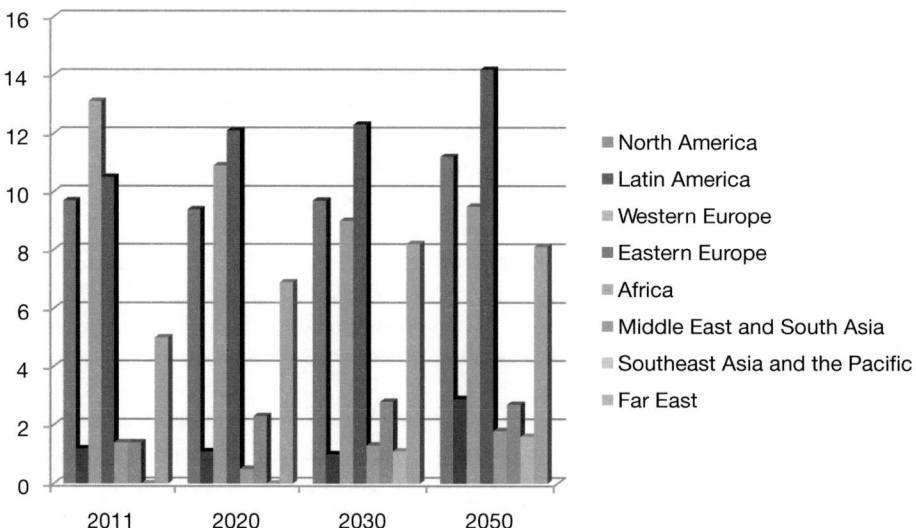

Figure 25.1 Estimates of nuclear as percentage of total electricity-generating capacity
Source: IAEA (2012).

Table 25.1 Major Asian nuclear power programmes, January 2013

	NPPs operable		NPPs under construction		NPPs planned		NPPs proposed	
	No.	MWe net	No.	MWe net	No.	MWe net	No.	MWe net
Bangladesh	0	0	0	0	2	2,000	0	0
China	16	12,918	29	29,990	51	59,800	120	123,000
India	20	4,385	7	5,300	18	15,100	39	45,000
Indonesia	0	0	0	0	2	2,000	4	4,000
Japan	50	44,396	3	3,036	10	13,772	3	4,000
Korea (South)	23	20,787	4	5,205	5	7,000	0	0
Malaysia	0	0	0	0	0	0	2	2,000
Pakistan	3	725	2	680	0	0	2	2,000
Thailand	0	0	0	0	0	0	5	5,000
Vietnam	0	0	0	0	4	4,000	6	6,700

Notes: MWe refers to megawatt electrical (as distinct from thermal).
Operable: NPPs connected to national power grid.
Under construction: First concrete for NPP poured or major refurbishment underway.
Planned: Approvals, funding or major commitment in place, mostly expected in operation within 8–10 years.
Proposed: Specific programme or site proposals, expected operation mostly within 15 years.

Source: World Nuclear Association (2013).

Table 25.2 New NPPs projected, 2012–2018

	Number of NPPs projected	Total MWe (net)
China	30	32,150
India	7	4,930
Japan	3	4,263
Korea (South)	5	6,400
Pakistan	2	600

Notes: MWe refers to megawatt electrical (as distinct from thermal).

Source: World Nuclear Association (2013).

by 2020 at 70 gigawatts (GW), up from the existing 10.8GW at the end of 2010 (*ET Net News* 2011). China's continued quest for nuclear energy also reflects its desire to compete with potential rivals in the nuclear industry sector and in the socioeconomic development realm. For example, the former head of China's National Energy Administration Zhang Guobao remarked in August 2011 that China needs to enact new and 'clear' policies to develop its nuclear energy sector to avoid falling behind other countries in this field of clean energy development, adding that the Fukushima disaster could provide China an opportunity to 'reach and overtake' the rest of the world (*Reuters News* 2011d). Despite having completed a country-wide post-Fukushima inspection of the safety standards of its existing NPPs by August 2011, the Chinese government appears to tread the issue of nuclear expansion more carefully. It took over a year for Beijing to lift the moratorium, imposed immediately after the Fukushima disaster, on new NPP construction (*Agence France Presse* 2012b). By early 2013, China appears more confident to expedite the expansion of its NPP project. It targeted a 20 percent growth within 2013 and looks set to become the third largest NPP operator after the United States and France by 2020 (*Xinhua News Agency* 2013).

India

A number of Indian government officials stepped out to emphasise that the country cannot abandon its nuclear energy programmes in the wake of the Fukushima disaster. For example, Principal Scientific Advisor to the Indian Government R. Chidambaram remarked that 'making technological choices requires foresight and forecasting as per the country's needs and in order to improve the quality of life and standard of living of people India cannot do without nuclear power' (*Financial Express* 2012; *Press Trust of India* 2011). The imperative for sustained economic growth appears to be the overriding consideration for the decision to remain fixed on nuclear power expansion. While advocating the utilisation of nuclear energy to its 'fullest extent', the chairman of India's Atomic Energy Commission, Dr Srikumar Banerjee stressed the need to consider the rather sterling safety record of NPP operation so far in India, by comparing the safety of NPP operation and traffic in New Delhi (Gupta 2011). According to former Indian president A.P.J. Abdul Kalam, India has plans to produce about 20GW nuclear power, in future leveraging on the country's rich thorium reserves (*Press Trust of India* 2012).

By August 2011, similar to the case of China, India completed a post-Fukushima nuclear safety review and concluded that its existing NPPs are capable of withstanding extreme climatic and seismic activities such as cyclones and earthquakes. New Delhi remains keen on acquiring Japanese nuclear technology and even struck an accord with Russia in December 2012 to proceed with the controversial Kudankulam NPP (KNPP) Units 1 and 2 construction plans, while at the same time expediting talks for Units 3 and 4 (*UNI (United News of India)* 2012). India's rather comprehensive NPP expansion plans were not without significant domestic backlashes. The KNPP programmes in Tamil Nadu state has been the focus of attention since Fukushima, with residents living in the vicinity of the NPP sites expressing concerns about the safety of those plants and, in the case of the fishing communities in the area, the prospect of NPP waste water being discharged into the sea to pollute the fishing grounds. Assurances by the Indian government that the KNPP project will be implemented in accordance to the 'highest safety standards' and promises of steps to be undertaken to allay the residents' concerns notwithstanding, villagers living close to the sites said they do not believe in the government assurances at all (*Associated Press Newswires* 2011d). KNPP-1 was originally slated to be commissioned in January 2013 but the schedule slipped to May 2013 instead amid allegations of radiation leaks at the plant (*Press Trust of India* 2013b). Nonetheless, India's nuclear power expansion plan appears to be on track, in view of governmental approval for KNPP Units 3 and 4 to be constructed in Tamil Nadu, with Units 5 and 6 under option (*Press Trust of India* 2013a).

Japan

The Fukushima disaster notwithstanding, the Japanese government continues to view nuclear energy as one of the three pillars of its long-term energy policy in the period 2030–2050, alongside renewable energy sources and expanding nationwide energy conservation efforts (*Associated Press Newswires* 2011b). The experience of Fukushima probably had a tumultuous influence on then Japanese Prime Minister Naoto Kan's remark in late May 2011 about possibly halting new NPP construction, reducing Japan's reliance on nuclear while targeting 20 percent renewables within the national energy mix under a revised energy plan by 2020 (*Kyodo News* 2011a). However, internal division within the Japanese government on the best way forward after the Fukushima disaster was apparent when, less than a month later, the Japanese Ministry of Economy, Trade and Industry was said to remain determined to keep nuclear energy as the primary power-generating source (*Kyodo News* 2011c). With the change of prime minister and his cabinet near the end of 2011, the direction appeared to be promoting the use of renewable energy (*Kyodo News* 2012c). The decision made throughout the bulk of 2012 was largely influenced by public opinion in Japan against nuclear energy, given that the effects of the Fukushima disaster continued to be felt by the people on the ground. Before another change of government by late 2012, there was even a suggestion made by the Japanese officials to explore a complete phaseout of nuclear energy by 2030 (*Reuters News* 2012b). However, this idea apparently did not garner unanimous agreement among all quarters within the Japanese government, with some high-ranking Japanese officials expressing doubts about immediately ending the country's nuclear reliance (*Kyodo News* 2012d). With the latest change of administration after the Democratic Party of Japan's electoral defeat, there are already plans to reinstate the operations of existing NPPs put off-line since the Fukushima debacle, while plans for complete nuclear phaseout in Japan have by now become a stillborn idea. Prime Minister Shinzo Abe in March 2013 stressed the need for nuclear power especially in the reconstruction of affected areas in the Fukushima disaster zone, noting that 'reconstruction will be hard without an inexpensive and stable source of power', apparently alluding to the continued importance of nuclear energy in ensuring Japan's energy security (*Kyodo News* 2013). Whether Japan will eventually restart new NPP projects or not, nuclear power seems to remain on the agenda even if emphasis is put into improving nuclear safety. For instance, a new nuclear regulatory body was established under the Ministry of Environment, replacing the Nuclear and Industrial Safety Agency under METI, which was widely criticised for cultivating close ties with the nuclear industry (*Associated Press Newswires* 2011c).

South Korea

South Korea, which is also geographically close to Japan and therefore also prone to the environmental effects of the Fukushima Daiichi fallout, is undeterred with its NPP projects. Strategic imperatives of economic growth and the lack of other alternative energy sources are cited by senior South Korean officials as reasons to continue with the nuclear quest (*Korea Times* 2011). As such, Seoul has decided to proceed with existing plans to increase nuclear power generation capacity. The Korea Electric Power Corporation in June 2011 announced that there was no change in its plan to install eight new NPPs in order to hike its existing nuclear power generation capacity to 11,200 megawatts (*Manila Bulletin* 2011). The decline of public support for nuclear energy after the Fukushima disaster as well as fervent civil opposition to new NPP construction plans notwithstanding, Seoul continued to seek expansion of its nuclear reliance. This was not without obstacles, however. In later December 2011, two sites – Yeongdeok in North Gyeongsang Province and Samcheok in Gangwon Province – were shortlisted by Korea Hydro and Nuclear Power Corporation as candidate locations for new NPPs, but the move was seriously protested by opposition parties and civic groups who accused the South Korean government of using the mass electrical blackout on 15 September 2011 as a justification to proceed with nuclear expansion plans (*Dong-A Ilbo Daily* 2011b). Undeterred, then South Korean Minister of Knowledge Economy Hong Suk-woo said that there was no change in the nuclear energy policy, which is to continue to build more NPPs, with a targeted 40 percent of the national power generating mix to be fulfilled by nuclear (*BBC Monitoring Asia Pacific* 2011).

The prevalent perspective of the South Korean government is that until non-nuclear alternative energy sources can feasibly and significantly replace hydrocarbon-based energy sources, nuclear energy will remain essential at least as an interim alternative measure. For instance, in October 2011 chairman of the Korea Nuclear Energy Promotion Agency Rhee Jae-wan remarked that nuclear energy remains an essential energy source and serves as a 'bridge' between the use of hydrocarbons and eventual future emergence of renewable energy sources as substitute (*Korea Herald* 2011). Probably similar to the case of China, South Korea's continued quest for nuclear energy has a competitive industrial aspect to it. Besides being seen as a solution to sustaining socioeconomic development before future widespread use of non-nuclear alternative energy sources, nuclear energy is possibly regarded as one of the centrepieces of South Korean high-technology global outreach. In December 2011 Seoul announced its aim of becoming one of the world's top three nuclear energy technology exporters and to capture 20 percent of the global NPP market by 2030 (Xie 2011). Prior to ending his term, former South Korean president Lee Myung-bak remarked that the domestic nuclear energy sector constitutes – just like automobiles, shipbuilding and electronics – a key export industry that powers national economic growth (*Yonhap English News* 2012).

Rest of Asia

Throughout the rest of Asia, nuclear remains an option at least in the foreseeable future. A handful of lesser NPP operators and aspirants have set into motion their national NPP programmes in varying degrees. In Pakistan, for example, a third NPP was formally put into operation in May 2011 and China was reported to be committed to expanding its South Asian neighbour's existing Chasma NPP using 1970s technology – the same type that was used in Fukushima Daiichi – raising safety concerns (*Reuters News* 2011a). Yet Islamabad remains keen on expanding the use of NPPs (also with Chinese assistance) to meet its growing energy needs. In February 2012, Islamabad and Beijing were reported to be in the final stage of negotiations for six more NPPs to be constructed in Pakistan by 2023 (*Kyodo News* 2012b). In all, Pakistan plans to produce up to 8.8GW nuclear power by 2030. A new South Asian nuclear energy aspirant, Bangladesh, also viewed nuclear power as a means to ensure energy security. In November 2011, Dhaka signed an accord with Russia to build an NPP at Rooppur, with the first of two 1,000MW units slated to go online by 2020 (*United News of Bangladesh Limited* 2011).

According to the Bangladesh government in September 2012, there is also a plan for a second NPP in southern Bangladesh after the Rooppur NPP project (*United News of Bangladesh Limited* 2012). Meanwhile, in Southeast Asia, the only country with the most concrete NPP plans to date is Vietnam, which remains undeterred by the Fukushima disaster and appears more concerned to meet growing energy and economic development needs. Plans for Vietnam's first NPP by 2020 remain well on track – a decision based on 'the long-term demand and supply outlook' according to Vietnam's Deputy Prime Minister Hoang Trung Hai, who also warned that if Hanoi does not push ahead with its NPP plans, it will become a net importer instead of maintaining its current position as net exporter by 2015 – a situation thought undesirable as it will make Vietnam's economic growth unsustainable. At the same time, Hanoi remains optimistic that it can uphold high safety standards through various measures, including paying attention to lessons of the Fukushima Daiichi nuclear disaster (*Bangkok Post* 2011). Not only are plans set into motion for the construction of the NPP at Ninh Thuan, Hanoi has also begun in earnest nuclear energy human resource development and capacity-building with Russian assistance. In addition, the Vietnamese government also invested over US$9.5 million in a public relations outreach project to disseminate information related to the benefits and safety measures of NPP development in the country (*Vietnam News Agency Bulletin* 2013).

While significant public sentiments against nuclear in the wake of the Fukushima disaster have stymied original plans, policymakers elsewhere in Southeast Asia do not appear to have shut their doors completely on nuclear energy. A good example is Indonesia whose NPP plan was put on hold without any concrete direction as Jakarta attempted to grapple with intense public opinion. However, it has not dismissed the nuclear option despite its possession of multiple renewable energy sources as alternatives to hydrocarbons. In June 2011, the country's director-general for oil and gas in the Ministry of Energy and Mineral Resources, Evita Herawati Legowo, remarked that nuclear power remains the most efficient way to generate electricity to improve public welfare, adding that 'nothing in the world is without risk' (*Jakarta Post* 2011). Since the Fukushima disaster to this day, a concrete decision on NPP remains pending. There has been a range of differing opinion, expressed openly in the public domain, among the Indonesian policymakers with some pushing for NPPs to stave off an anticipated energy crisis, while some others preferred to take the middle road of not dismissing the nuclear option outright yet not seeing it as an immediate priority. Some Indonesian officials have dismissed the nuclear option as viable for the country. One such vocal Indonesian official who publicly opposed the nuclear option was deputy minister for energy and mineral resources Widjajono Partowidagto who remarked on separate occasions that Indonesia is not ready to build an NPP due to the fact that the archipelagic country lies within a region of active seismic activity, and that the country is plagued by corruption and poor industrial safety and supervision records (*LKBN ANTARA* 2011).

In the Philippines, right after the Fukushima disaster, President Benigno Aquino decided not to back the revival of the defunct Bataan NPP (*Philippine Star* 2011). However, by 2012 the Aquino Administration expressed desire to carefully evaluate the possible use of nuclear energy, in particular for addressing a power shortage in Mindanao of the Southern Philippines (*BusinessWorld* 2012). The Philippine Department of Energy reportedly examined at the nuclear option as a possible long-term alternative energy source (*BBC Monitoring Asia Pacific* 2012). Another Southeast Asian nuclear aspirant, Malaysia, has put on hold plans to commission its first NPP by 2021 due to public concerns about nuclear energy (*Kyodo News* 2012a). Up north, Thailand has shelved its original plans for an NPP, particularly in the face of intense domestic public opposition to the project. Still, the Thai government did not dismiss the nuclear option, with clear intention to revive the plan in the near future once it completes its feasibility study (*Thai News Service* 2012). In the case of Singapore, another Southeast Asian country keen to explore alternative energy sources to wean off its reliance on foreign-sourced energy supplies, the initial enthusiasm to seriously look into nuclear energy has since the Fukushima disaster been replaced with a more cautious approach. A pre-feasibility study on nuclear energy commissioned in 2010 at the recommendation of the Economic Strategies Committee and completed in 2012 concluded that current nuclear energy technology is deemed unsuitable for Singapore's needs in view of the country's geographical size and population density, even though the report did not rule out nuclear energy totally (*Straits Times* 2012).

Observations and the road forward for Asia's nuclear quest

Judging from the official discourse in Asia especially, it would appear that even if nuclear energy opponents in the region would have liked the Fukushima disaster to mark the end of the region's nuclear quest, the reality may turn out to be otherwise. The key observation from the above survey examined here is that it is not possible to totally exclude nuclear power from the future energy mix, even if hydrocarbons may one day be totally supplanted. For many governments in Asia, socioeconomic development and energy supply security seem to remain the top priority even if there is growing awareness about reducing the environmental impact of hydrocarbon-driven growth. Emotional calls from certain interest groups, including academics, CSOs and the concerned general public and policymakers worldwide notwithstanding, nuclear power remains an attractive option to increasingly substitute hydrocarbons. This was well demonstrated during the World Economic Forum on East Asia held in Jakarta in June 2011, when many political leaders attending the event still regarded nuclear energy as relevant (*Straits Times* 2011).

National context in demographics, economy, society and culture, and political system differs from country to country, thus resulting in varying energy profiles for each, depending also on the availability, accessibility and applicability of alternative energy sources. Therefore it is not possible to expect the formulation of a one-size-fits-all energy profile for every government to adopt. For example, while Indonesia could have tapped geothermal, hydropower and biofuels as sources of clean electricity generation, the same cannot be said of Singapore, which lacks such resources, except perhaps solar power potential, and which is constrained by its limited geographical surface area. Seen in this light, while one country may not require nuclear power as a part of its energy mix, others cannot be expected to follow suit if the contextual factors surrounding energy resource availability, access, affordability and applicability are taken into consideration. Therefore, prudence dictates that no stone should be left unturned in the quest for alternative energy sources to sustain the world's population. Bearing this in mind, we propose the following policy recommendations for the path forward in Asia's continued nuclear quest.

Holistic methodology of energy cost evaluation

One consideration for nuclear operators is the importance of holism. Both nuclear and renewable energy possess drawbacks related to costs. For NPP projects, as Mark Diesendorf argues, the popular optimism about lower economic costs with new reactor designs may not be wholly accurate. He points out that most studies appear to portray nuclear technology as less expensive, and hence economically viable, while ignoring huge subsidies from governments or using accounting methods which shrink the capital cost component (Diesendorf 2012: 50–70). The paucity of data related to the cost of operating new generation reactors means that if one sticks to the current generation designs that are in widespread usage, the capital cost fails to compete favourably with renewables (*Nucleonics Week* 2011). This is not to say, however, that renewables are by default cheaper than nuclear energy. One still needs to factor in the availability of investments, sometimes in hitherto unexplored clean energy technology innovations, and also the considerable costs in terms of finance and time devoted to the research and development (R&D), followed by test bedding and commissioning processes. In short, nuclear and renewable energy may not necessarily have a clear cost advantage over one another. Any advantage would depend on the types of innovation involved, which entails some degree of tangible costs in terms of time and money. In the case of renewables, the prospect of long-term investments to enhance their share of the energy mix does have its own chequered path, especially in times of financial crisis. For instance, prior to the global economic recession in 2008–2009 and in the face of oil price spikes since 2004, there were tremendous interests worldwide in developing the potential of renewables for power generation. However, after the recession hit and until recently, there was a slowing of momentum in renewables development as oil prices dropped. What this means is that for renewables to have a clear, dominant advantage over hydrocarbons and nuclear power in the long run, persistent commitment in developing this particular sector is necessary. If not, it is hard to accept the claim that renewables are necessarily lower in economic costs compared to nuclear.

Besides economic costs, as Hooman Peimani points out, there are also non-economic, somewhat 'non-tangible' costs to consider when evaluating the suitability of particular types of alternative energy (Peimani 2012: 24–49). The environmental costs of using particular types of energy source are one such example of 'non-economic' costs to be taken into account as well. For future feasibility and evaluation studies of alternative energy types prior to decisions made on a particular optimum energy mix, calculations will need to be holistic. Simon Tay and Phir Paungmalit pointed out that, while all energy sources entail environmental costs, nuclear power throughout its entire lifecycle still emits greenhouse gases, although the amount emitted is small compared to fossil fuels (Tay & Paungmalit 2012: 90–111). A lifecycle analysis shows that a nuclear power plant emits about 25 grams of greenhouse gases per kilowatt hour of electricity generated while a coal power plant emits about 1,000 grams per kilowatt of electricity generated (Raadal et al. 2011: 3417–3422). Besides, unlike renewables, the problem is complicated in the case of nuclear power with respect to radioactive waste disposal and other environmental consequences resulting from a nuclear reactor accident, as the Chernobyl, Three Mile Island and Fukushima Daiichi episodes have amply shown. The environmental costs from using nuclear and renewable energy would therefore have to be considered within a holistic equation. To take this further, it is not sufficient to merely consider the cost of utilising particular energy types. The cost-benefit analysis also has to factor in energy efficiency and conservation measures (Diesendorf 2012: 62–67). A holistic methodology, tailored for each country in question, should therefore comprise the cost-benefit evaluation of various potential energy types, energy efficiency and conservation measures in order to derive a more accurate and realistic picture of an optimum energy profile for that particular country. Such an evaluation would not only look at the short term, but also at the long term, acknowledging the need to consider intergenerational equity rather than being preoccupied with the present and with short-term gains and risks (Tay & Paungmalit 2012: 92).

Pluralism in nuclear energy policy discourse

A holistic methodology for calculating the costs and benefits of a future, sustainable energy mix is not sufficient in itself if the public is not aware of the facts and how decisions are arrived at. The protests against government NPP plans in Asia, as witnessed in India, Indonesia, the Philippines, South Korea and Thailand, illustrate the problem of opacity in decision-making processes that breed scepticism among the general public that is both the beneficiary of energy plans and the victim of their negative consequences. In fact, as Tay and Paungmalit pointed out, the burden of proof in making development decisions lies with specific policy proponents, who must establish that any proposed activity is safe for the environment and human beings (ibid.: 92). Government policymakers and nuclear energy industry players need to be open to public scrutiny in their dealings and decision-making processes and transparent enough to address and dispel doubts, criticisms and scepticisms from multiple segments of the affected society at large. What this means is that there is a need to alter the manner business is being done among government policymakers and the nuclear energy industry in Asia – from a top-down, opaque, exclusive and sometimes heavy-handed approach to a bottom-up, transparent and inclusive one and eventually to a combination of the two approaches. Only through such a pluralistic form of participation among multiple stakeholders at the national level – the government, nuclear energy industry, non-government actors and the general public – will there be a comprehensive policy discourse that tolerates debates over the pros and contras of nuclear energy, enabled by openness and transparency in the dissemination and communication of opinions.

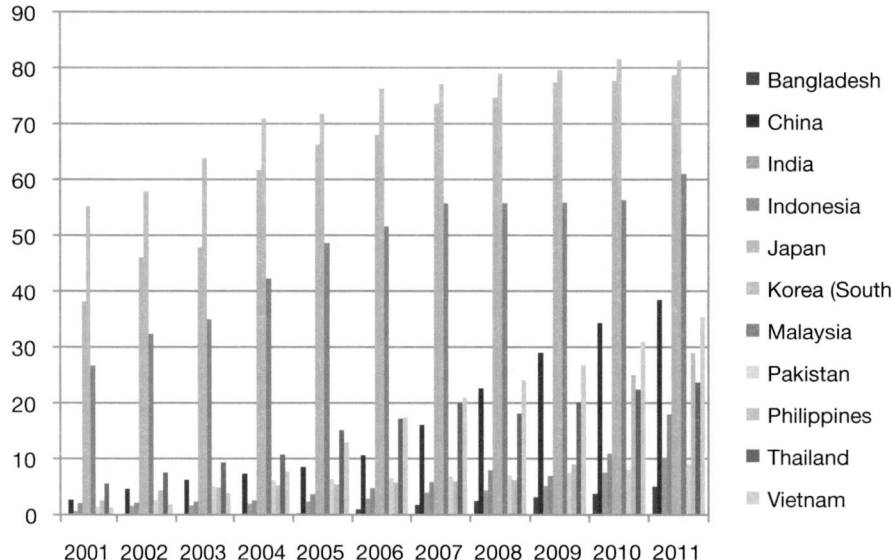

Figure 25.2 Number of internet users (per 100 people) in selected Asian countries
Source: World Bank Global Information and Communications Technologies Statistical database.

The traditional approach seen in much of Asia where policy discourse is concerned may no longer work in the present era due to a confluence of factors. The emergence of civil society organisations (CSOs) and the widespread proliferation of modern information and communication technologies such as internet-enabled social media networks, interplay with each other in a concoction that can potentially overthrow the 'old order' of how business is conducted in the policy sphere (Caballero-Anthony et al. 2012: 170–193). As seen in Figure 25.2, policymakers in Asia will have to contend with an increasingly better informed public using internet-based repositories of alternative viewpoints about and social media for mobilising public opinion towards nuclear energy policies. In Asia, the widespread proliferation of CSOs willing and eager to forward their agendas on nuclear energy, as well as the use of new media, means that future nuclear energy policy discourses can only be conducted in a 'business–as–usual' manner, especially after the Fukushima disaster, at their own peril. The failure of nuclear energy proponents in the government and industry sectors to communicate intent and decisions in an open, transparent manner through a pluralistic policy participation process may only engender distrust, impede policy implementation processes, and invite a backlash. As IAEA director Yukiya Amano remarked in March 2012, public acceptance remains important for the expansion of nuclear power generating capacity (*Business Standard* 2013). Policymakers of existing and aspiring nuclear power users in Asia will definitely have to devote more effort into this aspect.

Beyond the state in nuclear energy policy discourse

In an increasingly interdependent world, an event which takes place in one country can have significant transboundary impact on the others. The same goes for energy policies. In nuclear energy planning, policymakers no longer have to take into account only the domestic stakeholders but must encompass the views and interests of a wider range of participants. Environmental fallout from a nuclear reactor accident, for instance, again as seen in the

Fukushima debacle, has transnational ramifications. The building of even a single NPP in one country may become a cause of consternation within its neighbourhood. For example, China's plans to build four 1,250MW NPPs in the Jingyu County, Baishan City on the fringe of the Jilin Province some 100 kilometres from Mount Baekdu on the Korean peninsula have reportedly garnered some concern within South Korea regarding the potential radioactive fallout from these plants in the event of a nuclear accident due to an eruption of the dormant volcano and its impact on the Korean peninsula (*Dong-A Ilbo Daily* 2011a). In a geopolitically turbulent region where security risks abound against the safe and secure operation of an entire nuclear energy production cycle, there can be considerable cause for concern. Interstate and inter-agency cooperation at the regional and global levels is therefore even more pertinent for the use of nuclear energy. The world's continued quest for nuclear energy as a viable alternative energy option means that the associated safety and security risks of nuclear reactor operations will always be present (Choi 2012: 112–139; Pomper & Harvey 2012: 140–169). The danger of proliferation of nuclear materials among rogue states and terrorist organisations for the weaponisation of nuclear energy – which cannot yet be fully addressed by existing IAEA safeguards or by the relevant provisions of the Non-Proliferation Treaty (NPT) – remains an omnipresent security threat that cannot be ignored by the world community.

Pomper and Harvey call for states to subscribe to the NPT Additional Protocol as a means to enhance safeguards (Pomper & Harvey 2012: 158–159), but the onus is again on the state actors themselves to decide whether they wish to conform to such regulations that can be perceived as endangering national sovereignty. While new technologies have been invented or are currently under R&D for the proper handling and disposal of nuclear materials and wastes, they cannot be regarded as a panacea to all problems (Choi 2012: 125). While it is important for international nuclear safeguards and institutions to be strengthened against the safety and security risks of nuclear energy, these measures cannot provide holistic solutions. To make these safeguards work effectively, and to ensure that new technologies for proper disposal of nuclear materials can translate from their drawing boards to real, widespread acceptance and deployment, it is imperative for state actors – i.e., governments – to take steps to win confidence amongst their neighbours (and the world community at large) and to clearly express their intentions not merely to their domestic audiences but also to the wider regional or international community. What this implies is a further extension of the earlier argument for a pluralistic mode of policy participation in the nuclear energy discourse. Beyond the state, this framework has to work and interplay at all levels – domestic, regional and international – in order to offer a comprehensive, interstate and inter-agency network to ensure safe and secure nuclear energy operations. This is certainly much easier said than done, for the world still currently operates on a Westphalian model in which state sovereignty over critical policies of national interests remains sacrosanct.

The future scenario may not be as dire as one might envision. In fact, the Fukushima disaster did bring about noteworthy multinational efforts to improve interstate cooperation in the field of nuclear safety and security. In late 2011 for example, the nuclear safety authorities of Japan, China and South Korea agreed to adopt joint measures to allow for rapid sharing of information of nuclear accidents and other safety issues regarding NPP operations (*Kyodo News* 2011e). The participating countries at the Seoul Nuclear Security Summit in late March 2012 pledged to strengthen international cooperative efforts to enhance nuclear safety and security (*Seoul Communique* 2012). In that regard, perhaps the Fukushima disaster serves as a useful wake-up call for the international community to begin taking more resolute action to enhance interstate cooperation, particularly in the field of nuclear safety and security regulatory measures. Despite these observed post-Fukushima optimistic signs, unless further bold steps are undertaken to enhance interdependence and cooperation in the nuclear energy sphere, the international community will likely have to continue living with the uncertainties of security risks facing sustainable nuclear energy operations.

Concluding thoughts

If there is really a nuclear renaissance under way in Asia, then the Fukushima disaster has surely put a dent in it, far short of eradicating regional interest in the region. If the ultimate goal is to reconcile socioeconomic development and fight climate change, the ideal manner to proceed is through sustainable development. In this model, however, there is no energy source that can be deemed to be without drawbacks, and this applies to virtually all types of alternative energy source, none of which is wholly 'clean' since its use entails some environmental costs. The need for holism in evaluating various means of attaining sustainable development, in calculating the costs and benefits of deploying a particular energy mix alongside energy efficiency and conservation measures, has become all the more necessary in the wake of Fukushima, an event that is far from eradicating the nuclear option from the table, but one that certainly alters the nuclear energy calculus of policymakers. Nonetheless, viewing nuclear energy as part of a system of systems, not as a panacea for future energy woes but as a solution that might operate in conjunction with other alternative energy sources in the mix, is potentially a prudent way to go.

Judging from the post-Fukushima nuclear energy developments in Asia, it seems that the nuclear quest remains very much alive, albeit a cautious one at that. Asian governments are beginning to seriously examine the potential pitfalls and benefits of the nuclear energy option. In that sense, while it might not have put a total brake on the nuclear quest amongst aspiring Asian nuclear energy users, perhaps the Fukushima episode has successfully impressed upon them the need for balance not in the short run, but in the long term. The Fukushima debacle will not only occupy a prominent place in the annals of notable nuclear accidents but also provide useful lessons for existing users and potential aspirants for nuclear energy.

Bibliography

Agence France Presse (2011a) 'Japan Crisis Could Slow Global Nuclear Development: IEA', 16 March.
Agence France Presse (2011b) 'US still Wants to Expand Nuclear Energy: Chu', 16 March.
Agence France Presse (2011c) 'France Says not Ready to Give Up Nuclear', 31 May.
Agence France Presse (2011d) 'Nuclear Plant Firms Adopt Landmark Code of Conduct', 16 September.
Agence France Presse (2011e) 'Swiss Parliament Approves Nuclear Plant Phaseout', 28 September.
Agence France Presse (2012a) 'Nuclear Power Only Option Despite Fukushima: Industry', 23 March.
Agence France Presse (2012b) 'China Lifts Ban on Nuclear Plant Approvals: Official', 25 October.
Associated Press Newswires (2011a) 'China says Japan Crisis Won't Deter Expansion of its Domestic Nuclear Power Industry', 26 March.
Associated Press Newswires (2011b) 'Japan Won't Abandon Atomic Power in its Energy Policy despite Nuclear Crisis, Official Says', 9 May.
Associated Press Newswires (2011c) 'Japan to Set up New, More Independent Nuclear Safety Agency under Environment Ministry', 11 August.
Associated Press Newswires (2011d) 'Angry Villagers Block Highway to Demand Closure of Nuclear Plant in Southern India', 13 October.
Bangkok Post (2011) 'Vietnam Holds on to Nuclear Policy to Drive Growth', 6 June.
BBC Monitoring Asia Pacific (2011) 'South Korea Plans to Build more Nuclear Power Plants – Minister', 25 December.
BBC Monitoring Asia Pacific (2012) 'Nuclear Energy Back on Philippines' List of Power Source Alternatives', 25 December.
Business Standard (2013) 'Public Acceptance Critical for Nuclear Capacity Addition, Says IAEA Chief', 12 March.
BusinessWorld (2012) 'Palace Keen to Study Proposals on Nuclear Power Technology', 17 April.

Caballero-Anthony, M., Lina, A. & Punzalan, K.D.G. (2012) 'Civil Society Organizations and the Politics of Nuclear Energy in Southeast Asia: Exploring Processes of Engagement', in R. Basrur & K.S.L. Collin (Eds.), *Nuclear Power and Energy Security in Asia*, Routledge Security in Asia Pacific Series (Routledge).

Choi, J.-S. (2012) 'Nuclear Energy and Security Risks', in R. Basrur & K.S.L.Collin (Eds.), *Nuclear Power and Energy Security in Asia*, Routledge Security in Asia Pacific Series (Routledge).

Diesendorf, M. (2012) 'The Economics of Nuclear Energy', in R. Basrur & K.S.L. Collin (Eds.), *Nuclear Power and Energy Security in Asia*, Routledge Security in Asia Pacific Series (Routledge).

Dong-A Ilbo Daily (2011a) 'China to Build Nuclear Power Plants at Mount Baekdu', 4 April.

Dong-A Ilbo Daily (2011b) 'Opposition Parties, Civic Groups Protest Nuke Plant Project', 24 December.

ET Net News (2011) 'Beijing Planning Six-fold Increase in Nuclear Power – Press', 13 May.

Ewing, R.C. & Ritsema, J. (2011) 'Fukushima: What Don't We Know?', *Bulletin of the Atomic Scientists*, 3 May. Available at http://www.thebulletin.org/web-edition/roundtables/fukushima-what-dont-we-know.

Financial Express (2012) 'PM Reiterates Support for N-energy', 17 May.

Gupta, S. (2011) 'You Should Worry less about N-energy than about Walking on the Streets or Driving in Delhi', *Indian Express*, 22 March.

International Atomic Energy Agency (IAEA) (2011) 'IAEA International Fact Finding Expert Mission of the Fukushima Dai-Ichi NPP Accident following the Great East Japan Earthquake and Tsunami', 24 May–2 June. Available at http://www-pub.iaea.org/MTCD/meetings/PDFplus/2011/cn200/documentation/cn200_Final-Fukushima-Mission_Report.pdf.

International Atomic Energy Agency (IAEA) (2012) Energy, Electricity and Nuclear Power Estimates for the Period up to 2050, Reference Data Series No. 1, 2012 Edition, Vienna, International Atomic Energy Agency.

Jakarta Post (2011) 'Nuclear Energy Still Possible for Indonesia: Official', 13 June.

Jiji Press English News Service (2012) 'Interview: U.N. Chief Says Nuclear Energy Can Still Be Important', 9 March.

Korea Herald (2011) 'Nuclear Power is a Bridge for Cleaner Future', 11 October.

Korea Times (2011) 'PM says Korea Will Continue to Pursue Nuclear Energy', 11 April.

Kyodo News (2011a) 'UPDATE1: Kan Says 20% of Japan's Energy to be from Natural Resources in 2020s', 26 May.

Kyodo News (2011b) 'UPDATE1: Paris Seminar Agrees to Strengthen Nuclear Safety', 7 June.

Kyodo News (2011c) 'Japan to Retain Nuclear Power as Main Energy Source: Kaieda', 15 June.

Kyodo News (2011d) 'Some Asian Countries Delay Nuclear Programs in Wake of Fukushima Incident', 6 July.

Kyodo News (2011e) 'Japan, China, S. Korea to Share Nuclear Plant Info more Quickly', 29 November.

Kyodo News (2012a) 'Malaysia's Nuclear Power Plant Project Faces Delay', 15 January.

Kyodo News (2012b) 'Urgent: Pakistan, China Mull Ambitious Nuclear Power Deals: Sources', 13 February.

Kyodo News (2012c) 'Japan to Develop Maritime Renewable Energy', 25 May.

Kyodo News (2012d) 'Hosono Wary about Immediately Ending Japan's Nuclear Reliance', 21 August.

Kyodo News (2013) 'UPDATE1: Abe to Comprehensively Decide on Restart of Reactors in Japan', 24 March.

LKBN ANTARA (2011) 'Nuke Power Plant Idea not Applicable in RI: Deputy Minister', 7 November.

Manila Bulletin (2011) 'Unfazed by Fukushima, Kepco to Add 11,200-MW Nuclear Capacity', 14 June.

National Diet of Japan Fukushima Nuclear Accident Independent Investigation Commission (NAIIC) (2012) The Official Report of The Fukushima Nuclear Accident Independent Investigation Commission, The National Diet of Japan. Available at http://warp.da.ndl.go.jp/info:ndljp/pid/3856371/naiic.go.jp/en/report/.

Nucleonics Week (2011) 'Nuclear Industry Will be Changed by Fukushima but Will Survive, Companies Say', *Nucleonics Week* 52(9), 12 May.

Onishi, N. & Belson, K. (2011) 'Culture of Complicity Tied to Stricken Nuclear Plant', *New York Times*, 26 April.

Onishi, N. & Fackler, M. (2011a) 'Japanese Officials Ignored or Concealed Dangers', *New York Times*, 16 May.

Onishi, N. & Fackler, M. (2011b) 'In Nuclear Crisis, Crippling Mistrust', *New York Times*, 12 June.

Peimani, H. (2012) 'Is the Rapid Growth of Reliance on Nuclear Energy an Economically Viable Option?', in R. Basrur & K.S.L. Collin (Eds.), *Nuclear Power and Energy Security in Asia*, Routledge Security in Asia Pacific Series (Routledge).

Philippine Star (2011) 'Noy Won't Back BNPP Opening', 18 March.

Pomper, M.A. & Harvey, C.J. (2012) 'Nuclear Power and Proliferation: The Risk of the Nuclear Renaissance', in R. Basrur & K.S.L. Collin (Eds.), *Nuclear Power and Energy Security in Asia*, Routledge Security in Asia Pacific Series (Routledge).

Press Trust of India (2011) 'India Cannot do Without Nuclear Power: Prin Scientific Advisor', 29 March.

Press Trust of India (2012) 'India has Plans to Produce 20,000 MW of Nuclear Power: Kalam', 2 November.

Press Trust of India (2013a) 'CCS Nod for Two More N-plants at Kudankulam', 20 March 2013.

Press Trust of India (2013b) 'Kudankulam Nuclear Plant Commissioning Delayed Further', 20 March.

Raadal, H.L., Gagnon, L., Modahl, I.S. & Hanssen, O.J. (2011) 'Lifecycle Greenhouse Gas (GHG) Emissions from the Generation of Wind and Hydro Power', *Renewable and Sustainable Energy Reviews* 15(7), 3417–3422.

Reuters News (2011a) 'Update 2 – China Pushes ahead Pakistan Nuclear Plant Expansion', 24 March.

Reuters News (2011b) 'UPDATE 1 – Time to Rethink Global Nuclear Safety Regime – UN', 26 March.

Reuters News (2011c) 'German Upper House Unanimously Backs Nuclear Exit', 8 July.

Reuters News (2011d) 'China Needs to Act Fast to Develop Nuclear Sector – Official', 30 August.

Reuters News (2012a) 'Nuclear Power Champions Japan and France Turn Away', 14 September.

Reuters News (2012b) 'UPDATE2 – Japan Aims to Abandon Nuclear Power by 2030s', 14 September.

Seoul Communique (2012) 2012 Seoul Nuclear Security Summit, held in Seoul, 26–27 March. Available at http://www.thenuclearsecuritysummit.org/userfiles/Seoul%20Communique_ FINAL.pdf.

Straits Times (2011) 'Many at WEF Feel Nuclear Energy Still Relevant', 13 June.

Straits Times (2012) 'Parliament: Current Nuclear Technology not Suitable for S'pore: Study', 16 October.

Tay, S.S.C. & Paungmalit, P. (2012) 'Critical Environmental Questions: Nuclear Energy and Human Security in Asia', in R. Basrur & K.S.L. Collin (Eds.), *Nuclear Power and Energy Security in Asia*, Routledge Security in Asia Pacific Series (Routledge).

Tertrais, B. (2011) 'Black Swan over Fukushima', *Survival* 53(3), 91–100.

Thai News Service (2012) 'Thailand: Nuclear Power Study Planned: Minister', 2 November.

UNI (United News of India) (2012) 'India, Russia Decide to Go Ahead with their Nuclear Cooperation', 24 December.

United News of Bangladesh Limited (2011) 'Dhaka–Moscow Sign Cooperation Accord on Rooppur Nuclear Power Plant', 2 November.

United News of Bangladesh Limited (2012) 'Govt to Build Another Nuclear Power Plant in Southern Region: Hasina', 6 September.

Vietnam News Agency Bulletin (2013) 'Government to Promote Nuclear Power', 7 March.

Wallsten, P. & Yang, J.L. (2011) 'Obama Backs Nuclear Power', *Washington Post*, 19 March.

World Bank Global Information and Communications Technologies statistical database (n.d.) Available at http://data.worldbank.org/indicator/IT.NET.USER.P2/countries/1W?display=map.

World Nuclear Association (2012) 'Plans for New Reactors Worldwide', August 2012. Available at http://www.world-nuclear.org/info/Current-and-Future-Generation/Plans-For-New-Reactors-Worldwide/.

World Nuclear Association (2013) 'World Nuclear Power Reactors & Uranium Requirements', January 2013. Available at http://www.world-nuclear.org/info/Facts-and-Figures/World-Nuclear-Power-Reactors-and-Uranium-Requirements/.

Xie, Y. (2011) 'South Korea Unveils Plan to Rank Among World's Top Nuclear Exporters', *Nucleonics Week* 52(48), 1 December.

Xinhua News Agency (2013) 'China Targets 20-pct Growth in Installed Nuclear Power', 12 March.

Yonhap English News (2012) 'Lee: Nuclear Power Plants "Core Staple" for S. Korea', 26 November.

Part VII

Conclusion

26

Environmental change in Asian societies

Lessons learned and future directions

Graeme Lang and Paul G. Harris

In the introduction to this handbook, we previewed the work of the contributors. In this concluding chapter, we address some general conceptual issues raised or implied by their chapters, project some of their analyses into the future, and outline some general themes and proposals of our own. We do not claim that the authors of the previous chapters always agree with our themes and proposals. But a number of these ideas are implied or explicit in the preceding chapters.

We begin by noting again that, with nearly half of the human population of the planet, and some rapidly growing economies, Asian societies will determine much of the future of human impacts on the regional and global environment. At the same time, it is clear that Asia's environmental impacts are partly due to Asia's integration into the global capitalist system, with its highly interconnected cross-national networks for extractions from nature and for the production and distribution of goods. This globalized economic system has led to the transfer of much of the world's material production from non-Asian to Asian countries, mainly because of lower costs for land and labor but also to take advantage of minimal environmental regulation compared with the West. The consequence has been massive and growing extractions from nature and increasing flows of waste and toxic materials back into it. In short, some of the responsibility for Asia's environmental problems rests with the overseas populations which consume the products and services generated by Asian societies. But this is a choice that a number of Asian societies have embraced, and so they must deal with the inevitable and growing environmental impacts and problems in environmental management and governance.

The rapid growth in per capita incomes and consumption in many Asian societies, particularly among the rising and increasingly affluent middle classes in a number of these societies, has also contributed greatly to many types of adverse environmental impact in the region and around the world. The country with the biggest and still growing environmental impact in Asia, and indeed globally, is China. Environmental conditions, policies and politics within China have received much recent attention from environment-oriented scholars and journalists (e.g., Economy 2004; Harris 2011, 2012; Shapiro 2012; Simons 2013; Smil 2004; Watts 2010; Zhang & Barr 2013).The literature in English on these problems and processes in China is much richer and deeper than for any other country in the region. This work illuminates the very complex environmental politics in China, the limited but often ardent environmental activism and environment-related protests and confrontations there, and the country's evolution of environmental governance through both bottom-up and top-down processes and initiatives. (We recommend that readers consult that literature for more extensive investigations and analysis than we were able to include in this handbook.) It is important to point out that much more comparative research is needed to tell us more about human–environment relationships in all Asian societies.

The chapters in this handbook make substantial contributions to that kind of cross-national and comparative research for the region. In this chapter we offer some proposals for further work, beginning with the need for more future-oriented research, including 'models' and 'scenarios.' We end with an important admission: although this handbook is largely – and we hope adequately – academic in its treatment of these topics, it does not always convey the urgency of some of Asia's pressing environmental problems and the continuing depletions and degradations of key resources in Asia and in societies strongly impacted by the region's resource needs. In other words, the book is perhaps not sufficiently frightening. We provide some explanation toward the end of this chapter, and we offer some non-academic proposals to complement the further studies that we recommend in the sections below.

Models, scenarios and future-oriented research

For the topics covered in this handbook, we have focused mostly on the past 10 or 15 years. Where future developments were discussed, it is usually the near future – the next 10 or 15 years at most. Some problems are urgent, and researchers and policy analysts are forced to focus on immediate policy options in regard to forests, water contamination, air pollution, coral reefs, forests, and so on. Most of the environment-related political confrontations between aggrieved citizens and local authorities are focused on immediate problems, or NIMBY (not-in-my-backyard) reactions to local environmental impacts or threats. Most government responses and programs are also focused on these current problems, addressing conditions in the coming few years. The planning horizon is seldom longer than about a decade. This is typical of environmental governance in almost all contemporary societies. But this is far too short sighted. To the extent that we can see what is coming over the next few decades, we need to push the policy horizon farther out from the present, as others have argued (e.g., Diamond 2005: 522–523). To the extent that there are major uncertainties, we need more research and analysis to try to get a better grasp, not least for policy–advocacy, on where we are going.

Research should also be focused on the history of environmental conditions and changes in Asian societies up to the twentieth century, to illuminate what has been lost and to make ongoing degradation more striking for our contemporaries, and more poignant. It is important to try to counteract what ecologist Daniel Pauly (1995) has called the 'shifting baseline syndrome,' in which each generation gets accustomed to the increasingly poor environmental conditions because most of those conditions change slowly, leading each new generation of citizens and analysts to see and experience them as 'normal.' This phenomenon has also been called 'creeping normalcy,' which is even more likely where trends with potentially huge long-term implications are partly disguised by shorter-term 'noisy fluctuations' around the trend (Diamond 2005: 425).

Some past societies collapsed because they did not have the scientific capacity to study and understand on-going but slow-moving environmental degradation, or to see that they were undermining the ecological basis of their own society in a particular location (Diamond 2005: 422). We will not have that excuse. The scientific capacities of contemporary modern societies are unparalleled. The application of sufficient scientific research gives us reasonable projections and scenarios which are highly relevant for contemporary policy. If contemporary societies fail to fund and mobilize the scientific expertise for that kind of work, despite the many warnings and calls for more research by scientists who study those phenomena, this will be a major failure of our political and educational institutions. But we also need to distinguish the key types of environmental impacts, and allocate research resources accordingly.

'Inputs' and 'outputs'

Much environmentalist activism and discourse is focused on the 'outputs' from extraction, production, and consumption: pollution, environmental degradation, accumulating municipal and toxic waste, health impacts, and the related concepts of negative externalities, environmental justice, and mitigation. But the *inputs* are also part of the interaction between human societies and the natural world. Indeed, the concept of 'sustainability' is mostly about the *sustainability of inputs* – whether we can continue to operate a society in its current form, with current levels of consumption and with a similar or greater population, without completely depleting the resources on which that society depends. This is the 'hard' definition of sustainability. In many uses of 'sustainability' in contemporary discourse, the term is misused to refer only to goals such as increased energy efficiency in 'green' buildings, reducing and recycling waste or water, reducing consumption to save costs or to reduce the rate of depletion of resources, and so on. This kind of effort, while important and valuable to make better use of resources, does not deserve the label of 'sustainability' initiatives if we take the literal definition of the word.

Inputs include fossil fuels, metals, wood, foods, phosphates, fiber, and water. A comprehensive environmental policy analysis that is future oriented has to include the withdrawals from nature as well as the dumping of material back into it. Indeed, where we are withdrawing non-renewable but key resources which have no apparent substitutes, and where we can estimate the depletion of those resources within a human lifetime or less, the urgency of addressing the issue is obvious – at least to the scientists and analysts who have made those projections. The most important of these inputs are the resources used to generate energy, specifically, fossil fuels. Oil is the key resource for transportation in all contemporary societies, while coal and natural gas are used mainly for generating electricity and for heating and cooking. China and India both depend heavily on coal to produce the electricity that is driving their giant economies. Both countries pay a heavy price for their reliance on coal, particularly in the health impacts and increased mortality from the resulting pollution. China gets close to 70% of its primary energy and up to about 80% of its electricity from burning coal (Miao & Lang 2010). All of the countries in the region also depend heavily on oil for transportation and trade. Without oil, their economies would be severely crippled and would contract, resulting in a substantially lower standard of living. Without coal, their economies would collapse.

Up to the present, the economies of Asian societies have not begun to run up against the limits of these energy resources, even though the prices of oil and coal are much higher than in the 1990s. But there are a growing number of analysts who have studied the history of oil discoveries, extractions, and consumption, and have concluded that a 'peak' and eventual inevitable decline in global oil production is approaching, and that if we are not already in this 'peak oil' period, these analysts believe that it could occur during the coming one or two decades (e.g., Heinberg 2005). Demand for oil will continue to be strong for many years to come, especially with rapidly growing economies such as in China where rapid increases in car ownership, tourism, and air travel require its use. It is possible that the production of conventional oil has already reached a peak, and may be declining. Even if this is not the case, it is very likely that the cost of oil will continue to go up – perhaps to the point where supplies last longer than expected simply because societies cannot afford to use oil so freely – inevitably affecting those societies and economies that have come to rely on relatively cheap oil and have not already started to find alternatives.

The increasing scarcity of historically cheap oil has led to increasing reliance on so-called 'unconventional oil' from shale, tar sands, and deep-ocean oil drilling, most of which is much more expensive to extract and to process into useable hydrocarbons. The era of major oil discoveries seems to have come to an end decades ago. Recent discoveries, such as the Tupi field off the coast of Brazil, with possible reserves of 8 billion barrels of oil – albeit lying under 7,000 feet of water, 10,000 feet of sand and rock, and more than 6,000 feet of salt – would only supply the world economy for about three months if it were the only source of oil. Already, a number of oil-producing countries have seen a peak and substantial decline in their own conventional oil production, even as the price of oil has climbed to the range of US$80–110 per barrel. Shale oil and 'tight' oil may be unable to replace declining production in many oil-producing countries around the world, especially because wells drilled in shale deposits have very rapid rates of decline in output (Hughes 2013).

This potential 'peak oil' period – whether it is manifested in actual shortages of oil supplies or shortages in affordable oil – has huge implications for the global economy and for cities which depend on fossil fuels for electricity, trade, transportation, and mass tourism. This has already begun to provoke some longer term analysis for Asian societies, notably China and India (e.g., Li 2007). One of the most obvious implications is that if continual economic growth, combined with population growth, inevitably requires growing consumption of resources, even under the most optimistic scenarios of increased energy efficiency, then continued economic growth is not only impossible in the long term, but also that the period of rapid global economic growth, driven especially by economic growth in Asia, may be cut short by the growing costs of energy and the eventual depletion of the sources of that energy.

In response to these and related trends, there is now a growing body of research, especially in North America and Europe, which tries to envision a 'zero-growth' society and economy, and which argues that we must begin to make the transition to such an economy without delay (e.g., Heinberg 2005, 2011; Jackson 2009; Rubin 2012). Applying this kind of analysis to Asian societies is even more difficult in developing countries than in the rich countries of the West because most Asian governments are laboring under the political imperative that they must foster economic growth, as rapidly as possible, to bring even more of the population out of poverty and to satisfy the growing aspirations for affluent lifestyles of their rapidly growing middle classes. Nationalist citizens in some of these countries are particularly outraged at any suggestion by local or overseas analysts that they should restrict economic growth for any reason, including climate change and environmental impacts, and would condemn political leaders who echoed such ideas if that appeared to represent concessions to pressure from the rich countries of the 'West.'

But the projections in regard to the supply of oil (e.g., Murray & King 2012) and of coal (e.g., Heinberg & Fridley 2010) in the coming decades lead to the inevitable conclusion, for those who have studied the data and scenarios, that the transitions to low-carbon economies and cities have to get much more attention in public discussions and policy debates. Unfortunately, these calculations and perceptions are mostly confined to scientific circles inside and outside government, and followed closely only by environmental nongovernmental organizations (NGOs) and some concerned citizens. There is a scarcity of political leaders with the knowledge, vision, and courage to raise the issue of true 'sustainability' and genuine, rather than merely rhetorical, harmony with nature. At the very least, ministries of environmental protection in every country should provide this kind of vision and leadership, even if this challenges the usual combination of political and economic interests which are pursuing economic growth as the primary objective. Serious debate is needed in all societies.

Depletion of the 'commons'

Some ecological resources are 'commons' where depletion merely leads to competition to get as much as possible rather than to let others harvest or use the resource and gain all of the benefit. This is the classic 'tragedy of the commons.' The 'tragedy of the commons' in Asia can be observed especially in the decline of fisheries, forests, aquifers, rivers and lakes, the atmosphere, and many species of animals. Protecting commons where the resource is above the ground, and where exploitation can be more easily observed, is feasible, although it requires extensive monitoring and enforcement. Deforestation was largely halted in China by vigorous and determined state action after massive floods in 1998, which were attributed in part to deforestation along major rivers (Lang 2002a). Similar state interventions occurred in some other societies in the region, as for example in Thailand and the Philippines, following floods or landslides (Lang 2002b).

This illustrates a pattern that is characteristic of many polities on many issues: where solutions would be expensive, and problems are developing slowly, the state typically engages only in weak proactive policies in response to weak pressures from civil society, NGOs, and government critics. It is only in the aftermath of a disaster that the state takes strong and vigorous action, and only to the extent that the state's legitimacy is challenged by its failure to anticipate the disaster and prevent it or prepare adequately.

Unfortunately, this kind of disaster is especially likely where the depletion of the resource is not highly visible or easily monitored, and where the depletion does not cause immediate harmful impacts. The powerful political and economic imperatives that produce only weak proactive measures to protect these resources still operate in all of the societies in Asia, and hence we can observe continuing and unsustainable depletion of fisheries, forests, topsoil, and aquifers, and continuing massive emissions of pollutants into the air and atmosphere. Despite the well-articulated concerns of international organizations, NGOs, environment bureaus, and some politicians within these societies, the depletions continue. In some cases, such as the depletion of primary forests in Papua New Guinea to supply the trade in logs going to China, the rates of depletion are apparently accelerating (Simons 2013). In more than a few of these societies, this generic problem is closely linked to government corruption.

In this volume, we have not tried to outline generic solutions to the problem of unsustainable depletion or pollution of commons that are typically successful. Certification schemes such as by the Forest Stewardship Council (FSC) or the Marine Stewardship Council (MSC) have some effect. However, such schemes are beset by difficulties in monitoring the resources they aim to protect, and by the methods and high rates of extraction in many locations, and they are vulnerable to weak oversight and sometimes by conflicts of interest (see, for example, Jacquet et al. 2010, on MSC). If the depletion of a key local resource has already begun to affect or threaten a local community, that community may mobilize to protect its resources or obstruct their extraction, assuming local citizens have sufficient social capital and collective cunning to do so. Hence, there are some resource-depletion issues in Asia, focused especially on environmental 'commons,' which do generate much local interest, debate, and contention, even though these are not typically about energy. There are some forms of social organization at the local level that make a difference in depletion of some commons, leading in the best case situations to sustainability (e.g., Ostrom 2009), and local leadership and local social capital have also been shown to be important (Gutiérrez et al. 2011). But there are many 'commons' conditions that *facilitate* unsustainable extraction, particularly where local poverty, corruption, and weak monitoring of natural resources are endemic. Unsustainable depletion is particularly likely where the resource is also not visible to most people in the society.

Visibility, immediacy and environmental research

The public and political salience of environmental problems depends especially on three variables: the visibility, severity, and experience of the impacts in society. If visibility is low and people have not yet begun to experience the impacts, or they have not been able to prove that what they are experiencing is a result of the environmental problem, action is unlikely. This is the case even if the impacts over the longer term are likely to be severe according to longer term scientific analysis. Fisheries are classic examples of this phenomenon. It is only when fish stocks are collapsing and some fish species become increasingly expensive that most people become very engaged in the problem. In turn, governments fail to regulate adequately the taking of fish. Aquifers provide another good example: they are vital for agriculture in some parts of Asia, yet they are being depleted toward 'empty' in areas of China, India and beyond. The regions' aquifers are invisible. Apart from some scientists, the only people who are acutely aware of the depletion of aquifers are the farmers who have to dig their wells deeper each year to find water. The result is too little action by governments to preserve water and protect aquifers.

Most air pollution is highly visible, thus explaining why it is so often discussed in the media and a concern of publics and governments – which is not to say that the problem is being addressed sufficiently. In contrast to the obvious visibility of serious air pollution, carbon dioxide pollution, which is the chief cause of global warming and other manifestations of climate change, is invisible. It has no direct impact on people's lives or health because the consequences of today's carbon pollution will be felt far into the future (with today's climate change the result of pollution in past decades and centuries). Hence, carbon dioxide emissions from countries that rely heavily on coal, such as China, which contribute massively to future global warming, receive only a tiny fraction of the concern that particulate air pollution from power plants gets in China. This is simply because the particulate pollution is highly visible, the health impacts are personal, and the discomfort is immediate. In short, invisible pollution of the global commons (e.g., carbon dioxide in the atmosphere) is much less visible, immediate, and interesting than the visible pollution of the local commons (e.g., smog in the air around cities).

We can classify many of these issues using a simple typology of visibility and immediacy, as in Table 26.1.

Environmental impacts that are highly visible and immediately felt will receive the most attention from citizens and governments, weighted by the perceived severity of the problem.

Table 26.1 Visibility and immediacy of impacts

		Visibility of impacts	
		High	Low
Immediacy of impacts	High	* Air pollution * Industrial disasters	* Declining water table
	Low	* Deforestation	* Carbon dioxide emissions * Depletion of fossil fuels

In contrast, low-visibility, longer term problems will receive much less attention, fewer government resources, and lower citizen interest even if the longer term consequences are likely to be more severe than most of the high-visibility, immediate-impact problems, especially if the longer term consequences are in the 'indefinite future'.

More scientific research into long-term modeling and scenarios is needed, especially where resource depletion and its consequences are not well understood and where policymakers and activists do not yet receive sufficient guidance from the scientists. Examples of areas requiring much more attention are the long-term impacts of climate change, and depletion of key resources such as phosphates. To be sure, many scientists, academics, and environmental NGOs already attempt to raise public awareness for these and other low-visibility, longer term issues. In doing so, they play a very important role in environmental consciousness raising and in stimulating public discussion and political debate. Sometimes these efforts eventually do have some impacts. For example, in Hong Kong the 'Save our Seas' campaign promoted by WWF helped to prod the local government into declaring a ban on trawling in Hong Kong waters, thereby facilitating a recovery in local fish stocks. Another example is advocacy across the region, which may have contributed to the banning of the serving of sharks' fins at government-funded dinners in China. These sorts of efforts by NGOs, citizens' groups and environmental bureaus need to be assisted by much greater policy-oriented research.

At the same time, it must be acknowledged that not all research on environmental impacts and projections can be easily transferred into policy deliberations. All scientific papers on environmental impacts could include some reference to the question of how applicable the research is to policy (even if it just suggests trials or experiments) and, if the scientific consensus does not reach that level, what additional research would be needed to bring it to the level of well-supported policy applications.

One approach might be for the editors of academic journals and the organizers of academic conferences to require that policy relevance be made explicit in all articles and papers, with conclusions outlining what further research would be needed to bring the analysis to a reasonable level of confidence for policy applications. This kind of approach could make a substantial contribution to firmer scientific grounding for policies, and also for policy-advocacy by NGOs and international organizations, so that environmental issues with less obvious immediacy or impact move up the policy agenda.

The need for research on environmental externalities

In addition to more research and modeling on resource depletion and its consequences for Asian societies, we desperately need more research and analysis on the externalities that arise out of extraction, production, and consumption. These impacts on other sectors of the economy (e.g., the impacts of water pollution on fisheries), on health (e.g., the impact of particulate pollution on health and mortality), and on government costs (e.g., clean-up, repair of public facilities, public health expenditures) are not currently priced in the market for goods and services. Environmental governance would benefit from a much better analysis of these unpriced costs, because that analysis could be used by environmental bureaus within government, and by NGOs and citizens' groups outside government, to press for greater regulation, special taxes (i.e., bringing the externalities into the costs paid by consumers for those goods or services), or outright prohibitions for some of those externality-generating processes where the long-term costs are severe.

Bits and pieces of this kind of research have been done by a variety of scientists and research teams, particularly on the health impacts and increased mortality of particulate air pollution.

(Some of that research has been reviewed in this handbook.) But this kind of research must receive stronger government support with much greater funding. Additional research funds could be targeted at particular kinds of externalities, or at externalities from particular processes of production and consumption. This is urgent because so many of these processes are leading to increasing degradation of key resources – water, forests, fisheries, soil, air – and because some of the impacts are longer term and require prolonged research and comparisons of results, further research, and meta-analyses to arrive at the point where those results are useable for compelling advocacy and eventual public policy decisions. However, there are huge differences between countries in Asia in regard to their capacity to undertake research on their own environmental problems and resources issues. Wealthy societies with highly educated middle classes and elite universities, even in small polities such as Singapore, have much greater capacity than relatively poor countries to undertake the needed policy-relevant research and analysis.

Trials or pilot experiments have been conducted in China to test policy innovations, such as the Chinese government's recent trial of carbon trading in Shenzhen, or local innovations in Jiangsu that eventually led to a national scheme of National Model Cities of Environmental Protection (Li et al. 2011). This kind of experimental approach to policy innovations could be applied to many environmental problems in other countries in the region.

Inputs from international organizations and collaborations with researchers and research institutes outside the region can provide some of the needed knowledge, but these collaborations are often precarious and vulnerable to local political imperatives. Where a developing country is small or poor, lacking the resources to fund an extensive environmental bureaucracy or scientific institutes that have the freedom and resources to provide policy-relevant research, it can be argued that other countries, particularly in the region, could and should offer assistance with research on local conditions and possibilities, in collaboration with, and helping to train, local researchers. This can be politically sensitive if it appears that another more economically developed country is dictating research priorities or desired policies, especially if there are local suspicions that the research is designed to enable resource extraction by foreigners. It is even more questionable if research is targeted toward policies that are not implemented even in the developed countries, much less in the poor, developing ones.

However, Japan, China, South Korea, Thailand, and Singapore could and arguably should devote more research resources to helping their neighbors to get accurate longer term data and analysis to improve environmental governance. In forests, water management, eco-city experiments, energy efficiency, municipal waste management, recycling, and emissions caps or trading, substantial benefit could be accumulated by greater sharing of research and policy expertise, perhaps in collaboration with international organizations.

Some of these potential collaborations are highly contentious. For example, China's building of dams on rivers that run into Southeast Asia has raised the issue of river control and the possibility of limits to national sovereignty where rivers are shared among several countries. Supra-national bodies are needed in such circumstances, but nationalistic priorities interfere with efforts to set up or empower such bodies. Scientists and policy analysts in India and Pakistan are already concerned about the longer term supply of water to southern Asia from the glaciers in the Hindu Kush, Karakoram, and Himalayas, which are retreating as a result of global warming and the impact of black carbon. Some governments have begun to call for the kind of long-term modeling and analysis that is needed, but collaboration of Indian and Pakistani scientists is complicated by geopolitical and ethnic tensions and antagonisms, and by the importance for these countries of irrigation water from the same high mountain glaciers (Laghari 2013).

Nevertheless, much more research on the externalities of current economic activities is needed. It will be a crucial part of attempts to assess longer term environmental sustainability and to adjust public policies accordingly. It should also be brought into the education of each new generation of citizens so that they have a better understanding of the ecological impacts of their own activities, individually and collectively. It is probable that this is the only way by which public support for government initiatives to conserve and protect longer term environmental conservation can be increased. Nongovernmental organizations alone cannot have sufficient impact to really change the politics of Asian societies, despite their ability to mobilize people on some specific environmental issues.

Green GDP

Research on externalities can also feed into calculation of the so-called 'Green GDP', or the net GDP (gross domestic product) after subtracting the costs to the society and economy from pollution and other environmental impacts. Between 2004 and 2007 China undertook the most ambitious efforts to calculate a Green GDP. Although this effort met considerable resistance from local governments, and partly for that reason, was withdrawn into research institutes for further work, the effort is ongoing (see Li & Lang 2010). The ideas involved in the 'Green GDP' exercise continue to be raised in public discourse, including in other countries. Recently, Vietnam's government announced that it would attempt to undertake a similar analysis.

Green GDP cannot produce a comprehensive and full account of environmental costs because some of those costs are not quantifiable (e.g., suffering, psychological distress, reduced quality of life) or are difficult to calculate because the necessary research, such as on the depletion of aquifers in China and India, has not yet been done or must extend over decades because most impacts occur in the mid-term future, as is the case with air pollution effects on cancer mortality, impacts of aquifer depletion on agriculture and so forth. Nevertheless, the effort to quantify externalities has political value in calling attention to environment-related deficits and impacts from economic growth, and in introducing these considerations into political and economic debate. This kind of analysis could be greatly strengthened if there were much more research on the negative externalities of current forms of production and consumption.

Research on estimating the extent and costs of externalities is necessary in order to facilitate better environmental governance on specific issues, and eventually, to produce credible and politically useful Green GDP estimates. The link between scholarly research and government policy could be formalized with institutional exchanges and regular conferences among the researchers working on particular environmental externalities and the official teams attempting to estimate Green GDP (as for instance in China and Vietnam).

Further thoughts on key topics

A number of important topics that have come up in previous chapters warrant additional consideration. In this section, we briefly address questions of food security, say more about democratization and decentralization, and expand on earlier discussions of environment and culture, here focusing on the role of religion.

Food security

Food security will be a major challenge across much of East Asia in coming decades. The biggest problems of food security are likely to occur over the longer term in China. Serious and growing internal threats to China's long-term food supply include pollution of agricultural land through polluted irrigation water and toxic dust and emissions from factories and power plants, the erosion of topsoil into rivers and lakes, the overuse of fertilizers in much of the country's agriculture, affecting domestic fisheries, and the expansion of cities and towns into agricultural zones, which permanently destroys some of the best agricultural land in the region. China's government tries to restrain the loss of agricultural land around cities with various regulations about converting such land to other uses, but this is only partially effective. Even an attempt to prevent conversion of prime agricultural land into golf courses has had only limited effect, with several hundred golf courses being built since that regulation was promulgated in 2004. (The government was evidently forced to rely on satellite photos to try to keep track of the construction of new golf courses.)

In the future, those municipalities within China, such as Chengdu, which have managed to retain intensive agriculture in the rural hinterlands of the city, supplying much of the city's food supply, will be much better able to deal with rising energy costs than municipalities that have become heavily reliant on industry and the export of manufactured goods into the global economy, and which have largely abandoned local agriculture and converted most of their farmland into factories, malls, and housing estates (Lang & Miao 2013). The apparent Chinese strategy of trying to acquire agricultural land overseas in order to bolster China's food security, such as China's massive agricultural land deal with Ukraine (Zuo 2013), is risky and probably unsustainable in the longer term.

Climate change also poses long-term threats to food security for a number of countries in the region. In China, scientists have tried to estimate and project the impacts of climate change on water resources and agriculture within the country, but the current models are apparently not yet adequate to make firm predictions over such a wide range of geographical conditions and climate zones (Piao et al. 2010). Other countries in the region are more clearly facing food security problems related to climate change.

For example, agriculture and fisheries in the Philippines are highly vulnerable to climate change-related disruptions, and the Philippines has been importing large quantities of rice from Vietnam (nearly 2.5 million tons in 2010) to cover shortfalls in national supply, partly as a result of extreme weather events such as typhoons. The recent Typhoon Haiyan in November 2013, reportedly the strongest typhoon to hit an Asian coastline in the records of Pacific typhoons, has caused massive damage to towns and cities in its path, killing thousands of people and destroying innumerable houses and buildings and a large amount of crucial infrastructure. But its longer term economic impact includes the destruction of fishing boats and equipment, and major longlasting damage and disruptions to agricultural production.

A single extreme weather event cannot be conclusively linked to climate change, but all of the climate models predict more severe storms arising over warming oceans, and the severity of Typhoon Haiyan is very likely an indication of the damage which climate change will inflict on agriculture in the future, especially through extreme weather. With a still rapidly growing population, and a large number of people living in poverty in poor-quality houses highly susceptible to destruction by storms, this is a major problem for the Philippines (Lang & Chow 2011). But there will also be large impacts on food production and food security. We have already seen that the Philippines turned to Vietnam to make up shortfalls in rice.

But the rice-producing deltas in Vietnam in turn are vulnerable to rising sea levels and incursion of saltwater into agricultural districts (Newb et al. 2009). Some farmers in Vietnam have turned from rice to shrimp (which can be raised in saltwater ponds) to deal with the loss of income from rice, but this poses a longer term threat to food security in those countries which depend on rice imports from countries such as Vietnam, and countries facing local food supply shortages in the future are likely to restrict exports of food in order to retain sufficient supply for their own population.

Food security issues demonstrate the strong links between some environmental inputs and outputs: pollution, soil degradation, and the predicted longer term problems with energy undermine and, in the longer term, jeopardize an absolutely vital input: the supply of food. An entire scholarly paradigm of unsustainability could be developed around an analysis of the impacts of these factors on food security, especially for cities.

Democratization and decentralization

There have been notable debates over the past few decades about the political conditions that are most favorable for environmental conservation and best practice. Some argue that democratization facilitates better environmental governance, and that decentralization also has this effect because it puts decision making about local environmental conditions into the hands of people who are most familiar with those environments, and presumably have the most to lose if the local environment is seriously degraded. Some community-managed commons work because the conditions are right for local good governance (Ostrom 2009). Others argue that higher level central government authorities are best placed to provide environmental governance because of their greater resources and access to expertise, and their ability to create and enforce environmental regulations, and indeed that some degree of authoritarian environmental restrictions will sometimes be needed, for example during national crises or wars between societies. The luxury of 'democratic governance' succumbs to the necessities of quick and efficient action. Where key national leaders are scientifically illiterate or incompetent but have authoritarian powers (as in China's 'great leap forward' between 1958 and 1962), the results can be disastrous. But decentralization, such as in Indonesia after the fall of President Suharto in 1998, can actually facilitate increased unsustainable resource extraction if local elites are free to profit from such extraction after the waning of national regulation and control (e.g., Lang & Chan 2006).

These debates and studies do not provide clear answers to questions about the relation between types of governance and environmental conservation. However, it seems to be clear from a great deal of research that the processes that have been characterized as 'ecological modern-ization', that is, the progressive improvements in environmental good-governance by firms, cities, and states, depends at least in substantial part on the contributions of vigorous environmental NGOs and critical environmental reporting in the media to produce the pressures on corpora-tions and governments that lead to this kind of environment-conserving progress (e.g., Mol 2002, 2006). Whether that kind of flow of pressures and inputs leads eventually to sustainability is another issue. For example, some analysts argue that it is inadequate because it leaves the basic processes of a nature-exploiting capitalist system untouched (see, e.g., York & Rosa 2003). However, the struggles to conserve the natural environment and reduce the negative impacts on the health and well-being of local populations within and around some societies in Asia is also a struggle against attempts by local and national elites to suppress such protests and to control the information on impacts, externalities, and the sources and consequences of pollution. In general, an open society with environmentally literate citizens, free media, and vigorous NGOs is better placed to challenge environmental degradation than is a closed and repressive society controlled by environment-exploiting elites.

Culture and religion

The relationship between culture and environment is something that deserves close attention. Earlier in this handbook, Swan and Conrad argue that local cultures in Asia often take different approaches to wildlife than in other or Western societies, and that conservation measures are unlikely to be effective without taking account of these values and cultural practices. However, it could also be argued that 'culture' can include beliefs that are now known to be false, and practices which are now known to be environmentally destructive, and that 'culture' deserves no particular deference where these beliefs and practices are part of the reasons for a society's baleful impact on its own environment or on depletion of resources or species locally or in other parts of the world. For example, the belief that rhinoceros horns or tiger parts have medical benefits and potencies is false. Such beliefs are highly destructive because they generate the flow of such material to Asia and the further depletion of highly endangered species. Educators, scientists, NGOs, and officials should take account of local beliefs that fuel demand for some animal products and hence lead to poaching, smuggling, and further serious depletion of those species as citizens with such beliefs become rich enough to pay for those products regardless of cost. To the extent that the 'cultural' beliefs or practices are both false and destructive, they become part of the problem, and therefore they should be confronted and changed if there is any hope of solving the resulting problems. Indeed, this is happening in regard to the eating of shark's fin, which has led to steady depletion of shark species around the world. This 'cultural' practice is being confronted and challenged by media activists, NGOs, and even celebrities in some Asian societies (including China).

There has also been considerable attention given to the possible relationships between Asian religions, notably Buddhism and Confucianism, and environmentalist attitudes and practices (see, e.g., Tucker & Berthrong 1998; Tucker & Williams 1997; Weller 2006). Environmentalism among religious leaders in Asia is affected both by global international discourse and local environmentalist struggles. But does religious doctrine, independent of other outside secular engagements or educational experiences, have any separate effect on environmental activism? One view, which focuses on religion and environmentalism in Chinese societies, is that religion has little impact on environmentalism directly, and that most religious leaders who take up environmental themes in their preaching or advocacy or political engagements are doing so mostly because they are intellectually engaged with the environmentalist discourses and issues of their own societies, not because they are prompted to do so by their own religious texts or doctrines (Lang & Lu 2011). It is, in fact, rather common for religious leaders to scour their sacred texts for 'relevant' passages only after they have already become involved in thinking about environmental problems.

However, religious leaders, as *cultural* leaders, can have some important influences on believers if they find ways to articulate environmentalist or anti-environmentalist themes and to connect them to religious values. For example, if religious believers think that the end of the world is imminent for supernatural reasons, or that 'God' takes care of the natural environment and so humans do not have to worry about it, and hence that there is no need to worry about longer term environmental problems, this is clearly relevant to attempts to get support from those believers for pro-environmentalist policies. But the scholars who have looked most closely at possible linkages between religion and environmentalism have been ever more eager to find support in some religions for environmentalism and conservation. This is a difficult effort with not many positive results. The main effects seem to occur as a result of the diffusion of environmentalist concerns into the thinking and activism of some religious leaders.

Some practices associated with particular religions in Asia, particularly vegetarianism among many Buddhists and Hindus, have huge environment-conserving consequences in those societies when compared with other societies where meat consumption is high or growing as a result of increasing affluence. These and other relations between culture and the environment have been explored only briefly in this volume; they deserve much further discussion and debate.

Cultivating environmental 'urgency' through the sciences, arts and humanities

Finally, we must acknowledge that the academic nature of this handbook reflects the kinds of research and analysis that are common in environmental policy studies, environmental sociology, and the work of geographers and scientists who study particular resources and the methods of extracting them from nature, and the consequences of those extractions. We have also used the work of some journalists who have done outstanding work investigating and reporting on environmental issues. But it is difficult to convey in this kind of volume the sense of urgency that many people feel is needed to move public debate and political action forward in trying to address some of these problems. Fisheries and forests continue to decline in much of Asia and around the world, water tables and aquifers continue to fall, crucial topsoil is eroded and lost, the approaching energy crisis creeps up on all modern societies that rely heavily on fossil fuels, and meanwhile greenhouse gas emissions continue to increase despite devastatingly futile international meetings and discussions about restricting emissions (see, e.g., Harris 2005). Climate change threatens coastlines with rising seas and increasingly violent storms, particularly in the many Asian societies with coastlines exposed to typhoons (such as in the Philippines) (see Harris 2003). In the longer term, climate change is also shrinking a number of the glaciers and snow packs on the Tibetan plateau and depleting their function as a water tower for hundreds of millions of people in East, Southeast, and South Asia. Meanwhile, the rising appetite in Asia and particularly in China for increasingly threatened species of animals, as food or medicines, has already driven many of them toward crisis levels. There is reason for increasing alarm, as indeed many scientists admit privately and sometimes publicly.

Many environmental NGOs express and enhance this kind of concern and sense of urgency. This handbook covers a wide range of environment-related topics in Asia by scholars and researchers, based on research using primary and secondary sources, combined with analysis of environment-related policies and governance. Most of the contributors hold academic or policy-oriented positions in universities and research institutes, and follow the conventions and methods of academic publications in their fields of research. However, we should also acknowledge that some excellent and influential work on environmental problems is carried out by staff and volunteers in local and international NGOs, which publish reports through their websites, frequently based on intrepid field research, which feature useful and often striking pictures, graphs, maps, executive summaries, and recommendations. These reports can be found on the websites of organizations such as Greenpeace, WWF (World Wide Fund for Nature, also known as the World Wildlife Fund), the Environmental Investigation Agency (EIA), and Global Witness.

Some of these reports use investigative methods which can be locally risky for the investigators, and which could not be carried out by academic researchers using university-funded research grants, but which provide valuable information and insights. Examples include the investigations of corruption and military involvement in illegal logging in Cambodia's Aural Wildlife Sanctuary (Global Witness 2004), and of similar illegal logging and corruption in Tanjung Puting National Park in Indonesia (EIA & Telepak 2002). Some of these reports seem to be effective in raising awareness and in putting pressure on corporations, governments, and international organizations. Scientists are not usually as accomplished in campaigning for action as the environmental NGOs, and are arguably not the most qualified people to arouse the needed sense of urgency, although some have tried in various societies.

Journalists are as important as scientists and NGOs for this endeavor, and there are some outstanding environment-oriented journalists, but we need much more education of journalists in environmental impacts and policies. All journalism schools should include such courses and training. Artists and film-makers can also contribute, through documentaries and dramas that highlight the issues, and especially by portraying for the public what is normally hidden from them – the depletions and depredations of forests and fisheries, rivers and oceans, the depletion of many wild animal species, and the degradation and conversion of farmland. There are some outstanding examples of such productions, and film schools should encourage and support more of that kind of work. Such work could be applied to the past – showing what has been lost and how that occurred – and to the future – what will be lost if societies do not change their ways. People cannot easily imagine and would not voluntarily think about what a society might look like after it has lost its natural forests, or its fish, or its supply of drinkable water, but artists with imagination can show worlds like this, and make us think about what may be coming.

While these forms of public education are needed everywhere, this volume has shown that the need for them in Asia is profound. We have called on governments to fund more research, and on academics to make their research more clearly policy relevant. We now call on artists, writers, and filmmakers – in Asia and beyond – to join in, more often and more powerfully, with their own contributions to environmental consciousness and action in Asian societies. Hopefully this handbook will contribute toward that objective.

Bibliography

Diamond, J. (2005) *Collapse: How Societies Choose to Fail or Succeed* (Viking (Penguin)).

Economy, E.C. (2004) *The River Runs Black: The Environmental Challenge to China's Future* (Cornell University Press).

EIA (Environmental Investigation Agency) & Telepak (2002) *Above the Law: Corruption, Collusion, Nepotism, and the Fate of Indonesia's Forests*. Available at http://eia-global.org/news-media/above-the-law-corruption-collusion-nepotism-and-the-fate-of-indonesias-fore.

Global Witness (2004) *Taking a Cut: Institutionalised Corruption and Illegal Logging in Cambodia's Aural Wildlife Sanctuary – A Case Study*. Available at http://www.globalwitness.org/library/taking-cut.

Gutiérrez, N., Hillborn, R. & Defeo, O. (2011) 'Leadership, Social Capital, and Incentives Promote Successful Fisheries', *Nature* 470, 386–389.

Harris, PG. (Ed.) (2003) *Global Warming and East Asia: The Domestic and International Politics of Climate Change* (Routledge).

Harris, P.G. (Ed.) (2005) *Confronting Environmental Change in East and Southeast Asia: Eco-Politics, Foreign Policy, and Sustainable Development* (Earthscan).

Harris, P.G. (Ed.) (2011) *China's Responsibility for Climate Change: Ethics, Fairness, and Environmental Policy* (Policy Press).

Harris, P.G. (2012) *Environmental Policy and Sustainable Development in China* (Policy Press).

Heinberg, R. (2005) *The Party's Over: Oil, War and the Fate of Industrial Societies* (New Society Publishers).

Heinberg, R. (2011) *The End of Growth: Adapting to Our New Economic Reality* (New Society Publishers).

Heinberg, R. & Fridley, D. (2010) 'The End of Cheap Coal', *Nature* 468, 367–369.

Hughes, J.D. (2013) 'A Reality Check on the Shale Revolution', *Nature* 494, 307–308.

Jackson, T. (2009) *Prosperity Without Growth: Economics for a Finite Planet* (Earthscan).

Jacquet, J. Pauly, D., et al. (2010) 'Seafood Stewardship in Crisis', *Nature* 467, 28–29.

Laghari, J.R. (2013) 'Melting Glaciers Bring Energy Uncertainty', *Nature* 502, 617–618.

Lang, G. (2002a) 'Forests, Floods, and the Environmental State in China', *Organization and Environment* 15(2),109–130.

Lang, G. (2002b) 'Deforestation, Floods, and State Reactions in China and Thailand', in A. Mol & F. Buttel (Eds.), *The Environmental State Under Pressure* (Elsevier Science).

Lang, G. & Chan, C.H.W. (2006) 'China's Impact on Forests in Southeast Asia', *Journal of Contemporary Asia* 36(2),167–194.

Lang, G. & Chow, J. (2011) 'Climate Change, Energy, and Rural Livelihoods: Planning for Resilience', Proceedings of the 4th International Conference of the Asian Rural Sociology Association, 7–10 September, Legazpi City, Philippines.

Lang, G. & Lu, Y. (2011) 'Religion and Environmentalism in Chinese Societies', in F. Yang & G. Lang (Eds.), *Social Scientific Studies of Religion in China: Methodology, Theories and Findings* (Brill).

Lang, G. & Miao Bo (2013) 'Food Security for China's Cities', *International Planning Studies* 18(1), 1–16.

Lang, G. & Xu Ying (2013) 'Anti-incinerator Campaigns and the Evolution of Protest Politics in China', *Environmental Politics* 22(5), 832–848.

Li Minqi (2007) 'Peak Oil, the Rise of China and India, and the Global Energy Crisis', *Journal of Contemporary Asia* 37(4), 449–471.

Li, V. & Lang, G. (2010) 'China's "Green GDP" Experiment and the Struggle for Ecological Modernization', *Journal of Contemporary Asia* 40(1), 45–63.

Li Yu-wai, Miao Bo & Lang, G. (2011) 'The Local Environmental State in China: A Study of County-level Cities in Suzhou', *China Quarterly* 205, 115–132.

Lovins, A. (2012) 'A Farewell to Fossil Fuels: Answering the Energy Challenge, *Foreign Affairs* 91(2), 134–136.

Miao Bo & Lang, G. (2010) 'China's Emissions: Dangers and Responses', in C. Lever-Tracy (Ed.), *Routledge Handbook of Climate Change and Society* (Routledge).

Mol, A.P.J. (2002) 'Ecological Modernization and the Global Economy', *Global Environmental Politics* 2(2), 92–115.

Mol, A.P.J. (2006) 'Environment and Modernity in Transitional China: Frontiers of Ecological Modernization', *Development and Change* 37(1), 29–56.

Murray, J. & King, D. (2012) 'Oil's Tipping Point Has Passed', *Nature* 481, 433–435.

Newb, M., Ardiansyah, F. & Spector, E. (2009) *The Greater Mekong and Climate Change: Biodiversity, Ecosystem Services and Development at Risk* (WWF).

Ostrom, E. (2009) 'A General Framework for Analyzing Sustainability of Social-ecological Systems', *Science* 325(5939), 419–422.

Pauly, D. (1995) 'Anecdotes and the Shifting Baseline Syndrome of Fisheries'. *TREE* 10(10), 430.

Piao, S. et al. (2010) 'The Impacts of Climate Change on Water Resources and Agriculture in China', *Nature* 467, 43–51.

Rubin, J. (2012) *The End of Growth* (Random House Canada).

Shapiro, J. (2012) *China's Environmental Challenges* (Polity Press).

Simons, C. (2013) *The Devouring Dragon: How China's Rise Threatens Our Natural World* (St. Martin's Press).

Smil, V. (2004) *China's Past, China's Future* (RoutledgeCurzon Press).

Tucker, M.E. & William, D.R. (Eds.) (1997) *Buddhism and Ecology: The Interconnection of Darma and Deeds* (Harvard University Press).

Tucker, M.E. & Berthrong, J. (Eds.) (1998) *Confucianism and Ecology: The Interrelation of Heaven, Earth, and Humans* (Harvard University Press).

Wang Jianliang, Feng Kianyong, Zhao Lin, Snowden, S. & Wang Xu (2011) 'A Comparison of Two Typical Multicyclic Models used to Forecast the World's Conventional Oil Production', *Energy Policy* 39, 7616–7621.

Watts, J. (2010) *When a Billion Chinese Jump: How China Will Save Mankind – or Destroy It* (Faber & Faber).

Weller, R.P. (2006) *Discovering Nature: Globalization and Environmental Culture in China and Taiwan* (Cambridge University Press).

York, R. & Rosa, E. (2003) 'Key Challenges to Ecological Modernization Theory', *Organization & Environment* 16(3), 273–288.

Zhang, J.Y. & Barr, M. (2013) *Green Politics in China: Environmental Governance and State–Society Relations* (Pluto Press).

Zuo, M. (2013) 'Ukraine to be China's Largest Overseas Farmer', *South China Morning Post*, 22 September.

Index

Note: Page numbers in **bold** type refer to **figures**
Page numbers in *italic* type refer to *tables*